Lecture Notes in Computer Science 4121

Commenced Publication in 1973
Founding and Former Series Editors:
Gerhard Goos, Juris Hartmanis, and Jan van Leeuwen

T0226175

Armin Biere   Carla P. Gomes (Eds.)

Theory and Applications of Satisfiability Testing – SAT 2006

9th International Conference
Seattle, WA, USA, August 12-15, 2006
Proceedings

 Springer

Volume Editors

Armin Biere
Johannes Kepler University Linz, Institute for Formal Models and Verification
Altenbergerstr. 69, 4040 Linz, Austria
E-mail: biere@jku.at

Carla P. Gomes
Cornell University, Department of Computer Science
5133 Upson Hall, Ithaca, NY 14853, USA
E-mail: gomes@cs.cornell.edu

Library of Congress Control Number: 2006930253

CR Subject Classification (1998): F.4.1, I.2.3, I.2.8, I.2, F.2.2, G.1.6

LNCS Sublibrary: SL 1 – Theoretical Computer Science and General Issues

ISSN   0302-9743
ISBN-10   3-540-37206-7 Springer Berlin Heidelberg New York
ISBN-13   978-3-540-37206-6 Springer Berlin Heidelberg New York

Springer is a part of Springer Science+Business Media

springer.com

© Springer-Verlag Berlin Heidelberg 2006
Printed in Germany

Typesetting: Camera-ready by author, data conversion by Scientific Publishing Services, Chennai, India
Printed on acid-free paper   SPIN: 11814948   06/3142   5 4 3 2 1 0

Preface

This volume contains the papers presented at the 9th International Conference on Theory and Applications of Satisfiability Testing (SAT 2006).

The International Conference on Theory and Applications of Satisfiability Testing is the primary annual meeting for researchers studying the propositional satisfiability problem (SAT). SAT 2006 was part of FLoC 2006, the fourth Federated Logic Conference, which hosted, in addition to SAT, LICS, RTA, CAV, ICLP and IJCAR. SAT 2005 was held in St. Andrews, Scotland, and SAT 2004 in Vancouver, BC, Canada. This time SAT featured the SAT Race in spirit of the SAT Competitions, the first competitive QBF Evaluation, an Evaluation of Pseudo-Boolean Solvers and the Workshop on Satisfiability Solvers and Program Verification (SSPV).

Many hard combinatorial problems can be formulated as Boolean Satisfiability (SAT) problems. In fact, given the tremendous advances in the state of the art of SAT solvers in the last decade, many real-world applications are now being encoded as SAT problems. For example, many practical verification problems can be rephrased as SAT problems. This applies to verification problems in hardware and software. SAT is therefore becoming one of the most important core technologies to verify secure and dependable systems: Improvements in the theoretical and practical aspects of SAT will consequently apply to a range of real-world problems.

The topics of the conference spanned practical and theoretical research on SAT and its applications and included but were not limited to proof systems, proof complexity, search algorithms, heuristics, analysis of algorithms, hard instances, randomized formulae, problem encodings, industrial applications, solvers, simplifiers, tools, case studies and empirical results. SAT is interpreted in a rather broad sense: besides propositional satisfiability, it includes the domain of Quantified Boolean Formulae (QBF), Constraint Programming Techniques (CP) for word-level problems and their propositional encoding and particularly Satisfiability Modulo Theories (SMT).

There were 80 submissions including 75 regular papers with a page limit of 14 pages and 15 short papers with a page limit of 6 pages. Each submission was reviewed by at least three Programme Committee members. The committee decided to accept 26 regular papers and 11 short papers. Out of the 15 papers submitted as short papers, two papers were accepted. The other nine papers accepted as short papers had been submitted as regular paper.

The program also includes invited talks by Fahiem Bacchus and Karem Sakallah, the presentations of the results of the SAT Race, and the Evaluations of QBF, Pseudo-Boolean and MAX-SAT Solvers.

We would like to thank the organizers of FLoC for coordinating the different conferences. We thank Andrei Voronkov for his excellent EasyChair system.

It helped us streamline the reviewing progress tremendously and meet our decision deadlines. Last but not least we thank the Programme Committee and the additional external reviewers for their careful and thorough work, without which it would not have been possible for us to put together such an outstanding conference programme.

We would also like to acknowledge the support of our sponsors: Cadence, IBM, Microsoft Research, NEC, John von Neumann Minerva Center for the Development of Reactive Systems, and the Intelligent Information Systems Institute at Cornell University.

August 2006   Armin Biere and Carla P. Gomes

Organization

Programme Chairs

Armin Biere
Carla P. Gomes

FLoC Representative

Henry Kautz

Programme Committee

Dimitris Achlioptas
Carlos Ansótegui
Fahiem Bacchus
Paul Beame
Alessandro Cimatti
Niklas Eén
Enrico Giunchiglia
Holger Hoos
Henry Kautz
Hans Kleine Büning

James Kukula
Daniel Le Berre
Inês Lynce
Hans van Maaren
Sharad Malik
João Marques-Silva
Cristopher Moore
Jussi Rintanen
Ashish Sabharwal
Bart Selman

Carsten Sinz
Ewald Speckenmeyer
Ofer Strichman
Stefan Szeider
Allen Van Gelder
Miroslav Velev
Toby Walsh
Riccardo Zecchina
Lintao Zhang

External Reviewers

Johan Alfredsson
Anbulagan
Gilles Audemard
Ramón Béjar
Marco Benedetti
Wolfgang Blochinger
Maria Luisa Bonet
Marco Bozzano
Hajo Broersma
Uwe Bubeck
Amin Coja-Oghlan
Sylvie Coste-Marquis
Stefan Dantchev
Ivan Dotú
Anders Franzén
Zhaohui Fu
Roman Gershman
Eugene Goldberg
Dan Goldwasser
Emmanuel Hebrard

Marijn Heule
Jinbo Huang
Frank Hutter
Gabriel Istrate
Maya Koifman
Andrei Krokhin
Oliver Kullmann
Theodor Lettmann
Chu-Min Li
Jean Christophe Madre
Vasco Manquinho
Felip Manyà
Marco Maratea
Pierre Marquis
Massimo Narizzano
Arlindo Oliveira
Stefan Porschen
Steven Prestwich
Bert Randerath
Silvio Ranise

Marco Roveri
Marko Samer
Tian Sang
Vishal Sanwalani
Roberto Sebastiani
Andrew Slater
Greg Sorkin
Ted Stanion
Dominik Stoffel
Peter Stuckey
Christian Szegedy
Niklas Sörensson
John Thornton
Dave Tompkins
Michael Veksler
Dong Wang
Karen Yorav
Yinlei Yu

Table of Contents

Invited Talks

Session 1. Proofs and Cores

Session 2. Heuristics and Algorithms

Session 3. Applications

Session 4. SMT

Session 5. Structure

Session 6. MAX-SAT

Session 7. Local Search and Survey Propagation

Session 8. QBF

Session 9. Counting and Concurrency

From Propositional Satisfiability
to Satisfiability Modulo Theories

Hossein M. Sheini and Karem A. Sakallah

University of Michigan, Ann Arbor MI 48109, USA
{hsheini, karem}@umich.edu

Abstract. In this paper we present a review of SAT-based approaches for building scalable and efficient decision procedures for quantifier-free first-order logic formulas in one or more decidable theories, known as Satisfiability Modulo Theories (SMT) problems. As applied to different system verification problems, SMT problems comprise of different theories including fragments of elementary theory of numbers, the theory of arrays, the theory of list structures, etc. In this paper we focus on different DPLL-style satisfiability procedures for decidable fragments of the theory of integers. Leveraging the advances made in SAT solvers in the past decade, we introduce several SAT-based SMT solving methods that in many applications have outperformed classical decision methods. Aside from the classical method of translating the SMT formula to a purely Boolean problem, in recent methods, a SAT solver is utilized to serve as the "glue" that ties together the different theory atoms and forms the basis for reasoning and learning within and across them. Several methods have been developed to provide a combination framework for implications to flow through the theory solvers and to possibly activate other theory atoms based on the current assignments. Similarly, conflict-based learning is also extended to enable the creation of learned clauses comprising of the combination of theory atoms. Additional methods unique to one or more types of theory atoms have also been proposed that learn more expressive constraints and significantly increase the pruning power of these combination schemes. We will describe several combination strategies and their impact on scalability and performance of the overall solver in different settings and applications.

1 Introduction

The decision problem for quantifier-free first-order logic formulas arises quite naturally in a wide variety of applications including software and hardware verification, scheduling and planning [6,2,31,9,27]. Such formulas typically consist of logical combinations of atoms from different theories such as the theory of integer linear arithmetic, the theory of arrays, the theory of equality with uninterpreted functions, set theory, etc. Systematic procedures for deciding such formulas, based on equality propagation among the different theory solvers, were first described by Nelson and Oppen [20]. More recently, interest in this problem has increased dramatically, sparked in part by the phenomenal progress in the

A. Biere and C.P. Gomes (Eds.): SAT 2006, LNCS 4121, pp. 1–9, 2006.

capacity and speed of modern DPLL-based Boolean satisfiability (SAT) solvers. A significant number of researchers around the globe are actively exploring a variety of approaches for solving this problem by leveraging the computational power of modern SAT solvers. In this new incarnation, the problem has been dubbed Satisfiability Modulo Theories (SMT) [29] with particular emphasis on integrating the theory of linear (real and integer) arithmetic within a DPLL backtrack search framework[1].

In this paper we provide a brief survey of the SAT-based approaches for solving SMT problems. Specifically, after covering some preliminaries in Section 2, we describe methods for direct translation of SMT instances to SAT (Section 3), methods based on Abstraction/refinement (Section 4), online approaches (Section 5) and hybrid solutions methods (Section 6).

2   Preliminaries

Satisfiability Modulo Theories (SMT) is the problem of determining the satisfiability of a quantifier-free first-order logic (FOL) formula in one or more decidable theories. Quantifier-free first-order logic extends propositional logic with *terms*, *function symbols*, and *predicate symbols* defined according to the following rules:

- A variable is a term.
- $f(t_1, \cdots, t_n)$ is a *term*, where f is an n-arity function symbol (with $n \geq 0$) and $t_1, \cdots, t_n$ are terms. 0-arity functions are *constants*.
- $P(t_1, \cdots, t_n)$ is an *atom*, where P is an n-arity predicate symbol (with $n \geq 0$) and $t_1, \cdots, t_n$ are terms. 0-arity predicates are *propositional variables*.
- Quantifier-free FOL formulas are constructed by combining atoms according to the rules of propositional logic. In particular, a quantifier-free FOL formula in *conjunctive normal form* (CNF) is the conjunction of a set of *clauses* each of which is the disjunction of a set of *literals*, where a literal is either an atom or the negation of an atom.

Using quantifier-free FOL as an organizing framework, we can define several specialized "logics" or "theories" by specifying the domains of their variables, as well as the functions (terms) and predicates (atoms) they admit. We list below some of the most commonly-used theories in hardware and software verification applications. Note that in these theories all variables and constant are assumed to be integer-valued.

- **Propositional Logic ($\mathcal{P}$):** In this theory, there are no function symbols, and only 0-arity predicate symbols, i.e., propositional variables.
- **Equality Logic ($\mathcal{E}$):** This theory has no function symbols and only the equality predicate $t_i = t_j$ where t_i and t_j are terms.
- **Equality Logic with Successors ($\mathcal{ES}$):** This logic extends $\mathcal{E}$ logic by introducing the function $succ(t) = t + 1$ where t is a term.

---

[1] Independently, Hooker et al [13] and the OR community refer to similar problems as mixed logical linear programs.

- **Equality Logic with Uninterpreted Functions ($\mathcal{EUF}$):** This logic extends $\mathcal{E}$ with n-arity uninterpreted function and predicate symbols and enforces functional consistency.
- **Equality Logic with Successors and Uninterpreted Functions ($\mathcal{ESUF}$):** This logic extends $\mathcal{ES}$ with n-arity uninterpreted function and predicate symbols and enforces functional consistency.
- **Difference Logic ($\mathcal{DL}$):** This logic extends $\mathcal{ES}$ with the interpreted predicate of the form $t_i - t_j \leq d$ where t_i and t_j are terms and d is an integer constant.
- **Counter Arithmetic Logic with Lambda Expressions and Uninterpreted Functions ($\mathcal{CLU}$):** The set of functions and predicates are respectively the union of the functions and predicates of $\mathcal{EUF}$ and $\mathcal{DL}$.
- **Integer Unit-Two-Variable-Per-Inequality (UTVPI) Logic ($\mathcal{TVL}$):** This logic generalizes the interpreted predicates of $\mathcal{DL}$ to the form $a_i t_i + a_j t_j \leq d$ where $a_i, a_j \in \{\pm 1, 0\}$, t_i and t_j are terms and d is an integer constant.
- **Linear Integer Arithmetic ($\mathcal{LIA}$):** This is essentially the logic of integer linear inequalities. Note that $\mathcal{DL}$ and $\mathcal{TVL}$ are restrictions of $\mathcal{LIA}$.

The choice of which logic to apply in a particular situation depends on the expressiveness of the logic as well as the existence of efficient procedures for checking the satisfiability of conjunctions of atoms in that logic. Such procedures are referred to as *theory solvers* include congruence-closure for the logic of equality and its extensions [22], transitive-closure for the $\mathcal{DL}$ and $\mathcal{TVL}$ theories [15] and Simplex-based Branch-and-Bound algorithms for $\mathcal{LIA}$.

Propositional Satisfiability (SAT). Modern SAT solvers are based on the DPLL backtrack search algorithm [8] augmented with powerful techniques for search space pruning and efficient Boolean constraint propagation. These techniques include conflict-based learning and non-chronological backtracking [18], watched-literal schemes for Boolean Constraint Propagation and advanced variable ordering heuristics such as VSIDS [19]. For a survey of these methods, the reader is referred to [32].

The efficiency of the SMT solvers we describe on the remainder of this paper drive their power from these modern SAT solvers.

3   Translation to SAT

The earliest use of SAT solvers in the SMT context was through direct translation of an SMT instance to an equi-satisfiable Boolean formula. Such *eager* solution approaches were attractive because they did not require the development of specialized theory solvers or complex combination strategies. Their effectiveness derived from their reliance on the underlying SAT engine.

These approaches were most effective when applied to the $\mathcal{EUF}$ and $\mathcal{CLU}$ logics in [14,16,23,24,7]. Techniques to translate such formulas to propositional form include *Small Domain Instantiation* [23] and *Per-Constraint Encoding* [25].

Small Domain Instantiation. This method is based on the fact that for a formula whose atoms are only equalities between *input* variables, it is enough to give each variable the range $[1 \ldots n]$ (where n is the number of input variables) without affecting the satisfiability/unsatisfiability of the formula. Knowing the range enables us to replace each input variable in the formula with a bit-vector of Boolean variables of size $\lceil \log n \rceil$ yielding a purely Boolean instance. To apply this procedure to an $\mathcal{EUF}$ formula, the formula is first converted, using Ackermann reduction [1], to an equi-satisfiable formula that only involves equaliteis. Specifically, each occurrence of an uninterpreted function is replaced with a new variable and the formula is conjoined with constraints to preserve functional consistency.

Further enhancements to this method deal with reducing the $[1 \ldots n]$ range associated with each variable taking into account the structure of the formula. These methods include the equality graph and range analysis of [23], coloring the equality graph by analyzing the CNF representation of the formula in [12], positive equality method of [30] and the hybrid method of [24].

Per-Constraint Encoding. In this approach, also known as e_{ij} *encoding* [11], an $\mathcal{EUF}$ formula is first transformed, as above, into an equi-satisfiable formula, φ, whose atoms are only equalities. Each equality atom $t_i = t_j$ is replaced by a fresh unrestricted Boolean variable e_{ij}, yielding a Boolean abstraction, φ^{bool} that can be processed by a propositional SAT solver. To prevent false positives this abstraction is augmented with transitivity constraints of the form $(e_{ij} \wedge e_{jk}) \rightarrow e_{ik}$ for all e_{ij} variables within the formula. In the worst case the number of such transitively constraints grow exponentially [7].

4 Abstraction/Refinement

An SMT CNF formula whose propositional and theory atoms are denoted, respectively, by $\mathscr{P}$ and $\mathscr{T}$ can be generically expressed as

$$\varphi(\mathscr{P}, \mathscr{T}) = \bigwedge_{c \,\in\, \mathscr{C}} c \tag{1}$$

where $\mathscr{C}$ is a set of clauses whose elements have the form

$$\bigvee_{A \in\, \mathscr{P} \cup \mathscr{T}} A \,\vee \bigvee_{A \in\, \mathscr{P} \cup \mathscr{T}} \neg A \tag{2}$$

A Boolean abstraction of the SMT formula, φ^{bool}, can be constructed by introducing a mapping α that assigns a fresh Boolean *indicator* variable, $\alpha(A)$, to each theory atom, A. Thus the SMT CNF formula can be represented as

$$\varphi(\mathscr{P}, \mathscr{I}, \mathscr{T}) = \varphi^{bool}(\mathscr{P}, \mathscr{I}) \wedge \bigwedge_{A \in \mathscr{T}} (\alpha(A) \leftrightarrow A) \tag{3}$$

where $\mathscr{I}$ is the set of indicator variables. In the abstraction/refinement approach, a SAT solver is initially applied to φ^{bool}. If φ^{bool} is found to be unsatisfiable, then so is the SMT formula, φ and no theory solver needs to be invoked. If φ^{bool} is found to be satisfiable, however, a theory solver is applied to the conjunction of true theory literals in the solution returned by the SAT solver. Satisfiability of this conjunction establishes the satisfiability of the original formula φ; otherwise the abstracted formula, φ^{bool}, is *refined* to eliminate this solution and the process reiterates. In such a flow the theory solvers are sometimes referred to as *offline* solvers and their combination strategy within SAT is called *offline* integration.

An enhancement of this scheme was proposed in [5] whereby the non-Boolean literals in the SAT solution of φ^{bool} are categorized in terms of their logics and solved sequentially by increasing the expressiveness of their logic. For instance, if the solution to φ^{bool} contains $\mathcal{ES}$, $\mathcal{ESUF}$ and $\mathcal{TVL}$ literals, we can verify the consistency of only equality atoms first, and if satisfiable check the consistency of a more complete set of non-Boolean atoms, adding new types of atoms to the verified set in sequence. In this case the offline solvers are essentially layered in an increasing solving capability and each are applied to a satisfying solution found to the formula at the previous layer. Note that if a conflict at any layer is detected, this conflict is used to prune the search at the Boolean level and the system reiterates. In order to find conflicts earlier in the search, an *incremental layered approach* is suggested in [5] where instead of complete Boolean assignments, "partial" ones are passed through layers to be checked for consistency.

5   Online Solving Approach

In this approach [10,4,21,28], the consistency of the conjunction of theory literals is checked incrementally as soon as the SAT solver assigns those literals, i.e. adds them to the solution model. This combination strategy has been referred to as the *online* method in [4], DPLL($\mathcal{T}$) in [10] and Mixed Logical Integer Linear Programming (MLILP) in [28]. In this method the SAT solver is tightly integrated with one or more theory solvers. In [4], the authors integrate a $\mathcal{DL}$ solver into the SAT solver while in [10] an $\mathcal{EUF}$ solver is combined with the SAT solver. In [28], a combined method for solving systems of $\mathcal{TVL}$ literals using a SAT solver and a transitive-closure UTVPI solver is proposed. Considering that in this method a growing set of theory literals is incrementally checked for consistency, it is important that the theory solver is both incremental and backtrackable. This is the reason that only relatively cheap theory literals, i.e. $\mathcal{ESUF}$ or $\mathcal{TVL}$ are considered for this method.

Each theory solver, in this approach, maintains a set of "activated" theory atoms and incrementally updates this set, returning a conflict as soon an inconsistency is detected. Essentially the SAT solver treats such atoms similarly to the way it treats propositional atoms. The advantages of this approach over the layered and eager approaches can be summarized as follows:

- Compared to the eager approach, the online method only checks the consistency of theory atoms when they are absolutely required to establish the satisfiability of the formula. This avoids time- and memory-consuming "preprocessing" of all theory atoms.

- Compared to the layered approach, by processing each theory atom "on-demand", the online approach is able to detect conflicts among those atoms as they occur and not at a later layer. The advantage of early detection of conflicts is more pronounced in cases where the majority of conflicts are due to inconsistencies among theory atoms. To detect such conflicts, the layered approach requires solving the entire Boolean layer at each iteration.

6   Hybrid Solving Approach

The hybrid approach combines the online and layered approaches and is particularly effective when the SMT formula contains a large number $\mathcal{TVL}$ atoms relative to the number of $\mathcal{LIA}$ atoms. The efficiency of transitive-closure solver for UTVPI suggests an online integration with the SAT solver while the high computation requirements of simplex-based solvers for $\mathcal{LIA}$ atoms, suggests layering them or invoking them offline.

Combined Deduction Scheme. In Ario [3] a stand-alone deduction scheme outside the theory solvers is utilized that builds an implication graph taking into account all types of atoms. Implications in this graph are both due to unit-clause-propagation of the SAT solver and the linear combination of integer constraints. In this scheme all possible non-negative linear combinations of equality and UTVPI constraints are generated and implied. Non-UTVPI constraints on the other hand are combined with equality or UTVPI constraints only if a variable could be eliminated and they are not combined with other non-UTVPI constraints. Consequently the consistency of each implied integer constraint is verified within its respective theory solver.

Offline Refinement. An advanced refinement technique is introduced in [28] that further enhances the pruning power of the learned constraints. In this technique, instead of pruning the search space by adding a learned constraint in terms of indicator variables, by combining non-UTVPI and UTVPI constraints, a more powerful constraint including a conflicting combination of UTVPI constraints is generated and added to the SMT formula, taking into account that the UTVPI solver is tightly integrated within the SAT solver. For instance, let L_1, D_1, D_2, D_3 and D_4 (whose indicator variables are respectively P_{L1}, P_{D1}, P_{D2}, P_{D3} and P_{D4}) comprise the inconsistent set of integer constraints. This methods instead of simply learning $\neg(P_{L1} \wedge P_{D1} \wedge P_{D2} \wedge P_{D3} \wedge P_{D4})$, checks the linear non-negative combinations of L_1 and different permutations of the activated UTVPI constraints D_1, D_2, D_3 and D_4. In case, for instance, the linear combination of L_1, D_1 and D_2 results in a UTVPI constraint, $\hat{D}$, the following constraints is learned and is added to the SMT formula which is strictly stronger than the learned Boolean constraint, $P_{L1} \wedge D_1 \wedge D_2 \to \hat{D}$.

7   Conclusions and Future Work

The impressive advances in algorithms and implementations of propositional SAT solvers over the past decade have invigorated the quest for efficient SAT-based solution methods for the SMT problem. In this paper we briefly surveyed recently-proposed techniques for applying SAT technology to the SMT problem. While many issues are still unresolved as to how best architect an SMT solver, empirical evidence suggests that reasoning about the Boolean structure of the SMT formula is best handled by the SAT solver. Theory solvers in such a framework need only worry about conjunctions of theory literals.

On the application side, several SAT-based SMT solving methods have been adopted in verification problems such as equivalence checking of two versions of a hardware design or the input and output of a compiler. These methods have been recently utilized in several model checkers as well. Additionally, optimization in the context of SMT is also gaining importance [26] especially in planning and scheduling applications. An online SMT method [27] is applied to Disjunctive Temporal Problems with preferences [17], referred to as DTPPs in [26]. For particular applications, custom implementation of a SMT solver may also be helpful. We believe as more and more SMT solvers utilize SAT solvers as their deduction and reasoning framework, finding efficient integration techniques to combine different theories and/or logics within the SAT solver has become very significant and subject to further research.

Acknowledgement

This work was funded in part by the National Science Foundation (NSF) under ITR grant No. 0205288.

References

1. W. Ackermann. Solvable cases of the decision problem. In *Studies in Logic and the Foundations of Mathematics*. North-Holland, 1954.
2. Tod Amon, Gaetano Borriello, Taokuan Hu, and Jiwen Liu. Symbolic timing verification of timing diagrams using presburger formulas. In *DAC '97*, pages 226–231, 1997.
3. Ario SMT Solver. http://www.eecs.umich.edu/~ario/.
4. Sergey Berezin, Vijay Ganesh, and David L. Dill. An online proof-producing decision procedure for mixed-integer linear arithmetic. In *TACAS*, pages 521–536, 2003.
5. Marco Bozzano, Roberto Bruttomesso, Alessandro Cimatti, Tommi A. Junttila, Peter van Rossum, Stephan Schulz, and Roberto Sebastiani. An incremental and layered procedure for the satisfiability of linear arithmetic logic. In *TACAS*, pages 317–333, 2005.
6. Raik Brinkmann and Rolf Drechsler. RTL-datapath verification using integer linear programming. In *ASP-DAC '02*, pages 741–746, 2002.

7. Randal E. Bryant, Shuvendu K. Lahiri, and Sanjit A. Seshia. Deciding CLU logic formulas via boolean and pseudo-boolean encodings, 2002.
8. Martin Davis, George Logemann, and Donald Loveland. A machine program for theorem proving. *Communications of the ACM*, 5(7):394–397, July 1962.
9. Vinod Ganapathy, Sanjit A. Seshia, Somesh Jha, Thomas W. Reps, and Randal E. Bryant. Automatic discovery of API-level exploits. In *ICSE '05*, pages 312–321, 2005.
10. Harald Ganzinger, George Hagen, Robert Nieuwenhuis, Albert Oliveras, and Cesare Tinelli. DPLL(T): Fast decision procedures. In *Proceedings of the 16th International Conference on Computer Aided Verification, CAV'04*, pages 175–188, 2004.
11. Anuj Goel, Khurram Sajid, Hai Zhou, Adnan Aziz, and Vigyan Singhal. Bdd based procedures for a theory of equality with uninterpreted functions. *Form. Methods Syst. Des.*, 22(3):205–224, 2003.
12. Ramin Hojati, Adrian J. Isles, Desmond Kirkpatrick, and Robert K. Brayton. Verification using uninterpreted functions and finite instantiations. In *FMCAD '96: Proceedings of the First International Conference on Formal Methods in Computer-Aided Design*, pages 218–232, London, UK, 1996. Springer-Verlag.
13. John N. Hooker, Greger Ottosson, Erlendur S. Thorsteinsson, and Hak-Jin Kim. On integrating constraint propagation and linear programming for combinatorial optimization. In *AAAI '99/IAAI '99*, pages 136–141, 1999.
14. J. R. Burch and D. L. Dill. Automatic verification of pipelined microprocessors control. In *Proceedings of the sixth International Conference on Computer-Aided Verification CAV*, pages 68–80, 1994.
15. Joxan Jaffar, Michael J. Maher, Peter J. Stuckey, and Roland H. C. Yap. Beyond finite domains. In *Workshop on Principles and Practice of Constraint Programming*, pages 86–94, 1994.
16. Robert B. Jones, David L. Dill, and Jerry R. Burch. Efficient validity checking for processor verification. In *ICCAD '95: Proceedings of the 1995 IEEE/ACM international conference on Computer-aided design*, pages 2–6, Washington, DC, USA, 1995. IEEE Computer Society.
17. Lina Khatib, Paul Morris, Robert Morris, and Francesca Rossi. Temporal constraint reasoning with preferences. *17th International Joint Conference on Artificial Intelligence*, 1:322–327, 2001.
18. P. Marques-Silva and Karem A. Sakallah. GRASP: A search algorithm for propositional satisfiability. *IEEE Trans. Comput.*, 48(5):506–521, 1999.
19. Matthew W. Moskewicz, Conor F. Madigan, Ying Zhao, Lintao Zhang, and Sharad Malik. Chaff: engineering an efficient sat solver. In *38th Design Automation Conference*, pages 530–535, 2001.
20. Greg Nelson and Derek C. Oppen. Simplification by cooperating decision procedures. *ACM Trans. Program. Lang. Syst.*, 1(2):245–257, 1979.
21. Robert Nieuwenhuis and Albert Oliveras. DPLL(T) with exhaustive theory propagation and its application to difference logic. In *CAV*, pages 321–334, 2005.
22. Robert Nieuwenhuis and Albert Oliveras. Proof-Producing Congruence Closure. In *Proceedings of the 16th Int'l Conf. on Term Rewriting and Applications*, pages 453–468, 2005.
23. Amir Pnueli, Yoav Rodeh, Ofer Shtrichman, and Michael Siegel. Deciding equality formulas by small domains instantiations. In *CAV '99: Proceedings of the 11th International Conference on Computer Aided Verification*, pages 455–469, London, UK, 1999. Springer-Verlag.

24. Yoav Rodeh and Ofer Strichman. Building small equality graphs for deciding equality logic with uninterpreted functions. *Inf. Comput.*, 204(1):26–59, 2006.
25. Sanjit A. Seshia, Shuvendu K. Lahiri, and Randal E. Bryant. A hybrid SAT-based decision procedure for separation logic with uninterpreted functions. In *DAC '03: Proceedings of the 40th conference on Design automation*, pages 425–430, New York, NY, USA, 2003. ACM Press.
26. Hossein M. Sheini, Bart Peintner, Karem A. Sakallah, and Martha E. Pollack. On solving soft temporal constraints using SAT techniques. In *Proceedings of the Eleventh International Conference on Principles and Practice of Constraint Programming*, pages 607–621, 2005.
27. Hossein M. Sheini and Karem A. Sakallah. A SAT-based decision procedure for mixed logical/integer linear problems. In *CPAIOR*, pages 320–335, 2005.
28. Hossein M. Sheini and Karem A. Sakallah. A scalable method for solving satisfiability of integer linear arithmetic logic. In *SAT*, pages 241–256, 2005.
29. Cesare Tinelli. A DPLL-based calculus for ground satisfiability modulo theories. In Giovambattista Ianni and Sergio Flesca, editors, *Proceedings of the 8th European Conference on Logics in Artificial Intelligence (Cosenza, Italy)*, volume 2424 of *Lecture Notes in Artificial Intelligence*, pages 308–319. Springer, 2002.
30. Miroslav N. Velev and Randal E. Bryant. Exploiting positive equality and partial non-consistency in the formal verification of pipelined microprocessors. In *36th ACM/IEEE conference on Design automation*, pages 397–401, New York, NY, USA, 1999. ACM Press.
31. David Wagner, Jeffrey S. Foster, Eric A. Brewer, and Alexander Aiken. A first step towards automated detection of buffer overrun vulnerabilities. In *Network and Distributed System Security Symposium*, pages 3–17, February 2000.
32. Lintao Zhang and Sharad Malik. The quest for efficient boolean satisfiability solvers. In *CAV*, pages 17–36, 2002.

CSPs: Adding Structure to SAT

Fahiem Bacchus

University of Toronto, Canada
fbacchus@cs.toronto.edu

Abstract. One way of viewing the difference between SAT and CSPs is to think of programming in assembler vs programming in C. It can be considerably simpler to program in C than assembler. Similarly it can be considerably simpler to model real world problems in CSP than in SAT. On the other hand C's machine model is still rather close to the underlying hardware model accessed directly in assembler. Similarly, in CSPs the main method of reasoning, backtracking search, can be viewed as being an extension of DPLL, the main method of reasoning for SAT. Where the analogy breaks down is that unlike C and assembler whose machine models are computationally equivalent, some CSP techniques offer a considerable boost in inferential power over the resolution inferences preformed in DPLL. An intresting question is how to combine this additional inferential power with the more powerful forms of resolution preformed in modern DPLL solvers. One approach for achieving such a combination will be presented.

A. Biere and C.P. Gomes (Eds.): SAT 2006, LNCS 4121, p. 10, 2006.

Complexity of Semialgebraic Proofs with Restricted Degree of Falsity*

Arist Kojevnikov and Alexander S. Kulikov

St.Petersburg Department of Steklov Institute of Mathematics
27 Fontanka, 191023 St.Petersburg, Russia
http://logic.pdmi.ras.ru/{~arist,~kulikov}

Abstract. A weakened version of the Cutting Plane (CP) proof system with a restriction on the degree of falsity of intermediate inequalities was introduced by Goerdt. He proved an exponential lower bound for CP proofs with degree of falsity bounded by $\frac{n}{\log^2 n+1}$, where n is the number of variables. Hirsch and Nikolenko strengthened this result by establishing a direct connection between CP and Resolution proofs. This result implies an exponential lower bound on the proof length of the Tseitin-Urquhart tautologies, when the degree of falsity is bounded by cn for some constant c.

In this paper we generalize this result for extensions of Lovász-Schrijver calculi (LS), namely for LS^k+CP^k proof systems introduced by Grigoriev et al. We show that any LS^k+CP^k proof with bounded degree of falsity can be transformed into a $Res(k)$ proof. We also prove lower and upper bounds for the new system.

1 Introduction

The systematic study of propositional proof complexity was initiated by Cook and Reckhow in [1]. The motivation for this is the following: the NP≠co-NP assumption implies the existence of hard examples for any proof system. In this paper we are interested in semialgebraic proof systems, which restate a Boolean tautology as a set of inequalities and prove that this set has no solution in $\{0,1\}$-variables. No exponential lower bounds in these systems are known for tautologies that are hard for many other proof systems.

A weakened version of the Cutting Plane (CP) proof system with a restriction on the degree of falsity of intermediate inequalities was introduced by Goerdt [2]. He proved an exponential lower bound for CP proofs with the degree of falsity bounded by $\frac{n}{\log^2 n+1}$, where n is the number of variables. Hirsch and Nikolenko strengthened this result by establishing a direct connection between CP and Resolution proofs. This result implies an exponential lower bound on the proof

---

* Supported in part by INTAS (grants 04-77-7173, 04-83-3836, 05-109-5352), RFBR (grants 05-01-00932, 06-01-00502, 06-01-00584), RAS Program for Fundamental Research ("Modern Problems of Theoretical Mathematics"), and Russian Science Support Foundation.

A. Biere and C.P. Gomes (Eds.): SAT 2006, LNCS 4121, pp. 11–21, 2006.
© Springer-Verlag Berlin Heidelberg 2006

length of the Tseitin-Urquhart tautologies, when the degree of falsity is bounded by cn for some constant c.

In this paper we extend the notion of the degree of falsity to high degree semialgebraic proof systems and prove lower and upper bounds for the considered systems. We prove that an LS^k+CP^k proof with restricted degree of falsity can be transformed into a $Res(k)$ proof. We also provide exponential separations of the new proof system from CP and $Res(k)$ by giving short proofs of the Pigeon Hole Principle and the Weak Clique Coloring tautologies and prove.

Below we give the main ideas of transforming LS^k+CP^k proofs into $Res(k)$ proofs. Given an LS^k+CP^k proof Π we first linearize this proof, i.e., we replace each monomial by a new variable. This allows us to work with linear inequalities only. By a Boolean representation of a linear inequality we mean a CNF formula equivalent to this inequality. By bounding the degree of falsity of an inequality one bounds the size of this formula. We show that for any step of the proof Π it is possible to derive the Boolean representation of the conclusion from the Boolean representations of the premise(s). Thus, we transform an LS^k+CP^k proof into a Resolution proof of an auxiliary formula (with additional variables). This implies the existence of a $Res(k)$ proof of an initial formula.

The paper is organized as follows. Section 2 contains the necessary definitions. In Sect. 3 we show that any LS^k+CP^k proof with bounded degree of falsity can be transformed into a $Res(k)$ proof. Finally, in Sect. 4 we prove upper and lower bounds for considered systems.

2 General Setting

2.1 Proof Systems

A *proof system* [1] for a language L is a polynomial-time computable function mapping words (treated as proof candidates) to L (whose elements are considered as theorems).

A *propositional proof system* is a proof system for the co-NP-complete language TAUT of all Boolean tautologies in disjunctive normal form (DNF). Since this language is in co-NP, any proof system for a co-NP-hard language L can be considered as a propositional proof system. However, we need to fix a concrete reduction of TAUT to L before compare them.

The proof systems we consider are DAG-like derivation systems, i.e. a proof is a sequence of *lines* such that every line is either an axiom or is obtained by an application of a derivation rule to several previous lines. The proof finishes with a line called *goal*. Such a proof system is thus determined by notions of a line, a goal, a set of axioms and a set of derivation rules.

The *resolution* proof system [3] has clauses (disjunctions of literals) as its proof lines and an empty clause as its goal. Given a formula F in DNF, one takes clauses of $\neg F$ as the axioms and uses the following rules:

$$\text{Resolution:} \quad \frac{A \vee x \quad \neg x \vee B}{A \vee B} \quad , \qquad \text{Weakening:} \quad \frac{A}{A \vee l} \ .$$

The $Res(k)$ proof system [4] is a generalization of Resolution where one uses k-DNFs (disjunctions of terms, i.e. conjunctions of literals) as lines. The goal is to derive an empty clause. We use the clauses of the formula $\neg F$ as the axioms and the following inference rules:

Weakening: $\dfrac{A}{A \vee l}$

AND-introduction: $\dfrac{A \vee l_1 \cdots A \vee l_j}{A \vee \bigwedge_{i=1}^{j} l_i}$

Cut: $\dfrac{A \vee \bigwedge_{i=1}^{j} l_i \quad A \vee \bigvee_{i=1}^{j} \neg l_i}{A \vee B}$

AND-elimination: $\dfrac{A \vee \bigwedge_{i=1}^{j} l_i}{A \vee l_i}$

To define a propositional proof system dealing with inequalities, we translate each formula $\neg F$ in CNF with n variables into a system $\mathcal{D}$ of linear inequalities such that F is a tautology if and only if the system $\mathcal{D}$ has no solution in $\{0,1\}$-variables. For a given tautology F, we translate each clause C_i of $\neg F$ with variables $x_{j_1}, \ldots, x_{j_t}$, into the inequality

$$l_1 + \ldots + l_t \geq 1 , \qquad (1)$$

where $l_i = x_{j_i}$, if the variable x_{j_i} occurs positively in the clause and $l_i = 1 - x_{j_i}$, if x_{j_i} occurs negatively. For every variable x_i, $1 \leq i \leq n$, we also add the inequalities $0 \leq x_i \leq 1$ to the system $\mathcal{D}$.

The proof lines in the *Cutting Plane* proof system (CP) [5,6] are linear inequalities with integer coefficients. The goal is a contradiction $0 \geq 1$. We use as the axioms the system of linear inequalities $\mathcal{D}$ provided by the translation. The inference rules are

Addition: $\dfrac{f \geq 0 \quad g \geq 0}{\lambda_f f + \lambda_g g \geq 0}$,

Rounding: $\dfrac{af \geq c}{f \geq \lceil \frac{c}{a} \rceil}$,

where λ_f, λ_g are positive constants, a, c are constants, f, g are polynomials.

The extension of CP to higher degree was introduced in [7]. The proof system $LS^k + CP^k$ operates with inequalities of degree at most k with integer coefficients as lines, using the same set of axioms as CP extended by $x_i^2 - x_i \geq 0$ for $1 \leq i \leq n$ and the following rule:

Multiplication: $\dfrac{h \geq 0}{hx \geq 0}$, $\dfrac{h \geq 0}{h(1 - x) \geq 0}$,

where h is a polynomial of degree at most $k - 1$, x is a variable.

2.2   Proof Linearization

In order to transform an $LS^k + CP^k$ proof into a $Res(k)$ proof we transform an initial proof into a Resolution proof of an auxiliary formula. We show the connection between a $Res(k)$ proof of an initial formula and a Resolution proof of an auxiliary formula below.

For every set of literals $l_1, \ldots, l_m$ of a formula F, where $m \leq k$, we define a new variable $z(l_1, \ldots, l_m)$ denoting the conjunction of all these literals. This can be expressed by the following $m + 1$ clauses:

$$(z(l_1, \ldots, l_m) \vee \neg l_1 \vee \ldots \vee \neg l_m), \ (\neg z(l_1, \ldots, l_m) \vee l_1), \ \ldots, \ (\neg z(l_1, \ldots, l_m) \vee l_m) .$$

By $F(k)$ we denote the conjunction of F with all such clauses. We need the following property of $F(k)$:

Lemma 1 ([8]). *If $F(k)$ has a Resolution proof of size S, then F has a Res(k) proof of size $O(kS)$.*

For a variable $z_i = z(l_1, \ldots, l_s)$ of $F(k)$ and a variable x of F, by $z(z_i, x)$ we mean the variable $z(l_1, \ldots, l_s, x)$. For an inequality ι of degree at most k, by $lin(\iota)$ we denote a linear inequality obtained from ι by replacing each its monomial $x_1 \cdot \ldots \cdot x_m$ by a linear monomial $z(x_1, \ldots, x_m)$.

In the main theorem of this paper we also use the following simple lemma:

Lemma 2. *Let C be a clause containing variables of $F(k)$ and x be a variable of F. Let C' be a clause obtained from C by replacing each its variable z_i by $z(z_i, x)$. Then the clause $(C' \vee \neg x)$ can be obtained from C and clauses of $F(k)$ in at most $O(n^k)$ Resolution steps. If in addition C contains at least one negated variable, then one can also derive C'.*

Proof. For each variable $z_i = z(l_1, \ldots, l_s)$ we can derive clauses $(z_i \vee \neg z(z_i, x))$ and $(\neg z_i \vee z(z_i, x) \vee \neg x)$. The first clause is obtained by resolving $(z_i \vee \neg l_1 \vee \ldots \vee \neg l_s)$, $(\neg z(z_i, x) \vee l_1)$, ..., $(\neg z(z_i, x) \vee l_s)$, the second one — by resolving clauses $(z(z_i, x) \vee \neg l_1 \vee \ldots \vee \neg l_s \vee \neg x)$, $(\neg z_i \vee l_1)$, ..., $(\neg z_i \vee l_s)$.

For a literal z_i of the clause C, we resolve C with $(\neg z_i \vee z(z_i, x) \vee \neg x)$ and for a literal $\neg z_i$ — with $(z_i \vee \neg z(z_i, x))$. The result of these operations is either the clause C' or the clause $(C' \vee \neg x)$. In case the result is C', we derive $(C' \vee \neg x)$ by applying the Weakening rule. If C contains at least one negated variable, we resolve $(C' \vee \neg x)$ with $(\neg z(z_i, x) \vee x)$ for $\neg z_i \in C$.

The number of steps is as required, since the number of variables in C is $O(n^k)$. □

2.3 Degree of Falsity

The definition of the degree of falsity of a linear inequality was given by Goerdt [2].

Definition 1. *For a linear inequality ι of the form $\sum_{i=1}^{s} \alpha_i x_i \geq c$, $\mathrm{DGF}_1(\iota)$ is the difference of c and the minimal value of its left-hand side.*

Goerdt also gave a simpler definition of the degree of falsity.

Definition 2. *A literal form of a linear inequality is its representation in the form $\sum_{i=1}^{s} \alpha_i x_i + \sum_{i=s+1}^{s'} \alpha_i (1 - x_i) \geq c$, where $\alpha_i > 0$, for $1 \leq i \leq s'$. For an inequality ι, $\mathrm{DGF}_2(\iota)$ is the free coefficient of the literal form of ι.*

It is easy to see that these definitions are equivalent, i.e., for any linear inequality ι, $\mathrm{DGF}_1(\iota) = \mathrm{DGF}_2(\iota)$. Both these definitions can be extended naturally to inequalities of arbitrary degrees (one can just replace variables by monomials in both definitions). However, the new definitions would not be equivalent.

E.g., $\mathrm{DGF}_1(xy + xz - x \geq 2) = 3$, while $\mathrm{DGF}_2(xy + xz - x \geq 2) = 1$. From the other side, it is not difficult to show that for any inequality ι, $\mathrm{DGF}_1(\iota) \geq \mathrm{DGF}_2(\iota)$. So, if DGF_1 of a proof is bounded by a constant d, then DGF_2 is bounded by d too. For this reason, we use DGF_2 as the degree of falsity in this paper. The explicit definition is as follows.

Definition 3. *A literal form of an inequality is its representation in the form* $\sum_{i=1}^{s} \alpha_i m_i + \sum_{i=s+1}^{s'} \alpha_i(1 - m_i) \geq c$, *where* α_i*'s are positive constants,* m_i*'s are monomials. For an inequality* ι, $\mathrm{DGF}(\iota)$ *is the free coefficient of the literal form of* ι. *The degree of falsity of an* $\mathrm{LS}^k + \mathrm{CP}^k$ *proof is the maximal degree of falsity of intermediate inequalities of this proof.*

2.4   Boolean Representation of Linear Inequalities

By a Boolean representation of a linear inequality we mean a CNF formula that is equivalent to this inequality. Of course, such a formula is not unique. Below we describe the construction of a Boolean representation given in [9].

Let ι be a linear inequality of the form $\sum_{i=1}^{s} \alpha_i x_i + \sum_{i=s+1}^{s'} \alpha_i(1 - x_i) \geq c$, where $\alpha_i > 0$, for $1 \leq i \leq s'$. By satisfying a literal of ι we mean assigning either a value 1 to x_i, where $1 \leq i \leq s$, or a value 0 to x_i, where $s+1 \leq i \leq s'$. Let ι_0 be an inequality obtained from ι by satisfying some literals, such that no literal of ι_0 can be satisfied without trivializing ι_0. It is easy to see that ι_0 is equivalent to a clause (since it is not satisfied by exactly one assignment to its variables). By $\mathcal{B}(\iota)$ we denote the set of all such clauses. Moreover, in the rest of the paper by the Boolean representation of an inequality ι we mean exactly the set $\mathcal{B}(\iota)$. The following lemma shows that this construction is correct and provides an upper bound one the size of the constructed set.

Lemma 3 ([9]). *For any linear inequality* ι, $\mathcal{B}(\iota)$ *is equivalent to* ι. *Moreover, the number of clauses in* $\mathcal{B}(\iota)$ *is at most* $\binom{n}{d-1}$, *where* $d < n/2$ *is the degree of falsity of* ι.

We also use the following simple property of $\mathcal{B}(\iota)$, that follows immediately from the construction.

Lemma 4. *Let* ι *be a linear inequality of the form* $\sum_{i=1}^{s} \alpha_i x_i + \sum_{i=s+1}^{s'} \alpha_i(1 - x_i) \geq c$, *where* $\alpha_i > 0$, *for* $1 \leq i \leq s'$. *Then, the set of clauses of* $\mathcal{B}(\iota)$ *that do not contain the literal* x_i, *where* $1 \leq i \leq s$ *(or the literal* $\neg x_i$, *where* $s + 1 \leq i \leq s'$*) is exactly the set* $\mathcal{B}(\iota|_{x_i=1})$ *(respectively,* $\mathcal{B}(\iota|_{x_i=0})$*).*

3   Transforming $\mathrm{LS}^k + \mathrm{CP}^k$ Proofs with Restricted Degree of Falsity into $\mathrm{Res}(k)$ Proofs

Theorem 1. *For any* $\mathrm{LS}^k + \mathrm{CP}^k$ *proof* Π *of a CNF formula* F, *there exists a* $\mathrm{Res}(k)$ *proof of* F *of size* $O((\binom{n}{d-1})|\Pi|(n^k + 2^{6d}))$, *where* n *is the number of variables of* F *and* $d \leq n/2$ *is the degree of falsity of* Π.

Proof. We show that for any step $\frac{\iota_1 \ (\iota_2)}{\iota}$ of the proof Π, it is possible to derive all clauses of $\mathcal{B}(lin(\iota))$ from clauses of $\mathcal{B}(lin(\iota_1))$ (and $\mathcal{B}(lin(\iota_2))$) and clauses of $F(k)$ in at most $O(\binom{n}{d-1}(n^k + 2^{6d}))$ resolution steps. Note that if an inequality ι_0 is an axiom of the proof Π, then $\mathcal{B}(lin(\iota_0))$ is a clause of F. Observe also that $\mathcal{B}(lin(0 \geq 1))$ is an empty clause. Thus, the constructed proof is a Resolution proof of F.

For the Addition and Rounding rules this is shown by Hirsch and Nikolenko [9] (see lemma below). So, we only need to consider the Multiplication rule.

Lemma 5 ([9]).

- *The Rounding rule does not change the Boolean representation.*
- *If ι is an integer linear combination of linear inequalities ι_1 and ι_2, then every clause of $\mathcal{B}(\iota)$ can be derived from $\mathcal{B}(\iota_1) \cup \mathcal{B}(\iota_2)$ in at most 2^{6d} steps, where $\mathrm{DGF}(\iota_1), \mathrm{DGF}(\iota_2) \leq d$.*

Let ι_p be a premise of the Multiplication rule, ι_c be its conclusion, and x be a literal of this rule (so that ι_c is obtained from ι_p by multiplying by x). Let also the literal form of $lin(\iota_p)$ be $\sum_{i=1}^{s} \alpha_i z_i + \sum_{i=s+1}^{s'} \alpha_i(1 - z_i) \geq c$, where $\alpha_i > 0$, for $1 \leq i \leq s'$. The literal form of ι_c depends on the sign of $(\sum_{i=s+1}^{s'} \alpha_i - c)$. Consider two cases.

1. $(\sum_{i=s+1}^{s'} \alpha_i - c) \geq 0$. In this case, the literal form of $lin(\iota_c)$ is

$$\sum_{i=1}^{s} \alpha_i z(z_i, x) + \sum_{i=s+1}^{s'} \alpha_i(1 - z(z_i, x)) + (\sum_{i=s+1}^{s'} \alpha_i - c)x \geq \sum_{i=s+1}^{s'} \alpha_i \ .$$

Note that each clause of $\mathcal{B}(lin(\iota_c))$ contains a literal $\neg z(z_i, x)$ for some $s+1 \leq i \leq s'$ (since $lin(\iota_c)$ becomes trivial when all these literals are assigned the value 0). Each clause of $\mathcal{B}(lin(\iota_c))$ containing x can be obtained by the Weakening rule from the clause $(\neg z(z_i, x) \lor x)$.

Now consider all clauses of $\mathcal{B}(lin(\iota_c))$ that do not contain x, that is, the Boolean representation of $lin(\iota_c)|_{x=1}$. Observe that $lin(\iota_c)|_{x=1}$ can be obtained from $lin(\iota_p)$ just by replacing each variable z_i by $z(z_i, x)$, thus, we can apply Lemma 2.

2. $(\sum_{i=s+1}^{s'} \alpha_i - c) < 0$. In this case, the literal form of $lin(\iota_c)$ is

$$\sum_{i=1}^{s} \alpha_i z(z_i, x) + \sum_{i=s+1}^{s'} \alpha_i(1 - z(z_i, x)) + (c - \sum_{i=s+1}^{s'} \alpha_i)(1 - x) \geq c \ .$$

Consider all clauses of $\mathcal{B}(lin(\iota_c))$ that do not contain $\neg x$. By Lemma 4, these clauses form a Boolean representation of $lin(\iota_c)|_{x=0}$. As in the previous case, all these clauses contain a literal $\neg z(z_i, x)$ for some $s + 1 \leq i \leq s'$. Note that $lin(\iota_c)|_{x=0}$ can be obtained from $lin(\iota_p)$ by replacing each variable z_i by $z(z_i, x)$ and reducing the free coefficient from c to $\sum_{i=s+1}^{s'} \alpha_i$. Thus,

$\mathcal{B}(lin(\iota_c)|_{x=0})$ can be derived from $\mathcal{B}(lin(\iota_p))$ by applying the steps described in Lemma 2 and the Weakening rule.

Each clause C of $\mathcal{B}(lin(\iota_c))$ containing $\neg x$ corresponds to a clause C_0 of $\mathcal{B}(lin(\iota_p))$ resulting from C by removing $\neg x$ and replacing each variable $z(z_i, x)$ of C by z_i. All these clauses can be derived by Lemma 2.

The Boolean representation of $lin(\iota_c)$ contains at most $\binom{n}{d-1}$ clauses, so the number of steps is as required. □

4   Lower and Upper Bounds for $\mathrm{LS}^k{+}\mathrm{CP}^k$ with Restricted Degree of Falsity

In this section we give lower and upper bounds for $\mathrm{LS}^k{+}\mathrm{CP}^k$ with restricted degree of falsity. Namely, we give short proofs of the Pigeon Hole Principle (which is known to be hard for $\mathrm{Res}(k)$, when $k \leq \sqrt{\log n / \log\log n}$, [10]) and the Weak Clique-Coloring tautologies (which are known to be hard for CP [7]). This gives exponential separation of the new system from $\mathrm{Res}(k)$ and CP. We also prove an exponential lower bound for the $\mathrm{LS}^k{+}\mathrm{CP}^k$ with DGF bounded by cn for some constant c, provided that $\mathrm{Res}(k)$ has a strongly exponential lower bound.

4.1   Short Proof of the Pigeon Hole Principle

The M to N pigeon hole principle (PHP_N^M) is coded by the following set of clauses:

$$\bigvee_{1 \leq \ell \leq M} x_{k,\ell} , \qquad 1 \leq k \leq M , \tag{2}$$

$$\neg x_{k,\ell} \vee \neg x_{k',\ell} , \qquad 1 \leq k \neq k' \leq M, 1 \leq \ell \leq N . \tag{3}$$

This set of clauses is translated into the following set of inequalities:

$$\sum_{1 \leq \ell \leq N} x_{k,\ell} \geq 1 , \qquad 1 \leq k \leq M , \tag{4}$$

$$(1 - x_{k,\ell}) + (1 - x_{k',\ell}) \geq 1 , \qquad 1 \leq k \neq k' \leq M , \quad 1 \leq \ell \leq N . \tag{5}$$

Using similar to Goerdt [2] arguments we give a short proof of this contradiction in CP (and hence in $\mathrm{LS}^k{+}\mathrm{CP}^k$) with the degree of falsity bounded by $\sqrt{n}$.

Lemma 6. *Given a set of inequalities $x_i + x_j \leq 1$ for all $1 \leq i \neq j \leq M$ and an inequality $\sum_{i=1}^M x_i + A \geq 0$, where A is a polynomial not containing the variables x_i, $1 \leq i \leq M$, we can deduce an inequality $A + 1 \geq 0$ in $O(M^2)$ steps with the degree of falsity not exceeding DGF of the initial inequalities.*

Proof. We prove by induction on s that $A + \sum_{i=1}^{s} x_i - x_{s'} + 1 \geq 0$ for all $1 \leq s' \leq s$ can be deduced.

Base: an inequality $A + \sum_{i=1}^{M-1} x_i - x_j + 1 \geq 0$ is the sum of initial inequalities $A + \sum_{i=1}^{M} x_i \geq 0$ and $1 - x_M - x_j \geq 0$.

Induction step: for all $1 \leq s' \leq s - 1$ sum the following three inequalities $A + \sum_{i=1}^{s} x_i - x_s + 1 \geq 0$, $A + \sum_{i=1}^{s} x_i - x_{s'} + 1 \geq 0$, and $1 - x_s - x_{s'} \geq 0$, and apply the Rounding rule to the result to obtain

$$A + \sum_{i=1}^{s-1} x_i - x_{s'} + 1 \geq 0 \ . \qquad\qquad \square$$

Now, summing up all inequalities (4) we have

$$\sum_{j=1}^{N} \sum_{i=1}^{M} x_{i,j} \geq M \ . \qquad\qquad (6)$$

After then, step-by-step (for $i = M, \ldots, 1$) apply Lemma 6 to obtain A_{i-1} from A_i, where A_i is

$$\sum_{j=1}^{i} \sum_{i=1}^{M} x_{i,j} + (N - i) \geq M \ .$$

It is easy to see that A_0 is a contradiction.

4.2 Short Proof of the Weak Clique-Coloring Tautologies

First, we recall the definition of the Weak Clique-Coloring tautologies. Given a graph G with N vertices, we try to color it with $M - 1$ colors, while assuming the existence of a clique of size M in G. The set of variables of this tautology consists of the three following groups:

- for $1 \leq i, j \leq N$, $p_{ij} = 1$ iff there is an edge between i-th and j-th vertices of G
- for $1 \leq i \leq N, 1 \leq k \leq M$, $q_{ki} = 1$ iff the i-th vertex of G is the k-th vertex of the clique
- for $1 \leq i \leq N, 1 \leq \ell \leq M - 1$, $r_{i\ell} = 1$ iff the i-th vertex of G is colored by the color ℓ

Thus, the number of variables n is equal to $N^2 + NM + N(M - 1)$. The contradiction is given by the following set of inequalities.

$$(1 - p_{ij}) + (1 - r_{i\ell}) + (1 - r_{j\ell}) \geq 1 \ , \quad 1 \leq i < j \leq N \ , \quad 1 \leq \ell \leq M - 1 \ , \qquad (7)$$

$$\sum_{\ell=1}^{M-1} r_{i\ell} \geq 1 \ , \quad 1 \leq i \leq N \ , \qquad\qquad (8)$$

$$\sum_{i=1}^{N} q_{ki} \geq 1 \ , \quad 1 \leq k \leq M \ , \qquad\qquad (9)$$

$$(1 - q_{ki}) + (1 - q_{k',i}) \geq 1 \ , \quad 1 \leq k \neq k' \leq M \ , \tag{10}$$

$$p_{ij} + (1 - q_{ki}) + (1 - q_{k',j}) \geq 1 \ , \quad 1 \leq i < j \leq N \ , \ \ 1 \leq k \neq k' \leq M \ . \tag{11}$$

Grigoriev et al. [7] added to them one more family of inequalities, such that any CP refutation of the new system still requires at least $2^{\Omega((n/\log n)^{1/3})}$ steps:

$$(1 - q_{kj}) + (1 - q_{ki}) \geq 1 \ , \quad 1 \leq k \leq M \ , \ \ 1 \leq i \neq j \leq N \ . \tag{12}$$

Now let us give a short proof of this contradiction with degree of falsity bounded by $\sqrt{n}$ (we just rewrite the proof of [7] by putting each inequality into its literal form in order to show that the degree of falsity is as required). First, for each i, we multiply (8) by q_{ki} and sum the resulting inequalities over i to obtain

$$\sum_{i=1}^{N} \sum_{\ell=1}^{M-1} q_{ki} r_{i\ell} + \sum_{i=1}^{N} (1 - q_{ki}) \geq N \ ,$$

Adding (9) to this inequality yields

$$\sum_{i=1}^{N} \sum_{\ell=1}^{M-1} q_{ki} r_{i\ell} \geq 1 \ . \tag{13}$$

Next, we eliminate p_{ij} from (7) and (11) and obtain

$$(1 - q_{ki}) + (1 - q_{k',j}) + (1 - r_{i\ell}) + (1 - r_{j\ell}) \geq 1 \ , \tag{14}$$

for $1 \leq i < j \leq N, \leq k \neq k' \leq M$.

Then, we sum (14) with axioms $(1-q_{ki}) r_{i\ell} \geq 0$, $q_{ki}(1-r_{i\ell}) \geq 0$, $q_{k',j}(1-r_{j\ell}) \geq 0$ and $(1 - q_{k',j}) r_{j\ell} \geq 0$ and apply the Rounding rule:

$$(1 - q_{ki} r_{i\ell}) + (1 - q_{k',j} r_{j\ell}) \geq 1 \ , \quad 1 \leq i < j \leq N \ , \ \ 1 \leq k \neq k' \leq M \ , \tag{15}$$

Using $q_{ki}(1 - r_{i\ell}) \geq 0$, $q_{kj}(1 - r_{j\ell}) \geq 0$ and (12), we obtain

$$(1 - q_{ki} r_{i\ell}) + (1 - q_{kj} r_{j\ell}) \geq 1 \ , \quad 1 \leq \ell \leq M - 1 \ , \ \ 1 \leq k \leq M \ . \tag{16}$$

Multiplying every (10) by $r_{i\ell}$ and adding $(1 - r_{i\ell}) \geq 0$ to the result, we obtain

$$(1 - q_{ki} r_{i\ell}) + (1 - q_{k',i} r_{i\ell}) \geq 1 \ . \tag{17}$$

Relations (15)–(17) imply that any length 2 sub-sum of monomials in the the sum

$$\sum_{k=1}^{M} \sum_{i=1}^{N} q_{ki} r_{i\ell} \ , \quad 1 \leq \ell \leq M - 1 \ ,$$

is bounded by 1.

The proof of the Weak Clique-Coloring tautologies is as follows. Sum (13) for all $1 \leq k \leq M$ to obtain

$$\sum_{k=1}^{M} \sum_{i=1}^{N} \sum_{\ell=1}^{M-1} q_{ki} r_{i\ell} \geq M \ . \tag{18}$$

Then, apply Lemma 6 to (15)–(17) and (18) for $s = M - 1, \ldots, 1$ to obtain

$$\sum_{k=1}^{M} \sum_{i=1}^{N} \sum_{\ell=1}^{s-1} q_{ki} r_{i\ell} + M - 1 - s \geq M \ . \tag{19}$$

4.3 An Exponential Lower Bound for $LS^k + CP^k$ with Bounded Degree of Falsity

The following lemma extends the Corollary 1, [9].

Lemma 7. *If a formula F with n variables has no $Res(k)$ proof containing less then $\exp(cn)$, $c > 0$ clauses, then for sufficiently large n this formula does not have an $LS^k + CP^k$ proof of size less than $\exp(\epsilon n)$ and degree of falsity bounded by dn for every choice of positive constants $\epsilon < c/2$ and $d < 1/2$ such that*

$$2\epsilon + 6d - d \log_2 d - (1 - d) \log_2(1 - d) \leq c \ . \tag{20}$$

Proof. By Theorem 1 any $LS^k + CP^k$ proof of size $2^{\epsilon n}$ can be transformed into a $Res(k)$ proof of size

$$\binom{n}{dn - 1} 2^{\epsilon n + 6dn + k \log_2(n)} = o(2^{(\epsilon + 6d + k \log_2(n)/n - d \log_2 d - (1-d) \log_2(1-d))n}) \ ,$$

by Stirling's formula. This is $o(2^{cn})$, since for sufficiently large n, $k \log_2(n)/n < \epsilon$. Note that $f(x) = 6x - x \log_2 x - (1 - x) \log_2(1 - x)$ decreases to 0 as x decreases from $1/2$ to 0, thus, for every $\epsilon < c/2$ there is d that satisfies (20). □

Below we show that this lemma implies an exponential lower bound on the size of $LS^k + CP^k$ proofs with bounded degree of falsity for a class of formulas that encode a linear system $Ax = b$ that has no solution over $\mathbb{GF}_2$, where the matrix A is a "good" expander.

Recall the definition of hard formulas based on expanders matrices [11] which is a generalization of Tseitin-Urquhart tautologies. For a set of strings I of a matrix $A \in \{0, 1\}^{m \times n}$, we define its *boundary* ∂I as the set of all columns J of A such that there is exactly one string $i \in I$ such that $a_{ij} = 1$ for some $j \in J$ and for all other $i' \in I$, $i' \neq i$, is true that $a_{i',j} = 0$. We say that A is an (r, s, c)-*boundary expander* if

1. Each string contains at most s ones.
2. For all set of strings I of size at most r, $|\partial I| \geq c \cdot |I|$.

Let b be a vector from $\{0, 1\}^n$. Then $\Phi(A, b)$ is a formula expressing the equality $Ax = b$ modulo 2, namely, every equation $\oplus_{l=1}^{s} a_{ij_l} x_{j_l} = b_i$ is transformed into the 2^s clauses on $x_{j_1}, \ldots, x_{j_s}$ satisfying all its solutions.

Lemma 8. *There exists a positive constant δ such that formulas $\Phi(A, b)$ with respect to $(n/2, 3, c)$-expander A have only $\exp(\Omega(n))$-size $LS^k + CP^k$ proofs with degree of falsity bounded by δn.*

Proof (sketch). The proof follows from Lemma 7 and the proof of the following theorem by Alekhnovich.

Theorem 2 (Theorem 4.1, [12]). *For any constant Δ with probability $1 - o(1)$ every $Res(k)$ refutation of a random 3-CNF formula with Δn clauses and n variables has size $\exp(n^{1-o(1)})$.*

To prove this theorem Alekhnovich showed that a random 3-CNF formula with Δn clauses and n variables with good probability is an $\Phi(A, b)$ formula for an $(r, 3, c)$-expander matrix A and proved an exponential lower bound for it.   □

Acknowledgment

The authors are very grateful to their supervisor Edward A. Hirsch for helpful comments.

References

1. Cook, S.A., Reckhow, R.A.: The Relative Efficiency of Propositional Proof Systems. The Journal of Symbolic Logic **44**(1) (1979) 36–50
2. Goerdt, A.: The Cutting Plane Proof System with Bounded Degree of Falsity. In: Proceedings of CSL 1991. Volume 626 of Lecture Notes in Computer Science., Springer (1991) 119–133
3. Robinson, J.A.: The generalized resolution principle. Machine Intelligence **3** (1968) 77–94
4. Krajíček, J.: On the weak pigeonhole principle. Fundamenta Mathematicæ **170**(1-3) (2001) 123–140
5. Gomory, R.E.: An algorithm for integer solutions of linear programs. In Graves, R.L., Wolfe, P., eds.: Recent Advances in Mathematical Programming. McGraw-Hill (1963) 269–302
6. Cook, W., Coullard, C.R., Turán, G.: On the complexity of cutting-plane proofs. Discrete Applied Mathematics **18**(1) (1987) 25–38
7. Grigoriev, D., Hirsch, E.A., Pasechnik, D.V.: Complexity of semialgebraic proofs. Moscow Mathematical Journal **2**(4) (2002) 647–679
8. Atserias, A., Bonet, M.L.: On the automatizability of resolution and related propositional proof systems. Information and Computation **189**(2) (2004) 182–201
9. Hirsch, E.A., Nikolenko, S.I.: Simulating Cutting Plane proofs with restricted degree of falsity by Resolution. In: Proceedings of SAT 2005. Volume 3569 of Lecture Notes in Computer Science., Springer-Verlag (2005) 135–142
10. Nathan Segerlind, Samuel R. Buss, Russell Impagliazzo: A Switching Lemma for Small Restrictions and Lower Bounds for k-DNF Resolution. SIAM Journal on Computing **33**(5) (2004) 1171–1200
11. Alekhnovich, M., Ben-Sasson, E., Razborov, A.A., Wigderson, A.: Pseudorandom generators in propositional proof complexity. SIAM Journal on Computing **34**(1) (2004) 67–88
12. Alekhnovich, M.: Lower bounds for k-DNF resolution on random 3-CNFs. In: STOC '05: Proceedings of the thirty-seventh annual ACM symposium on Theory of computing, New York, NY, USA, ACM Press (2005) 251–256

Categorisation of Clauses in Conjunctive Normal Forms: Minimally Unsatisfiable Sub-clause-sets and the Lean Kernel

Oliver Kullmann[1,*], Inês Lynce[2,**], and João Marques-Silva[3]

[1] Computer Science Department
University of Wales Swansea
Swansea, SA2 8PP, UK
O.Kullmann@Swansea.ac.uk
http://cs-svr1.swan.ac.uk/~csoliver
[2] Departamento de Engenharia Informática
Instituto Superior Técnico / INESC-ID
Universidade Técnica de Lisboa
ines@sat.inesc-id.pt
http://sat.inesc-id.pt/~ines
[3] School of Electronics and Computer Science
University of Southampton
Highfield, Southampton SO17 1BJ, UK
jpms@soton.ac.uk
http://www.ecs.soton.ac.uk/~jpms

Abstract. Finding out that a SAT problem instance F is unsatisfiable is not enough for applications, where *good reasons* are needed for explaining the inconsistency (so that for example the inconsistency may be repaired). Previous attempts of finding such good reasons focused on finding some minimally unsatisfiable sub-clause-set F' of F, which in general suffers from the non-uniqueness of F' (and thus it will only find *some reason*, albeit there might be others).

In our work, we develop a fuller approach, enabling a more fine-grained analysis of necessity and redundancy of clauses, supported by meaningful semantical and proof-theoretical characterisations. We combine known techniques for searching and enumerating minimally unsatisfiable sub-clause-sets with (full) autarky search. To illustrate our techniques, we give a detailed analysis of well-known industrial problem instances.

1 Introduction

Explaining the causes of unsatisfiability of Boolean formulas is a key requirement in a number of practical applications. A paradigmatic example is SAT-based model checking, where analysis of unsatisfiability is an essential step ([7,22]) for ensuring completeness of bounded model checking ([3]). Additional examples

---

* Supported by grant EPSRC GR/S58393/01.
** Supported by FCT under research project POSC/EIA/61852/2004.

A. Biere and C.P. Gomes (Eds.): SAT 2006, LNCS 4121, pp. 22–35, 2006.

include fixing wire routing in FPGAs ([24]), and repairing inconsistent knowledge from a knowledge base ([21]).

Existing work on finding the causes of unsatisfiability can be broadly organised into two main categories. The first category includes work on obtaining a *reasonable* unsatisfiable sub-formula, with no guarantees with respect to the size of the sub-formula ([5,11,28,4]). The second category includes work that provides some *guarantees* on the computed sub-formulas ([10,20,23]). Most existing work has focused on computing one minimally unsatisfiable sub-formula or all minimally unsatisfiable sub-formulas. Thus also relevant here is the literature on minimally unsatisfiable clause-sets, for example the characterisation of minimally unsatisfiable clause-sets of small deficiency ([1,9,6,12]), where [12] might be of special interest here since it provides an algorithm (based on matroids) searching for "simple" minimally unsatisfiable sub-clause-sets.

In this paper now we seek to obtain a more differentiated picture of the (potentially many and complicated) causes of unsatisfiability by a characterisation of (single) clauses based on their contribution to the causes of unsatisfiability. The following subsection gives on overview on the this categorisation of clauses.

From necessary to unusable clauses. The problem is to find some "core" in an unsatisfiable clause-set F: Previous attempts were (typically) looking for some minimally unsatisfiable sub-clause-set $F' \subseteq F$, that is, selecting some element $F' \in \mathrm{MU}(F)$ from the set of all minimally unsatisfiable sub-clause-sets of F. The problem here is that $\mathrm{MU}(F)$ in general has many elements, and thus it is hard to give meaning to this process. So let us examine the role the elements of F play for the unsatisfiability of F.

At the base level we have *necessary clauses*, which are clauses whose removal renders F satisfiable. These clauses can also be characterised by the condition that they must be used in every resolution refutation of F, and the set of all necessary clauses is $\bigcap \mathrm{MU}(F)$ (the intersection of all minimally unsatisfiable sub-clause-sets). Determining $\bigcap \mathrm{MU}(F)$ is not too expensive (assuming the SAT decision for F and sub-clause-sets is relatively easy), and every "core analysis" of F should determine these clauses as the core parts of F. It is $\bigcap \mathrm{MU}(F)$ itself unsatisfiable if and only if F has exactly one minimally unsatisfiable sub-clause-set (that is, $|\mathrm{MU}(F)| = 1$ holds), and in this case our job is finished. However, in many situations we do not have a unique minimally unsatisfiable core, but $\bigcap \mathrm{MU}(F)$ has to be "completed" in some sense to achieve unsatisfiability.

At the next level we consider *potentially necessary clauses*, which are clauses which can become necessary clauses when removing some other (appropriately chosen) clauses. The set of all potentially necessary clauses is $\bigcup \mathrm{MU}(F)$ (the union of all minimally unsatisfiable sub-clause-sets); $\bigcup \mathrm{MU}(F)$ is unsatisfiable and seems to be the best choice for a canonical unsatisfiable core of F. However, it is harder to compute than $\bigcap \mathrm{MU}(F)$, and the best method in general seems to consist in enumerating in some way all elements of $\mathrm{MU}(F)$. Clauses which are potentially necessary but which are not necessary are called *only potentially necessary*; these are clauses which make an essential contribution to the unsatisfiability of F, however not in a unique sense (other clauses may play this role as well).

$\bigcup \mathrm{MU}(F)$ is the set of all clauses in F which can be forced to be used in every resolution refutation by removing some other clauses. Now at the third and weakest level of our categorisation of "core clauses" we consider all *usable clauses*, that is, all clauses which can be used in *some* resolution refutation (without dead ends); the set of all usable clauses of F is $\mathrm{N_a}(F)$ (see below for an explanation for this notation). Clauses which are usable but not potentially necessary are called *only usable*; these clauses are superfluous from the semantical point of view (if C is only usable in F, and $F' \subseteq F$ is unsatisfiable, then also $F' \setminus \{C\}$ is unsatisfiable), however their use may considerably shorten resolution refutations of F, as can be seen by choosing F as a pigeonhole formula extended by appropriate clauses introduced by Extended Resolution: Those new clauses are only usable, but without them pigeonhole formulas require exponential resolution refutations, while with them resolution refutations become polynomial.

Dual to these three categories of "necessity" we have the corresponding degrees of "redundancy", where a SAT solver might aim at removing redundant clauses to make its life easier; however this also can backfire (by making the problem harder for the solver and harder even for non-deterministic proof procedures). The weakest notion is given by *unnecessary clauses*; the set of all unnecessary clauses is $F \setminus \bigcap \mathrm{MU}(F)$. Removing such a clause still leaves the clause-set unsatisfiable, but in general we cannot remove two unnecessary clauses simultaneously (after removal of some clauses other clauses might become necessary).

At the next (stronger) level we have *never necessary clauses*, that is, clauses which are not potentially necessary; the set of all never necessary clauses is $F \setminus \bigcup \mathrm{MU}(F)$. Here now we can remove several never necessary clauses at the same time, and still we are guaranteed to maintain unsatisfiability; however it might be that after removal of never necessary clauses the resolution complexity is (much) higher than before.

For necessary clauses we have a "proof-theoretical" characterisation, namely that they must be used in any resolution refutation, and an equivalent "semantical" characterisation, namely that removal of them renders the clause-set satisfiable. Now for unnecessary clauses we also have a semantical criterion, namely a clause is never necessary iff it is contained in every maximal satisfiable sub-clause-set.

Finally the strongest notion of redundancy is given by *unusable clauses*; the set of unusable clauses is $F \setminus \mathrm{N_a}(F)$. These clauses can always be removed without any harm (that is, at least for a non-deterministic resolution-based SAT algorithm). As shown in [16], a clause $C \in F$ is unusable if and only if there exists an *autarky* for F satisfying C. This enables a non-trivial computation of $\mathrm{N_a}(F)$ (as discussed in Section 4), which is among the categorisation algorithms considered here the least expensive one, and thus can be used for example as a preprocessing step.

Organisation of the paper. The paper is organised as follows. The next section introduces the notations used throughout the paper. Section 3 develops the proposed clause categorisation for unsatisfiable clause sets. A discussion on the computation of the lean kernel is included in Section 4. Section 5 presents results for the well-known Daimler-Chrysler's [27] problem instances. Finally, Section 6 concludes the paper and outlines future research work.

2 Preliminaries

Clause-sets and autarkies. We are using a standard environment for (boolean) clause-sets, partial assignments and autarkies; see [16,17] for further details and background. Clauses are complement-free (i.e., non-tautological) sets of literals, clause-sets are sets of clauses. The application of a partial assignment φ to a clause-set F is denoted by $\varphi * F$. An autarky for a clause-set F is a partial assignment φ such that every clause $C \in F$ touched by φ (i.e., $\mathrm{var}(\varphi) \cap \mathrm{var}(C) \neq \emptyset$) is satisfied by φ.[1] Applying autarkies is a satisfiability-equivalent reduction, and repeating the process until no further autarkies are found yields the (uniquely determined) *lean kernel* $\mathrm{N_a}(F) \subseteq F$.

Hypergraphs. A hypergraph here is a pair $G = (V, E)$, where V is a (finite) set of vertices and $E \subseteq \mathbb{P}(V)$ is a set of subsets. Let $\mathsf{C}(G) := (V(G), \{V(G) \setminus E : E \in E(G)\})$ be the *complement hypergraph* of G. Obviously we have $\mathsf{C}(\mathsf{C}(G)) = G$. A *transversal* of G is a subset $T \subseteq V(G)$ such that for all $E \in E(G)$ we have $T \cap E \neq \emptyset$; the hypergraph with vertex set V and hyperedge set the set of all minimal transversals of G is denoted by $\mathbf{Tr}(G)$; we have the well-known fundamental fact (see for example [2]) $\mathrm{Tr}(\mathrm{Tr}(G)) = \min(G)$, where $\min(G)$ is the hypergraph with vertex set $V(G)$ and hyperedges all inclusion minimal elements of G (the dual operator is $\max(G)$). An *independent set* of G is a subset $I \subseteq V(G)$ such that $V(G) \setminus I$ is a transversal of G; in other words, the independent sets of G are the subsets $I \subseteq V(G)$ such that no hyperedge $E \in E(G)$ with $E \subseteq I$ exists. Let $\mathbf{Ind}(G)$ denote the hypergraph with vertex set G and as hyperedges all maximal independent sets of G. By definition we have $\mathrm{Ind}(G) = \mathsf{C}(\mathrm{Tr}(G))$.

Sub-clause-sets. For a clause-set F let $\boldsymbol{\mathcal{USAT}(F)}$ be the hypergraph with vertex set F and hyperedges the set of all unsatisfiable sub-clause-sets of F, and let $\mathbf{MU}(F) := \min(\mathcal{USAT}(F))$. Thus $\mathrm{MU}(F)$ has as hyperedges all minimally unsatisfiable sub-clause-sets of F, and $\mathrm{MU}(F) = \emptyset \Leftrightarrow F \in \mathcal{SAT}$. And let $\boldsymbol{\mathcal{SAT}(F)}$ be the hypergraph with vertex set F and hyperedges the set of all satisfiable sub-clause-sets of F, and $\mathbf{MS}(F) := \max(\mathcal{SAT}(F))$. Thus $\mathrm{MS}(F)$ has as hyperedges all maximal satisfiable sub-clause-sets of F, and $F \in \mathrm{MS}(F) \Leftrightarrow F \in \mathcal{SAT}$; we always have $\mathrm{MS}(F) \neq \emptyset$. Finally let $\mathbf{CMU}(F) := \mathsf{C}(\mathrm{MU}(F))$ and

---

[1] Equivalently, φ is an autarky for F iff for all $F' \subseteq F$ we have $\varphi * F' \subseteq F'$.

CMS(F) $:= \mathbf{C}(\mathrm{MS}(F))$. In [20] the observation of Bailey and Stuckey has been used that for every clause-set F we have

$$\mathrm{MU}(F) = \mathrm{Tr}(\mathrm{CMS}(F)). \tag{1}$$

This can be shown as follows: By definition we have $\mathrm{MS}(F) = \mathrm{Ind}(\mathrm{MU}(F))$, whence $\mathrm{MS}(F) = \mathbf{C}(\mathrm{Tr}(\mathrm{MU}(F)))$, and thus $\mathbf{C}(\mathrm{MS}(F)) = \mathrm{Tr}(\mathrm{MU}(F))$; applying Tr to both sides we get $\mathrm{Tr}(\mathbf{C}(\mathrm{MS}(F))) = \mathrm{Tr}(\mathrm{CMS}(F)) = \mathrm{MU}(F)$.

3   Classification

Let $F \in \mathcal{USAT}$ be an unsatisfiable clause-set for this section. When we speak of a resolution refutation "using" a clause C then we mean the refutation uses C as an axiom (and we consider here only resolution refutations without "dead ends"; since we are not interested in resolution *complexity* here this can be accomplished most easily by only considering *tree* resolution refutations).

3.1   Necessary Clauses

The highest degree of necessity is given by "necessary clauses", where a clause $C \in F$ is called **necessary** if every resolution refutation of F must use C. By completeness of resolution, a clause C is necessary iff there exists a partial assignment φ satisfying $F \setminus \{C\}$. So we can compute all necessary clauses by running through all clauses and checking whether removal renders the clause-set satisfiable. The set of all necessary clauses of F is $\bigcap \mathrm{MU}(F)$. Clause-sets with $F = \bigcap \mathrm{MU}(F)$, that is, clause-sets where every clause is necessary, are exactly the minimally unsatisfiable clause-sets. So the complexity of computing $\bigcap \mathrm{MU}(F)$ is closely related to deciding whether a clause-set F is minimally unsatisfiable, which is a D^P-complete decision problem (see [25]). The corresponding (weakest) notion of redundancy is that of clauses which are **unnecessary**, which are clauses $C \in F$ such that $F \setminus \{C\}$ still is unsatisfiable, or, equivalently, clauses for which resolution refutations of F exist not using this clause.

3.2   Potentially Necessary Clauses

$C \in F$ is called **potentially necessary** if there exists an unsatisfiable $F' \subseteq F$ with $C \in F'$ such that C is necessary for F'. In other words, potentially necessary clauses become necessary (can be forced into every resolution refutation) by removing some other clauses. Obviously the set of potentially necessary clauses is $\bigcup \mathrm{MU}(F)$ (and every necessary clause is also potentially necessary). The class of (unsatisfiable) clause-sets F with $F = \bigcup \mathrm{MU}(F)$ (unsatisfiable clause-sets, where every clause is potentially necessary) has been considered in [16], and it is mentioned that these clause-sets are exactly those clause-sets obtained from minimally unsatisfiable clause-sets by the operation of crossing out variables: The operation of crossing out a set of variables V in F is denoted by $V * F$. That if F is minimally unsatisfiable, then $V * F$ is the union of minimally unsatisfiable clause-sets, has been shown in [26]. For the converse direction consider the characteristic case of two minimally unsatisfiable clause-sets F_1, F_2. Choose a new variable v and let $F := \{C \cup \{v\} : C \in F_1\} \cup \{C \cup \{\overline{v}\} : C \in F_2\}$; obviously F is minimally unsatisfiable and $\{v\} * F = F_1 \cup F_2$.

So given (unsatisfiable) F with $F = \bigcup \text{MU}(F)$, we have a (characteristic) representation $F = V * F_0$ for some minimally unsatisfiable F_0; it is conceivable but not known to the authors whether such a representation might be useful (considering "good" F_0). The complexity of deciding whether for a clause-set F we have $F = \bigcup \text{MU}(F)$ is not known to the authors; by definition the problem is in PSPACE, and it seems to be a very hard problem. See below for the computation of $\bigcup \text{MU}(F)$.

Clauses which are potentially necessary, but which are not necessary (i.e., the clauses in $\bigcup \text{MU}(F) \setminus \bigcap \text{MU}(F)$), are called **only potentially necessary**. By Lemma 4.3 in [12] we have $\bigcup \text{MU}(F) = F \setminus \bigcap \text{MS}(F)$, i.e., a clause is potentially necessary iff there exists a maximally satisfiable sub-clause-set not containing this clause, or, in other words, a clause is not potentially necessary iff the clause is in every maximally satisfiable sub-clause-set. Thus for computing $\bigcup \text{MU}(F)$ we see two possibilities:

1. Enumerating $\text{MU}(F)$ and computing $\bigcup \text{MU}(F)$.
2. Enumerating $\text{MS}(F)$ and computing $\bigcup \text{MU}(F) = F \setminus \bigcap \text{MS}(F)$ (this is more efficient than using (1), since for applying (1) we must store all elements of $\text{MS}(F)$, and furthermore it is quite possible that while $\text{MS}(F)$ is a small set, $\text{MU}(F)$ is a big set).

The corresponding (medium) degree of redundancy is given by clauses which are **never necessary** (not potentially necessary), that is, clauses which can not be forced into resolution refutations by removing some other clauses, or equivalently, clauses which are contained in every maximally satisfiable sub-clause-set. A clause which is never necessary is also unnecessary. Blocked clauses (see [14]), and, more generally, clauses eliminated by repeated elimination of blocked clauses, are never necessary; an interesting examples for such clauses are clauses introduced by extended resolution (see [15]).

3.3   Usable Clauses

The weakest degree of necessity if given by "usable clauses", where $C \in F$ is called **usable** if there exists some tree resolution refutation of F using C. Obviously every potentially necessary clause is a usable clause. By Theorem 3.16 in [16] the set of usable clauses is exactly the lean kernel $\text{N}_a(F)$. The set of F with $\text{N}_a(F) = F$, which are called *lean clause-sets* (every clause is usable) has been studied in [17], and the decision problem whether a clause-set is lean has been shown to be co-NP complete. In Section 4 we discuss the computation of the lean kernel. The corresponding strongest degree of redundancy is given by **unusable clauses**, clauses $C \in F$ which are not used in any resolution refutation, which are exactly the clauses for which an autarky φ for F exists satisfying C. An unusable clause is never necessary. Clauses which are never necessary but are which are usable are called **only usable**, and are given for example by clauses (successfully) introduced by Extended Resolution: They are never necessary as discussed before, but they are usable (since we assumed the introduction to be "successful"), and actually these clauses can exponentially speed up the resolution refutation as shown in [8]. *

3.4 Discussion

Figure 1 relates the concepts introduced above. Consider a formula with 9 clauses (represented with bullets). These clauses can be partitioned into necessary clauses (nc) and unnecessary clauses (un). The unnecessary clauses can be partitioned into only potentially necessary clauses (opn) and never necessary clauses (nn). The (disjoint) union of the only potentially necessary clauses with the necessary clauses gives the potentially necessary clauses (pn). In addition, the never necessary clauses can be partitioned into only usable clauses (ou) and unusable clauses (uu). The (disjoint) union of the potentially necessary clauses with the only usable clauses gives the usable clauses (us).

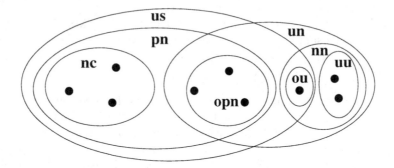

Fig. 1. Clause classification: an example

3.5 Finding the Cause

Given is an unsatisfiable clause-set F, which is partitioned into $F = F_s \cup F_u$, where F_s come from "system axioms", while F_u comes from a specific "user requirements". The unsatisfiability of F means that the user requirements together with the system axioms are inconsistent, and the task now is to find "the cause" of this problem.

First if F_u is already unsatisfiable, then the user made a "silly mistake", while if F_s already is unsatisfiable, then the whole system is corrupted. So we assume that F_u as well as F_s is satisfiable. The natural first step now is to consider $\bigcap \mathrm{MU}(F)$. The best case is that $\bigcap \mathrm{MU}(F)$ is already unsatisfiable (i.e., F has a unique minimally unsatisfiable sub-clause-set). Now $F_u \cap \bigcap \mathrm{MU}(F)$

are the critical user requirements, which together with the system properties $F_{\mathrm{s}} \cap \bigcap \mathrm{MU}(F)$ yield the (unique) contradiction. So assume that $\bigcap \mathrm{MU}(F)$ is satisfiable in the sequel.

That $F_{\mathrm{s}} \cap \bigcap \mathrm{MU}(F) \neq \emptyset$ is the case typically does not reveal much; it can be a very basic requirement which when dropped (or when "some piece is broken out of it") renders the whole system meaningless (if for example numbers would be used, then we could have "if addition wouldn't be addition, then there would be no problem"). However if $F_{\mathrm{u}} \cap \bigcap \mathrm{MU}(F) \neq \emptyset$ holds, then this could contain valuable information: These clauses could also code some very basic part of the user requirement, where without these requirements the whole user requirement breaks down, and then (again) we do not know much more than before; if however at least some clauses code some very specific requirement, then perhaps with their identification already the whole problem might have been solved.

In general the consideration of $\bigcap \mathrm{MU}(F)$ is not enough to find "the cause" of the unsatisfiability of F. Finding some $F' \in \mathrm{MU}(F)$ definitely yields some information: F' will contain some system clauses and some user clauses which together are inconsistent, however this inconsistency might not be the only inconsistency. Also if $F \setminus F'$ is satisfiable (which is guaranteed if $\bigcap \mathrm{MU}(F) \neq \emptyset$) we do not gain much, again because some very fundamental pieces might now be missing. So what really is of central importance here is $\bigcup \mathrm{MU}(F)$. The clauses $F_{\mathrm{u}} \cap \bigcup \mathrm{MU}(F)$ are exactly all (pieces) of user requirements which can cause trouble, while the clauses $F_{\mathrm{s}} \cap \bigcup \mathrm{MU}(F)$ are exactly all pieces of basic requirements needed (under certain circumstances) to complete the contraction. The clauses in $F \setminus \bigcup \mathrm{MU}(F)$, the unnecessary clauses, might be helpful to see some contradiction with less effort, but they are never really needed.

So what now is the role of $\mathrm{N}_{\mathrm{a}}(F)$ (the lean kernel, or, in other words, the set of usable clauses) here?! To identify the causes of inconsistency the clauses in $\mathrm{N}_{\mathrm{a}}(F) \setminus \bigcup \mathrm{MU}(F)$ (the only usable clauses) are not needed. One role of $\mathrm{N}_{\mathrm{a}}(F)$ is as a stepping stone for the computation of $\bigcup \mathrm{MU}(F)$, since the computation of $\mathrm{N}_{\mathrm{a}}(F)$ is easier than the computation of $\bigcup \mathrm{MU}(F)$, and removing the "fat" helps to get faster to the potentially necessary clauses. Another, quite different role now is, that the set $F \setminus \mathrm{N}_{\mathrm{a}}(F)$ of unusable clauses are the clauses satisfied by a maximal autarky φ; and this φ can be considered as the largest "conservative model", which doesn't remove any possibilities to satisfy further clauses.

Satisfying any clause from $\mathrm{N}_{\mathrm{a}}(F)$ necessarily implies that some other clause is touched but not satisfied. Trying to satisfy these touched clauses will lead to an element $F' \in \mathrm{MS}(F)$, characterised by the condition that every satisfying assignment φ for F' must falsify all clauses in $F \setminus F'$, whence these satisfying assignments are normally not useful here. In a certain sense a maximal autarky φ for F is the largest generally meaningful model for some part of F. Finding such a model yields a fulfilment of the "really harmless" user requirements. So with the set of potentially necessary clauses we covered all causes of the unsatisfiability, while with the set of unusable clauses we covered everything what can be "truly satisfied" (without remorse).

4   Computing the Lean Kernel

In Section 6 of [13] the following procedure for the computation of a "maximal autarky" φ for F (that is, an autarky φ for F with $\varphi * F = N_a(F)$) has been described, using a SAT solver $\mathcal{A}$ which for a satisfying input F returns a satisfying assignment φ with $\mathrm{var}(\varphi) \subseteq \mathrm{var}(F)$, while for an unsatisfiable input F a set $V \subseteq \mathrm{var}(F)$ of variables is returned which is the set of variables used in some (tree) resolution refutation of F:

1. Apply $\mathcal{A}(F)$; if F is satisfiable then return φ.
2. Otherwise let $F := F[V]$, and go to Step 1.

Here $F[V]$ is defined as $(V * F) \setminus \{\bot\}$, where $V * F$ denotes the operation of removing all literals x from F with $\mathrm{var}(x) \in V$, while $\bot$ is the empty clause. So the above procedure can be outlined as follows: Apply the given SAT solver $\mathcal{A}$ to F. If we obtain a satisfying assignment, then φ is a maximal autarky for the original input (and applying it we obtain the lean kernel). Otherwise we obtain a set V of variable used in a resolution refutation of F; cross out all these variables from F, remove the (necessarily obtained) empty clause, and repeat the process.

Correctness follows immediately with Theorem 3.16 in [16] together with Lemma 3.5 in [16]. More specifically, Lemma 3.5 in [16] guarantees that if by iterated reduction $F \to F[V]$ for arbitrary sets V of variables at the end we obtain some satisfiable F^* then any satisfying assignment φ for F^* with $\mathrm{var}(\varphi) \subseteq \mathrm{var}(F^*)$ is an autarky for F (thus the above process returns only autarkies). For the other direction (the non-trivial part) Theorem 3.16 guarantees that by using such V coming from resolution refutations we don't loose any autarky.

The computation of V by a SAT solver can be done following directly the correspondence between tree resolution refutations and semantic trees (for a detailed treatment see [18]). Since the set of used variables needs to be maintained only on the active path, the space required by this algorithm is (only) quadratic in the input size; the only implementation of this algorithm we are aware of is in OKsolver (as participated in the SAT 2002 competition), providing an implementation of "intelligent backtracking" without learning; see [19] for a detailed investigation.

By heuristical reasoning, a procedure computing some unsatisfiable $F' \subseteq N_a(F)$ for unsatisfiable F has been given in [28], also based on computing resolution refutations. Compared to the autarky approach, F' is some set of usable clauses, while $N_a(F)$ is the set of *all* usable clauses. Furthermore $N_a(F)$ comes with an autarky (a satisfying assignment for all the other clauses, not touching $N_a(F)$), and the computation of $N_a(F)$ can be done quite space-efficient (as outlined above), while [28] computes the whole resolution tree, and thus the space requirements can be exponential in the input size.

5   Experimental Results

The main goal of this section is to analyse a set of problem instances with respect to the concepts described above. To achieve this goal, we have selected 38 problem instances from the DC family ([27])[2]. These instances are obtained from the validation and verification of automotive product configuration data and encode different consistency properties of the configuration data base which is used to configure Daimler Chrysler's Mercedes car lines. For example, some instances refer to the stability of the order completion process (SZ), while others refer to the order independence of the completion process (RZ) or to superfluous parts (UT). We have chosen these instances because they are well known for having small minimal unsatisfiable cores and usually more than one minimal unsatisfiable core [20]. Hence, they provide an interesting testbed for the new concepts introduced in the paper.

The size of DC problem instances analysed in this paper ranges from 1659 to 1909 variables and 4496 to 8686 clauses. However, and as mentioned in [20], these formulas have a few repeated clauses and also repeated literals in clauses. Also, there are some variable codes that are not used. Consequently, we have performed a preprocessing step to eliminate the repeated clauses and literals, as well as non-used variables. In the resulting formulas the number of variables ranges from 1513 to 1805 and the number of clauses ranges from 4013 to 7562.

Table 1 gives the number of variables, the number of clauses and the average clause size for each of the 38 problem instances from the DC family. Table 1 also gives the number of minimal unsatisfiable sub-clause-sets (#MU) contained in each formula, the number of maximal satisfiable sub-clause-sets (#MS) (recall (1)), the percentage of necessary clauses (nc) and the percentage of the number of clauses in the smallest (min) and largest (max) minimal unsatisfiable sub-clause-set. Furthermore Table 1 shows the percentages of only potentially necessary clauses (opn) and the percentage of only potentially necessary clauses (pn), as well as the percentage of only usable clauses (ou) and the percentage of usable clauses (us). Then redundant clauses are considered: the percentage of unusable clauses (un), the percentage of never necessary clauses (nn) and the percentage of unnecessary clauses (un). Recall that uu stands for the clauses which can be satisfied by some autarky; in the final column we give the percentage of the uu-clauses which can be covered by (iterated) elimination of pure literals alone.[3]

These results have been obtained using a tool provided by the authors of [20], and also a Perl script for computing the lean kernel that iteratively invokes a SAT solver ([28]) which identifies variables used in a resolution refutation. From this table some conclusions can be drawn. As one would expect in general to be the case, as the number of mus's increases, the relative number of necessary clauses decreases. Regarding the number of mus's we see of lot of variation:

---

[2] Available from http://www-sr.informatik.uni-tuebingen.de/~sinz/DC/.

[3] Since all instances contain necessary clauses, the maximum (size) maximal satisfiable sub-clause-sets are always as large as possible (only one clause missing); the minimum (size) maximal satisfiable sub-clause-sets here are never much smaller, so we considered these number negligible.

Table 1. The structure of MU(F) for formulas F from the DC family, the number of variables, the number of clauses, the average clause size, the number of minimally unsatisfiable and maximal satisfiable sub-clause-sets, and as percentages the number of necessary clauses, the minimal and maximal sizes of the minimally unsatisfiable sub-clause-sets, the only potentially necessary clauses, the potentially necessary clauses, the only usable clauses, the usable clauses, the unusable clauses, the never necessary clauses and the unnecessary clauses; finally the percentage of uu (autarky reduction) achievable by pure literal elimination (iterated) alone

Bench	#vars	#cls	μ w	#MU	#MS	nc	min	max	opn	pn	ou	us	uu	m	un	pl
C208_FC_RZ_70	1513	4468	2.14	1	212	4.74	4.74	4.74	0	4.74	4.39	9.13	90.87	95.26	95.26	75.91
C208_FC_SZ_127	1513	4469	2.14	1	34	0.76	0.76	0.76	0	0.76	2.48	3.24	96.76	99.24	99.24	70.54
C208_FC_SZ_128	1513	4469	2.14	1	32	0.72	0.72	0.72	0	0.72	2.17	2.89	97.11	99.28	99.28	70.28
C208_FA_RZ_64	1516	4246	2.04	1	212	4.99	4.99	4.99	0	4.99	4.62	9.61	90.39	95.01	95.01	76.65
C208_FA_SZ_120	1516	4247	2.04	2	34	0.78	0.8	0.8	0.05	0.82	2.52	3.34	96.66	99.18	99.22	70.89
C208_FA_SZ_121	1516	4247	2.04	2	32	0.73	0.75	0.75	0.05	0.78	2.19	2.97	97.03	99.22	99.27	70.61
C208_FA_SZ_87	1516	4255	2.04	12884	139	0.31	0.42	0.63	0.8	1.1	1.06	2.16	97.84	98.9	99.69	69.59
C170_FR_RZ_32	1528	4067	2.28	32768	242	5.21	5.58	5.61	0.76	5.97	23.29	29.26	70.74	94.03	94.79	100
C170_FR_SZ_95	1528	4068	2.28	6863389	175	0.71	0.93	1.62	1.5	2.21	2.8	5.01	94.99	97.79	99.29	75.57
C170_FR_SZ_58	1528	4083	2.28	218692	177	0.91	1.13	1.54	1.44	2.35	30.69	33.04	66.96	97.65	99.09	100
C170_FR_SZ_92	1528	4195	2.28		131	3.12	3.12	3.12		3.12	31.97	35.09	64.91	96.88	96.88	100
C220_FV_RZ_14	1530	4013	2.13	80	20	0.12	0.27	0.45	0.5	0.62	16.97	17.59	82.14	99.11	99.88	95.19
C220_FV_RZ_13	1530	4014	2.13	6772	76	0.07	0.25	0.67	1.07	1.15	17.61	18.76	81.24	98.85	99.93	95.89
C220_FV_SZ_65	1530	4014	2.15	103442	198	0.45	0.57	1	2.12	2.57	15.94	18.51	81.49	97.43	99.55	95.93
C220_FV_RZ_12	1530	4017	2.13	80272	150	0.07	0.27	0.87	1.32	1.39	18.28	19.67	80.33	98.61	99.93	96.22
C220_FV_SZ_121	1530	4035	2.18	1	102	1.34	1.44	1.61	0.45	1.78	3.15	4.93	95.07	98.22	98.66	80.92
C202_FS_SZ_122	1556	5385	2.79		33	0.61	0.61	0.61	0.11	0.61	7.97	8.58	91.42	99.39	99.39	78.65
C202_FS_SZ_121	1556	5387	2.79	4	24	0.37	0.41	0.45	1.57	0.48	6.11	6.59	93.41	99.52	99.63	79.09
C202_FS_SZ_95	1556	5388	2.78	3415443	59307	0.45	0.65	0.91	2.02	2.02	1.69	3.71	96.29	97.98	99.55	80.53
C202_FS_SZ_44	1556	5399	2.79	4589596	2658	0.22	0.33	0.98	2.13	2.35	0.74	3.09	96.91	97.65	99.78	80.28
C202_FS_SZ_104	1556	5405	2.81	1626843	3593	0.48	0.48	1.72	2.09	2.57	18.1	20.67	79.33	97.43	99.52	97.07
C202_FW_RZ_57	1561	7434	3.32	1	213	2.87	2.87	2.87	0	2.87	2.63	5.5	94.5	97.13	97.13	58.33
C202_FW_SZ_124	1561	7435	3.32	1	33	0.44	0.44	0.44	0	0.54	30.41	30.85	69.15	99.56	99.56	86.5
C202_FW_SZ_123	1561	7437	3.32	4	38	0.46	0.48	0.51	0.08	0.54	30.56	31.1	68.9	99.46	99.54	86.46
C202_FW_SZ_118	1561	7562	3.3	4194235	257	0.53	0.54	1.75	1.52	2.05	27.89	29.94	70.06	97.95	99.47	98.74
C210_FS_RZ_40	1607	4894	2.87	15	212	2.74	2.86	3.54	1.53	4.27	3.89	8.16	91.84	95.73	97.26	67.76
C210_FS_SZ_129	1607	4894	2.87		33	0.67	0.67	0.67		0.67	26.87	27.54	72.46	99.33	99.33	97.67
C210_FS_SZ_130	1607	4894	2.87	15	31	0.63	0.63	0.63	1.18	0.63	26.91	27.54	72.46	99.37	99.37	87.96
C210_FW_RZ_59	1628	6381	3.75	1	212	2.1	2.19	2.71		3.28	2.97	6.25	93.75	96.72	97.9	61.43
C210_FW_SZ_135	1628	6384	3.75	15	33	0.52	0.52	0.52	0	0.52	34.47	34.99	65.01	99.48	99.48	90.36
C210_FW_SZ_136	1628	6384	3.75	1	31	0.49	0.49	0.49	0	0.49	34.5	34.99	65.01	99.51	99.51	90.36
C210_FW_SZ_129	1628	6595	3.93	3731050	584	1.88	1.9	4.26	2.5	4.38	35.13	39.51	60.49	95.62	98.12	87.96
C168_FW_UT_855	1804	6752	4.09	102	30	0.09	0.12	0.24	0.36	0.44	54.31	54.75	45.25	99.56	99.91	95.06
C168_FW_UT_854	1804	6753	4.09	102	30	0.09	0.12	0.24	0.36	0.44	54.32	54.76	45.24	99.56	99.91	95.06
C168_FW_UT_852	1804	6756	4.09	102	30	0.09	0.12	0.24	0.36	0.44	54.34	54.78	45.22	99.56	99.91	95.06
C168_FW_UT_851	1804	6758	4.08	102	30	0.09	0.12	0.24	0.36	0.44	54.35	54.79	45.21	99.56	99.91	95.06
C208_FA_UT_3254	1805	6153	4.38	17408	155	0.47	0.65	1.2	1.12	1.59	54.45	56.04	43.96	98.41	99.53	99.26
C208_FA_UT_3255	1805	6156	4.38	52736	155	0.47	0.65	1.2	1.19	1.66	54.38	56.04	43.96	98.34	99.53	99.26

Although half of the problem instances have only a few mus's, there are also many problems with many mus's. In addition, there seems to be no relation between the number of mus's and the number of mss's.

Looking at the levels of necessity, we may observe the following. For all instances the percentage of clauses in the smallest mus is quite small (in most cases less than 1%) and the largest mus is usually not much larger than the smallest one. The number of potentially necessary clauses is typically somewhat bigger than the size of the largest mus, but for all instances the set of potentially necessary clauses is still fairly small. The percentage of usable clauses is typically substantially larger, but only for the UT-family more than half of all clauses are usable. Looking at the levels of redundancy, we see that in many cases autarky reduction to a large part boils down to elimination of pure literals. In most cases most never necessary clauses are already unusable, with the notable exceptions of the UT- and (to a somewhat lesser degree) the SZ-family, while almost all unnecessary clauses are already never necessary.

6   Conclusions

This paper proposes a categorisation of clauses in unsatisfiable instances of SAT, with the objective of developing new insights into the structure of unsatisfiable formulas. The paper also addresses which sets of clauses are relevant when dealing with unsatisfiable instances of SAT. Finally, the paper evaluates the proposed categorisation of clauses in well-known unsatisfiable problem instances, obtained from industrial test cases [27].

We see the following main directions for future research:

- Regarding the industrial test cases considered, we were mainly interested in them as proof of concept, and likely there are many more interesting relations hidden in the data (especially when combining them with special insights into the structure of these formulas).
- In Subsection 3.5 we outlined a general approach for finding causes of unsatisfiability in a scenario motivated by [27]; it would now be interesting to see how helpful these considerations are in practice.
- Obviously there are many non-trivial problems regarding the complexity of the algorithms involved. A main problem here, which according to our knowledge has not been tackled until now, is the complexity of the computation of $\bigcup MU(F)$.
- For the computation of the lean kernel in this paper we considered an algorithm exploiting the "duality" between resolution proofs and autarkies. It would be interesting to compare this approach with a direct approach (directly searching for autarkies).
- Finally, it would be interesting to perform an analysis as in Table 1 on many other classes of SAT problems and to see how useful these statistics are for the categorisation of classes of problem instances.

References

1. Ron Aharoni and Nathan Linial. Minimal non-two-colorable hypergraphs and minimal unsatisfiable formulas. *Journal of Combinatorial Theory*, A 43:196–204, 1986.
2. Claude Berge. *Hypergraphs: Combinatorics of Finite Sets*, volume 45 of *North-Holland Mathematical Library*. North Holland, Amsterdam, 1989. ISBN 0 444 87489 5; QA166.23.B4813 1989.
3. Armin Biere, Alessandro Cimatti, Edmund M. Clarke, Ofer Strichman, and Y. Zhu. *Highly Dependable Software*, volume 58 of *Advances in Computers*, chapter Bounded model checking. Elsevier, 2003. ISBN 0-12-012158-1.
4. Renato Bruni. On exact selection of minimally unsatisfiable subformulae. *Annals for Mathematics and Artificial Intelligence*, 43:35–50, 2005.
5. Renato Bruni and Antonio Sassano. Restoring satisfiability or maintaining unsatisfiability by finding small unsatisfiable subformulae. In *LICS Workshop SAT 2001*, volume 9 of *ENDM*.
6. Hans Kleine Büning. On subclasses of minimal unsatisfiable formulas. *Discrete Applied Mathematics*, 107:83–98, 2000.
7. P. Chauhan, E. Clarke, J. Kukula, S. Sapra, H. Veith, and D. Wang. Automated abstraction refinement for model checking large state spaces using SAT based conflict analysis. In *International Conference on Formal Methods in Computer-Aided Design*, 2002.
8. Stephen A. Cook. A short proof of the pigeonhole principle using extended resolution. *SIGACT News*, pages 28–32, October-December 1976.
9. Gennady Davydov, Inna Davydova, and Hans Kleine Büning. An efficient algorithm for the minimal unsatisfiability problem for a subclass of CNF. *Annals of Mathematics and Artificial Intelligence*, 23:229–245, 1998.
10. M. G. de la Banda, P. J. Stuckey, and J. Wazny. Finding all minimal unsatisfiable sub-sets. In *International Conference on Principles and Practice of Declarative Programming*, 2003.
11. E. Goldberg and Y. Novikov. Verification of proofs of unsatisfiability for CNF formulas. In *Design, Automation and Test in Europe Conference*, pages 10886–10891, March 2003.
12. Oliver Kullmann. An application of matroid theory to the SAT problem. In *Fifteenth Annual IEEE Conference on Computational Complexity (2003)*, pages 116–124.
13. Oliver Kullmann. On the use of autarkies for satisfiability decision. In *LICS Workshop SAT 2001*, volume 9 of *ENDM*.
14. Oliver Kullmann. New methods for 3-SAT decision and worst-case analysis. *Theoretical Computer Science*, 223(1-2):1–72, July 1999.
15. Oliver Kullmann. On a generalization of extended resolution. *Discrete Applied Mathematics*, 96-97(1-3):149–176, 1999.
16. Oliver Kullmann. Investigations on autark assignments. *Discrete Applied Mathematics*, 107:99–137, 2000.
17. Oliver Kullmann. Lean clause-sets: Generalizations of minimally unsatisfiable clause-sets. *Discrete Applied Mathematics*, 130:209–249, 2003.
18. Oliver Kullmann. Upper and lower bounds on the complexity of generalised resolution and generalised constraint satisfaction problems. *Annals of Mathematics and Artificial Intelligence*, 40(3-4):303–352, March 2004.

19. Oliver Kullmann. Modelling the behaviour of a DLL SAT solver on random formulas. In preparation, 2006.
20. Mark H. Liffiton and Karem A. Sakallah. On finding all minimally unsatisfiable subformulas. In *Theory and Applications of Satisfiability Testing (SAT 2005)*, pages 173–186.
21. B. Mazure, L. Sais, and E. Grégoire. Boosting complete techniques thanks to local search methods. *Annals of Mathematics and Artificial Intelligence*, 22(3-4):319–331, 1998.
22. K. L. McMillan. Interpolation and SAT-based model checking. In *International Conference on Computer-Aided Verification*, 2003.
23. M. N. Mneimneh, I. Lynce, Z. S. Andraus, K. A. Sakallah, and J. P. Marques-Silva. A branch and bound algorithm for extracting smallest minimal unsatisfiable formulas. In *Theory and Applications of Satisfiability Testing (SAT 2005)*, pages 467–474.
24. G.-J. Nam, K. A. Sakallah, and R. A. Rutenbar. Satisfiability-based layout revisited: Detailed routing of complex FPGAs via search-based boolean SAT. In *International Symposium on Field-Programmable Gate Arrays*, February 1999.
25. Christos H. Papadimitriou and M. Yannakakis. The complexity of facets (and some facets of complexity). *Journal on Computer and System Sciences*, 28:244–259, 1984.
26. Ma Shaohan and Liang Dongmin. A polynomial-time algorithm for reducing the number of variables in MAX SAT problem. *Science in China (Series E)*, 40(3):301–311, June 1997.
27. Carsten Sinz, Andreas Kaiser, and Wolfgang Küchlin. Formal methods for the validation of automotive product configuration data. *Artificial Intelligence for Engineering Design, Analysis and Manufacturing*, 17(1):75–97, January 2003.
28. Lintao Zhang and Sharad Malik. Extracting small unsatisfiable cores from unsatisfiable Boolean formula. In *Sixth International Conference on Theory and Applications of Satisfiability Testing (SAT 2003)*, pages 239–249.

A Scalable Algorithm for
Minimal Unsatisfiable Core Extraction[*]

Nachum Dershowitz[1], Ziyad Hanna[2], and Alexander Nadel[1,2]

[1] School of Computer Science, Tel Aviv University, Ramat Aviv, Israel
{nachumd, ale1}@post.tau.ac.il
[2] Design Technology Solutions Group, Intel Corporation, Haifa, Israel
{ziyad.hanna, alexander.nadel}@intel.com

Abstract. We propose a new algorithm for minimal unsatisfiable core extraction, based on a deeper exploration of resolution-refutation properties. We provide experimental results on formal verification benchmarks confirming that our algorithm finds smaller cores than suboptimal algorithms; and that it runs faster than those algorithms that guarantee minimality of the core. (A more complete version of this paper may be found at arXiv.org/pdf/cs.LO/0605085.)

1 Introduction

Many real-world problems, arising in formal verification of hardware and software, planning and other areas, can be formulated as constraint satisfaction problems, which can be translated into Boolean formulas in conjunctive normal form (CNF). When a formula is unsatisfiable, it is often required to find an *unsatisfiable core*—that is, a small unsatisfiable subset of the formula's clauses. Example applications include functional verification of hardware, field-programmable gate-array (FPGA) routing, and abstraction refinement. An unsatisfiable core is a *minimal unsatisfiable core (MUC)*, if it becomes satisfiable whenever any one of its clauses is removed.

In this paper, we propose an algorithm that is able to find a minimal unsatisfiable core for large "real-world" formulas. Benchmark families, arising in formal verification of hardware (such as [8]), are of particular interest to us.

The folk algorithm for MUC extraction, which we dub *Naïve*, works as follows: For every clause C in an unsatisfiable formula F, Naïve checks if it belongs to the minimal core by invoking a propositional satisfiability (SAT) solver on F, but without clause C. Clause C does not belong to a minimal core if and only if the solver finds that $F \setminus \{C\}$ is unsatisfiable, in which case C is removed from F. In the end, F contains a minimal unsatisfiable core.

There are four more practical approaches for unsatisfiable core extraction in the current literature: adaptive core search [2], AMUSE [7], MUP [5] and a

---

[*] We thank Jinbo Huang and Zaher Andraus for their help in providing MUP and AMUSE, respectively. The work of Alexander Nadel was carried out in partial fulfillment of the requirements for a Ph.D. This research was supported in part by the Israel Science Foundation (grant no. 250/05).

A. Biere and C.P. Gomes (Eds.): SAT 2006, LNCS 4121, pp. 36–41, 2006.

resolution-based approach [9,4]. MUP is the only one guaranteeing minimality of the core, whereas the only algorithm that scales well for large formal verification benchmarks is the resolution-based approach. We refer to the latter method as the *EC (Empty-clause Cone)* algorithm.

EC exploits the ability of modern SAT solvers to produce a resolution refutation, given an unsatisfiable formula. Most state-of-the-art SAT solvers, beginning with GRASP [6], implement a DPLL backtrack search enhanced by a failure-driven assertion loop. These solvers explore the variable-assignment tree and create new *conflict clauses* at the leaves of the tree, using resolution on the initial clauses and previously created conflict clauses. This process stops when either a satisfying assignment for the given formula is found or when the empty clause (□)—signifying unsatisfiability—is derived. In the latter case, SAT solvers are able to produce a *resolution refutation* in the form of a directed acyclic graph (dag) $\Pi(V, E)$, whose vertices V are associated with clauses, and whose edges describe resolution relations between clauses. The vertices $V = V^i \cup V^c$ are composed of a subset V^i of the initial clauses and a subset V^c of the conflict clauses, including the empty clause □. The empty clause is the sink of the refutation graph, and the sources are V^i. Here, we understand a refutation to contain those clauses connected to □. The sources of the refutation comprise the unsatisfiable core returned by EC. Invoking EC until a fixed point is reached [9], allows one to reduce the unsatisfiable core even more. We refer to this algorithm as *EC-fp*. However, the resulting cores are still not guaranteed to be minimal and can be further reduced.

The basic flow of the algorithm for minimal unsatisfiable core extraction proposed in this paper is composed of the following steps:

1. Produce a resolution refutation Π of a given formula using a SAT solver.
2. For every initial clause C in Π, check whether it belongs to a MUC in the following manner:
   (a) Remove C from Π, along with all conflict clauses for which C was required to derive them. Pass all the remaining clauses (including conflict clauses) to a SAT solver.
   (b) If they are satisfiable, then C belongs to a MUC, so continue with another initial clause.
   (c) If the clauses are unsatisfiable, then C does not belong to a MUC, so replace Π by a new valid resolution refutation not containing C.
3. Terminate when all the initial clauses remaining in Π comprise a MUC.

Our basic *Complete Resolution Refutation (CRR)* algorithm is described in Sect. 2, and a pruning technique, enhancing CRR and called *Resolution Refutation-based Pruning (RRP)*, is described in Sect. 3. Experimental results are presented and analyzed in Sect. 4. This is followed up by a brief conclusion.

2   The Complete Resolution Refutation (CRR) Algorithm

One says that a vertex D is *reachable* from vertex C in graph Π if there is a path (of 0 or more edges) from C to D. The sets of all vertices that are reachable and unreachable from C in Π are denoted $Re(\Pi, C)$ and $UnRe(\Pi, C)$, respectively. The *relative hardness* of a resolution refutation is the ratio between the total number of clauses and the number of initial clauses.

Our goal is to find a minimal unsatisfiable core of a given unsatisfiable formula F. The proposed *CRR* method is displayed as Algorithm 1.

Algorithm 1 (CRR). Returns a MUC, given an unsatisfiable formula F.

1: Build a refutation $\Pi(V^i \cup V^c, E)$ using a SAT solver
2: **while** unmarked clauses exist in V^i **do**
3:     $C \leftarrow PickUnmarkedClause(V^i)$
4:     Invoke a SAT solver on $G = UnRe(\Pi, C)$
5:     **if** $UnRe(\Pi, C)$ is *satisfiable* **then**
6:         Mark C as a MUC member
7:     **else**
8:         Let $\Pi'(V_G^i \cup V_G^c, E_G)$ be the refutation built by the solver
9:         $V^i \leftarrow V^i \cap V_G^i; \ V^c \leftarrow (V_G^i \cup V_G^c) \setminus V^i; \ E \leftarrow E_G$
10: **return** V^i

First, CRR builds a resolution refutation $\Pi(V^i \cup V^c, E)$. CRR checks, for every unmarked clause C left in V^i, whether C belongs to a minimal core. Initially, all clauses are unmarked. At each stage of the algorithm, CRR maintains a valid refutation of F.

By construction of Π, the $UnRe(\Pi, C)$ clauses were derived independently of C. To check whether C belongs to a minimal core, we provide the SAT solver with $UnRe(\Pi, C)$, including the conflict clauses. We are trying to *complete the resolution refutation* without using C as one of the sources. Observe that $\square$ is always reachable from C; thus $\square$ is never passed as an input to the SAT solver. We let the SAT solver try to derive $\square$, using $UnRe(\Pi, C)$ as the input formula, or else prove that $UnRe(\Pi, C)$ is satisfiable.

In the latter case, we conclude that C must belong to a minimal core, since we found a model for an unsatisfiable subset of initial clauses minus C. Hence, if the SAT solver returns *satisfiable*, the algorithm marks C (line 6) and moves to the next initial clause. Otherwise, the SAT solver returns a valid resolution refutation $\Pi'(V_G^i \cup V_G^c, E_G)$, where $G = UnRe(\Pi, C)$. We cannot use Π' as is, as the refutation for the subsequent iterations, since the sources of the refutation may only be initial clauses of F. The necessary adjustments to the refutation are shown on line 9.

3   Resolution-Refutation-Based Pruning

In this section, we propose an enhancement of Algorithm CRR by developing resolution refutation-based pruning techniques for when the SAT solver is invoked on $UnRe(\Pi, C)$ to check whether it is possible to complete a refutation without C. We refer to the suggested technique as *Resolution Refutation-based*

Pruning (RRP). (We presume that the reader is familiar with the functionality of a modern SAT solver.)

An assignment σ *falsifies* a clause C if every literal of C is *false* under σ; it *falsifies* a set of clauses P if every clause $C \in P$ is falsified by σ. We claim that a model for $UnRe(\Pi, C)$ can only be found under a partial assignment that falsifies every clause in some path from C to the empty clause in $Re(\Pi, C)$. The reason is that otherwise there would exist a satisfiable vertex cut U in Π, contradicting the fact that the empty clause is derivable from U. (We omit a formal proof due to space limitations.)

Denote a subtree connecting C and $\square$ by $\Pi\!\restriction_C$. The RRP technique is integrated within the decision engine of the SAT solver. The solver receives $\Pi\!\restriction_C$, together with the input formula $UnRe(\Pi, C)$. The decision engine of the SAT solver explores $\Pi\!\restriction_C$ in a depth-first manner, picking unassigned variables in the currently explored path as decision variables and assigning them *false*. As usual, Boolean Constraint Propagation (BCP) follows each assignment. Backtracking in $\Pi\!\restriction_C$ is tightly coupled with backtracking in the assignment space. Both happen when a satisfied clause in $\Pi\!\restriction_C$ is found or when a new conflict clause is discovered during BCP. After a particular path in $\Pi\!\restriction_C$ has been falsified, a general-purpose decision heuristic is used until the SAT solver either finds a satisfying assignment or proves that no such assignment can be found under the currently explored path. This process continues until either a model is found or the decision engine has completed exploring $\Pi\!\restriction_C$. In the latter case, one can be sure that no model for $UnRe(\Pi, C)$ exists. However, the SAT solver should continue its work to produce a refutation. (Refer to the full version of this paper for details.)

4   Experimental Results

We have implemented CRR and RRP in the framework of the VE solver. VE, a simplified version of the industrial solver Eureka, is similar to Chaff [3]. We used benchmarks from four well-known unsatisfiable families, taken from bounded model checking (*barrel, longmult*) [1] and microprocessor verification (*fvp-unsat.2.0, pipe_unsat_1.0*) [8]. The instances we used appear in the first column of Table 1. The experiments on Families *barrel* and *fvp-unsat.2.0* were carried out on a machine with 4Gb of memory and two Intel Xeon CPU 3.06 processors. A machine with the same amount of memory and two Intel Xeon CPU 3.20 processors was used for the other experiments.

Table 1 summarizes the results of a comparison of the performance of two algorithms for suboptimal unsatisfiable core extraction and five algorithms for minimal unsatisfiable core extraction in terms of execution time and core sizes.

First, we compare algorithms for minimal unsatisfiable core extraction, namely, Naïve, MUP, plain CRR, and CRR enhanced by RRP. In preliminary experiments, we found that invoking suboptimal algorithms for trimming down the sizes of the formulas prior to MUC algorithm invocation is always useful. We used Naïve, combined with EC-fp and AMUSE, and MUP, combined with

Table 1. Comparing algorithms for unsatisfiable core extraction. Columns **Instance**, **Var** and **Cls** contain instance name, number of variables, and clauses, respectively. The next seven columns contain execution times (in seconds) and core sizes (in number of clauses) for each algorithm. The cut-off time was 24 hours (86,400 sec.). Column **Rel. Hard.** contains the relative hardness of the final resolution refutation, produced by CRR+RRP. Bold times are the best among algorithms guaranteeing minimality.

Instance	Var	Cls	Subopt. EC	EC-fp	CRR RRP	plain	Naïve EC-fp	AMUSE	MUP EC-fp	Rel. Hard.
4pipe	4237		9	171	**3527**	4933	24111	time-out	time-out	1.4
		80213	23305	17724	17184	17180	17182			
4pipe_1_ooo	4647		10	332	**4414**	10944	25074	time-out	mem-out	1.7
		74554	24703	14932	12553	12515	12374			
4pipe_2_ooo	4941		13	347	**5190**	12284	49609	time-out	mem-out	1.7
		82207	25741	17976	14259	14192	14017			
4pipe_3_ooo	5233		14	336	**6159**	15867	41199	time-out	mem-out	1.6
		89473	30375	20034	16494	16432	16419			
4pipe_4_ooo	5525		16	341	**6369**	16317	47394	time-out	mem-out	1.6
		96480	31321	21263	17712	17468	17830			
3pipe_k	2391		2	20	**411**	493	2147	12544	mem-out	1.5
		27405	10037	6953	6788	6786	6784	6790		
4pipe_k	5095		8	121	**3112**	3651	15112	time-out	time-out	1.5
		79489	24501	17149	17052	17078	17077			
5pipe_k	9330		16	169	**13836**	17910	83402	time-out	mem-out	1.4
		189109	47066	36571	36270	36296	36370			
barrel5	1407		2	19	93	**86**	406	326	mem-out	1.8
		5383	3389	3014	2653	2653	2653	2653		
barrel6	2306		35	322	**351**	423	4099	4173	mem-out	1.8
		8931	6151	5033	4437	4437	4437	4437		
barrel7	3523		124	1154	**970**	1155	6213	24875	mem-out	1.9
		13765	9252	7135	6879	6877	6877	6877		
barrel8	5106		384	9660	**2509**	2859	time-out	time-out	mem-out	1.8
		20083	14416	11249	10076	10075				
longmult4	1966		0	0	8	**7**	109	152	13	2.6
		6069	1247	1246	972	972	972	976	972	
longmult5	2397		0	1	74	**31**	196	463	35	3.6
		7431	1847	1713	1518	1518	1518	1528	1518	
longmult6	2848		2	13	**288**	311	749	2911	5084	5.6
		8853	2639	2579	2187	2187	2187	2191	2187	
longmult7	3319		17	91	6217	**3076**	6154	32791	68016	14.2
		10335	3723	3429	2979	2979	2979	2993	2979	

EC-fp. CRR performs best when combined with EC, rather than EC-fp. The sizes of the cores do not vary much between MUC algorithms, so we concentrate on a performance comparison. One can see that the combination of EC-fp and Naïve outperforms the combination of AMUSE and Naïve, as well as MUP. Plain CRR outperforms Naïve on every benchmark, whereas CRR+RRP outperforms Naïve on 15 out of 16 benchmarks (the exception being the hardest instance of *longmult*). This demonstrates that our algorithms are justified practically. Usually, the speed-up of these algorithms over Naïve varies between 4 and 10x, but it can be as large as 34x (for the hardest instance of *barrel* family) and as small as 2x (for the hardest instance of *longmult*). RRP improves performance on most instances. The most significant speed-up of RRP is about 2.5x, achieved on hard instances of Family *fvp-unsat.2.0*. The only family for which RRP is usually unhelpful is *longmult*, a family that is hard for CRR, and even harder for RRP due to the hardness of the resolution proofs of its instances.

Comparing CRR+RRP on one side and EC and EC-fp on the other, we find that CRR+RRP always produce smaller cores than both EC and EC-fp. The average gain on all instances of cores produced by CRR+RRP over cores produced by EC and EC-fp is 53% and 11%, respectively. The biggest average gain of CRR+RRP over EC-fp is achieved on Families *fvp-unsat.2.0* and *longmult* (18% and 17%, respectively). Unsurprisingly, both EC and EC-fp are usually much faster than CRR+RRP. However, on the three hardest instances of the barrel family, CRR+RRP outperforms EC-fp in terms of execution time.

5 Conclusions

We have proposed an algorithm for minimal unsatisfiable core extraction. It builds a resolution refutation using a SAT solver and finds a first approximation of a minimal unsatisfiable core. Then it checks, for every remaining initial clause, if it belongs to a minimal unsatisfiable core. The algorithm reuses conflict clauses and resolution relations throughout its execution. We have demonstrated that the proposed algorithm is faster than currently existing ones for minimal unsatisfiable cores extraction by a factor of 6 or more on large problems with non-overly hard resolution proofs, and that it finds smaller unsatisfiable cores than suboptimal algorithms.

References

1. A. Biere, A. Cimatti, E. M. Clarke, and Y. Zhu. Symbolic model checking without BDDs. In *Proc. Fifth Intl. Conf. on Tools and Algorithms for the Construction and Analysis of Systems (TACAS'99)*, pages 193–207, 1999.
2. R. Bruni. Approximating minimal unsatisfiable subformulae by means of adaptive core search. *Discrete Applied Mathematics*, 130(2):85–100, 2003.
3. Z. Fu, Y. Mahajan, and S. Malik. ZChaff2004: An efficient SAT solver. In *Proc. Seventh Intl. Conf. on Theory and Applications of Satisfiability Testing (SAT'04)*, pages 360–375, 2004.
4. E. Goldberg and Y. Novikov. Verification of proofs of unsatisfiability for CNF formulas. In *Proc. Design, Automation and Test in Europe Conference and Exhibition (DATE'03)*, pages 10886–10891, 2003.
5. J. Huang. MUP: A minimal unsatisfiability prover. In *Proc. Tenth Asia and South Pacific Design Automation Conference (ASP-DAC'05)*, pages 432–437, 2005.
6. J. P. Marques-Silva and K. A. Sakallah. GRASP: A search algorithm for propositional satisfiability. *IEEE Transactions on Computers*, 48(5):506–521, 1999.
7. Y. Oh, M. N. Mneimneh, Z. S. Andraus, K. A. Sakallah, and I. L. Markov. AMUSE: A minimally-unsatisfiable subformula extractor. In *Proc. 41st Design Automation Conference (DAC'04)*, pages 518–523, 2004.
8. M. N. Velev and R. E. Bryant. Effective use of Boolean satisfiability procedures in the formal verification of superscalar and VLIW microprocessors. In *Proc. 38th Design Automation Conference (DAC'01)*, pages 226–231, 2001.
9. L. Zhang and S. Malik. Extracting small unsatisfiable cores from unsatisfiable Boolean formula. In *Prelim. Proc. Sixth Intl. Conf. on Theory and Applications of Satisfiability Testing (SAT'03)*, 2003.

Minimum Witnesses for Unsatisfiable 2CNFs

Joshua Buresh-Oppenheim and David Mitchell

Simon Fraser University
jburesho@cs.sfu.ca, mitchell@cs.sfu.ca

Abstract. We consider the problem of finding the smallest proof of un-satisfiability of a 2CNF formula. In particular, we look at Resolution refutations and at minimum unsatisfiable subsets of the clauses of the CNF. We give a characterization of minimum tree-like Resolution refutations that explains why, to find them, it is not sufficient to find shortest paths in the implication graph of the CNF. The characterization allows us to develop an efficient algorithm for finding a smallest tree-like refutation and to show that the size of such a refutation is a good approximation to the size of the smallest general refutation. We also give a polynomial time dynamic programming algorithm for finding a smallest unsatisfiable subset of the clauses of a 2CNF.

1 Introduction

Two important areas of SAT research involve identification of tractable cases, and the study of minimum length proofs for interesting formulas. Resolution is the most studied proof system, in part because it is among the most amenable to analysis, but also because it is closely related to many important algorithms. The two most important tractable cases of SAT, 2-SAT and Horn-SAT, have linear time algorithms that can be used to produce linear-sized Resolution refutations of unsatisfiable formulas. However, for Horn formulas it is not possible even to approximate the minimum refutation size within any constant factor, unless P=NP [1]. Here, we consider the question of finding minimum-size Resolution refutations, both general and tree-like, for 2-SAT.

The linear-time 2-SAT algorithm of [2] is based on the implication graph, a directed graph on the literals of the CNF. It seems plausible that finding a minimum tree-like Resolution refutation would amount to finding shortest paths in the implication graph of the CNF. This approach is proposed in [3], but is incorrect. Hence, while [3] correctly states that finding a minimum tree-like refutation can be done in polytime, the proof is flawed. We show that a different notion of shortest path is needed, and give an $O(n^2(n + m))$-time algorithm based on BFS. We also show that such a refutation is at most twice as large as the smallest general Resolution refutation and that there are cases where this bound is tight. This contrasts with the above-mentioned inapproximability in the Horn case.

Since 2-SAT is linear time, the formula itself, or any unsatisfiable subset of its clauses, is an efficiently checkable certificate of unsatisfiability. For the question of finding a minimum unsatisfiable subset of a set of 2-clauses, analysis

A. Biere and C.P. Gomes (Eds.): SAT 2006, LNCS 4121, pp. 42–47, 2006.

of certain types of paths in the implication graph again allows us to develop a polytime algorithm. This is interesting in light the fact that finding a maximum satisfiable subset of the clauses of a 2CNF is NP-hard, even to approximate. Perhaps surprisingly, a minimum tree-like Resolution refutation of a 2CNF is not necessarily a refutation of a minimum unsatisfiable subformula. This also seems to be the case with minimum general Resolution refutations.

2   Preliminaries and Characterization

Throughout, let $\mathcal{C}$ be a collection of 2-clauses over the variables $\{x_1, ..., x_n\}$. Say $|\mathcal{C}| = m$. As first suggested by [2], $\mathcal{C}$ can be represented as a directed graph $G_{\mathcal{C}}$ on $2n$ nodes, one for each literal. If $(a \vee b) \in \mathcal{C}$ for literals a, b, then the edges $(\bar{a}, b)$ and $(\bar{b}, a)$ appear in $G_{\mathcal{C}}$ (note that literals a and b can be the same). Both of these edges are labelled by the clause $(a \vee b)$. For an edge $e = (a, b)$, let $dual(e)$, the dual edge of e, be the edge $(\bar{b}, \bar{a})$.

Consider a directed path P in $G_{\mathcal{C}}$ (that is, a sequence of not-necessarily-distinct directed edges). Note that in $G_{\mathcal{C}}$ even a simple path may contain two edges with the same clause label. Let $set(P)$ denote the set of clause-labels underlying the edges of P. We define $|P|$, the *size* of the path P, to be $|set(P)|$. In contrast, let $length(P)$ denote the length of P as a sequence. Call a path P *singular* if it does not contain two edges that have the same clause label. For any singular path P, $|P| = length(P)$.

For literals a, b, define $\mathcal{P}_{ab}$ to be the set of all simple, directed paths from a to b in $G_{\mathcal{C}}$. If c is also a literal, let $\mathcal{P}_{abc}$ be the set of all simple, directed paths that start at a, end at c and visit b at some point. Let $P \in \mathcal{P}_{ab}$. We say P is *minimum* if it has minimum size among all paths in $\mathcal{P}_{ab}$.

Proposition 1 ([2]). *If $\mathcal{C}$ is unsatisfiable, then there is a variable x such that there is a path from x to $\bar{x}$ and a path from $\bar{x}$ to x in $G_{\mathcal{C}}$. Furthermore, for any Resolution derivation of the clause $(\bar{a} \vee b)$ ($\bar{a}$ and b need not be distinct) there must a path $P \in \mathcal{P}_{ab}$ whose labels are contained in the axioms of this derivation.*

Let $P \in \mathcal{P}_{ab}$. Let $IR(P)$ be the Input Resolution derivation that starts by resolving the clauses labelling the first two edges in P and then proceeds by resolving the latest derived clause with the clause labelling the next edge in the sequence P. This is a derivation of either $(\bar{a} \vee b)$ or simply (b). It is not hard to see that the size of the derivation $IR(P)$ is $2 \cdot length(P) - 1$.

For a path $P = (e_1, ..., e_k) \in \mathcal{P}_{ab}$, let $dual(P) \in \mathcal{P}_{\bar{b}\bar{a}}$ be the path $(dual(e_k), ..., dual(e_1))$. Let $suf(P)$ be the maximal singular suffix of P (as a sequence). Similarly, let $pre(P)$ be the maximal singular prefix of P. For a simple path $P \in \mathcal{P}_{ab\bar{b}}$, let $extend(P)$ be the following path in $\mathcal{P}_{a\bar{a}}$: let P' be the portion of P that starts at a and ends at b. Then $extend(P)$ is the sequence P concatenated with the sequence $dual(P')$. If $P \in \mathcal{P}_{a\bar{a}b}$, then $extend(P) \in \mathcal{P}_{\bar{b}b}$ is defined similarly.

Proposition 2. *Let x be a literal and let $P \in \mathcal{P}_{x\bar{x}}$. There is some literal a (possibly equal to x) such that $suf(P) \in \mathcal{P}_{a\bar{a}\bar{x}}$, $pre(P) \in \mathcal{P}_{xa\bar{a}}$. If P is minimum, then $extend(suf(P))$ and $extend(pre(P))$ are minimum.*

Lemma 1. *Assume a clause* (a)*, for some literal* a*, has a Resolution derivation from* $\mathcal{C}$*. Then the size of the smallest Resolution derivation of* (a) *is* $2\ell - 1$*, where* $\ell = \min_{P \in \mathcal{P}_{\bar{a}a}} |P|$*. Moreover, if* P *is the minimum such path, then* $IR(suf(P))$ *is a smallest derivation.*

Proof. We first show that there is an input derivation of size at most $2\ell - 1$. Let P be a minimum path from $\bar{a}$ to a. Then $length(suf(P)) = |suf(P)| = \ell$ and, by Proposition 2, there is some b such that $suf(P) \in \mathcal{P}_{b\bar{b}a}$. Let P' be the prefix of $suf(P)$ that ends at literal $\bar{b}$. Then $IR(P')$ is a derivation of the singleton clause $(\bar{b})$ and $IR(suf(P))$ is a derivation of (a). This derivation has size $2 \cdot length(suf(P)) - 1 = 2\ell - 1$.

To see that any Resolution derivation of (a) has size at least $2\ell - 1$, assume otherwise. Any Resolution derivation that uses k axioms has size at least $2k - 1$, so (a) is derivable from $\ell' < \ell$ axioms of $\mathcal{C}$. These axioms cannot form a path from $\bar{a}$ to a by minimality, so (a) cannot be derived from them by Proposition 1.

3   Finding Minimum Tree-Like Refutations

Lemma 1 gives us the size of a minimum tree-like Resolution refutation of any contradictory $\mathcal{C}$ and suggests a way to find one. Let $size_{gen}(\mathcal{C})$ $(size_{tree}(\mathcal{C}))$ be the size of a smallest general (tree-like) Resolution refutation of $\mathcal{C}$. Then, $size_{tree}(\mathcal{C})$ is $2\min_{i \in [n]} \left(\min_{P \in \mathcal{P}_{x_i \bar{x}_i}} |P| + \min_{P \in \mathcal{P}_{\bar{x}_i x_i}} |P| \right) - 1$. That is, any minimum tree-like refutation of $\mathcal{C}$ consists of minimum derivations of x_i and $\bar{x}_i$, for some x_i, plus the empty clause. Such derivations of x_i and $\bar{x}_i$ come from input derivations along the suffix of minimum paths from x_i to $\bar{x}_i$ and vice versa. We search for such suffixes by doing BFS from x_i, avoiding already-used clause labels, until either we reach $\bar{x}_i$ or, for some literal y, both y and $\bar{y}$ are visited along the same path (the latter case constitutes the prefix of a minimum path in $\mathcal{P}_{x_i \bar{x}_i}$, which defines the suffix to be used in the minimum derivation of $\bar{x}_i$).

The algorithm proceeds as follows. For each literal x, perform a modified BFS starting at x, except: (1) Whenever y is reached from x, store a list $L_1(y)$ of all clause-labels on the path from x to y and a list $L_2(y)$ of all literals on the path from x to y; (2) If $\bar{y}$ appears in $L_2(y)$, set $path(x, \bar{x})$ to $extend(path(x, y))$. Terminate BFS at this point; Otherwise, (3) when continuing from y, avoid all edges labelled with clauses in $L_1(y)$. When BFS is completed for each literal, find a literal x such that $|path(x, \bar{x})| + |path(\bar{x}, x)|$ is minimum. The tree-like refutation is $IR(suf(path(x, \bar{x})))$, $IR(suf(path(\bar{x}, x)))$ and the empty clause.

BFS, runs in time $O(n + m)$; Doing it for each literal takes time $O(n(n + m))$. Adding the time to check lists L_1 and L_2, the algorithm takes time $O(n^2(n + m))$.

Theorem 1. *For any contradictory 2CNF* $\mathcal{C}$*,* $size_{tree}(\mathcal{C}) < 2\, size_{gen}(\mathcal{C})$*.*

Proof. Let π be the minimum General Resolution refutation of $\mathcal{C}$. Assume π ends by resolving variable x with $\bar{x}$. Assume, wlog, that the minimum Resolution derivation of x is at least as big as the minimum derivation of $\bar{x}$, and let ℓ be the size of this derivation. Clearly $size(\pi) \geq \ell$ since π contains a derivation of

x. In fact, $size(\pi) \geq \ell + 1$ since π also contains the empty clause (which is not used in the derivation of x). On the other hand, there is a tree-like refutation of size at most $2\ell + 1$: use the minimum derivations of x and $\bar{x}$, which are tree-like by Lemma 1, and then resolve the two.

Hence, the algorithm for finding the shortest tree-like refutation is an efficient 2-approximation for computing $size_{gen}$. In fact, this algorithm cannot do better than a 2-approximation in the worst-case.

Theorem 2. *For any $\epsilon > 0$, there exists a contradictory 2CNF C_n such that $size_{tree}(C) \geq (2 - \epsilon) \cdot size_{gen}(C)$.*

Proof. Choose n such that $2\epsilon n \geq 9$. C will be a formula over $n + 1$ variables $\{a, x_1, \ldots, x_n\}$ with the following clauses: $(\bar{a} \vee x_1), \{(\bar{x}_i \vee x_{i+1})\}_{i=1}^{n-1}, (\bar{x}_n \vee \bar{a}), (a \vee x_1), (\bar{x}_n \vee a)$. It is not hard to verify that $\forall y, P \in \mathcal{P}_{y\bar{y}}, P' \in \mathcal{P}_{\bar{y}y}, |P| + |P'| \geq 2n + 2$. Any refutation must consist of a derivation of y, a derivation of $\bar{y}$ and the empty clause, for some variable y. By Lemma 1, the size of a derivation for y plus the size of a derivation for $\bar{y}$ must be at least $2(2n + 2) - 2 = 4n + 2$, so any tree-like refutation has size at least $4n + 3$.

On the other hand, there is a general Resolution refutation that proceeds as follows: derive the clause $(\bar{x}_1 \vee x_n)$ using an input derivation of size $2(n-1)-1 = 2n - 3$. Using also $(\bar{a} \vee x_1)$ and $(\bar{x}_n \vee \bar{a})$, derive $\bar{a}$. Likewise, using $(a \vee \bar{x}_n)$ and $(x_1 \vee a)$ and the already-derived $(\bar{x}_1 \vee x_n)$, derive a. Finally derive the empty clause. This derivation has size $2n - 3 + 4 + 4 + 1 = 2n + 6$. Certainly $4n + 3 \geq (2 - \epsilon)(2n + 6)$.

4   Finding Minimum Unsatisfiable Subformulas

Any unsatisfiable subformula of C must have a variable x for which there is a path from x to $\bar{x}$ and a path from $\bar{x}$ to x in G_C. However, each of these paths might use the same clause twice and the two paths may share clauses. Therefore, we are searching for the set of clauses that comprise the paths that minimize the expression $\min_x \min_{P_1 \in \mathcal{P}_{x\bar{x}}, P_2 \in \mathcal{P}_{\bar{x}x}} |set(P_1) \cup set(P_2)|$. Call two such paths *joint-minimum*. Define the *cost* of any two paths P_1 and P_2 to be $|set(P_1) \cup set(P_2)|$.

Proposition 2 states that if P is minimum path, then $extend(suf(P))$ is minimum. We can say a similar thing about joint-minimum paths: If P_1 and P_2 are joint-minimum, then $extend(suf(P_1))$ and $extend(suf(P_2))$ are joint-minimum, and $cost(suf(P_1), suf(P_2)) = cost(P_1, P_2)$. Therefore, we need to find not-necessarily distinct literals x, a, b and singular paths $P_1 \in \mathcal{P}_{a\bar{a}\bar{x}}$ and $P_2 \in \mathcal{P}_{b\bar{b}x}$ of minimum cost.

A *segment* of a path is a consecutive subsequence of the path's sequence. For two singular paths P_1 and P_2, a *shared segment* is a maximal common segment. A *dual shared segment* of P_1 with respect to P_2 is a maximal segment t of P_1 such that $dual(t)$ is a segment of P_2. For two disjoint segments s and t of P, say $s \prec_P t$ if s appears before t in P.

Consider the following properties of two paths P_1 and P_2.

Property I: Let $s_1 \prec_{P_1} \cdots \prec_{P_1} s_k$ be the shared segments of P_1 and P_2. Then $s_k \prec_{P_2} \cdots \prec_{P_2} s_1$.

Property II: Let $t_1 \prec_{P_1} \cdots \prec_{P_1} t_\ell$ be the dual shared segments of P_1 with respect to P_2. Then $dual(t_1) \prec_{P_2} \cdots \prec_{P_2} dual(t_\ell)$.

Property III: Let $s_1 \prec_{P_1} \cdots \prec_{P_1} s_k$ be the shared segments of P_1 and P_2 and let $t_1 \prec_{P_1} \cdots \prec_{P_1} t_\ell$ be the dual shared segments of P_1 with respect to P_2. For any i, j, $t_i \prec_{P_1} s_j$ if and only if $dual(t_i) \prec_{P_2} s_j$.

Lemma 2. *There are joint-minimum paths P_1 and P_2 such that $suf(P_1)$ and $suf(P_2)$ satisfy Properties I-III.*

Proof. Consider Property I. If $suf(P_1)$ and $suf(P_2)$ violate the property, then there is some $i < j$ such that $s_i \prec_{P_2} s_j$. Let P_1' be the segment of P_1 starting at the beginning of s_i and ending at the end of s_j. Likewise, let P_2' be the segment of P_2 that starts at the beginning of s_i and ends at the end of s_j. Assume, wlog, that $length(P_1') \le length(P_2')$. Let P_2'' be the path P_2 with P_2' replaced by P_1'. Certainly P_1 and P_2'' are still joint-minimum. Property II follows in the same way by looking at P_1 and $dual(P_2)$.

Consider Property III. If $suf(P_1)$ and $suf(P_2)$ violate the property, then there is some i, j such that, wlog, $t_i \prec_{P_1} s_j$, but $s_j \prec_{P_2} dual(t_i)$. Let a, b be the endpoints of t_i. Then there is a cycle that includes a and $\bar{a}$ that uses a strict subset of the edges of P_1 and P_2.

The algorithm will search for the suffixes guaranteed by Lemma 2. More generally, given two pairs of endpoints (and possibly two intermediate points), we will find a pair of (not necessarily singular) paths P_1 and P_2 that obey Properties I-III, that have the specified endpoints (and perhaps intermediate points) and that have minimum cost over all such pairs of singular paths. The fact that P_1 and P_2 themselves may not be singular is not a problem since they will achieve the same optimum that singular paths achieve.

The algorithm uses dynamic programming based on the following idea. The reason joint-minimum paths P_1 and P_2 may not each be of minimum length is that, while longer, they benefit by sharing more clauses. If we demand that P_1 and P_2 have a shared segment with specified endpoints, then that segment should be as short as possible; likewise, for any segment of, say, P_1 with specified endpoints that is guaranteed not to overlap any shared segment. By doing this, we isolate segments of P_1 and P_2 that we can locally optimize and then concentrate on the remainder of the paths.

We will compute a table $A[(a_1, b_1, c_1), (a_2, b_2, c_2), k, \ell]$ which stores the minimum of $cost(P_1, P_2)$ over all paths $P_1 \in \mathcal{P}_{a_1 b_1 c_1}$ and $P_2 \in \mathcal{P}_{a_2 b_2 c_2}$ such that: **(1)** We recognize at most k shared segments between P_1 and P_2; **(2)** We recognize at most ℓ dual shared segments of P_1 with respect to P_2; and **(3)** P_1, P_2 obey Properties I-III. By "recognizing" k shared segments, we mean that if there are more shared segments, their lengths are added twice to the cost of P_1 and P_2, with no benefit from sharing. If we omit b_1, respectively b_2, as a parameter in $A[\ ]$, then P_1, respectively P_2, comes from $\mathcal{P}_{a_1 c_1}$.

To begin, for all literals a, b, set $B[a, b]$ to the length of a shortest path in $\mathcal{P}_{ab}$. Likewise, set $B[a, b, c]$ to the length of a shortest path in $\mathcal{P}_{abc}$. For all $a_1, b_1, c_1, a_2, b_2, c_2$, set $A[(a_1, b_1, c_1), (a_2, b_2, c_2), 0, 0]$ equal to $B[a_1, b_1, c_1] + B[a_2, b_2, c_2]$. Set $A[((a_1, c_1), (a_2, c_2), 0, 0]$ to $B[a_1, c_1] + B[a_2, c_2]$.

To compute a general entry in A where ℓ is nonzero, let P_1 and P_2 be the paths that achieve the minimum corresponding to the entry in question. By Properties II and III, there are two cases. **(1)** The first shared segment of any kind in P_1 (in order of appearance) is a dual shared segment t_1 and $dual(t_1)$ is the first shared segment of any kind in P_2. **(2)** The last shared segment of any kind in P_1 is a dual shared segment t_k and $dual(t_k)$ is the last shared segment of any kind in P_2.

Suppose we are in Case 1 (Case 2 is similar). We try placing b_1 before, in, or after t_1 in P_1 (likewise for $b_2, dual(t_1), P_2$) and we try all endpoints for t_1. For example, in the case where we try placing b_1 before t_1 in P_1 and b_2 before $dual(t_1)$ in P_2, we take the minimum over all literals u, v, of

$$B[a_1, b_1, u] + B[a_2, b_2, \bar{v}] + B[u, v] + A[(v, c_1), (\bar{u}, c_2), k, \ell - 1].$$

Then we assign A the minimum over all nine placements of b_1 and b_2. Finally, if $A[(a_1, b_1, c_1), (a_2, b_2, c_2), k, \ell - 1]$ is less than the calculated value, we replace the current entry with that.

If $\ell = 0$ and k is nonzero, we proceed similarly except that the first shared segment in P_1 is the last shared segment in P_2 by Property I. Therefore (placing b_1 before s_1 and b_2 before s_k), we take the minimum over all u, v of

$$B[a_1, b_1, u] + B[v, c_2] + B[u, v] + A[(v, c_1), (a_2, b_2, u), k - 1, 0].$$

Again, minimize over all b_1 and b_2, then check $A[(a_1, b_1, c_1), (a_2, b_2, c_2), k - 1, 0]$.

The size of the joint minimum paths will finally be stored in $A[(a, \bar{a}, x), (b, \bar{b}, \bar{x}), n, n]$ for some literals a, b, x; we simply find the smallest such entry. We can recover the actual set of edges comprising these paths using the standard dynamic-programming technique of remembering which other entries of A were used to compute the current entry. The algorithm is clearly polynomial time, since there are polynomially-many entries in A and each one is computed as the minimum of polynomially-many expressions.

References

1. A. Alekhnovich, S. Buss, S. Moran, and T. Pitassi. Minimal propositional proof length is NP-hard to linearly approximate. In *Proc., MFCS'98*, pages 176–184, 1998. (Also LNCS 1450).
2. Bengt Aspvall, Michael F. Plass, and Robert Endre Tarjan. A linear-time algorithm for testing the truth of certain quantified boolean formulas. *Information Processing Letters*, 8(3):121–123, March 1979.
3. K. Subramani. Optimal length tree-like resolution refutations for 2SAT formulas. *ACM Transactions on Computational Logic*, 5(2):316–320, April 2004.

Preliminary Report on Input Cover Number as a Metric for Propositional Resolution Proofs

Allen Van Gelder

University of California, Santa Cruz CA 95060, USA
http://www.cse.ucsc.edu/~avg

Abstract. Input Cover Number (denoted by κ) is introduced as a metric for difficulty of propositional resolution derivations. If $\mathcal{F} = \{C_i\}$ is the input CNF formula, then $\kappa_{\mathcal{F}}(D)$ is defined as the minimum number of clauses C_i needed to form a superset of (i.e., cover) clause D. Input Cover Number provides a refinement of the clause-width metric in the sense that it applies to families of formulas whose clause width grows with formula size, such as pigeon-hole formulas $\mathrm{PHP}(m, n)$ and $\mathrm{GT}(n)$. Although these two families have much different general-resolution complexities, it is known that both require $\Theta(n)$ clause width (after transforming to 3-CNF). It is shown here that κ is $\Theta(n)$ for pigeon-hole formulas and is $\Theta(1)$ for $\mathrm{GT}(n)$ formulas and variants of $\mathrm{GT}(n)$.

1 Introduction

Ben-Sasson and Wigderson showed that, if the minimum-length general resolution refutation for a CNF formula $\mathcal{F}$ has S steps, and if the minimum-length tree-like refutation of $\mathcal{F}$ has S_T steps, then there is a (possibly different) refutation of $\mathcal{F}$ using clauses of width at most:

$$w(\mathcal{F} \vdash \bot) \le w(\mathcal{F}) + c \sqrt{n \ln S}; \tag{1}$$
$$w(\mathcal{F} \vdash \bot) \le w(\mathcal{F}) + \lg S_T. \tag{2}$$

where $\mathcal{F}$ has n variables and $w(\mathcal{F} \vdash \bot)$ denotes resolution-refutation width. The $w(\mathcal{F})$ terms were omitted from their statement in the introduction, but appear in the theorems [3].

Our first results essentially eliminate the $w(\mathcal{F})$ terms in the Ben-Sasson and Wigderson theorems, and replace resolution width by $\kappa_{\mathcal{F}}(\pi)$, the *input cover number*, as defined below.

Our interest in input cover number stems from the indications that it separates polynomial families from super-polynomial families for a wide class of formulas that represent SAT encodings of *constraint satisfaction* problems.

Two prototypical and widely studied examples are the pigeon-hole family $\mathrm{PHP}(n + 1, n)$ and the $\mathrm{GT}(n)$ family. Both families have a similar appearance: $\Theta(n)$ clause width, $\Theta(n^2)$ propositional variables, $\Theta(n^3)$ clauses, and $\Theta(n^3)$ overall formula length. However, the pigeon-hole family has minimum resolution length in $\Omega(2^n)$ [6,3], whereas the $\mathrm{GT}(n)$ family has minimum resolution length

A. Biere and C.P. Gomes (Eds.): SAT 2006, LNCS 4121, pp. 48–53, 2006.

in $O(n^3)$ [9,4]. The clause-width metric does not distinguish between these two families: after the standard transformations into 3-CNF, giving $\text{EPHP}(n+1, n)$ and $\text{MGT}(n)$, they both have lower bounds for $w(\mathcal{F} \vdash \perp)$ in $\Omega(n)$ [3,4]. The input distance metric [10] also does not distinguish them. We show that the input-cover-number metric distinguishes sharply between them: $\kappa(\text{PHP}(n+1, n) \vdash \perp)$ is in $\Theta(n)$, whereas $\kappa(\text{GT}(n) \vdash \perp)$ is in $\Theta(1)$.

Definition 1.1. (input cover number) All clauses mentioned are nontautologous sets of literals. Let D be a clause; let C be a clause of formula $\mathcal{F}$. The *input cover number* of D w.r.t. $\mathcal{F}$, denoted $\kappa_\mathcal{F}(D)$, is the minimum number of clauses $C_i \in \mathcal{F}$ such that $D \subseteq \bigcup_i C_i$, i.e., the cardinality of the minimum set cover.

For a resolution proof π $\kappa_\mathcal{F}(\pi)$ is the maximum over $D \in \pi$ of the *input cover numbers* of D w.r.t. $\mathcal{F}$.

When $\mathcal{F}$ is understood from the context, $\kappa(D)$ and $\kappa(\pi)$ are written. $\kappa(\mathcal{F} \vdash D)$ denotes the minimum of $\kappa_\mathcal{F}(\pi)$ over all π that are derivations of D from $\mathcal{F}$.

The theorems shown in the full paper[1] are that, if π is a resolution refutation of $\mathcal{F}$ and π uses all clauses of $\mathcal{F}$ and the length of π is S, then there is a refutation of $\mathcal{F}$ using clauses that have *input cover number* w.r.t. $\mathcal{F}$ that is at most:

$$\kappa(\mathcal{F} \vdash \perp) \leq c\sqrt{n \ln S}; \tag{3}$$

$$\kappa(\mathcal{F} \vdash \perp) \leq \lg S_T. \tag{4}$$

Proofs and additional details may be found in full paper.

Also, we show that the pigeon-hole family of formulas $\text{PHP}(m, n)$ require refutations with input cover number $\Omega(n)$, although they contain clauses of width n. This result suggests that input cover number provides a refinement of the clause-width metric as a measure of resolution difficulty. That is, when a family of formulas with increasing clause-width, such as $\text{PHP}(m, n)$, is transformed into a bounded-width family, such as $\text{EPHP}(m, n)$, and the bounded-width family has large resolution width, this is not simply because they rederive the wide clauses of the original family, then proceed to refute the original family. Rather, it is the case that wide clauses substantially different from those in the original family must be derived.

Although the results are promising in some cases, the input-cover-number metric has an inherent fragility. Although $\kappa(\text{GT}(n) \vdash \perp)$ is in $\Theta(1)$ for the natural encoding of $\text{GT}(n)$, for 3-CNF variant, $\kappa(\text{MGT}(n) \vdash \perp)$ is necessarily the same order of magnitude as the clause-width lower bound, $w(\text{MGT}(n) \vdash \perp)$, i.e., in $\Omega(n)$.

Recall that Bonet and Galesi showed that $w(\text{MGT}(n) \vdash \perp)$ is in $\Omega(n)$, yet $\text{MGT}(n)$ has a refutation in $\Theta(n^3)$ [4]. Due to the fragility of κ mentioned in the previous paragraph, the following attractive conjecture must *fail*: If a family has κ in $\Omega(n)$, its refutation length must be super-polynomial in n. The full paper discusses fragilities of κ in greater length.

---

[1] See http://www.cse.ucsc.edu/~avg/Papers/cover-number.{pdf,ps}.

Table 1. Summary of notations

$a, \ldots, z$	Literal; i.e., propositional variable or negated propositional variable.
$A, \ldots, Z$	Disjunctive clause, or set of literals, depending on context.
$\mathcal{A}, \ldots, \mathcal{H}$	CNF formula, or set of literals, depending on context.
π	Resolution derivation DAG.
$[p_1, \ldots, p_k]$	Clause consisting of literals $p_1, \ldots, p_k$.
$\bot, \top$	*empty clause, tautologous clause.*
$\alpha, \ldots, \delta$	Subclause, in the notation $[p, q, \alpha]$.
C^-	Read as "C, or some clause that subsumes C".
$\mathbf{res}(q, C, D)$	Resolvent of C and D, where q and $\neg q$ are the clashing literals (see Definition 2.1).
$C\|\mathcal{A}, \quad \mathcal{F}\|\mathcal{A},$ $\pi\|\mathcal{A}$	C (respectively $\mathcal{F}$, π) *restricted* by $\mathcal{A}$ (see Definition 2.3).

2 Preliminaries

Notations are summarized in Table 1. Although the general ideas of resolution and derivations are well known, there is no standard notation for many of the technical aspects, so it is necessary to specify our notation in detail.

Defining resolution as a total function removes the need to include the weakening rule in the proof system. Numerous proof complexity papers include the weakening rule as a crutch to handle "life after restrictions" [3,4,1]. However, according to Alasdair Urquhart, the weakening rule might add power to some resolution strategies, such as linear resolution. See Table 1 for the notation of the resolution operator, which satisfies these symmetries:

$$\mathbf{res}(q, C, D) = \mathbf{res}(q, D, C) = \mathbf{res}(\neg q, C, D) = \mathbf{res}(\neg q, D, C).$$

Definition 2.1. (resolution, tautologous) A clause is *tautologous* if it contains complementary literals. All tautologous clauses are considered to be indistinguishable and are denoted by $\top$.

Fix a total order on the clauses definable with the n propositional variables such that $\bot$ is smallest, $\top$ is largest, and wider clauses are "bigger" than narrower clauses. Other details of the total order are not important. The following table, in which α and β denote clauses that do not contain q or $\neg q$, extends resolution to a total function:

C	D	$\mathbf{res}(q, C, D)$
$[q, \alpha]$	$[\neg q, \beta]$	$[\alpha, \beta]$
$[\gamma]$	$\top$	$[\gamma]$
$[\alpha]$	$[\neg q, \beta]$	$[\alpha]$
$[\alpha]$	$[\beta]$	smaller of α, β

$\square$

With this generalized definition of resolution, we have an algebra, and the set of clauses (including $\top$) is a lattice. Now resolution "commutes up to subsumption" with *restriction* (see Definition 2.3), so restriction can be applied to any resolution derivation to produce another derivation.

Definition 2.2. (derivation, refutation) A *derivation* (short for *propositional resolution derivation*) from formula $\mathcal{F}$ is a *rooted*, directed acyclic graph (DAG) in which each vertex is labeled with a clause and, unless it is a *leaf* ($C \in \mathcal{F}$), it is also labeled with a clashing literal and has two out-edges.   □

Definition 2.3. (restricted formula, restricted derivation) Let $\mathcal{A}$ be a partial assignment for formula $\mathcal{F}$. Let π be a derivation from $\mathcal{F}$. Read "$|\mathcal{A}$" as "restricted by $\mathcal{A}$".

1. $C|\mathcal{A} = \top$, if C contains any literal that occurs in $\mathcal{A}$, otherwise $C|\mathcal{A} = C - \{\neg q \mid q \in \mathcal{A}\}$.
2. $\mathcal{F}|\mathcal{A}$ results by applying restriction to each clause in $\mathcal{F}$.
3. $\pi|\mathcal{A}$ is defined differently from most previous papers. It is the same DAG as π structurally, but the clauses labeling the vertices are changed as follows. If a leaf (input clause) of π contains C, then the corresponding leaf of $\pi|\mathcal{A}$ contains $C|\mathcal{A}$. Each derived clause of $\pi|\mathcal{A}$ uses resolution on the same clashing literal as the corresponding vertex of π.   □

Lemma 2.4. Given formula $\mathcal{F}$, and a restriction literal p,

$$\mathbf{res}(q, D_1|p, D_2|p) \subseteq \mathbf{res}(q, D_1, D_2)|p.$$

Lemma 2.5. Given formula $\mathcal{F}$, and a restriction literal p, if π is a derivation of C from $\mathcal{F}$, then $\pi|p$ is a derivation of $(C|p)^-$ (a clause that subsumes $C|p$) from $\mathcal{F}|p$.

Lemma 2.6. Let C be a clause of $\mathcal{F}$ and let $\mathcal{A}$ be a partial assignment. If $C|\mathcal{A} \neq \top$ (i.e., $\mathcal{A}$ does not satisfy C), then $\kappa_{\mathcal{F}}(C|\mathcal{A}) = 1$.

Lemma 2.7. Let D be a clause of $\mathcal{F}$, let $\mathcal{A}$ be a partial assignment, and let $\mathcal{G} = \mathcal{F}|\mathcal{A}$. If $D|\mathcal{A} \neq \top$ (i.e., $\mathcal{A}$ does not satisfy D), then $\kappa_{\mathcal{F}}(D) \leq \kappa_{\mathcal{G}}(D|\mathcal{A}) + |\mathcal{A}|$.

3 Size vs. Input Cover Number Relationships

Ben-Sasson and Wigderson [3] derived size-width relationships that they describe as a "direct translation of [CEI96] to resolution derivations." Their informal statement, "if $\mathcal{F}$ has a *short* resolution refutation then it has a refutation with a small *width*," applies only when $\mathcal{F}$ has no wide clauses.

This section shows that by using input cover number rather than clause width, the restriction on the width of $\mathcal{F}$ can be removed. That is, the relationships are strengthened by removing the additive term, $width(\mathcal{F})$.

The use of restriction for recursive construction of refutations with special properties originates with Anderson and Bledsoe [2], and has been used by numerous researchers subsequently [5,3,10]. We use it to construct resolution refutations of small input cover number.

Lemma 3.1. Let $\mathcal{G} = \mathcal{F}|p$. If derivation π_1 derives clause D from $\mathcal{G}$ with $\kappa_{\mathcal{G}}(\pi_1) = (d-1)$, then there is a derivation π_2 that derives $(D + \neg p)^-$ from $\mathcal{F}$ with $\kappa_{\mathcal{F}}(\pi_2) \leq d$.

Lemma 3.2. Let $\mathcal{G} = \mathcal{F}|p$ and $\mathcal{H} = \mathcal{F}|\neg p$. If derivation π_1 derives $\perp$ from $\mathcal{G}$ with $\kappa_{\mathcal{G}}(\pi_1) = d-1$, and derivation π_2 derives $\perp$ from $\mathcal{H}$ with $\kappa_{\mathcal{H}}(\pi_2) = d$, then there is a derivation π_3 that derives $\perp$ from $\mathcal{F}$ with $\kappa_{\mathcal{F}}(\pi_3) \leq d$.

Theorem 3.3. Let $\mathcal{F}$ be an unsatisfiable formula on $n \geq 1$ variables and let $d \geq 0$ be an integer. Let S_T be the size of the shortest tree-like refutation of $\mathcal{F}$. If $S_T \leq 2^d$, then $\mathcal{F}$ has a refutation π with $\kappa_{\mathcal{F}}(\pi) \leq d$.

Corollary 3.4. $S_T(\mathcal{F}) \geq 2^{\kappa(\mathcal{F} \vdash \perp)}$.

Theorem 3.5. Let $\mathcal{F}$ be an unsatisfiable formula on $n \geq 1$ variables and let $d \geq 0$ be an integer. Let $S(\mathcal{F})$ be the size of the shortest refutation of $\mathcal{F}$. If $S(\mathcal{F}) \leq e^{(d^2/8n)}$, then $\mathcal{F}$ has a refutation π_1 with $\kappa_{\mathcal{F}}(\pi_1) \leq d$.

Corollary 3.6. $S(\mathcal{F}) \geq e^{(\kappa(\mathcal{F} \vdash \perp)^2/8n)}$.

4   Pigeon-Hole Formulas

The well-known family of Pigeon-Hole formulas for m pigeons and n holes $(\text{PHP}(m,n))$ is defined by these clauses:

$$C_i = [x_{i,1}, \ldots, x_{i,n}] \qquad \text{for } 1 \leq i \leq m$$
$$B_{ijk} = [\neg x_{i,k}, \neg x_{j,k}] \qquad \text{for } 1 \leq i \leq m, 1 \leq j \leq m, 1 \leq k \leq n.$$

Theorem 4.1. Any refutation of $\text{PHP}(m,n)$ with $m > n$ has input cover number at least $n/6$.

5   The GT(n) Family

The GT(n) family was conjectured to require exponential length refutations [7], but Stålmarck demonstrated the first polynomial solution, then Bonet and Galesi found another [9,4]. Both of these solutions produce derived clauses of width about double that of the input and have input cover numbers of two. The full paper describes a refutation with input cover number 3, which also has no derived clause wider than an input clause. This new refutation is half as long as those previously published.

Definition 5.1. The clauses of GT(n) are named as follows for indexes indicated.

$$C_n(j) \equiv [\langle 1,j \rangle, \ldots, \langle j-1,j \rangle, \langle j+1,j \rangle, \ldots, \langle n,j \rangle] \qquad 1 \leq j \leq n$$
$$B(i,j) \equiv [\neg \langle i,j \rangle, \neg \langle j,i \rangle] \qquad 1 \leq i < j \leq n$$
$$A(i,j,k) \equiv [\neg \langle i,j \rangle, \neg \langle j,k \rangle, \langle i,k \rangle] \qquad 1 \leq i,j,k \leq n \text{ and } i,j,k \text{ distinct.}$$

We recursively construct a refutation with input cover number 3, which also limits derived clause width to that of the input. The base case is $GT(1)$, in which $C_1(1) = \bot$. For $GT(n)$, where $n > 1$, the refutation begins by deriving $C_{n-1}(m)$ for $1 \leq m \leq n - 1$. Then $GT(n - 1)$ is refuted. The subderivation of $C_{n-1}(m)$ from $C_n(m)$, $C_n(n)$, $B(i, j)$ and $A(i, j, k)$ begins by resolving $C_n(n)$ with $B(m, n)$. This is the key difference from earlier published refutations, and introduces $\neg\langle n, m\rangle$ in place of $\langle m, n\rangle$. Then $\langle i, n\rangle$ are replaced one by one with $\langle i, m\rangle$ by resolving with $A(i, n, m)$. Finally, $\neg\langle n, m\rangle$ is removed by subsumption resolution with $C_n(m)$.

6   Conclusion

We proposed the *input cover number* metric (κ) as a refinement of clause width and input distance for studying the complexity of resolution. For families with wide clauses, the trade-off between resolution refutation size and κ is sharper than the trade-off between resolution refutation size and clause width.

The $GT(n)$ family has exponential tree-like refutations [4], and can be modified so that regular refutations are also exponential [1]. The original and modified families have $\kappa = 3$. These results suggest (very tentatively) that κ might be the sharper metric for general resolution, while clause-width is sharper for tree-like resolution.

References

1. Alekhnovich, M., Johannsen, J., Pitassi, T., Urquhart, A.: An exponential separation between regular and unrestricted resolution. In: Proc. 34th ACM Symposium on Theory of Computing. (2002) 448–456
2. Anderson, R., Bledsoe, W.W.: A linear format for resolution with merging and a new technique for establishing completeness. Journal of the ACM **17** (1970) 525–534
3. Ben-Sasson, E., Wigderson, A.: Short proofs are narrow — resolution made simple. JACM **48** (2001) 149–168
4. Bonet, M., Galesi, N.: Optimality of size-width tradeoffs for resolution. Computational Complexity **10** (2001) 261–276
5. Clegg, M., Edmonds, J., Impagliazzo, R.: Using the Groebner basis algorithm to find proofs of unsatisfiability. In: Proc. 28th ACM Symposium on Theory of Computing. (1996) 174–183
6. Haken, A.: The intractability of resolution. Theoretical Computer Science **39** (1985) 297–308
7. Krishnamurthy, B.: Short proofs for tricky formulas. Acta Informatica **22** (1985) 253–274
8. Letz, R., Mayr, K., Goller, C.: Controlled integration of the cut rule into connection tableau calculi. Journal of Automated Reasoning **13** (1994) 297–337
9. Stålmarck, G.: Short resolution proofs for a sequence of tricky formulas. Acta Informatica **33** (1996) 277–280
10. Van Gelder, A.: Lower bounds for propositional resolution proof length based on input distance. In: Eighth International Conference on Theory and Applications of Satisfiability Testing, St. Andrews, Scotland (2005)

Extended Resolution Proofs for
Symbolic SAT Solving with Quantification

Toni Jussila, Carsten Sinz, and Armin Biere

Institute for Formal Models and Verification
Johannes Kepler University Linz, Austria
{toni.jussila, carsten.sinz, armin.biere}@jku.at

Abstract. Symbolic SAT solving is an approach where the clauses of a CNF formula are represented using BDDs. These BDDs are then conjoined, and finally checking satisfiability is reduced to the question of whether the final BDD is identical to false. We present a method combining symbolic SAT solving with BDD quantification (variable elimination) and generation of extended resolution proofs. Proofs are fundamental to many applications, and our results allow the use of BDDs instead of—or in combination with—established proof generation techniques like clause learning. We have implemented a symbolic SAT solver with variable elimination that produces extended resolution proofs. We present details of our implementation, called EBDDRES, which is an extension of the system presented in [1], and also report on experimental results.

1 Introduction

Propositional logic decision procedures [2,3,4,5,6] lie at the heart of many applications in hard- and software verification, artificial intelligence and automatic theorem proving [7,8,9,10,11], and have been used to successfully solve problems of considerable size. In many practical applications it is not sufficient to obtain a yes/no answer from the decision procedure, however. Either a model, representing a sample solution, or a justification why the formula possesses none is required. In the context of model checking proofs are used, e.g., for abstraction refinement [11] or approximative image computations through interpolants [12]. Proofs are also important for certification by proof checking [13], in declarative modeling [9], or product configuration [10].

Using BDDs for SAT is an active research area [14,15,16,17,18,19]. It turns out that BDD and search based techniques are complementary [20,21,22]. There are instances for which one works better than the other. Therefore, combinations have been proposed [15,16,19] to obtain the benefits of both, usually in the form of using BDDs for preprocessing. However, in all these approaches where BDDs have been used, proof generation has not been possible so far.

In [1], we presented a method for symbolic SAT solving that produces extended resolution proofs. However, in that paper the only BDD operation considered is conjunction. Here, we address the problem of existential quantification left open in [1]. In particular, we demonstrate how BDD quantification can be combined with the construction of extended resolution proofs for unsatisfiable instances. Using quantification allows to build algorithms that have an exponential run-time only in the width of the elimination order

A. Biere and C.P. Gomes (Eds.): SAT 2006, LNCS 4121, pp. 54–60, 2006.

used [17,21]. It can therefore lead to much faster results on appropriate instances and hence produce shorter proofs, which is also confirmed by our experiments. For instance, we can now generate proofs for some of the Urquhart problems [23].

2   Theoretical Background

We assume that we are given a formula in CNF that we want to refute by an extended resolution proof. In what follows, we largely use an abbreviated notation for clauses, where we write $(l_1 \ldots l_k)$ for the clause $l_1 \vee \cdots \vee l_k$.

We assume that the reader is familiar with the resolution calculus [24]. Extended resolution [25] enhances the ordinary resolution calculus by an *extension rule*, which allows introduction of definitions (in the form of additional clauses) and new (defined) variables into the proof. Additional clauses must stem out of the CNF conversion of definitions of the form $x \leftrightarrow F$, where F is an arbitrary formula and x is a new variable, i.e. a variable neither occurring in the formula we want to refute nor in previous definitions nor in F. In this paper—besides introducing variables for the Boolean constants true and false—we only define new variables for if-then-else (*ITE*) constructs. $ITE(x, a, b)$ is the same as x ? a : b (for variables x, a, b), which is an abbreviation for $(x \rightarrow a) \wedge (\neg x \rightarrow b)$. So introducing a new variable w as an abbreviation for $ITE(x, a, b)$ results in the additional clauses $(\bar{w}\bar{x}a)$, $(\bar{w}xb)$, $(w\bar{x}\bar{a})$ and $(wx\bar{b})$, which may then be used in subsequent resolution steps. Extended resolution is among the strongest proof systems available and equivalent in strength to extended Frege systems [26].

Binary Decision Diagrams (BDDs) [27] are used to compactly represent Boolean functions as directed acyclic graphs. In their most common form as reduced ordered BDDs (that we also adhere to in this paper) they offer the advantage that each Boolean function is uniquely represented by a BDD, and thus all semantically equivalent formulae share the same BDD. BDDs are based on the Shannon expansion $f = ITE(x, f_1, f_0)$, decomposing f into its *co-factors* f_0 and f_1 (w.r.t variable x). The co-factor f_0 (resp. f_1) is obtained by setting variable x to false (resp. true) in formula f and subsequent simplification.

In [1], we presented a symbolic SAT solver that conjoins all the BDDs representing the clauses. This approach has the potential hurdle that the intermediate BDDs may grow too large. If memory consumption is not a problem, however, the BDD approach can be orders of magnitude faster than DPLL-style implementations [17,18,20]. Using existential quantification can speed up satisfiability checking even more and, moreover, improve memory consumption considerably by eliminating variables from the formula and thus produce smaller BDDs.

If the formula is a conjunction, rules of quantified logic allow existential quantification of variable x to be restricted to those conjuncts where x actually appears, formally:

$$\exists x(f(x, Y) \wedge g(Z)) = (\exists x f(x, Y)) \wedge g(Z)$$

where Y and Z are sets of variables not containing x. This suggests the following satisfiability algorithm [17]. First, choose a total order $\pi = (x_1, \ldots, x_n)$ of the variables X of formula F. Then, build for each variable x_i a *bucket*. The bucket B_i for x_i initially contains the BDD representations of all the clauses where x_i is the first variable according to π. Start from bucket B_1 and build the conjunction BDD b of all its elements. Then, compute $\exists x_1 b$ and put the resulting BDD to the bucket of its first variable according to π. Then, the computation proceeds to B_2 and continues until all buckets have been processed. If for any bucket, the conjunction of its elements is the constant false, we know that F is unsatisfiable. If the instance is satisfiable we get the true BDD after processing all the buckets.

3   Proof Construction

As above, we assume that we are given a formula F in CNF and that F contains the variables $\{x_1, \ldots, x_n\}$. Furthermore, we assume a given variable ordering π and that the BDD representation of clauses are initially divided into buckets $B_1, \ldots, B_n$ according to π and that variables in the BDDs are ordered according to π (the first variable of π is the root etc.). The details of how clauses are converted to BDDs are given in [1].

Our computation builds intermediate BDDs for the buckets one by one in the order mandated by π. Assume that we process a bucket that contains the BDDs $b_1, \ldots, b_m$. We construct intermediate BDDs h_i corresponding to partial conjunctions of $b_1 \wedge \cdots \wedge b_i$ until, by computing h_m, we have computed a BDD for the entire bucket. Finally, we compute a BDD $\exists h_m$ corresponding to h_m where its root variable has been existentially quantified, and add the BDD $\exists h_m$ to the (so far unprocessed) bucket of its root variable. Assuming that the children of h_m are called h_{m0} and h_{m1}, respectively, these intermediate BDDs can be computed recursively by the equations:

$$h_2 \leftrightarrow b_1 \wedge b_2, \quad h_i \leftrightarrow h_{i-1} \wedge b_i \quad \text{for } 3 \leq i \leq m \quad \text{and} \quad \exists h_m \leftrightarrow h_{m0} \vee h_{m1}$$

If it turns out that h_m is the false BDD, F is unsatisfiable and the construction of the proof can start. For this construction, we introduce new variables (using the extension rule) for each BDD node that is generated during the BDD computation, i.e. for all b_i, h_i, and $\exists h_m$ as well as for the nodes of the BDDs of the original clauses. Let f be such an internal node with the children f_0 and f_1 (leaf nodes are handled according to [1]). Then we introduce a variable (also called f) based on Shannon expansion as follows:

$$f \leftrightarrow (x ? f_1 : f_0) \qquad (\bar{f}\bar{x}f_1)(\bar{f}xf_0)(f\bar{x}\bar{f}_1)(fx\bar{f}_0)$$

On the right, we have also given the clausal representation of the definition. In order to prove F, we have to construct proofs of the following formulas for all buckets:

$$F \vdash b_i \qquad\qquad\qquad \text{for all } 1 \leq i \leq m \qquad\qquad \text{(ER-1)}$$
$$F \vdash b_1 \wedge b_2 \rightarrow h_2 \qquad\qquad\qquad\qquad\qquad\qquad \text{(ER-2a)}$$
$$F \vdash h_{i-1} \wedge b_i \rightarrow h_i \qquad \text{for all } 3 \leq i \leq m \qquad\qquad \text{(ER-2b)}$$
$$F \vdash h_{m0} \vee h_{m1} \rightarrow \exists h_m \qquad\qquad\qquad\qquad\qquad \text{(ER-3a)}$$
$$F \vdash h_m \rightarrow \exists h_m \qquad\qquad\qquad\qquad\qquad\qquad \text{(ER-3b)}$$
$$F \vdash \exists h_m \qquad\qquad\qquad\qquad\qquad\qquad\qquad\qquad \text{(ER-4)}$$

Here, the elements b_i can either be (initially present) clauses or results of an existential quantification. For clauses, the proof is straightforward (see [1]). For non-clauses, the proof is ER-4 (shown below). The proofs of ER-2a, and ER-2b are also given in [1] and we now concentrate on proving ER-3a, ER-3b, and ER-4. For the proof of ER-3a, we use the fact that $\exists h_m$ is the disjunction of the children (we call them h_{m0} and h_{m1}) of h_m. We first prove that $h_{m0} \vee h_{m1} \rightarrow \exists h_m$, in clausal form $(\bar{h}_{m0}\exists h_m)(\bar{h}_{m1}\exists h_m)$. For representational purposes, assume $h_{m0} = f$, $h_{m1} = g$, and $\exists h_m = h$, and that the root variable of f, g and h is x. We know that:

$$
\begin{array}{ll}
f \leftrightarrow (x\,?\,f_1 : f_0) & (\bar{f}\bar{x}f_1)(\bar{f}xf_0)(f\bar{x}\bar{f}_1)(fx\bar{f}_0) \\
g \leftrightarrow (x\,?\,g_1 : g_0) & (\bar{g}\bar{x}g_1)(\bar{g}xg_0)(g\bar{x}\bar{g}_1)(gx\bar{g}_0) \\
h \leftrightarrow (x\,?\,h_1 : h_0) & (\bar{h}\bar{x}h_1)(\bar{h}xh_0)(h\bar{x}\bar{h}_1)(hx\bar{h}_0) \ .
\end{array}
$$

We now recursively construct an ER proof for $f \vee g \rightarrow h$, where in the recursive step we assume that proofs for both $f_0 \vee g_0 \rightarrow h_0$ and $f_1 \vee g_1 \rightarrow h_1$ are already given. We prove $f \vee g \rightarrow h$ by generating separate proofs for $(\bar{f}h)$ and $(\bar{g}h)$. The proof for $(\bar{f}h)$ is as follows.

$$
\cfrac{
  \cfrac{(h x \bar{h}_0) \quad \cfrac{(\bar{f} x f_0) \quad (\bar{f}_0 h_0)}{(\bar{f} x h_0)}}{(\bar{f} x h)}
  \qquad
  \cfrac{\cfrac{(\bar{f}_1 h_1) \quad (\bar{f}\bar{x}f_1)}{(\bar{f}\bar{x}h_1)} \quad (h\bar{x}\bar{h}_1)}{(\bar{f}\bar{x}h)}
}{(\bar{f}h)}
$$

The recursive process stops when we arrive at the leaf nodes resp. the base case of the recursive *BDD-or* algorithm. The proof for $(\bar{g}h)$ is the same, except that f, f_0, and f_1 are replaced with g, g_0, and g_1, respectively.

The case ER-3b, in clausal form $(\bar{h}_m\exists h_m)$, is not recursive but consists of just three simple steps. The proof uses the results of ER-3a, i.e. $(\bar{h}_{m0}\exists h_m)$ and $(\bar{h}_{m1}\exists h_m)$. The root variables of h_m and $\exists h_m$ are different. To illustrate this we use w instead of x.

$$
\cfrac{
  \cfrac{(\bar{h}_m w h_{m0}) \quad (\bar{h}_{m0}\exists h_m)}{(\bar{h}_m w \exists h_m)}
  \qquad
  \cfrac{(\bar{h}_{m1}\exists h_m) \quad (\bar{h}_m \bar{w} h_{m1})}{(\bar{h}_m \bar{w}\exists h_m)}
}{(\bar{h}_m\exists h_m)}
$$

The proof of ER-4 is just a combination of parts one to three. First, having unit clauses b_1 and b_2, we resolve h_2 (using ER-2a), then all the h_i up to h_m (using ER-2b) and finally $\exists h_m$ (using ER-3b). The so-produced proofs may contain tautological clauses. As stated in [1] for the case of conjunction, careful analysis is needed in order to remove them, but it is clearly possible, also in case of existential quantification (disjunction). The full details will be given in an extended version.

4   Implementation and Experimental Result

We have implemented our approach in the SAT solver EBDDRES. It takes as input a CNF formula in DIMACS format and computes the bucket elimination algorithm. The

Table 1. Comparison of Trace generation with MINISAT and with EBDDRES

1	2	3	4	5	6	7	8	9	10	11	12	13	14	15	16	17	18
	MINISAT			EBDDRES							EBDDRES, quantification						
	solve	trace		solve		trace				bdd	solve		trace				bdd
	resources	size		resources		gen	ASCII	bin	chk	nodes	resources		gen	ASCII	bin	chk	nodes
	sec	MB	MB	sec	MB	sec	MB	MB	sec	$\times 10^3$	sec	MB	sec	MB	MB	sec	$\times 10^3$
ph7	0	0	0	0	0	0	1	0	0	3	0	5	0	12	4	1	60
ph8	0	4	1	0	0	0	3	1	0	15	1	14	1	49	15	4	236
ph9	6	4	11	0	0	0	3	1	0	8	6	52	4	186	59	14	864
ph10	44	4	63	1	17	1	30	10	2	136	20	214	16	683	*	*	2974
ph11	884	6	929	1	13	1	21	8	2	35	-	*	-	-	-	-	-
ph12	*	-	-	2	22	1	33	12	3	31	-	*	-	-	-	-	-
ph13	*	-	-	10	126	7	260	92	20	850	-	*	-	-	-	-	-
ph14	*	-	-	9	111	7	204	74	18	166	-	*	-	-	-	-	-
mutcb8	0	0	0	0	0	0	2	1	0	10	0	0	0	3	1	0	16
mutcb9	0	4	0	0	5	0	5	2	0	27	0	4	0	6	2	0	35
mutcb10	0	4	1	0	8	0	11	4	1	58	0	5	0	11	4	1	59
mutcb11	1	4	4	1	17	1	31	10	2	153	1	8	1	23	7	2	123
mutcb12	8	4	22	2	32	2	69	22	5	320	1	13	1	38	12	3	198
mutcb13	112	5	244	7	126	5	181	61	13	817	2	24	2	70	22	5	347
mutcb14	488	8	972	14	250	10	393	132	27	1694	4	37	3	127	40	8	621
mutcb15	*	-	-	36	498	26	1009	*	*	4191	6	52	5	211	67	14	1012
mutcb16	*	-	-	-	*	-	-	-	-	-	12	104	9	391	126	26	1821
urq35	95	4	218	2	22	1	37	13	3	24	0	0	0	1	0	0	5
urq45	*	-	-	-	*	-	-	-	-	-	0	0	0	1	0	0	10
urq55	*	-	-	-	*	-	-	-	-	-	0	0	0	2	1	0	15
urq65	*	-	-	-	*	-	-	-	-	-	0	4	0	6	2	0	34
urq75	*	-	-	-	*	-	-	-	-	-	0	4	0	7	2	0	39
urq85	*	-	-	-	*	-	-	-	-	-	0	5	0	10	3	1	59
fpga108	0	2		6	47	4	135	47	11	186	8	92	6	239	77	18	1088
fpga109	0	0		3	44	2	70	24	6	83	10	114	8	323	105	9	1434
fpga1211	0	0		53	874	37	1214	*	*	1312	-	*	-	-	-	-	-
add16	0	0	0	0	4	0	6	2	0	30	0	3	0	4	2	0	26
add32	0	0	0	1	9	1	24	8	2	122	1	7	1	19	6	1	106
add64	0	0	0	12	146	9	338	112	23	1393	12	95	9	393	127	26	1839
add128	0	4	0	-	*	-	-	-	-	-	-	*	-	-	-	-	-

The first column lists the name of the instance (see [1] for descriptions of the instances). Columns 2-4 contain data for MINISAT, first the time taken to solve the instance including the time to produce the trace, then the memory used, and in column 4 the size of the generated trace. The data for EBDDRES takes up the rest of the table, columns 5-11 for the approach only conjoining BDDs [1] and 12-18 for variable elimination. Column 5 (12) shows the time taken to solve the instance with EBDDRES including the time to generate and dump the trace. The latter is shown separately in column 7 (14). The memory used by EBDDRES, column 6 (13), is linearly related to the number of BDD nodes shown in column 11 (18). Column 8 (15) shows the size of the trace files in ASCII format. Column 9 (16) shows the size in a binary format comparable to that used by MINISAT (column 4). Finally, column 10 (17) shows the time needed to check that the trace is correct. The * denotes either *time out* (> 1000 seconds) or *out of memory* (> 1GB of main memory). The table shows that quantification performs worse than conjoining on pigeonhole formulas (ph*). We assume that this could be improved if we used separate variable orderings for BDDs and elimination. On the other hand, quantification is faster on the mutilated checker board instances (mutcb*) and Urquhart formulas (urq*).

result is either the false BDD or the true BDD. In the latter case, a satisfying assignment is created by traversing the intermediate BDDs right before existential quantification (called h_m above) from the last eliminated variable to the first. For the last eliminated variable, a truth value is chosen based on which branch of the BDD leads to the sink true. For all the previous BDDs, the value for the root variable is chosen based on seeking from its children a path to the sink true. Notice that for all the variables below the root, the truth value is already fixed. Therefore, at maximum two paths have to be traversed for each root h_m. The length of the traversed paths grow from one to the number of variables in the worst case. Thus, the algorithm to find a satisfiable valuation is quadratic in the number of variables. In practise with our test cases, this has not been a problem. Finally, for unsatisfiable cases a proof trace (deduction of the empty clause) can be generated.

For the experiments we used a cluster of Pentium IV 3.0 GHz PCs with 2GB of main memory running Debian Sarge Linux. The time limit was set to 1000 seconds and the memory limit to 1GB main memory. No limit was imposed on the generated traces. The experimental results are presented in Table 1.

5   Conclusions

Resolution proofs are used in many practical applications. Our results enable the use of BDDs for these purposes instead—or in combination with—already established methods based on DPLL with clause learning. This paper extends work in [1] by presenting a practical method to obtain extended resolution proofs for symbolic SAT solving with existential quantification. Our experiments confirm that on appropriate instances we are able to outperform both a fast search based approach as well as our symbolic approach only conjoining BDDs.

References

1. C. Sinz and A. Biere. Extended resolution proofs for conjoining BDDs. In *Proc. CSR'06*, 2006.
2. M. Davis and H. Putnam. A computing procedure for quantification theory. *JACM*, 7, 1960.
3. J. P. Marques-Silva and K. A. Sakallah. GRASP — a new search algorithm for satisfiability. In *Proc. ICCAD'96*.
4. M. Moskewicz, C. Madigan, Y. Zhao, L. Zhang, and S. Malik. Chaff: Engineering an efficient SAT solver. In *Proc. DAC'01*.
5. E. Goldberg and Y. Novikov. BerkMin: A fast and robust SAT-solver. In *Proc. DATE'02*.
6. N. Eén and N. Sörensson. An extensible SAT-solver. In *Proc. SAT'03*.
7. A. Biere, A. Cimatti, E. Clarke, and Y. Zhu. Symbolic model checking without BDDs. In *Proc. TACAS'99*.
8. M. Velev and R. Bryant. Effective use of boolean satisfiability procedures in the formal verification of superscalar and VLIW microprocessors. *J. Symb. Comput.*, 35(2), 2003.
9. I. Shlyakhter, R. Seater, D. Jackson, M. Sridharan, and M. Taghdiri. Debugging overconstrained declarative models using unsatisfiable cores. In *Proc. ASE'03*.
10. C. Sinz, A. Kaiser, and W. Küchlin. Formal methods for the validation of automotive product configuration data. *AI EDAM*, 17(1), 2003.

11. K. McMillan and N. Amla. Automatic abstraction without counterexamples. In *Proc. TACAS'03*.
12. K. McMillan. Interpolation and SAT-based model checking. In *Proc. CAV'03*, volume 2725 of *LNCS*.
13. L. Zhang and S. Malik. Validating SAT solvers using an independent resolution-based checker: Practical implementations and other applications. In *Proc. DATE'03*.
14. D. Motter and I. Markov. A compressed breath-first search for satisfiability. In *ALENEX'02*.
15. J. Franco, M. Kouril, J. Schlipf, J. Ward, S. Weaver, M. Dransfield, and W. Fleet. SBSAT: a state–based, BDD–based satisfiability solver. In *Proc. SAT'03*.
16. R. Damiano and J. Kukula. Checking satisfiability of a conjunction of BDDs. In *DAC'03*.
17. J. Huang and A. Darwiche. Toward good elimination orders for symbolic SAT solving. In *Proc. ICTAI'04*.
18. G. Pan and M. Vardi. Search vs. symbolic techniques in satisfiability solving. In *SAT'04*.
19. H.-S. Jin, M. Awedh, and F. Somenzi. CirCUs: A hybrid satisfiability solver. In *Proc. SAT'04*.
20. T. E. Uribe and M. E. Stickel. Ordered binary decision diagrams and the Davis-Putnam procedure. In *Proc. Intl. Conf. on Constr. in Comp. Logics*, volume 845 of *LNCS*, 1994.
21. I. Rish and R. Dechter. Resolution versus search: Two strategies for SAT. *J. Automated Reasoning*, 24(1–2), 2000.
22. J. F. Groote and H. Zantema. Resolution and binary decision diagrams cannot simulate each other polynomially. *Discrete Applied Mathematics*, 130(2), 2003.
23. A. Urquhart. Hard examples for resolution. *J. ACM*, 34(1), 1987.
24. J. A. Robinson. A machine-oriented logic based on the resolution principle. *JACM*, 12, 1965.
25. G. Tseitin. On the complexity of derivation in propositional calculus. In *Studies in Constructive Mathematics and Mathematical Logic*, 1970.
26. A. Urquhart. The complexity of propositional proofs. *Bulletin of the EATCS*, 64, 1998.
27. R. Bryant. Graph-based algorithms for Boolean function manipulation. *IEEE Trans. on Comp.*, 35(8), 1986.

Encoding CNFs to Empower
Component Analysis*

Mark Chavira and Adnan Darwiche

Computer Science Department
University of California, Los Angeles
{chavira, darwiche}@cs.ucla.edu

Abstract. Recent algorithms for model counting and compilation work
by decomposing a CNF into syntactically independent components
through variable splitting, and then solving the components recursively
and independently. In this paper, we observe that syntactic component analysis can miss decomposition opportunities because the syntax
may hide existing semantic independence, leading to unnecessary variable splitting. Moreover, we show that by applying a limited resolution
strategy to the CNF prior to inference, one can transform the CNF to
syntactically reveal such semantic independence. We describe a general
resolution strategy for this purpose, and a more specific one that utilizes
problem–specific structure. We apply our proposed techniques to CNF
encodings of Bayesian networks, which can be used to answer probabilistic queries through weighted model counting and/or knowledge compilation. Experimental results demonstrate that our proposed techniques
can have a large effect on the efficiency of inference, reducing time and
space requirements significantly, and allowing inference to be performed
on many CNFs that exhausted resources previously.

1   Introduction

Recent algorithms for model counting [17,6] and compilation [13] work by decomposing a CNF into syntactically independent components through variable
splitting, and then solving the components recursively and independently. Critical to the efficiency of these *search with decomposition* algorithms is the early
identification of independent components, which would minimize the amount of
variable splitting required (a typical source of exponential behavior).

Search–with–decomposition algorithms consider two CNFs independent when
they do not have variables in common, a condition which we call *syntactic independence*. Note, however, that even though two CNFs α and β may share
variables (and are hence syntactically dependent), they may still be capable of
being solved separately in two circumstances. First, there may exist CNFs α'
and β' that encode the same semantics as α and β, respectively, and which do
not have variables in common. This happens when one of the the CNFs α or

* This work has been partially supported by Air Force grant #FA9550-05-1-0075-
P00002 and JPL/NASA grant #1272258.

A. Biere and C.P. Gomes (Eds.): SAT 2006, LNCS 4121, pp. 61–74, 2006.
© Springer-Verlag Berlin Heidelberg 2006

β mentions irrelevant variables. Second, it may be that values of shared variables are implied by α and β, but removing subsumed clauses and performing unit resolution is insufficient to recognize this situation. If the information were known, then the variables could be set accordingly and the CNFs would thus become syntactically independent. Both of these situations cause decomposition algorithms to perform unnecessary splitting on the variables common to α and β. The phenomenon is not only present at the first level of decomposition, but can be exhibited at any level in the search tree, leading to compounded inefficiencies. As we demonstrate in this paper, the gap between syntactic independence and what we will call semantic independence can be bridged considerably by applying limited forms of resolution to the CNF, leading to major improvements to search–with–decomposition algorithms. In fact, the effect of such pre–processing can be much more dramatic if one pays attention to where the CNF originated.

Table 1. Cachet model count times for ISCAS89 circuits using three different CNF encodings. Timeout was four hours on a 2.40 GHz Intel Xeon CPU with 4GB of memory.

Circuit	SYNTAX 1 Time (s)	SYNTAX 2 Time (s)	Improvement	RESOLUTION STRATEGY 1 Time (s)	Improvement
s510	0.09	0.06	1.55	0.06	1.50
s444	0.12	0.07	1.69	0.07	1.74
s382	0.12	0.07	1.78	0.07	1.73
s400	0.12	0.07	1.80	0.07	1.78
s420	0.21	0.07	3.19	0.07	3.19
s344	0.25	0.07	3.44	0.07	3.31
s349	0.25	0.07	3.47	0.08	3.29
s386	0.29	0.07	4.26	0.07	4.26
s838	0.86	0.19	4.55	0.14	6.22
s1238	7.48	0.99	7.59	1.03	7.29
s713	4.66	0.50	9.37	0.40	11.61
s526n	2.28	0.18	12.73	0.18	12.32
s1196	11.89	0.93	12.81	0.97	12.29
s526	2.27	0.18	12.82	0.18	12.61
s953	5.34	0.34	15.48	0.33	16.13
s641	5.20	0.32	16.30	0.33	15.67
s1488	2.78	0.13	21.24	0.13	21.24
s1494	2.89	0.13	22.43	0.13	22.25
s832	3.15	0.10	31.21	0.11	29.19
s838.1	2.95	0.08	38.80	0.08	38.80
s1423	timeout	63.78	n/a	49.94	n/a
s13207.1	timeout	186.61	n/a	199.49	n/a
s35932	timeout	2.83	n/a	3.09	n/a

To demonstrate the effect of initial CNF syntax on the performance of search–with–decomposition algorithms, consider Table 1 which depicts the result of running a state–of–the–art model counter Cachet [17,1] on two CNF encodings of

ISCAS89 benchmark circuits (e.g., [2]).[1] We will have more to say about the two encodings later, but for now, suffice it to say that SYNTAX 1 is chosen carefully to worsen the gap between syntactic and semantic independence, and that SYNTAX 2 is chosen to bridge this gap. The first four columns (other columns will be discussed in Section 3) of Table 1 illustrate the dramatic performance difference between these two encodings, where SYNTAX 2 model count times range from 1.5 times faster on easy problems to over 38 times faster on harder ones, and where three networks could not be processed in four hours using SYNTAX 1 but required only minutes or less using SYNTAX 2.

The primary goal of this paper is to demonstrate that CNFs can be pre-processed, or carefully encoded, to better bridge the gap between syntactic and semantic independence. The approach we propose is to apply a limited resolution strategy to the CNF prior to execution of search. We first identify a general resolution strategy that can be applied to any CNF. As we shall see, for example, this strategy matches the performance of SYNTAX 2 when applied to CNFs encoded according to SYNTAX 1 from Table 1. We also show that by paying attention to where a CNF originates, and by bringing to bear structure that exists in the system being modeled, it is possible to define a more effective *structured* resolution strategy. We demonstrate this on CNF encodings of Bayesian networks, which have been used as inputs to both model counters [18] and knowledge compilers [12]. Using this structured strategy, we achieve significant improvements in the time and space efficiency of inference compared to unprocessed CNFs. Moreover, we are able to perform inference on some models that proved too difficult without applying the resolution strategy.

This paper is organized as follows. In Section 2, we review search with decomposition and demonstrate the importance of syntax. In Section 3, we define semantic independence, and describe a resolution strategy that is meant to bridge the gap between syntactic and semantic independence. Section 4 then reviews CNF encodings of Bayesian networks. Section 5 presents a technique that utilizes structure in a Bayesian network to guide the encoding of the corresponding CNF. In Section 6, we provide experimental results that show the benefits of this strategy. Finally, we conclude with a few remarks in Section 7.

2   The Effect of Syntax on the Syntactic Identification of Components

In this section, we review how search with decomposition works and then demonstrate the effect that syntax can have on the ability of the algorithm to identify components. Consider the problem of counting the models in the CNF at the top of Figure 1. Because all of the clauses in this CNF contain variable A, we cannot syntactically decompose the CNF, and so we must split on some variable. Splitting on variable A and performing unit resolution generates the two CNFs

---

[1] Sequential circuits have been converted into combinational circuits in a standard way, by cutting feedback loops into flip-flops, treating a flip-flop's input as a circuit output and its output as a circuit input.

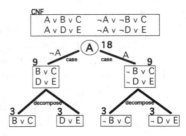

Fig. 1. An example of a search algorithm (with decomposition) that performs model counting

at the middle of Figure 1. At this point, we solve each of the two subproblems recursively and independently. We see that each subproblem decomposes into two sets of syntactically independent clauses. The four resulting sets are shown at the bottom of the figure. The CNFs at the bottom represent base cases in the recursion, each having a count of 3, as indicated. From these counts, we can compute counts for the CNFs in the middle, both 9 in this case. And from the middle counts, we compute the count for the CNF at the top of the figure, 18, which is the answer to the original problem. Although we have illustrated the search in a breadth first–fashion, it is normally performed depth–first [6]. In addition, advanced techniques such as clause learning, component caching and non–chronological backtracking, are used to improve efficiency, but we do not detail them here; see [17,11,13,5].

We next present an example which reveals the effect that syntax can have on the identification of components. Consider Figure 2(a) which depicts two fragments of a CNF: fragment α which includes, among other things, an encoding of an AND gate g with output D and inputs A, B, and C, and fragment β which includes clauses that mention variables A, B and C. Suppose further that the clauses for gate g are the only ones that mention variables A, B and C within α. These two fragments are then syntactically dependent as they share common variables, and cannot be solved in isolation. Suppose now that we decide to split on variable A by setting it to false. Under this setting, the output D of the gate must become false, and the inputs B and C are no longer relevant to fragment α. Semantically, fragments α and β are now independent and can be solved in isolation. However, depending on how we encode the gate g, this semantic independence may or may not be revealed syntactically! In particular, Figures 2(b) and 2(c) depict two different encodings of g, which we shall call SYNTAX 1 and SYNTAX 2, respectively. Either of these encodings could form the part of fragment α pertaining to gate g. The figures also show the result of simplifying (by performing unit resolution and removing subsumed clauses) these encodings when setting variable A to false. As is clear from this example, variables B and C continue to appear in the clauses of SYNTAX 1 even though they are irrelevant. These variables, however, cease to appear in the clauses of SYNTAX 2. Therefore, SYNTAX 2 enables decomposition, but SYNTAX 1 will probably require splitting on variables B and C.

A different situation occurs when we set the output D to true. In this case, all gate inputs must be true. Setting them accordingly is sufficient to sever the dependency between the two fragments. In the case of SYNTAX 1, simplifying is insufficient to discover that the inputs can be set, but in the case of SYNTAX 2, simplifying does indeed tell us the values of the inputs. As a result, SYNTAX 2 once again enables decomposition, but SYNTAX 1 requires more splitting (or a more powerful inference than unit resolution). In fact, the two different encodings of Table 1 are based on the encodings of gates shown in Figure 2, which encode each gate in isolation, in the two ways described. We have seen the significant discrepancy in performance that search with decomposition can have on these two different encodings.

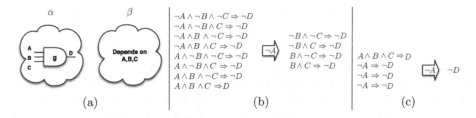

(a)   (b)   (c)

Fig. 2. (a) A depiction of two sets of clauses; (b) the AND gate encoded according to SYNTAX 1; and (c) the AND gate encoded according to SYNTAX 2

3   Semantic Independence

In this section, we define the notions of syntactic and semantic independence and discuss the encoding of CNFs to reduce the gap between them. Two CNFs are *syntactically independent* if they do not have variables in common. Two CNFs are *semantically independent* if each variable is irrelevant to either CNF (or both). More formally, two CNFs α and β are semantically independent iff for every variable V, $\alpha|V \equiv \alpha|\neg V$ or $\beta|V \equiv \beta|\neg V$, where $\alpha|V$ is the result of setting variable V to true in α, and $\alpha|\neg V$ is the result of setting V to false.

Given a logical theory Δ on which we must perform inference, there are many CNFs that specify Δ, any one of which may be supplied to the search with decomposition to obtain a correct answer. However, to make inference efficient, the goal will be to supply a CNF that makes semantic independence visible syntactically throughout the search. That is, whenever two subsets of the clauses are semantically independent, one should strive to also make them syntactically independent.

Given a CNF for Δ, we now describe a general method that produces another CNF for Δ that may better reveal semantic independence. The idea is to perform a limited type of resolution on the CNF prior to invoking the search. In particular, the strategy, which we will call RESOLUTION STRATEGY 1, specifies that whenever there are two clauses of the form $\alpha \vee \beta \vee X$ and $\alpha \vee \neg X$, where α and β are clauses and X is a variable, replace the former clause with $\alpha \vee \beta$.

RESOLUTION STRATEGY 1 makes semantic independence more visible within the CNF, as demonstrated by the following example. Consider again the AND gate with inputs A, B, and C and output D. As we have seen, encoding this gate into CNF according to SYNTAX 1 results in the following clauses:

$$\neg A \wedge \neg B \wedge \neg C \Rightarrow \neg D \qquad A \wedge \neg B \wedge \neg C \Rightarrow \neg D$$
$$\neg A \wedge \neg B \wedge C \Rightarrow \neg D \qquad A \wedge \neg B \wedge C \Rightarrow \neg D$$
$$\neg A \wedge B \wedge \neg C \Rightarrow \neg D \qquad A \wedge B \wedge \neg C \Rightarrow \neg D$$
$$\neg A \wedge B \wedge C \Rightarrow \neg D \qquad A \wedge B \wedge C \Rightarrow D$$

Applying RESOLUTION STRATEGY 1 transforms the clauses as follows:

$$A \wedge B \wedge C \Rightarrow D \qquad \neg A \Rightarrow \neg D \qquad \neg B \Rightarrow \neg D \qquad \neg C \Rightarrow \neg D$$

These reduced clauses correspond to SYNTAX 2's encoding of the AND gate.

For ISCAS89 circuits, applying RESOLUTION STRATEGY 1 to SYNTAX 1 is very efficient. The last two columns of Table 1 demonstrate what happens to model count times using Cachet [1]. The most important point is that RESOLUTION STRATEGY 1 matches SYNTAX 2's performance, even though SYNTAX 2 had the advantage of utilizing structure from the source domain (gate types), which was unavailable to RESOLUTION STRATEGY 1.

It will help at this point to describe two types of structure that can exist in a circuit: local and global. One approach to encoding a circuit is to construct a truth table over all variables in the circuit, and for each term that corresponds to falsehood, generate a clause that outlaws the term. This approach utilizes no structure and is clearly impractical in most cases. SYNTAX 1 described earlier represents an improvement that makes use of structure that can be inferred from the topology of the circuit. In particular, the topology implies a *factorization* of the global truth table into many smaller truth tables, one for each gate, that allows us to encode each smaller truth table in isolation. We refer to this type of structure as *global structure*. Utilizing global structure makes many problems practical that would not be otherwise. SYNTAX 2 goes even further, paying attention to gate type during the encoding of a specific gate. We refer to this type of structure as *local structure*. As we have seen, harnessing local structure can uncover additional semantic independence, making a large difference in how efficiently search with decomposition runs. Benefits that arise from exploiting global and local structure have long been realized in the domain of logical circuits, as SYNTAX 2 is the standard way of encoding such circuits. However, these benefits may also exist in other domains, where they are not always fully exploited. To demonstrate further how both global and local structure can be utilized to reveal semantic independence, we now turn to a specific application where CNFs correspond to encodings of Bayesian networks. Although RESOLUTION STRATEGY 1 is very efficient when applied to logical circuits, when dealing with Bayesian networks, more can be gained by paying attention to where the CNF originated.

4   CNF Encodings of Bayesian Networks

The encoding of Bayesian networks into CNFs was proposed in [12], which called for compiling these CNFs into a tractable form, d-DNNF, allowing probabilistic inference to be performed in time linear in the size of resulting compilation (through weighted model counting on the compiled form [10]). More recently, [18] proposed a similar approach, but using a different CNF encoding and applying a model counter directly on the CNF, instead of compiling the CNF first. Both approaches, however, use search with decomposition as the core algorithm, yet the compilation approach keeps a trace of the search [15].

We will now review the CNF encoding of a Bayesian network as given in [12] as the specific encoding will play a role in the remainder of the paper. A Bayesian network is a directed acyclic graph (DAG) and a set of tables called conditional probability tables (CPTs), one table for each node in the DAG. The CPTs are analogous to the truth tables of gates in a circuit. Two major differences are that variables in a CPT can be multi–valued and instead of mapping each row to truth or falsehood, a CPT maps each row to a real–number called a *parameter*.[2] Figure 3(a) depicts an example CPT, where variable A and B each have two values and variable C has three values. When encoding gates of a circuit, global structure allowed SYNTAX 1 to encode each gate separately. In a similar way, each CPT in a network can be encoded in isolation. When encoding a truth table for a particular logic gate, local structure allowed SYNTAX 2 to tailor its encoding to the particular gate type. It can be more difficult to utilize local structure in a Bayesian network. Tables are not normally associated with a type, so local structure must be inferred from parameter values.

A	B	C	$Pr(c\|a,b)$	
a_1	b_1	c_1	0.7	(θ_1)
a_1	b_1	c_2	0.0	(false)
a_1	b_1	c_3	0.3	(θ_2)
a_1	b_2	c_1	0.4	(θ_3)
a_1	b_2	c_2	0.3	(θ_2)
a_1	b_2	c_3	0.3	(θ_2)
a_2	b_1	c_1	0.333	(θ_4)
a_2	b_1	c_2	0.333	(θ_4)
a_2	b_1	c_3	0.333	(θ_4)
a_2	b_2	c_1	0.2	(θ_5)
a_2	b_2	c_2	0.3	(θ_2)
a_2	b_2	c_3	0.5	(θ_6)

(a)

(b):
$$\lambda_{a_1} \wedge \lambda_{b_1} \wedge \lambda_{c_1} \rightarrow \theta_1$$
$$\neg\lambda_{a_1} \vee \neg\lambda_{b_1} \vee \neg\lambda_{c_2}$$
$$\lambda_{a_1} \wedge \lambda_{b_1} \wedge \lambda_{c_3} \rightarrow \theta_2$$
$$\lambda_{a_1} \wedge \lambda_{b_2} \wedge \lambda_{c_1} \rightarrow \theta_3$$
$$\lambda_{a_1} \wedge \lambda_{b_2} \wedge \lambda_{c_2} \rightarrow \theta_2$$
$$\lambda_{a_1} \wedge \lambda_{b_2} \wedge \lambda_{c_3} \rightarrow \theta_2$$
$$\lambda_{a_2} \wedge \lambda_{b_1} \wedge \lambda_{c_1} \rightarrow \theta_4$$
$$\lambda_{a_2} \wedge \lambda_{b_1} \wedge \lambda_{c_2} \rightarrow \theta_4$$
$$\lambda_{a_2} \wedge \lambda_{b_1} \wedge \lambda_{c_3} \rightarrow \theta_4$$
$$\lambda_{a_2} \wedge \lambda_{b_2} \wedge \lambda_{c_1} \rightarrow \theta_5$$
$$\lambda_{a_2} \wedge \lambda_{b_2} \wedge \lambda_{c_2} \rightarrow \theta_2$$
$$\lambda_{a_2} \wedge \lambda_{b_2} \wedge \lambda_{c_3} \rightarrow \theta_6$$

(c):
$$\lambda_{a_1} \wedge \lambda_{b_1} \wedge \lambda_{c_1} \rightarrow \theta_1$$
$$\neg\lambda_{a_1} \vee \neg\lambda_{b_1} \vee \neg\lambda_{c_2}$$
$$\lambda_{a_1} \wedge \lambda_{c_3} \rightarrow \theta_2$$
$$\lambda_{b_2} \wedge \lambda_{c_2} \rightarrow \theta_2$$
$$\lambda_{a_1} \wedge \lambda_{b_2} \wedge \lambda_{c_1} \rightarrow \theta_3$$
$$\lambda_{a_2} \wedge \lambda_{b_1} \rightarrow \theta_4$$
$$\lambda_{a_2} \wedge \lambda_{b_2} \wedge \lambda_{c_1} \rightarrow \theta_5$$
$$\lambda_{a_2} \wedge \lambda_{b_2} \wedge \lambda_{c_3} \rightarrow \theta_6$$

Fig. 3. (a) A CPT over three variables, (b) Clauses generated by the encoding from [8] for the CPT, and (c) an equivalent encoding.

---

[2] There are other restrictions on the CPTs of a Bayesian network that are not important to the current discussion.

The encoding that will serve as our starting point, which we will refer to as BASELINE ENCODING, captures a large amount of local structure and was consequently shown in [8] to vastly improve compilation performance on many benchmark networks. This encoding begins by looking at the network variables. For each value x of each network variable X, we create in the CNF an *indicator variable* λ_x. For example, for network variable C with values c_1, c_2, and c_3, the encoding would generate CNF variables λ_{c_1}, λ_{c_2}, and λ_{c_3}. Next, for each network variable, we generate *indicator clauses*, which assert that in each model, exactly one of the corresponding indicator variables is true. For variable C, these clauses are as follows: $\lambda_{c_1} \vee \lambda_{c_2} \vee \lambda_{c_3}$, $\neg\lambda_{c_1} \vee \neg\lambda_{c_2}$, $\neg\lambda_{c_1} \vee \neg\lambda_{c_2}$, and $\neg\lambda_{c_1} \vee \neg\lambda_{c_1}$. The encoding then looks at each CPT in isolation. For each non–zero parameter value that is unique within its CPT, the encoding generates a CNF *parameter variable*. For example, the parameters in rows 7–9 in the CPT in Figure 3(a), all equal to 0.333, might generate parameter variable θ_4. Finally, for each row in the CPT, the encoding generates a *parameter clause*. A parameter clause asserts that the conjunction of the corresponding indicators implies θ, where θ is the parameter variable for the row, or falsehood if the row's parameter is zero. For example, the seventh row in Figure 3(a) generates the clause $\lambda_{a_2} \wedge \lambda_{b_1} \wedge \lambda_{c_1} \Rightarrow \theta_4$, and the second row generates the clause $\neg\lambda_{a_1} \vee \neg\lambda_{b_1} \vee \neg\lambda_{c_2}$. The encoding in [8] uses a few additional optimizations, which are unimportant for the current discussion. The complete set of clauses for the rows of the CPT in Figure 3(a) is shown in Figure 3(b).

5   Encoding with Local Structure

BASELINE ENCODING capitalizes on *determinism* (zero probabilities) and equal parameters in the network by omitting the generation of parameter variables for certain parameters. The effect on inference can be dramatic, as was shown in [8]. However, BASELINE ENCODING does not go as far as possible to capitalize on local structure. In this section, we introduce a new encoding method that retains the advantages of BASELINE ENCODING while further harnessing local structure to uncover semantic independence and improve component analysis.

Consider again Figure 3(a) and observe that given values for certain variables, other variables sometimes become irrelevant. For example, given $A = a_2$ and $B = b_1$, the probability no longer depends upon C (C has a uniform probability). Moreover, given values $A = a_1$ and $C = c_3$, variable B becomes irrelevant to the probability of variable C. This phenomenon is similar to context–specific–independence (CSI) [7] and can be very powerful. CSI is normally taken to mean that given values of certain parents (A or B in this case), some other parent becomes irrelevant to the probability of the child (C). The phenomenon described here is a more powerful generalization as it also captures cases where (1) setting one or more parents causes the distribution on the child to become uniform or (2) when setting the child to a certain value makes a parent irrelevant. This type of structure allows the clauses in Figure 3(b) to be simplified to the clauses in Figure 3(c), which will tend to have fewer occurrences of irrelevant variables as we set variables in search process.

Before defining a general procedure for simplifying the clauses of a given CPT, we observe that because we are working with multi–valued variables, it makes sense to use a multi–valued form of resolution. We therefore define a logic over multi–valued variables $\mathbf{X}$. The syntax of the logic is identical to that of standard propositional logic, except that an *atom* is an assignment to a variable in $\mathbf{X}$ of a value in its domain. For example, $C = c_2$ is an atom. The semantics is also like that of standard propositional logic, except that a *world*, which consists of an atom for each variable, satisfies an atom iff it assigns the common variable the same value. Within this logic, a *term* over $\mathbf{X}' \subseteq \mathbf{X}$ is a conjunction of atoms, one for each variable in $\mathbf{X}'$. Let Γ be a disjunction of terms over $\mathbf{X}$. An *implicant* γ of Γ is a term over $\mathbf{X}' \subseteq \mathbf{X}$ that implies Γ. A *prime implicant* γ of Γ is an implicant that is minimal in the sense that the removal of any atom would result in a term that is no longer an implicant of Γ.

Algorithm 1. EncodeCPT(ϕ: CPT) Generates a set of clauses for ϕ.

Partition the rows of ϕ into groups so that all rows with the same parameter are in the same group
for each encoding group Γ **do**
    $M \leftarrow$ terms of Γ
    $\theta \leftarrow$ consequent of Γ
    $P \leftarrow$ the prime implicants of M
    for p in P **do**
        $I \leftarrow$ encoding of p
        if $\theta = 0$ **then**
            assert clause $\neg I$
        else
            assert clause $I \Rightarrow \theta$
        end if
    end for
end for

Given these definitions, we can encode the network by generating the same CNF variables and indicator clauses as in BASELINE ENCODING and by generating clauses for each CPT according to Algorithm 1. This algorithm encodes a CPT ϕ over variables $\mathbf{X}$ by first partitioning the CPT into *encoding groups*, which are sets of rows that share the same parameter value. Note that each row in the CPT induces a term over variables $\mathbf{X}$ and so each encoding group induces a set of terms. Moreover, the terms within an encoding group will share a common parameter variable or all correspond to falsehood. We refer to this variable (or falsehood) as the consequent of the encoding group. To process encoding group Γ, we find the prime implicants of Γ's terms, and for each prime implicant p, we assert a clause $I \Rightarrow \theta$, where I is conjunction of indicators corresponding to p, and θ is the consequent of the encoding group. If the parameter θ equals 0, we simply generate the clause $\neg I$. Figure 4 demonstrates this algorithm for the CPT in Figure 3(a).

The algorithm we use to find prime implicants is an extension of the venerable Quine-McCluskey (QM) algorithm (e.g., [14]). QM works only for binary variables, so we extend it to multi–valued variables in a straightforward manner. Extensions of the QM algorithm for multi–valued variables are common, some of them defining a prime implicant differently (e.g., [16]). The definition given here was found effective for the purpose at hand.

Encoding Group	Param Value	Terms	Consequent	Prime Implicants	Encoding
Γ_1	.7	$a_1 b_1 c_1$	θ_1	$a_1 b_1 c_1$	$\lambda_{a_1} \wedge \lambda_{b_1} \wedge \lambda_{c_1} \Rightarrow \theta_1$
Γ_2	0	$a_1 b_1 c_2$	false	$a_1 b_1 c_2$	$\neg \lambda_{a_1} \vee \neg \lambda_{b_1} \vee \neg \lambda_{c_2}$
Γ_3	.3	$a_1 b_1 c_3, a_1 b_2 c_2, a_1 b_2 c_3, a_2 b_2 c_2$	θ_2	$a_1 c_3, b_2 c_2$	$\lambda_{a_1} \wedge \lambda_{c_3} \Rightarrow \theta_2,$ $\lambda_{b_2} \wedge \lambda_{c_2} \Rightarrow \theta_2$
Γ_4	.4	$a_1 b_2 c_1$	θ_3	$a_1 b_2 c_1$	$\lambda_{a_1} \wedge \lambda_{b_2} \wedge \lambda_{c_1} \Rightarrow \theta_3$
Γ_5	.333	$a_2 b_1 c_1, a_2 b_1 c_2, a_2 b_1 c_3$	θ_4	$a_2 b_1$	$\lambda_{a_2} \wedge \lambda_{b_1} \wedge \Rightarrow \theta_4$
Γ_6	.2	$a_2 b_2 c_1$	θ_5	$a_2 b_2 c_1$	$\lambda_{a_2} \wedge \lambda_{b_2} \wedge \lambda_{c_1} \Rightarrow \theta_5$
Γ_7	.5	$a_2 b_2 c_3$	θ_6	$a_2 b_2 c_3$	$\lambda_{a_2} \wedge \lambda_{b_2} \wedge \lambda_{c_3} \Rightarrow \theta_6$

Fig. 4. Encoding a CPT using prime implicants

The new encoding method described defines a structured resolution strategy which we will refer to as RESOLUTION STRATEGY 2. The strategy is structured in the sense that rather than working on a set of clauses, the strategy works on a partition of clauses, and restricts resolution to clauses within the same element of the partition. Each element in the partition corresponds to clauses belonging to the same CPT and having the same consequent.

We close this section with a few observations. First, even though computing prime implicants can be expensive in general, RESOLUTION STRATEGY 2 adds little overhead to BASELINE ENCODING. This efficiency stems from the small number of variables that are involved in the computation (those appearing in a CPT). This is to be contrasted with our first resolution strategy, which is applied to variables in the whole CNF. Second, there is a strong similarity between the two strategies. In particular, both are capable of removing occurrences of literals, transforming a set of terms into a more minimal set, and in this way revealing semantic independence. Third, the main idea presented might be applied more generally to other domains where a CNF is encoded from a set of functions over finitely valued variables. As we have seen, two examples are truth tables and Bayesian networks. Other examples are Markov networks and influence diagrams. Finally, CNFs created using RESOLUTION STRATEGY 2 will be smaller than those created using BASELINE ENCOING. A natural question is how much of any gains achieved arise from smaller CNFs as opposed to increased decomposability? It is not clear how one would conduct an analysis to answer this question, but the magnitude of improvements obtained clearly demonstrate that reduction in size could not be solely responsible.

6   Experimental Results

In this section, we examine a number of Bayesian networks. For each, we generate a CNF according to BASELINE ENCODING and another using RESOLUTION STRATEGY 2. We compile the CNFs into d-DNNF using the c2d compiler [3,13] and compare performance. Table 2 shows five sets of networks. The first set consists of ISCAS89 circuits converted to Bayesian networks by placing uniform probability distributions on inputs and encoding other gates with deterministic CPTs (all parameters 0 or 1). The blockmap (bm) networks were generated from relational probabilistic models and were first used in [9] to demonstrate the effectiveness of the compilation approach to networks of this type. The OR and grid (gr) networks were used in [18] to also show the effectiveness of weighted model counting for probabilistic inference, this time using search rather than compilation. From the large number of OR and grid networks, which are divided into sets of ten, we selected sets that provided a challenge for c2d, while still possible to compile within 2GB of memory using RESOLUTION STRATEGY 2. Finally, the last set consists of benchmark networks from various sources that have long been used to compare probabilistic inference algorithms. Experiments ran on a 1.6Ghz Pentium M with 2GB of memory. The implementation of the encoding and compiling algorithms have been packaged in the publicly available tools Ace 1.1 [4] (RESOLUTION STRATEGY 2 for encoding Bayesian networks) and c2d 2.2 [3] (RESOLUTION STRATEGY 1 for general CNFs).

For each network, Table 2 first lists the maximum cluster size as computed by a minfill heuristic. This measure is important because inference algorithms that do not use local structure run in time that is exponential in this number. We next list encoding times for the two encoding algorithms. The main point is that the resolution taking place in RESOLUTION STRATEGY 2 is not adding significant time to the encoding. Compile times then reveal the extent to which RESOLUTION STRATEGY 2 helps. In particular, we see that, except for one case, compile times improve anywhere from 1.45 times to over 17 times. Moreover, many of the grid networks and also barley caused the compiler to run out of memory (as indicated by dashes) when applied to BASELINE ENCODING but compiled successfully using RESOLUTION STRATEGY 2. The last three columns show the improvement to the size (number of edges) of the resulting compilations. This size is important to demonstrate space requirements and also because online inference, which may be repeated a great many times for a given application, runs in time that is linear in this size. Here, we see that on networks where BASELINE ENCODING was successful, sizes were sometimes comparable and otherwise significantly reduced.

Before closing this section, we place these results into a broader perspective. The first critical point is that on many of these networks, inference approaches that do not utilize local structure would simply fail, because of large cluster sizes. The second point is that the gains that RESOLUTION STRATEGY 2 achieves are particularly noteworthy, since they are being compared to a state–of–the–art technique for utilizing local structure [8]. The approach described in [18] harnesses local structure within the Bayesian network, applies the Cachet model

Table 2. Results for compiling a number of networks using BASELINE ENCODING and the RESOLUTION STRATEGY 2 encoding. All times are in seconds.

Network	Max. Clst. Size	BASELINE Enc. Time	RS 2 Enc. Time	Baseline Comp. Time	RS 2 Comp. Time	Imp-rove-ment	Baseline Comp. Size	RS 2 Comp. Size	Imp-rove-ment
s1238	61.0	0.91	0.86	11.32	1.83	6.19	853,987	263,786	3.24
s713	19.0	0.80	0.79	1.40	0.35	4.00	67,428	37,495	1.80
s526n	18.0	0.73	0.73	0.23	0.12	1.92	10,088	10,355	0.97
s1196	54.0	0.86	0.83	6.16	1.33	4.63	685,254	189,381	3.62
s526	18.0	0.69	0.68	0.22	0.14	1.57	13,352	14,143	0.94
s953	70.0	0.79	0.80	13.88	2.19	6.34	691,220	205,043	3.37
s641	19.0	0.84	0.79	2.54	0.38	6.68	78,071	36,555	2.14
s1488	46.0	0.89	0.88	1.65	0.56	2.95	333,629	125,739	2.65
s1494	48.0	0.88	0.90	1.82	0.44	4.14	419,274	85,469	4.91
s832	27.0	0.75	0.76	1.38	0.24	5.75	62,756	32,715	1.92
s838.1	13.0	0.79	0.80	0.43	0.20	2.15	49,856	30,899	1.61
s1423	24.0	0.91	0.91	56.62	14.54	3.89	3,010,821	994,518	3.03
bm-05-03	19.0	1.04	1.10	0.29	0.20	1.45	19,190	10,957	1.75
bm-10-03	51.0	2.90	3.03	19.57	4.97	3.94	938,371	275,089	3.41
bm-15-03	62.0	7.76	7.39	254.96	44.07	5.79	7,351,823	1,460,842	5.03
bm-20-03	90.0	17.96	17.40	1,505.24	388.65	3.87	37,916,087	6,195,000	6.12
bm-22-03	107.0	26.26	25.62	4,869.64	748.13	6.51	72,169,022	14,405,730	5.01
or-60-20-1	24.0	0.69	0.77	338.48	54.47	6.21	6,968,339	7,777,867	0.90
or-60-20-3	25.0	1.04	0.69	1.40	0.77	1.82	104,275	119,779	0.87
or-60-20-5	27.0	0.74	0.70	728.36	118.17	6.16	17,358,747	14,986,497	1.16
or-60-20-7	26.0	1.08	0.71	250.72	97.13	2.58	11,296,613	12,510,488	0.90
or-60-20-9	25.0	0.73	0.70	19.58	7.17	2.73	1,011,193	1,060,217	0.95
gr-50-16-1	24.0	0.76	0.75	137.25	43.95	3.12	14,692,963	5,739,854	2.56
gr-50-16-2	25.0	0.86	4.52	-	292.42	-	-	35,473,955	-
gr-50-16-3	24.0	0.92	0.74	65.03	40.45	1.61	7,755,318	5,280,027	1.47
gr-50-16-4	24.0	1.21	0.80	407.60	46.83	8.70	35,950,912	6,128,859	5.87
gr-50-16-5	25.0	0.88	0.82	-	26.70	-	-	3,431,139	-
gr-50-16-6	25.0	0.85	0.79	44.40	22.99	1.93	4,598,373	3,159,007	1.46
gr-50-16-7	24.0	0.85	0.84	51.68	2.99	17.28	6,413,897	421,060	15.23
gr-50-16-8	24.0	0.84	0.81	86.19	32.29	2.67	10,341,755	4,280,261	2.42
gr-50-16-9	24.0	0.84	0.94	-	60.55	-	-	7,360,872	-
gr-50-16-10	24.0	0.84	0.83	133.70	287.08	0.47	15,144,602	33,561,672	0.45
gr-50-18-1	27.0	1.02	0.87	411.45	48.36	8.51	39,272,847	6,451,916	6.09
gr-50-18-2	28.0	0.94	0.92	-	172.13	-	-	19,037,468	-
gr-50-18-3	27.0	0.91	0.86	362.90	29.18	12.44	32,120,267	2,507,215	12.81
gr-50-18-4	28.0	1.62	0.98	-	139.81	-	-	15,933,651	-
gr-50-18-5	27.0	1.26	1.07	-	158.13	-	-	18,291,116	-
gr-50-18-6	28.0	1.05	0.86	403.96	52.55	7.69	37,411,619	7,111,893	5.26
gr-50-18-7	27.0	0.98	0.98	-	79.97	-	-	9,439,318	-
gr-50-18-8	28.0	0.93	0.89	-	42.17	-	-	5,036,670	-
gr-50-18-9	27.0	0.96	0.87	-	68.51	-	-	7,890,645	-
gr-50-18-10	28.0	1.00	1.00	-	188.66	-	-	22,387,841	-
water	20.8	1.04	0.95	2.81	1.73	1.62	101,009	103,631	0.97
pathfinder	15.0	2.97	1.86	12.45	2.86	4.35	36,024	33,614	1.07
diabetes	18.2	10.76	7.77	6,281.23	3,391.18	1.85	15,426,793	15,751,044	0.98
mildew	20.7	13.37	8.16	6,245.45	1,869.92	3.34	1,693,750	1,696,139	1.00
barley	23.4	4.36	6.75	-	14,722.19	-	-	37,321,497	-

Table 3. Median times for Cachet search and for c2d compilation using BASELINE
ENCODING and RESOLUTION STRATEGY 2

Network Set	Cachet Search Time (s)	BASELINE ENCODING Compile Time (s)	RESOLUTION STRATEGY 2 Compile Time (s)
grid-50-16	890	135	42
grid-50-18	13,111	592	74
or-60-5	1.7	3.9	1.9
or-60-10	3.9	24.9	8.7
or-60-20	54	294.6	64.8

counter to a different CNF encoding, and has been shown to be successful on
some of the networks considered here. Table 3 repeats some of the results reported in [18] with regards to networks in Table 2. In particular, for each of
several sets of networks, search times running on a dual 2.8GHz processor with
4GB of memory are shown. Each time represents the median over ten networks.
Also shown in the table are median compile times we achieved for the two encodings considered in this paper. As can be seen from the table, the times are
comparable for the OR networks, but both BASELINE ENCODING and RESOLU-
TION STRATEGY 2 allow compilation to run more efficiently on grid networks
(even though compilation would normally require much more overhead than
search). We note here that the grid networks in Table 3 were chosen from a
large number of such networks because they represent some of the hardest of
the group (they contain the least amount of determinism and any larger grids
having the same degree of determinism cause Cachet to fail).

7 Conclusion

We observe in this paper that the particular syntax of a CNF can be critical for
the performance of search–with–decomposition algorithms, as it can lead to a gap
between semantic and syntactic independence that can hinder the identification
of semantically independent components. We provide two resolutions strategies,
one general and one more structured, for pre–processing a CNF with the aim of
reducing the gap between syntactic and semantic independence. We apply our
proposed techniques to general CNF encodings, and to more specific ones corresponding to Bayesian networks. Experimental results show large improvements
when applying state of the art search–with–decomposition algorithms, including the Cachet model counter and the c2d compiler, allowing us to solve some
problems that have previously exhausted available resources.

References

1. The cachet model counter, http://www.cs.washington.edu/homes/kautz/Cachet.
2. ISCAS89   Benchmark   Circuits,   http://www.cbl.ncsu.edu/www/CBL_Docs/
   iscas89.html.

3. The c2d compiler, http://reasoning.cs.ucla.edu/c2d.
4. The Ace compiler, http://reasoning.cs.ucla.edu/ace.
5. Fahiem Bacchus, Shannon Dalmao, and Toniann Pitassi. Dpll with caching: A new algorithm for #SAT and Bayesian inference. *Electronic Colloquium on Computational Complexity (ECCC)*, 10(003), 2003.
6. R. Bayardo and J. Pehoushek. Counting models using connected components. In *AAAI*, pages 157–162, 2000.
7. Craig Boutilier, Nir Friedman, Moisés Goldszmidt, and Daphne Koller. Context–specific independence in Bayesian networks. In *Proceedings of the 12th Conference on Uncertainty in Artificial Intelligence (UAI)*, pages 115–123, 1996.
8. Mark Chavira and Adnan Darwiche. Compiling Bayesian networks with local structure. In *Proceedings of the 19th International Joint Conference on Artificial Intelligence (IJCAI)*, pages 1306–1312, 2005.
9. Mark Chavira, Adnan Darwiche, and Manfred Jaeger. Compiling relational Bayesian networks for exact inference. In *Proceedings of the Second European Workshop on Probabilistic Graphical Models (PGM)*, pages 49–56, 2004.
10. Adnan Darwiche. On the tractability of counting theory models and its application to belief revision and truth maintenance. *Journal of Applied Non-Classical Logics*, 11(1-2):11–34, 2001.
11. Adnan Darwiche. A compiler for deterministic, decomposable negation normal form. In *Proceedings of the Eighteenth National Conference on Artificial Intelligence (AAAI)*, pages 627–634, Menlo Park, California, 2002. AAAI Press.
12. Adnan Darwiche. A logical approach to factoring belief networks. In *Proceedings of KR*, pages 409–420, 2002.
13. Adnan Darwiche. New advances in compiling CNF to decomposable negational normal form. In *Proceedings of European Conference on Artificial Intelligence*, pages 328–332, 2004.
14. John P. Hayes. *Introduction to Digital Logic Design*. Addison Wesley, 1993.
15. Jinbo Huang and Adnan Darwiche. Dpll with a trace: From sat to knowledge compilation. In *Proceedings of the 19th International Joint Conference on Artificial Intelligence (IJCAI)*, pages 156–162, 2005.
16. M. M. Mirsalehi and T. K. Gaylord. Logical minimization of multilevel coded functions. *Applied Optics*, 25:3078–3088, September 1986.
17. Tian Sang, Fahiem Bacchus, Paul Beame, Henry A. Kautz, and Toniann Pitassi. Combining component caching and clause learning for effective model counting. In *SAT*, 2004.
18. Tian Sang, Paul Beame, and Henry Kautz. Solving Bayesian networks by weighted model counting. In *Proceedings of the Twentieth National Conference on Artificial Intelligence (AAAI-05)*, volume 1, pages 475–482. AAAI Press, 2005.

Satisfiability Checking of Non-clausal Formulas Using General Matings*

Himanshu Jain, Constantinos Bartzis, and Edmund Clarke

School of Computer Science, Carnegie Mellon University, Pittsburgh, PA

Abstract. Most state-of-the-art SAT solvers are based on DPLL search and require the input formula to be in clausal form (cnf). However, typical formulas that
arise in practice are non-clausal. We present a new non-clausal SAT-solver based
on General Matings instead of DPLL search. Our technique is able to handle
non-clausal formulas involving ∧, ∨, ¬ operators without destroying their structure or introducing new variables. We present techniques for performing search
space pruning, learning, non-chronological backtracking in the context of a General Matings based SAT solver. Experimental results show that our SAT solver
is competitive to current state-of-the-art SAT solvers on a class of non-clausal
benchmarks.

1 Introduction

The problem of propositional satisfiability (SAT) is of central importance in various
areas of computer science, including theoretical computer science, artificial intelligence, hardware design and verification. Most state-of-the-art SAT procedures are
variations of the Davis-Putnam-Logemann-Loveland (DPLL) algorithm and require the
input formula to be in conjunctive normal form (cnf). Typical formulas generated by
the previously mentioned applications are not necessarily in cnf. As argued by Thiffault
et al. [17] converting a general formula to cnf introduces overhead and may destroy the
initial structure of the formula, which can be crucial in efficient satisfiability checking.

We propose a new propositional SAT-solving framework based on the General Matings technique due to Andrews [6]. It is closely related to the Connection method discovered independently by Bibel [8]. Theorem provers based on these techniques have
been used successfully in higher order theorem proving [5]. To the best of our knowledge, General Matings has not been used in building SAT-solvers for satisfiability
problems arising in practice. This paper presents techniques for building an efficient
SAT-solver based on General Matings.

When applied to propositional formulas the General Matings approach can be summarized as follows [7]. The input formula is translated into a 2-dimensional format
called *vertical-horizontal path form (vhpform)*. In this form disjuncts (operands of ∨)
are arranged horizontally and conjuncts (operands of ∧) are arranged vertically. The

* This research was sponsored by the Gigascale Systems Research Center (GSRC), the Semicon-
ductor Research Corporation (SRC), the Office of Naval Research (ONR), the Naval Research
Laboratory (NRL), the Army Research Office (ARO), and the General Motors Collaborative
Research Lab at CMU.

A. Biere and C.P. Gomes (Eds.): SAT 2006, LNCS 4121, pp. 75–89, 2006.

formula is *satisfiable* if and only if there exists a *vertical path* through this arrangement that does not contain two opposite literals (l and $\neg l$). The input formula is not required to be in cnf.

We have designed a SAT procedure for non-clausal formulas based on the General Matings approach. At a high level our search algorithm enumerates all possible vertical paths in the vhpform of a given formula until a vertical path is found which does not contain two opposite literals. If every vertical path contains two opposite literals, then the given formula is unsatisfiable. The number of vertical paths can be exponential in the size of a given formula. Thus, the key challenge in obtaining an efficient SAT solver is to prevent the enumeration of vertical paths as much as possible. We present several novel techniques for preventing the enumeration of vertical paths. Our contributions can be summarized as follows:

- The vhpform of a given formula succinctly encodes: 1) disjunctive normal form (dnf) of a given formula as a set of vertical paths 2) conjunctive normal form (cnf) of a given formula as a set of *horizontal paths*. Our solver employs a combination of both vertical and horizontal path exploration for efficient SAT solving. The choice of which variable to assign next (*decision making*) is made using the vertical paths which are similar to the terms (conjunction of literals) in the dnf of a given formula. Conflict detection is aided by the use of horizontal paths which are similar to the clauses (disjunction of literals) in the cnf of a given formula.
- We show how to adapt the techniques found in the current state-of-the-art SAT solvers in our algorithm. We describe how to perform *search space pruning, conflict driven learning, non-chronological backtracking* by using the vertical paths and horizontal paths present in the vhpform of a given formula.
- We present graph based representations of the set of vertical paths and the set of horizontal paths which makes it possible to implement our algorithms efficiently.

Related Work: Many SAT solvers have been developed, most employing some combination of two main strategies: the DPLL search and heuristic local search. Heuristic local search techniques [12] are not guaranteed to be complete, that is, they are not guaranteed to find a satisfying assignment if one exists or prove unsatisfiability. As a result, complete SAT solvers (such as GRASP [11], SATO [18], zChaff [14], BerkMin [10], Siege [4], MiniSat [2]) are based almost exclusively on the DPLL search. While most DPLL based SAT solvers operate on cnf, there has been work on applying DPLL directly to circuit [9] and non-clausal [17] representations. The key differences between existing work and our approach are as follows:

- Unlike heuristic local search based techniques, we propose a complete SAT solver.
- Unlike DPLL based SAT solvers (operating on either cnf, circuit or non-clausal representation), the basis of our search procedure is General Matings. There is a crucial difference between the two techniques. In DPLL the search space is the set of all possible assignments to the propositional variables, whereas in General Matings the search space is the set of all possible vertical paths in the vertical-horizontal path form of a given formula. We give an example illustrating the difference in Section 2. In contrast to current cnf SAT solvers which produce a complete satisfying assignment (all variables are assigned), our solver produces partial satisfying assignments when possible.

– The General Matings technique is designed to work on non-clausal forms. In particular, any arbitrary propositional formula involving $\wedge, \vee, \neg$ is handled naturally, without introduction of new variables or loss of structural information.

Semantic Tableaux [16] is a popular theorem proving technique. The basic idea is to expand a given formula in the form of a tree, where nodes are labeled with formulas. If all the branches in the tree lead to contradiction, then the given formula is unsatisfiable. The tableau of a given propositional formula can blowup in size due to repetition of subformulas along the various paths. In contrast, when using General Matings a vertical-horizontal path form of a given formula is built first. This representation is a directed acyclic graph (DAG) and polynomial in the size of the given formula.

2   Preliminaries

A propositional formula is in *negation normal form (nnf)* iff it contains only the propositional connectives $\wedge$, $\vee$ and $\neg$ and the scope of each occurrence of $\neg$ is a propositional variable. It is known that every propositional formula is equivalent to a formula in nnf. Furthermore, a negation normal form of a formula can be much shorter than any dnf or cnf of that formula. The internal representation in our satisfiability solver is nnf. More specifically, we use a two-dimensional format of a nnf formula, called a *vertical-horizontal path form (vhpform)* as described in [7][1]. In this form disjunctions are written horizontally and conjunctions are written vertically. For example Fig. 1(a) shows the formula $\phi = (((p \vee q) \wedge \neg r \wedge \neg q) \vee (\neg p \wedge (r \vee \neg s) \wedge q))$ in vhpform.

Vertical path: A vertical path through a vhpform is a sequence of literals in the vhpform that results by choosing either the left or the right scope for each occurrence of $\vee$. For the vhpform in Fig. 1(a) the set of vertical paths is $\{\langle p, \neg r, \neg q \rangle, \langle q, \neg r, \neg q \rangle, \langle \neg p, r, q \rangle, \langle \neg p, \neg s, q \rangle\}$.

Horizontal path: A horizontal path through a vhpform is a sequence of literals in the vhpform that results by choosing either the left or the right scope for each occurrence of $\wedge$. For the vhpform in Fig. 1(a) the set of horizontal paths is $\{\langle p, q, \neg p \rangle, \langle p, q, r, \neg s \rangle, \langle p, q, q \rangle, \langle \neg r, \neg p \rangle, \langle \neg r, r, \neg s \rangle, \langle \neg r, q \rangle, \langle \neg q, \neg p \rangle, \langle \neg q, r, \neg s \rangle, \langle \neg q, q \rangle\}$.

The following are two important results regarding satisfiability of negation normal formulas from [7]. Let F be a formula in negation normal form and let σ be an assignment (σ can be a partial truth assignment).

Theorem 1. *σ satisfies F iff there is a vertical path P in the vhpform of F such that σ satisfies every literal in P.*

Theorem 2. *σ falsifies F iff there is a horizontal path P in the vhpform of F such that σ falsifies every literal in P.*

---

[1] In [7] the term *vertical path form (vpform)* is used in place of vertical-horizontal path form (vhpform). We use vertical-horizontal path form (vhpform) in this paper for clarity.

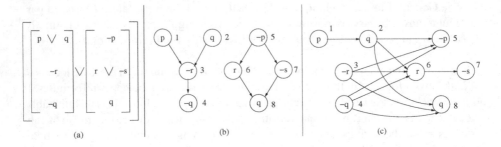

Fig. 1. We show the negation of a variable by a − sign. (a) vhpform for the formula $(((p \vee q) \wedge \neg r \wedge \neg q) \vee (\neg p \wedge (r \vee \neg s) \wedge q))$ (b) the corresponding vpgraph (c) the corresponding hpgraph.

The vhpform in Fig. 1(a) has a vertical path $\langle p, \neg r, \neg q \rangle$ whose every literal can be satisfied by an assignment σ which sets p to true and r, q to false. It follows from Theorem 1 that σ satisfies ϕ. Thus, ϕ is satisfiable. An example of a vertical path whose every literal cannot be satisfied by any assignment is $\langle q, \neg r, \neg q \rangle$ (due to opposite literals q and $\neg q$). An assignment σ' which sets p, r to true, falsifies every literal in the horizontal path $\langle \neg r, \neg p \rangle$ in the vhpform of ϕ. Thus, from Theorem 2 it follows that σ' falsifies ϕ.

Let $\mathcal{VP}(\phi)$ and $\mathcal{HP}(\phi)$ denote the set of vertical paths and the set of horizontal paths in the vhpform of ϕ, respectively. We use $l \in \pi$ to denote the occurrence of a literal l in a vertical/horizontal path π. The following result from [7] states that the set of vertical paths encodes the dnf and the set of horizontal paths encodes the cnf of a given formula.

Theorem 3. *(a) ϕ is equivalent to the dnf formula $\bigvee_{\pi \in \mathcal{VP}(\phi)} \bigwedge_{l \in \pi} l$. (b) ϕ is equivalent to the cnf formula $\bigwedge_{\pi \in \mathcal{HP}(\phi)} \bigvee_{l \in \pi} l$.*

Theorem 1 forms the basis of a General Matings based SAT procedure. The idea is to check the satisfiability of a given nnf formula by examining the vertical paths in its vhpform. For the vhpform in Fig. 1(a) the search space is $\{\langle p, \neg r, \neg q \rangle, \langle q, \neg r, \neg q \rangle, \langle \neg p, r, q \rangle, \langle \neg p, \neg s, q \rangle\}$. In contrast, the search space for a DPLL-based SAT solver is the set of all possible truth assignments to the variables p, q, r, s. We use Theorem 2 for efficient Boolean constraint propagation in two ways: 1) For detecting when the current candidate for a satisfying assignment falsifies the given formula (*conflict detection*). 2) For obtaining a *unit literal rule* (Section 3) similar to the one used in cnf SAT solvers.

3   Graph Representations

Our SAT procedure operates on the graph based representations of the vhpform of a given formula. These graph based representations are described below.

Graphical encoding of vertical paths (vpgraph): A graph containing all vertical paths present in the vhpform of a nnf formula is called a *vpgraph*. Given a nnf formula ϕ, we define the vpgraph $G_v(\phi)$ as a tuple (V, R, L, E, Lit), where V is the set of nodes corresponding to all occurrences of literals in ϕ, $R \subseteq V$ is a set of root nodes, $L \subseteq V$

is a set of leaf nodes, $E \subseteq V \times V$ is the set of edges, and $Lit(n)$ denotes the literal associated with node $n \in V$. A node $n \in R$ has no incoming edges and a node $n \in L$ has no outgoing edges.

The vpgraph containing all vertical paths in the vhpform of Fig. 1(a) is shown in Fig. 1(b). For the vpgraph in Fig. 1(b), we have $V = \{1, 2, 3, 4, 5, 6, 7, 8\}$, $R = \{1, 2, 5\}$, $L = \{4, 8\}$, $E = \{(1, 3), (2, 3), (3, 4), (5, 6), (5, 7), (6, 8), (7, 8)\}$ and for each $n \in V$, $Lit(n)$ is shown inside the node labeled n in Fig. 1(b). Each path in the vpgraph $G_v(\phi)$, starting from a root node and ending at a leaf node, corresponds to a vertical path in the vhpform of ϕ. For example, path $\langle 1, 3, 4 \rangle$ in Fig. 1(b) corresponds to the vertical path $\langle p, \neg r, \neg q \rangle$ in Fig. 1(a) (obtained by replacing node n on path by $Lit(n)$). Using this correspondence one can see that vpgraph contains all vertical paths present in the vhpform shown in Fig. 1(a).

Given nnf formula ϕ, we can construct the vpgraph $G_v(\phi) = (V, R, L, E, Lit)$ directly without constructing the vhpform of ϕ. This is done inductively as follows:

- If ϕ is a literal l, then we create a graph containing just one node fv, where fv is a fresh identifier. The literal stored inside fv is set to l.

  $G_v(\phi) = (\{fv\}, \{fv\}, \{fv\}, \emptyset, Lit)$ and $Lit(fv) = l$, fv is a fresh identifier.
- If $\phi = \phi_1 \vee \phi_2$, then the vpgraph for ϕ is obtained by taking the union of the vp-graphs of ϕ_1 and ϕ_2. Let $G_v(\phi_1) = (V_1, R_1, L_1, E_1, Lit_1)$ and $G_v(\phi_2) = (V_2, R_2, L_2, E_2, Lit_2)$. Then $G_v(\phi)$ is the union of $G_v(\phi_1)$ and $G_v(\phi_2)$.

  $G_v(\phi) = (V_1 \cup V_2, R_1 \cup R_2, L_1 \cup L_2, E_1 \cup E_2, Lit_1 \cup Lit_2)$
- If $\phi = \phi_1 \wedge \phi_2$, then the vpgraph for ϕ is obtained by concatenating the vpgraph of ϕ_1 with the vpgraph of ϕ_2. Let $G_v(\phi_1) = (V_1, R_1, L_1, E_1, Lit_1)$ and $G_v(\phi_2) = (V_2, R_2, L_2, E_2, Lit_2)$. Then $G_v(\phi)$ contains all the nodes and edges in $G_v(\phi_1)$ and $G_v(\phi_2)$. But $G_v(\phi)$ has additional edges connecting leaves of $G_v(\phi_1)$ with the roots of $G_v(\phi_2)$. The set of additional edges is denoted as $L_1 \times R_2$ below. The set of roots of $G_v(\phi)$ is R_1, while the set of leaves is L_2.

  $G_v(\phi) = (V_1 \cup V_2, R_1, L_2, E_1 \cup E_2 \cup (L_1 \times R_2), Lit_1 \cup Lit_2)$

Graphical encoding of horizontal paths (hpgraph): A graph containing all horizontal paths present in the vhpform of a nnf formula is called a *hpgraph*. We use $G_h(\phi)$ to denote the hpgraph of a formula ϕ. The procedure for constructing a hpgraph is similar to the above procedure for constructing the vpgraph. The difference is that the hpgraph for $\phi = \phi_1 \wedge \phi_2$ is obtained by taking the union of hpgraphs for ϕ_1 and ϕ_2 and the hpgraph for $\phi = \phi_1 \vee \phi_2$ is obtained by concatenating the hpgraphs of ϕ_1 and ϕ_2.

The hpgraph containing all horizontal paths in the vhpform in Fig. 1(a) is shown in Fig. 1(c). For the hpgraph in Fig. 1(c), we have $V = \{1, 2, 3, 4, 5, 6, 7, 8\}$, $R = \{1, 3, 4\}$, $L = \{5, 7, 8\}$, $E = \{(1, 2), (2, 5), (2, 6), (2, 8), (3, 5), (3, 6), (3, 8), (4, 5), (4, 6), (4, 8), (6, 7)\}$ and for each $n \in V$, $Lit(n)$ is shown inside the node labeled n.

Using vpgraph/hpgraph: We use $G(\phi)$ to refer to either a vpgraph or hpgraph of ϕ. It can be shown by induction that the vpgraph and hpgraph of a nnf formula are directed acyclic graphs (DAGs). This fact allows obtaining more efficient versions of standard graph algorithms (such as shortest path computation) for vpgraph/hpgraph. The construction of vpgraph/hpgraph takes $O(k^2)$ time in the worst case where k is the size of the given formula. This is mainly due to the $L_1 \times R_2$ term in the handling of $\phi_1 \wedge \phi_2$ (for vpgraph construction) and $\phi_1 \vee \phi_2$ (for hpgraph construction).

Given a vpgraph or hpgraph $G(\phi) = (V, R, L, E, Lit)$, the following definitions will be used in subsequent discussion.

r-path: A path $\pi = \langle n_0, \ldots, n_k \rangle$ in $G(\phi)$ is said to be a r-path (rooted path) iff it starts with a root node ($n_0 \in R$). In Fig. 1(b), $\langle 2, 3 \rangle$ is a r-path while $\langle 3, 4 \rangle$ is not a r-path.
rl-path: A path $\pi = \langle n_0, \ldots, n_k \rangle$ in $G(\phi)$ is said to be a rl-path iff it starts at a root node and ends at a leaf node ($n_0 \in R$ and $n_k \in L$). In Fig. 1(b), both $\langle 2, 3, 4 \rangle$, $\langle 5, 6, 8 \rangle$ are rl-paths, but $\langle 3, 4 \rangle$ is not a rl-path.
Conflicting nodes: Two nodes $n_1, n_2 \in V$ are said to be conflicting iff $Lit(n_1) = \neg Lit(n_2)$. In Fig. 1(b), nodes 2,4 are conflicting.

- We say an assignment σ *satisfies (falsifies)* a node $n \in V$ iff σ satisfies (falsifies) $Lit(n)$. An assignment which sets q to true satisfies nodes 2, 8 and falsifies node 4 in Fig. 1(b).
- We say an assignment σ *satisfies (falsifies)* a path $\pi \in G(\phi)$ iff σ satisfies (falsifies) every node on π. For example, in Fig. 1(b) path $\langle 5, 6, 8 \rangle$ is satisfied by an assignment which sets p to false and r, q to true. The same path is falsified by an assignment which sets p to true and r, q to false. We say that a path $\pi \in G$ is *satisfiable* iff there exists an assignment which satisfies π. In Fig. 1(b), path $\langle 5, 6, 8 \rangle$ is satisfiable, while the path $\langle 2, 3, 4 \rangle$ is not satisfiable due to conflicting nodes 2,4.

Recall, that an rl-path in a vpgraph $G_v(\phi)$ corresponds to a vertical path in the vhpform of ϕ. Similarly, an rl-path in a hpgraph $G_h(\phi)$ corresponds to a horizontal path in the vhpform of ϕ. The following corollaries adapt Theorem 1 and Theorem 2 to the graph representations of the vhpform of a given formula ϕ.

Corollary 1. *An assignment σ satisfies ϕ iff there exists a rl-path π in $G_v(\phi)$ such that σ satisfies π.*

Corollary 2. *An assignment σ falsifies ϕ iff there exists a rl-path π in $G_h(\phi)$ such that σ falsifies π.*

The following corollary is a re-statement of corollary 1.

Corollary 3. *ϕ is satisfiable iff there exists a rl-path π in $G_v(\phi)$ which is satisfiable.*

The following corollary connects the notion of conflicting nodes with the satisfiability of a path.

Corollary 4. *A path π in $G(\phi)$ is satisfiable iff no two nodes on π are conflicting.*

Discovery of unit literals from hpgraph: Modern SAT solvers operating on a cnf representation employ a *unit literal rule* for efficient Boolean constraint propagation. The unit literal rule states that if all but one literal of a clause are set to false, then the un-assigned literal in the clause must be set to true under the current assignment. In our context the input formula is not necessarily represented in cnf, however, it is still possible to obtain the unit literal rule via the use of the hpgraph of a given formula. The following claim states the unit literal rule for the non-clausal formulas.

Input: vpgraph $G_v(\phi) = (V, R, L, E, Lit)$ and hpgraph $G_h(\phi) = (V', R', L', E', Lit')$
Output: If $G_v(\phi)$ has a satisfiable rl-path return SAT, else return UNSAT
Algorithm:

1: $st \leftarrow R$	//push all roots in $G_v(\phi)$ on stack st
2: $\sigma \leftarrow \emptyset$	//initial truth assignment is empty
3: $\forall n \in V : mrk(n) \leftarrow false$	//all nodes are un-marked to start with
4: while $(st \neq \emptyset)$	//stack st is not empty
5: $m \leftarrow st.top()$	// top element of stack st
6: if $(mrk(m) == false)$	//can we extend current r-path CRP with m
7: if (conflict == prune())	//check if taking m causes conflict
8: learn()	//compute reason for conflict and learn
9: backtrack()	//non-chronological backtracking
10: continue	//goto while loop (line 4)
11: end if	
12: $mrk(m) \leftarrow true$	//extend current satisfiable r-path with m
13: $\sigma \leftarrow \sigma \cup \{Lit(m)\}$	//add $Lit(m)$ to current assignment
14: if $(m \in L)$	//node m is a leaf
15: return SAT	//we found a satisfiable rl-path in $G_v(\phi)$
16: else	
17: push all children of m on st	//try extending CRP$\langle m \rangle$ to reach a leaf
18: end if	
19: else	//backtracking mode
20: backtrack()	//non-chronological backtracking
21: end if	
22: end while	
23: return UNSAT	//no satisfiable rl-path exists in $G_v(\phi)$

Fig. 2. Searching a vpgraph for a satisfiable rl-path

Corollary 5. *Let an assignment σ falsify all but a subset of nodes V_s on an rl-path π in $G_h(\phi)$. If all nodes in V_s contain the same literal l and l is not already assigned by σ, then l must be set to true under σ in order to obtain a satisfying assignment.*

The above corollary follows from Theorem 3(b). Intuitively, each rl-path in the hpgraph corresponds to a clause in the cnf of a given formula. Thus, at least one literal from each rl-path in $G_h(\phi)$ must be satisfied in order to obtain a satisfying assignment.

Example: Consider the hpgraph shown in Fig. 1(c) and an assignment σ which sets p, q to false and s to true. σ falsifies all but node 6 on the rl-path $\langle 1, 2, 6, 7 \rangle$ in the hpgraph. It follows from Corollary 5 that $Lit(6)$ which is r must be set to true under σ.

4   Top Level Algorithm

In order to check the satisfiability of a nnf formula ϕ, we obtain a vpgraph $G_v(\phi)$. From Corollary 3 it follows that ϕ is satisfiable iff $G_v(\phi)$ has a satisfiable rl-path. At a high level our search algorithm enumerates all possible rl-paths until a satisfiable rl-path is found. If no satisfiable rl-path is found, then ϕ is unsatisfiable. For dnf (or dnf-like) formulas the number of rl-paths in vpgraph is small, linear in the size of the

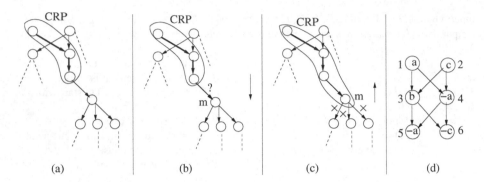

Fig. 3. (a) Current r-path or CRP in a vpgraph. (b) Can CRP be extended by node m? (c) Backtracking from node m. (d) vpgraph for formula $(a \vee c) \wedge (b \vee \neg a) \wedge (\neg a \vee \neg c)$.

formula, and therefore the basic search algorithm is efficient. However, for formulas that are not in dnf form, the algorithm of just enumerating all rl-paths in $G_v(\phi)$ does not scale. We have adapted several techniques found in modern SAT solvers such as *search space pruning, conflict driven learning, non-chronological backtracking* to make the search efficient. Search space pruning and conflict driven learning will be described in detail in the following sections. Due to space restriction we present non-chronological backtracking in a detailed version of this paper available at [3].

The high level description of the algorithm is given in Fig. 2. The input to the algorithm is a vpgraph $G_v(\phi) = (V, R, L, E, Lit)$ and a hpgraph $G_h(\phi) = (V', R', L', E', Lit')$ corresponding to a formula ϕ. If $G_v(\phi)$ contains a satisfiable rl-path, then the algorithm returns SAT as the answer. Otherwise, ϕ is unsatisfiable and the algorithm returns UNSAT. The algorithm uses the hpgraph $G_h(\phi)$ in various sub-routines such as prune and learn. The following data structures are used:

- st is a stack. It stores a subset of nodes from V that need to be explored when searching for a satisfiable rl-path in $G_v(\phi)$. Initially, the roots in $G_v(\phi)$ are pushed on the stack st (line 1). Let $st.top()$ return the top element of st. We write st as $[n_0, \ldots, n_k]$ where the top element is n_k and the bottom element is n_0.
- σ stores the current truth assignment as a set. Each element of σ is a literal which is true under the current assignment. It is ensured that σ is *consistent*, that is, it does not contain contradictory pairs of the form l and $\neg l$. Initially, σ is the empty set (line 2). For example, an assignment which sets variables a, b to true and c to false will be denoted as $\{a, b, \neg c\}$.
- mrk maps a node in V to a Boolean value. It identifies an r-path in $G_v(\phi)$ which is currently being considered by the algorithm to obtain a satisfiable rl-path (see Fig. 3(a)). We refer to this r-path as the *current r-path* (CRP for short). Intuitively, $mrk(n)$ is true for nodes that lie on CRP ($n \in$ CRP) and false for all other nodes in $G_v(\phi)$. More precisely, the CRP is obtained by removing every node n from the stack st for which $mrk(n)$ is false. The remaining nodes constitute the CRP. Initially, $mrk(n)$ is set to false for every node n (line 3), thus, CRP is empty.

Example: The vpgraph for the formula $\phi = (a \vee c) \wedge (b \vee \neg a) \wedge (\neg a \vee \neg c)$ is shown in Fig. 3(d). Initially, we have st as $[2, 1]$ where the top element of the stack is 1, $\sigma = \emptyset$, $mrk(n) = false$ for all $n \in \{1, 2, 3, 4, 5, 6\}$. Suppose during the execution of the algorithm we have st as $[2, 1, 4, 3, 6, 5]$, and $mrk(1), mrk(3)$ are $true$ and $mrk(n) = false$ for $n \in \{2, 4, 5, 6\}$. Thus, CRP is $\langle 1, 3 \rangle$. Observe that CRP is an r-path. Intuitively, the algorithm tries to extend CRP by one node at a time, to obtain a satisfiable rl-path. In this case CRP can be extended to obtain two rl-paths $\pi_1 = \langle 1, 3, 5 \rangle$ or $\pi_2 = \langle 1, 3, 6 \rangle$. However, only π_2 is satisfiable (by $\sigma = \{a, b, \neg c\}$) and is enough to show that ϕ is satisfiable.

The main part of the algorithm is the while loop (lines 4-22) which executes as long as st is not empty and the algorithm has not returned SAT on line 15. The algorithm maintains the following loop invariant.

Loop invariant: At the beginning of iteration number i of the while loop: let the current r-path (CRP) be $\langle n_0, \ldots, n_k \rangle$. Then the assignment σ is equal to $\{Lit(n_i) | n_i \in$ CRP$\}$. That is, σ satisfies each node on CRP and thus, σ satisfies CRP. For example, suppose CRP is $\langle 1, 3 \rangle$ in the vpgraph shown in Fig. 3(d), then σ will be $\{a, b\}$.

If st is not empty, then the top element of the stack (denoted by m) is considered in line 5. There are two possibilities for node m according to the if statement in line 6.

• $mrk(m)$ is $false$: In this case the algorithm checks if the current r-path CRP can be extended by node m as shown in Fig. 3(b). This check is carried out by a call to prune (line 7). If prune returns conflict, then the current r-path extended by node m cannot lead to a satisfiable rl-path. Thus, the solver needs to backtrack from node m, and if possible extend CRP by some other node. This is done by calling backtrack on line 9 and going back to while loop (line 4) by using continue (line 10). Before backtracking a call to learn (line 8) is made which summarizes the reason for the conflict when CRP is extended by m. This reason is learned in form of a clause and is used later to avoid similar conflicts. We denote CRP concatenated with m as CRP$\langle m \rangle$. Depending upon the reason why there is no satisfiable rl-path with CRP$\langle m \rangle$ as prefix, the backtrack routine can pop several nodes from st (non-chronological backtracking) instead of just popping m from st.

If a call to prune results in no−conflict (line 7), then m can extend CRP. In this case execution reaches line 12. At line 12 $mrk(m)$ is set to true, which means that the new current r-path is CRP concatenated with m, that is, CRP$\langle m \rangle$. The algorithm maintains the loop invariant that the assignment σ satisfies the current r-path. In order to maintain this invariant σ now needs to satisfy node m which is on the current r-path CRP$\langle m \rangle$. This is done by adding $Lit(m)$ to σ (line 13). If m is a leaf in the vpgraph, then CRP$\langle m \rangle$ is a satisfiable rl-path. In this case SAT is returned (lines 14-15). If m is not a leaf, then the children of m are pushed on the stack (line 17). The algorithm will next attempt to extend the current r-path CRP$\langle m \rangle$.

• $mrk(m)$ is $true$: This happens when the current r-path is of the form $\langle n_0, \ldots, n_k, m \rangle$. Intuitively, the algorithm has explored all possible rl-paths with $\langle n_0, \ldots, n_k, m \rangle$ as prefix, but none of them leads to a satisfiable rl-path as shown in Fig. 3(c). The algorithm now backtracks from node m by calling backtrack on line 20 . Depending upon the

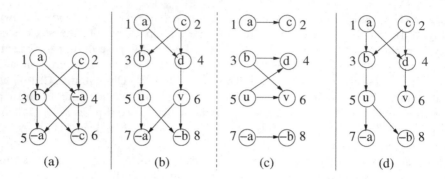

Fig. 4. (a) vpgraph for formula $(a \vee c) \wedge (b \vee \neg a) \wedge (\neg a \vee \neg c)$. (b,c) vpgraph and hpgraph for formula $(a \vee c) \wedge ((b \wedge u) \vee (d \wedge v)) \wedge (\neg a \vee \neg b)$, respectively. (d) vpgraph for formula $(a \vee c) \wedge ((b \wedge u \wedge (\neg a \vee \neg b)) \vee (d \wedge v))$.

reason why there is no satisfiable rl-path with $\langle n_0, \ldots, n_k, m \rangle$ as prefix, the algorithm can pop several nodes from st instead of just popping m.

For each node n removed from the stack during backtracking (lines 9, 20) $mrk(n)$ is set to false again. This enables the removed nodes to be examined again on rl-paths which have not yet been explored.

5   Search Space Pruning

This section describes the procedure `prune` called in the non-clausal SAT algorithm shown in Fig. 2 (line 7). A call to `prune` checks if the current r-path CRP can be extended by node m or not, as shown in Fig. 3(b). Intuitively, `prune` returns `conflict` if there cannot be a satisfiable rl-path in vpgraph $G_v(\phi)$ with $\text{CRP}\langle m \rangle$ as prefix. When `prune` is called, the current r-path CRP is satisfied by assignment σ, which is equal to $\{Lit(n) | n \in \text{CRP}\}$ (maintained as a `while` loop invariant in the top level algorithm shown in Fig. 2). The three cases when `conflict` is returned are as follows:

Case 1: When $\text{CRP}\langle m \rangle$ is not satisfiable. This happens when there is a node n on CRP such that $Lit(n) = \neg Lit(m)$. In this case no assignment can satisfy the r-path $\text{CRP}\langle m \rangle$ (Corollary 4). For example, in the vpgraph shown in Fig. 4(a) this conflict arises when the CRP is $\langle 1, 3 \rangle$ and m is node 5.

Otherwise, $\text{CRP}\langle m \rangle$ is satisfiable and $\sigma' = \sigma \cup \{Lit(m)\}$ satisfies $\text{CRP}\langle m \rangle$. However, it is still possible that there is no satisfiable rl-path in $G_v(\phi)$ with $\text{CRP}\langle m \rangle$ as prefix. These cases are described below.

Case 2 (Global conflict): When σ' falsifies ϕ. In this case no satisfiable rl-path in $G_v(\phi)$ can be obtained with $\text{CRP}\langle m \rangle$ as prefix. We prove this claim by contradiction. Assume that there is an rl-path π in $G_v(\phi)$ which has $\text{CRP}\langle m \rangle$ as prefix and is satisfiable. By definition there exists an assignment σ'' which satisfies π. From Corollary 1 we know that σ'' satisfies ϕ. In order to satisfy π, σ'' must satisfy $\text{CRP}\langle m \rangle$. That is, σ'' must contain $Lit(n)$ for every $n \in \text{CRP}\langle m \rangle$. Since $\sigma' = \{Lit(n) | n \in \text{CRP}\langle m \rangle\}$, it follows that $\sigma' \subseteq \sigma''$. But σ' falsifies ϕ and hence σ'' must falsify ϕ. This leads to a contradiction.

Example: In Fig. 4(b) vpgraph for formula $\phi := (a \vee c) \wedge ((b \wedge u) \vee (d \wedge v)) \wedge (\neg a \vee \neg b)$ is given. Consider the case when CRP is $\langle 1 \rangle$ and $\sigma = \{a\}$. The algorithm checks if CRP can be extended by node 3 ($m = 3$). Using our notation $\sigma' = \{a, b\}$. Observe that σ' falsifies ϕ by substituting $a = true, b = true$ in ϕ. There are two rl-paths $\pi_1 := \langle 1, 3, 5, 7 \rangle, \pi_2 := \langle 1, 3, 5, 8 \rangle$ in the vpgraph shown in Fig. 4(b) which have $\langle 1, 3 \rangle$ as prefix. Neither of these rl-paths is satisfiable: π_1 is not satisfiable due to conflicting nodes 1, 7 and π_2 is not satisfiable due to conflicting nodes 3, 8.

Detection of global conflict: We use Corollary 2 to check if σ' falsifies ϕ. We check if there is an rl-path π in $G_h(\phi)$ such that σ' falsifies π. Continuing the above example, the hpgraph corresponding to ϕ is shown in Fig. 4(c). Observe that $\sigma' = \{a, b\}$ falsifies the rl-path $\langle 7, 8 \rangle$ in Fig. 4(c). Thus, using Corollary 2, it follows that σ' falsifies ϕ.

If there is no global conflict, then the set of implied assignments can be found by the application of unit literal rule on $G_h(\phi)$ as described in Corollary 5.

Case 3 (Local conflict): This conflict arises when every rl-path in $G_v(\phi)$ with CRP$\langle m \rangle$ as prefix contains two nodes which are conflicting and one of the conflicting nodes lies on CRP$\langle m \rangle$. Formally, this conflict arises when for every rl-path π in $G_v(\phi)$ with CRP$\langle m \rangle$ as prefix there exist two nodes $k, l \in \pi$ and $k \in$ CRP$\langle m \rangle$ such that $Lit(k) = \neg Lit(l)$. From Corollary 4, it follows that any rl-path π containing conflicting nodes is not satisfiable. Thus, when a local conflict occurs no rl-path in $G_v(\phi)$ with CRP$\langle m \rangle$ as prefix is satisfiable. Whenever there is a global conflict (case 2 above) there is also a local conflict, however, the reverse need not hold as shown by the example below.

Example: In Fig. 4(d) the vpgraph for formula $\phi := (a \vee c) \wedge ((b \wedge u \wedge (\neg a \vee \neg b)) \vee (d \wedge v))$ is shown. Consider the case when CRP is $\langle 1 \rangle$ and m is node 3 ($m = 3$). Using our earlier notation $\sigma' = \{a, b\}$. Note that σ' does not falsify ϕ, which means there is no global conflict. There are two rl-paths $\langle 1, 3, 5, 7 \rangle, \langle 1, 3, 5, 8 \rangle$ in the vpgraph shown in Fig. 4(d) which have $\langle 1, 3 \rangle$ as prefix. Both of these rl-paths contain two conflicting nodes, nodes 1,7 are conflicting on $\langle 1, 3, 5, 7 \rangle$ and nodes 3,8 are conflicting on $\langle 1, 3, 5, 8 \rangle$. Thus, there is a local conflict and the solver needs to backtrack from node $m = 3$.

Detection of global and local conflicts can be done in linear time as described in a more detailed version of this paper available at [3]. Depending upon the type of conflict (global or local) we perform global or local learning as described below.

6 Learning

Learning records the cause of a conflict. This enables the preemption of similar conflicts later on in the search. In the following, a *clause* will refer to a disjunction of literals. A clause C is *conflicting* under an assignment σ iff all literals in C are falsified by σ. If a clause C is not conflicting under an assignment σ, we say C is *consistent* under σ. We distinguish between two types of learning:

Global learning: A *globally learned* clause is a clause whose consistency must be maintained irrespective of the current search state, which is given by the current r-path CRP (and assignment $\sigma = \{Lit(n)|n \in CRP\}$). That is, whenever a globally learned clause becomes conflicting under σ the solver abandons the current search state and backtracks. A globally learned clause is generated from a conflicting clause. A conflicting clause C arises in two cases as described below.

1) When analyzing global conflicts as described in the previous section. When a global conflict occurs there is an rl-path π in hpgraph $G_h(\phi)$ which is falsified by the assignment σ currently under consideration. The set of literals corresponding to the nodes on π gives us a clause $C := \bigvee_{n \in \pi} (Lit(n))$. Observe that C is a conflicting clause, that is, all literals occurring in C are set to false under the current assignment.

Example: The hpgraph corresponding to $\phi := (a \vee c) \wedge ((b \wedge u) \vee (d \wedge v)) \wedge (\neg a \vee \neg b)$ is shown in Fig. 4(c). A global conflict occurs when the current assignment is $\sigma = \{a, b\}$, that is, σ falsifies ϕ. In this case the rl-path in the hpgraph which is falsified by σ is $\langle 7, 8 \rangle$. Thus the required conflicting clause is $\neg a \vee \neg b$.

2) When all literals of an existing globally learned clause C become false.

Once a conflicting clause C is obtained, we perform a 1-UIP (*first unique implication point*) analysis [19] to obtain a learned clause C'. Clause C' is added to the database of globally learned clauses. In order to perform 1-UIP analysis we maintain a notion of a decision level. We associate a decision level $dec(n)$ with each node n in the current r-path CRP. We also maintain a set of implied literals at each node (or decision level) along with the reason (set of variable assignments) which led to the implication. We follow the same algorithm as in [19] to perform the 1-UIP learning.

Local learning: A *locally learned* clause is associated to a node n in the vpgraph when a local conflict occurs at n. Suppose C is a locally learned clause at node n. Then the consistency of C needs to be maintained only when n is part of the current search state, that is, $n \in CRP$. If n does not lie on CRP, then the consistency of C is irrelevant. This is in contrast to a globally learned clause whose consistency must always be maintained.

Example: Consider the local conflict which occurs in the vpgraph in Fig. 4(d) when CRP is $\langle 1 \rangle$ and it is checked if CRP can be extended by $m = 3$. In this case every rl-path in vpgraph with $\langle 1, 3 \rangle$ as prefix contains two conflicting nodes one of which lies on $\langle 1, 3 \rangle$. The rl-path $\langle 1, 3, 5, 7 \rangle$ has conflicting nodes 1,7 and the rl-path $\langle 1, 3, 5, 8 \rangle$ has conflicting nodes 3,8. In this case a clause $Lit(7) \vee Lit(8) = \neg a \vee \neg b$ can be learned at node 3. Intuitively, when we consider extending the CRP with node m the (locally) learned clauses at node m must be consistent with the assignment $\sigma = \{Lit(n)|n \in CRP\langle m \rangle\}$. Otherwise, a local conflict will occur at m causing the solver to backtrack. Having learned clauses at node m avoids repeating the work done in detecting the same local conflict. For the vpgraph in Fig. 4(d), when CRP is $\langle 2 \rangle$ and $m = 3$, $\sigma = \{c, b\}$ is consistent with the learned clause $\neg a \vee \neg b$ at node 3, thus, the solver cannot get the same local conflict at node 3 as before (when CRP was $\langle 1 \rangle$ and $m = 3$).

If a local conflict occurs when extending CRP by node m, then a clause is learned at node m as follows: For each rl-path π having CRP$\langle m \rangle$ as prefix let $\omega_1(\pi), \omega_2(\pi)$ denote the pair of conflicting nodes on π. Without loss of generality assume that $\omega_1(\pi)$ lies on CRP$\langle m \rangle$. Then the learned clause C at node m is given by $\bigvee_\pi Lit(\omega_2(\pi))$. Consistency of C must be maintained only when considering rl-paths passing through m.

Table 1. Comparison between SatMate, MiniSat, BerkMin, Siege, zChaff. "Time" gives total time in seconds and "Solved" gives #problems solved within timeout of 600 seconds/problem.

Bench -mark	#Probs	SatMate		MiniSat		BerkMin		Siege		zChaff	
		Time	Solved	Time	Solved	Time	Solved	Time	Solved	Time	Solved
QG6	256	23266	235	49386	179	46625	184	46525	184	47321	180
QG6*	256	23266	235	37562	211	15975	239	30254	225	45557	186
Mboard	19	4316	12	4331	12	4947	11	4505	12	5029	11
Pigeon	19	5110	11	6114	9	5459	10	6174	9	5483	11

7 Experimental Results

The experiments were performed on a 1.5 GHZ AMD machine with 3 GB of memory running Linux. The techniques described in the paper have been implemented in a SAT solver called SatMate [3]. The non-clausal input formula is given in EDIMACS [1] or ISCAS format. SatMate also accepts cnf inputs in DIMACS format. We compare Sat-Mate against four state-of-the-art cnf SAT solvers MiniSat version 1.14 [2], BerkMin version 561 [10], Siege version 4 [4], and zChaff version 2004.5.13 [14].

QG6 benchmarks: The authors of [13] provided us with a benchmark set called QG6 which consists of 256 non-clausal formulas of varying difficulty. These benchmarks were generated during the construction of classification theorems for quasigroups [13]. The cnf version of these problems was also made available to us by the authors of [13]. The cnf version was obtained by directly expressing the problem of classifying quasigroups into cnf as opposed to the translation of non-clausal formulas into cnf. The non-clausal versions of these benchmarks have 300 variables and 7500 gates (AND, OR gates) on average, while the cnf versions have 1700 variables and 7500 clauses on average. We ran SatMate on the non-clausal formulas and cnf SAT solvers on the corresponding cnf formulas from QG6 suite.

QG6* benchmarks: We translated the non-clausal formulas from the QG6 suite into cnf by introducing new variables [15]. The cnf formulas obtained after translation have 7500 variables and 30000 clauses on average. We ran cnf SAT solvers on the cnf formulas obtained after translation. Note that we still ran SatMate on the non-clausal formulas.

Mboard benchmarks: encode the mutilated-checkerboard problem.

Pigeon benchmarks: encode the pigeon hole principle with n holes and $n + 1$ pigeons.

Both QG6 and QG6* benchmarks contain a mixture of satisfiable and unsatisfiable problems. All problems in the Mboard and Pigeon benchmarks are unsatisfiable.

The experimental results are summarized in Table 1. The column "#Probs" gives the number of problems in each benchmark set. There was a timeout of 10 minutes per problem per solver. For each solver we report two quantities: 1) "Time" is the total time spent in seconds when solving problems in a given benchmark, including the time spent (= timeout) for each instance not solved within timeout. 2) "Solved" gives the total number of problems that were solved within timeout.

Table 2. Comparison on individual benchmarks. Timeout is 1 hour per problem per solver. "Time" sub-column gives time taken in seconds.

Problem	SatMate			MiniSat	BerkMin	Siege	zChaff
	Time	Local confs	Global confs	Time	Time	Time	Time
dnd02	**174**	23500	15588	1308	1085	1238	TO
brn13	**181**	20699	20062	1441	1673	1508	TO
icl39	**200**	22683	14069	TO	TO	2629	TO
icl45	TO	4850	72106	TO	2320	**1641**	TO
q2.14	237	113	15863	**23**	24	34	88
cache.inv12	58	659	7131	**1**	**1**	**1**	2

Summary of results in Table 1: On QG6 benchmarks SatMate solves around 50 more problems and it is approximately 2 times faster than the cnf SAT solvers MiniSat, BerkMin, Siege, and zChaff. On QG6* benchmarks SatMate performs better than MiniSat, zChaff, Siege. However, BerkMin outperforms SatMate on QG6* benchmarks. The difference in the performance of cnf SAT solvers on QG6 and QG6* benchmarks shows how the differences in the encoding of a given problem to cnf can significantly impact the performance of cnf SAT solvers. The performance of SatMate on Mboard and Pigeon benchmarks is slightly better than the cnf SAT solvers.

Table 2 summarizes the performance of SatMate and four cnf SAT solvers on various individual problems. Problems dnd02, brn13, icl39, icl45 are from QG6 benchmark suite. Problems q2.14, cache.inv12 are generated by a verification tool. The sub-column "Time" gives the time required for SAT solving (in seconds). For SatMate we report the number of local conflicts and the number of global conflicts (Section 5) in the "Local confs" and "Global confs" sub-columns, respectively. A timeout of 1 hour was set per problem. We denote timeout by "TO". In case of timeout we report the number of conflicts just before the timeout for SatMate.

Performance of SatMate is correlated with the number of local conflicts and global conflicts. A local conflict is a conflict that occurs in a part of a formula and it depends on the structure of the vpgraph. There is no equivalent of local conflict in cnf SAT solvers. In cnf SAT solvers a conflict arises when the current assignment falsifies an original/learned clause which is equivalent to a global conflict. As shown in Table 2 the number of local conflicts is usually comparable to the number of global conflicts on the benchmarks where SatMate outperforms cnf SAT solvers. Indeed the performance of SatMate degrades if no local conflict detection and local learning is done.

8 Conclusion

We presented a new non-clausal SAT solver based on the General Matings approach. This approach involves the search for a vertical path which does not contain opposite literals in the vertical-horizontal path form (vhpform) of a given negation normal form formula. The main challenge in obtaining an efficient SAT solver based on the General Matings approach is to prevent the enumeration of vertical paths. We presented techniques for preventing the enumeration of vertical paths and graph based representations of the vhpform for efficient implementation of these ideas. Experimental results show that on a class of non-clausal benchmarks our SAT solver has a performance comparable to the current state-of-the-art cnf SAT solvers. Overall, our results show the promise of the General Matings approach in building SAT solvers.

Acknowledgment. We thank Peter Andrews for his useful comments and Malay Ganai, Guoqiang Pan, Sanjit Seshia, Volker Sorge for providing us with benchmarks.

References

1. Edimacs format, www.satcompetition.org/2005/edimacs.pdf.
2. Minisat sat solver, http://www.cs.chalmers.se/cs/research/formalmethods/minisat/.
3. SatMate website, http://www.cs.cmu.edu/~modelcheck/satmate.
4. Siege (version 4) sat solver, http://www.cs.sfu.ca/~loryan/personal/.
5. TPS and ETPS, http://gtps.math.cmu.edu/tps-papers.html.
6. Peter B. Andrews. Theorem Proving via General Matings. *J. ACM*, 28(2):193–214, 1981.
7. Peter B. Andrews. *An Introduction to Mathematical Logic and Type Theory: to Truth through Proof*. Kluwer Academic Publishers, Dordrecht, second edition, 2002.
8. Wolfgang Bibel. On Matrices with Connections. *J. ACM*, 28(4):633–645, 1981.
9. M. K. Ganai, P. Ashar, A. Gupta, L. Zhang, and S. Malik. Combining Strengths of Circuit-based and CNF-based Algorithms for a High-performance SAT solver. In *DAC*, 2002.
10. E. Goldberg and Y. Novikov. BerkMin: A Fast and Robust Sat-Solver. In *DATE*, 2002.
11. Joao P. Marques-Silva and Karem A. Sakallah. GRASP - A New Search Algorithm for Satisfiability. In *ICCAD*, pages 220–227, 1996.
12. David McAllester, Bart Selman, and Henry Kautz. Evidence for invariants in local search. In *AAAI*, pages 321–326, Providence, Rhode Island, 1997.
13. Andreas Meier and Volker Sorge. A new set of algebraic benchmark problems for sat solvers. In *SAT*, pages 459–466, 2005.
14. M. W. Moskewicz, C. F. Madigan, Y. Zhao, L. Zhang, and S. Malik. Chaff: Engineering an efficient SAT solver. In *DAC*, pages 530–535, June 2001.
15. David A. Plaisted and Steven Greenbaum. A structure-preserving clause form translation. *J. Symb. Comput.*, 2(3), 1986.
16. R. M. Smullyan. *First Order Logic*. Springer-Verlag, 1968.
17. Christian Thiffault, Fahiem Bacchus, and Toby Walsh. Solving Non-clausal Formulas with DPLL Search. In *SAT*, 2004.
18. H. Zhang. Sato: An efficient propositional prover. In *CADE-14*, pages 272–275, 1997.
19. Lintao Zhang, Conor F. Madigan, Matthew W. Moskewicz, and Sharad Malik. Efficient conflict driven learning in boolean satisfiability solver. In *ICCAD*, pages 279–285, 2001.

Determinization of Resolution by an Algorithm Operating on Complete Assignments

Eugene Goldberg

Cadence Berkeley Labs, USA
egold@cadence.com

Abstract. "Determinization" of resolution is usually done by a DPLL-like procedure that operates on partial assignments. We introduce a resolution-based SAT-solver operating on complete assignments and give a theoretical justification for determinizing resolution this way. This justification is based on the notion of a point image of resolution proof. We give experimental results confirming the viability of our approach. The complete version of this paper is given in [2].

1 Introduction

The resolution proof system has achieved an outstanding popularity in practical applications. Since resolution is a non-deterministic proof system, any SAT-solver based on resolution, one way or another, has to perform its "determinization". In the state-of-the-art SAT-solvers this determinization is based on using the DPLL procedure [1] that operates on partial assignments. The current partial assignment is extended until a clause is falsified. Then, the DPLL procedure backtracks to the last decision assignment and flips it. The search performed by the DPLL procedure can be simulated by so-called tree-like resolution (a special type of general resolution).

The reason for using partial rather than complete assignments is that by rejecting a partial assignment the DPLL procedure may "simultaneously" reject an exponential number of complete assignments. The premise of such an approach is that to prove that a CNF formula F is unsatisfiable one has to show that F evaluates to 0 for all complete assignments.

In this paper, we introduce the notion of a point image of a resolution proof that questions the premise above. A point image of a resolution proof can be viewed as an "encryption" of this resolution proof. Given a resolution proof R, one can always build its point image whose size is at most twice the size of R (measured in the number of resolution operations). On the other hand, given a set of points T and a CNF formula F one can use a simple procedure to test if T is a point image of a resolution proof. If it is, this procedure builds a resolution proof "specified" by T. This result implies that a resolution proof that F is unsatisfiable can be "guided" by testing the value of F in a sequence of points. Moreover, if a CNF formula F has a short resolution proof, the number of "guiding" points is negligible with respect to the size of the entire search space.

A. Biere and C.P. Gomes (Eds.): SAT 2006, LNCS 4121, pp. 90–95, 2006.

We introduce a SAT-solver operating on complete assignments that is inspired by the notion of a point image of resolution proof. The complete version of this paper is given in [2].

2   Main Definitions

Let F be a CNF formula over a set X of Boolean variables. The satisfiability problem (SAT) is to find a complete assignment p (called a **satisfying assignment**) to the variables of X such that $F(p) = 1$ or to prove that such an assignment does not exist. If F has a satisfying assignment, F is called **satisfiable**. Otherwise, F is **unsatisfiable**. A disjunction of literals is further referred to as a **clause**. A complete assignment to variables of X will be also called a **point** of the Boolean space $B^{|X|}$ where $B=\{0,1\}$. A point p **satisfies** clause C if $C(p)=1$. If $C(p)=0$, p is said to **falsify** C.

Let C_1 and C_2 be two clauses that have opposite literals of a variable x_i. Then the clause consisting of all the literals of C_1,C_2 except those of x_i is called the **resolvent** of C_1,C_2. The resolvent of C_1,C_2 is said to be obtained by the **resolution operation**. Given an unsatisfiable CNF formula F, one can always generate a sequence of resolution operations resulting in an empty clause. This sequence of operations is called a **resolution proof**. The resolution proof system is very important from a practical of view because many successful SAT-algorithms for solving "industrial" CNF formulas are based on resolution.

3   Justification of Our Approach

In this section, we give a theoretical justification of our approach.

Let R be a resolution proof that a CNF formula F is unsatisfiable. Let T be a set of points that has the following property. For any resolvent C of R, obtained from parent clauses C' and C'' there are two points p' and p'' of T such that

1. $C'(p') = 0$ and $C''(p'') = 0$
2. Points p' and p'' are different only in the variable in which clauses C' and C'' are resolved.

Then the set T is called a **point image of resolution proof R**. The points p' and p'' are called **a point image of the resolution operation** over clauses C' and C''.

Building a point image of a resolution proof. Given a resolution proof R that F is unsatisfiable, a point image T of R can be built as follows. We start with an empty set T. Then for every resolution operation from R over clauses C' and C'' we add to T two points p' and p'' forming a point image of this operation (unless p' and/or p'' have been added to T before). Clearly, the size of a set T built this way is at most twice the size of R.

Checking if a set of points is a point image. Given a set of points T and a CNF formula F, one can test if T is a point image of a resolution proof by the following procedure. Let S be a set of clauses that initially consists of the clauses of F. At each step of the procedure, we pick a pair of clauses C' and C'' of S such that a point image of the resolution operation over C' and C'' is in T and add the resolvent to S unless it is subsumed by a clause of S. This procedure has three termination conditions. 1) If a point of T satisfies F, then clearly F is satisfiable and T is not a point image. 2) No new clause can be added to S at a step of the procedure. This means that T is not a point image of a resolution proof (because T does not have "enough" points) and so one can not say yet whether F is satisfiable or not. 3) An empty clause is derived at a step of the procedure. This means that T is a point image of a resolution proof that F is unsatisfiable.

The procedure above implies that one can use complete assignments to "guide" a resolution proof. The size of the "guiding" set T is at most twice as large as the size of the proof R the set T "guides".

A proof has a huge number of point images. Let Res be the resolution operation over clauses C' and C'' used in a proof that CNF formula F is unsatisfiable. The operation Res, in general, has a huge number of point images because points p' and p'' forming a point image of Res are specified only for the variables of C' and C''. For the variables of F that are not in C' and C'', points p' and p'' may have arbitrary (but identical) values (because, by definition, p' and p'' are different only in the variable in which C' and C'' are resolved).

Since a point image of a resolution proof R is essentially the union of point images of resolution operations comprising R, the latter has a huge number of point images. However, not all point images of R are equivalent in the sense that some images are more regular and so can be more easily built by a deterministic algorithm. Resolution, being a non-deterministic proof system, does not distinguish between different point images of R. On the other hand, the fact that the "space" of images of R is very rich (and that some images are easier to find than others) implies that an algorithm operating on complete assignments can be used for finding resolution proofs.

4   Algorithm Operating on Complete Assignments

In this section, we introduce a resolution-based SAT-solver called **FI** (Find Image). Although *FI* is inspired by the ideas of Section 3 it does not look for a point image of a resolution proof "directly". Instead, *FI* implements a DPLL-like procedure that operates on complete assignments. Here, we give only a very high-level picture of *FI*. A detailed description can be found in [2]. Besides, [2] explains the relation between the set of points "visited" by *FI* and the proof *FI* builds.

FI operation. Operationally, *FI* can be viewed as a regular resolution-based SAT-solver that uses a complete assignment p as an "oracle". An initial assignment p can be generated randomly or using some heuristic. Let F be the current CNF formula (consisting of initial and conflict clauses) and $M(p)$ be the set of clauses of F falsfiied by p. Then only a variable of a clause from $M(p)$ can be assigned a value during decision making. If a variable x_i is assigned a value $b \in \{0, 1\}$ (either during decision making or BCP) and the value of x_i in p is $\bar{b}$, then value of x_i in p is flipped to b. In other words, one can view FI as a regular SAT-solver in which the choice of variables for decision making is controlled by a complete assignment that dynamically changes.

Interpretation in terms of complete assignments. The interpretation of FI above is convenient for "historical" reasons. However, we believe that a much more fruitful interpretation of FI is as follows. FI operates on complete assignments and so at any step of FI, every variable of the formula is assigned. Instead of making an assignment to a free variable x_i as in DPLL , FI fixes the assignment of x_i in p. This fixing means that in all the points p visited later the value of x_i stays the same until the time it is "unfixed". Unfixing x_i in FI corresponds to unassigning x_i in a DPLL-like procedure and making it free again. Only variables of clauses from the set $M(p)$ can be fixed. FI either fixes the value of a variable x_i that agrees with current point p or it first changes the value of x_i in p and only then fixes it. In the first case, the set $M(p)$ of falsified clauses stays the same, in the second case it has to be recomputed.

Note, that while a DPLL-like procedure can reproduce the decision-making of FI, the opposite is not true. Namely, the "overwhelming majority" of search trees that can be built by DPLL are out of reach for FI because at every step, FI is limited in the choice of variables that can be used for decision-making. (In a sense, this extra power of DPLL is due to the fact that DPLL is not an algorithm, but rather a proof system still containing a lot of non-determinism.)

In [2] we list reasons why FI should be interpreted in terms of complete assignments. Here we mention only one of them. The underlying semantics of a DPLL-like procedure operating on partial assignments is that it covers all the points of the search space. If one considers FI just as a regular DPLL-like procedure with a particular decision-making "heuristic", it is hard to explain why this "heuristic" works. As we mentioned above, the decisions of FI are extremely limited being controlled by a complete assignment, that is by $1/2^n$ of the search space. On the other hand, such an explanation can be easily done in terms of point images of resolution proofs.

5   Advantages of Using Complete Assignments

In Section 3 we gave a very "abstract" justification of using complete assignments in a resolution based algorithm. In this section, we list more concrete arguments (that are, in a sense, consequences of this abstract justification) in favor of our approach. In [2] we substantiate our claims experimentally.

Identifying small unsatisfiable sub-formulas. Current SAT-solvers are often used for solving huge CNF formulas e.g. in bounded model checking. The fact that SAT-solvers can efficiently prove the unsatisfiability of a CNF formula F of, say, 1 million variables usually means that there is an unsatisfiable sub-formula of a relatively small size (like 10-20 thousand variables). So identifying an unsatisfiable sub-formula is of great practical importance. Let G be an unsatisfiable sub-formula of F. The advantage of using complete assignments is that any complete assignment p falsifies at least one clause of G. So at least one clause of G is always present in $M(p)$ and so clauses from G are always on the "radar" of FI. On the other hand, a resolution based SAT-solver operating on partial assignments may spend a lot of time trying to satisfy clauses of F that are not in G.

Finding clauses that can be resolved. Let C be a clause that contains variable x_i and is falsified by the current point p. Let C' be a new clause falsified by the point p' obtained from p by flipping the value of x_i. Then clauses C and C' can be resolved in x_i. So, FI takes into account the "resolution nature" of the underlying proof system.

Efficient decision making. When solving large formulas it is very important to have a decision making procedure that is fast and at the same time manages to avoid branching on "irrelevant" variables. Making "redundant" decision assignments increases time spent on decision-making and may lead to the increase of redundant assignments made by the BCP procedure. Since the size of $M(p)$ is much smaller than that of the formula, picking the next variable to be fixed can be done efficiently. On the other hand, as we mentioned above, new clauses that appear in the set $M(p')$ (where p' is obtained from p by flipping the value of a variable) can be resolved with clauses of $M(p)$ satisfied by p'. So, by reducing our choice of variables to those of the clauses falsified by the current point, one reduces the probability of making assignments to "irrelevant" variables.

Successful use of frequent restarts. FI employs restarts. Instead of generating a new initial point randomly, FI starts with the last point visited in the previous iteration. This makes the resolution proof more "coherent" and allows one to use more frequent restarts successfully.

6   Experimental Results

In this section, we give results of experiments with an implementation of FI. We compare FI's results (in terms of the number of conflicts i.e. backtracks) with those of *Forklift* and *Minisat*. (Many more experimental results can be found in [2].) Experiments show that although FI is extremely limited in its decision-making, it is competitive in the number of conflicts with *Forklift* and *Minisat*.

7   Conclusions

We introduce a resolution based SAT-solver FI that operates on complete assignments (points). FI is inspired by the fact that a resolution proof can be specified by a small set of points. Experimental results show the viability of our

Table 1. Some experimental results

Name	# formulas	Forklift #conflicts	Minisat #conflicts (#aborted)	FI #conflicts
Dimacs formulas				
aim	72	3,303	3,587	**3,256**
bf	4	774	383	**379**
dubois	13	**3,062**	4,904	3,260
hanoi	2	**26,156**	65,428	223,040
hole	5	227,102	1,538,350	**56,884**
ii	41	6,505	4,088	**1,254**
jnh	49	2,151	2,096	**2,069**
par16	10	**42,934**	47,568	70,915
par8	10	304	162	**83**
pret	8	4,342	6,892	**2,942**
ssa	8	744	367	**348**
Velev's formulas				
vliw-sat.1.0	100	679,827	1,413,027	**527,416**
fvp-unsat.1.0	4	101,991	180,240	**92,333**
3pipe	4	**24,738**	66,567	33,856
4pipe	5	**125,850**	538,932	154,321
5pipe	6	268,463	1,261,229	**231,975**
6pipe	2	218,461	>470,779(1)	**176,067**
Some other known formulas				
Beijing	16	494,534	> 721,258(1)	**106,896**
blocksworld	7	**2,116**	4,732	8,209
bmc	13	54,098	**44,195**	48,568
bmc1	31	**1,033,434**	1,326,812	1,568,729
planning	6	29,415	**17,153**	24,426

approach. Determinization of resolution by operating on complete assignments seems to be a promising way to design resolution-based SAT-solvers. In [2], we give the complete version of this paper that contains a more detailed description of *FI* and more experimental results.

References

1. M.Davis, G.Longemann, D.Loveland. *A Machine program for theorem proving.* Communications of the ACM. -1962. -V.5. -P.394-397.
2. E.Goldberg. *Determinization of resolution by an algorithm operating on complete assignments.* Technical report, CDNL-TR-2006-0110, January 2006, available at http://eigold.tripod.com/papers/fi.zip

A Complete Random Jump Strategy with Guiding Paths

Hantao Zhang[*]

Department of Computer Science
University of Iowa Iowa City, IA 52242, U.S.A.
hzhang@cs.uiowa.edu

Abstract. The restart strategy can improve the effectiveness of SAT solvers for satisfiable problems. In 2002, we proposed the so-called random jump strategy, which outperformed the restart strategy in most experiments. One weakness shared by both the restart strategy and the random jump strategy is the ineffectiveness for unsatisfiable problems: A job which can be finished by a SAT solver in one day cannot not be finished in a couple of days if either strategy is used by the same SAT solver. In this paper, we propose a simple and effective technique which makes the random jump strategy as effective as the original SAT solvers. The technique works as follows: When we jump from the current position to another position, we remember the skipped search space in a simple data structure called "guiding path". If the current search runs out of search space before running out of the allotted time, the search can be recharged with one of the saved guiding paths and continues. Because the overhead of saving and loading guiding paths is very small, the SAT solvers is as effective as before for unsatisfiable problems when using the proposed technique.

1 Introduction

Modern SAT solvers based on the DPLL method can handle instances with hundreds of thousands of variables and several million clauses. To improve the chance of solving these problems, in [16], we proposed another technique called the "random jump strategy" which solves the same problem as the restart strategy [11]: When the procedure stuck at a region of the search space along a search tree, the procedure jumps to another region of the search space. One major advantage of the random jump strategy over the restart strategy is that there is no danger of visiting any region of the search space more than once (except those nodes appearing in a guiding path).

However, the strategy proposed in [16] may (with a small chance) destroy the completeness of a SAT solver because of skipped search space: A job which can be finished by a SAT solver in one day might not be finished in one or two days if either this strategy or the restart strategy is used by the same SAT solver. In this paper, we propose a simple technique which makes the random jump strategy

[*] Supported in part by NSF under grant CCR-0098093.

A. Biere and C.P. Gomes (Eds.): SAT 2006, LNCS 4121, pp. 96–101, 2006.

effective for both satisfiable and unsatisfiable problems. The technique works as follows: When we jump from the current search position to another position, we remember the skipped search space as a "guiding path" [18]. If we have no more search space left in the current run before running out of the allotted time, we may load one of the saved guiding paths and start another run of the search. The procedure stops when either a model is found or when no time left or when no more guiding paths left. Because the overhead of saving and loading guiding paths is very small, the SAT solvers are as effective as before for unsatisfiable problems when using this version of the random jump strategy. While the proposed technique can be used for any combinatorial backtrack search procedures, to keep the presentation simple, in this paper we will limit the discussion of our idea on the DPLL method.

We wish to point out that our random jump strategy is different from the general random jump strategy proposed in [10] in that when we backtrack to a previously selected literal, we require that the selected literal be set to its opposite value. This will prevent the selected literal from being assigned the same value. The strategy proposed in [10] cannot be complete with this requirement; they require that the literal be unassigned. Since the literal may take the same value in the next step, some portion of the search will be repeated. Another difference is that the completeness of their strategy is based on keeping the lemmas learned from the conflict analysis. The completeness of our strategy does not depend on the use of lemmas.

2   Random Jump in the DPLL Method

To use the random jump strategy, we divide the allotted search time into, say eight, equal time intervals and set up a checkpoint at the end of each time slot. At a checkpoint, we look at (the first few nodes of) the path from the root to the current node to estimate the percentage of the remaining space. If the remaining space is sufficiently large, we may jump up along the path, skipping some open branches along the way. After the jump, we wish that the remaining space is still sufficiently large.

2.1   Search Space of the DPLL Method

The space explored by a backtrack search procedure can be represented by a tree, where an internal node represents a backtrack point, and a leaf node represents either a solution or no solution. The search space explored by the DPLL method is a binary tree where each internal node is a decision point and the two outgoing branches are marked, respectively, by the two literals of the same variable with opposite sign. The case-splitting rule creates an internal node with two children in the search space. Without loss of any generality, if L is the literal picked at an internal node for splitting, we assume that the link pointing to the left child is marked with $\langle L, 1 \rangle$ (called *open link*) while the link pointing to the right child is labeled with $\langle L, 0 \rangle$ (called *closed link*). A leaf node in the search space is marked with either an empty clause (a conflict) or an empty set of clauses (a model).

We can record the path from the root of the tree to a given node, called *guiding path* in [18], by listing the links on the path. Thus, a *guiding path* is a list of literals each of which is associated with a Boolean flag [18]. The Boolean flag tells if the chosen literal has been assigned true only (1) or both true and false (0). According to [9], the concept of "guiding path" was first introduced in [6]. However, the name of "guiding path" does not appear in [6] and the literals used in [6] do not have an associated Boolean flag. For the implementation of the DPLL method, the current guiding path is stored in a stack.

In addition to the input formula, a DPLL algorithm can also take a guiding path as input. The DPLL algorithm will use the guiding path for decision literals until all the literals in the guiding path are used. If a literal is taken from a closed link, the DPLL algorithm will treat it as having only one (right) child. We can use guiding paths to avoid repeated search so that the search effort can be accumulated. For instance, the guiding path $(\langle x_1, 1 \rangle \langle \neg x_5, 0 \rangle \langle x_3, 0 \rangle)$ tells us that the literals are selected in the order of $x_1, \neg x_5, x_3$, and the subtree starting with the path $(\langle x_1, 1 \rangle \langle \neg x_5, 1 \rangle)$ is already complete; so is the subtree starting with $(\langle x_1, 1 \rangle \langle \neg x_5, 0 \rangle \langle x_3, 1 \rangle)$.

For the standard DPLL method, the backtrack is done by looking bottom-up for the first open link in the guiding path. If an open link is found, we replace it by its corresponding closed link and continue the search from there. To implement the jump, we may skip a number of open links in the guiding path as long as the remaining search space is sufficiently large.

2.2   Random Jump in the DPLL Method

The random jump strategy is proposed in [16] with the following goals in mind.

- Like the restart strategy, the new strategy should allow the search to jump out of a "trap" when it appears to explore a part of the space far from a solution.
- The new strategy will never cause any repetition of the search (except the few nodes appearing in guiding paths) performed by the original search algorithm.
- The new strategy will not demand any change to the branching heuristic, so that any powerful heuristic can be used without modification.
- If a search method can exhaust the space without finding a solution, thus showing unsatisfiability, the same search method with the new strategy should be able to do the same using the same amount of time.

Suppose the root node is at *level* 0 and the two children of a level i node are at *level* $i + 1$. To facilitate the presentation in this section, we assume that each path of the search tree is longer than three. Under this assumption, there are eight internal nodes at level 3 and we number the subtrees rooted by these nodes as $T_1, T_2, ..., T_8$. To check if the current search position in one of these subtrees, we just need to check the first three Boolean flags in the current guiding path: If they are $(1, 1, 1)$, then it is in T_1; if they are $(1, 1, 0)$, then it is in T_2; ...; if they are $(0, 0, 0)$, then it is in T_8.

The main idea of our new strategy can be described as follows: Suppose a search program is allotted t hours to run on a problem. We divide t into eight time intervals. At the end of time interval i, where $1 \leq i < 8$, suppose the current search position is in T_j, we say the search is "on time" if $j \geq i$ and is "late" if $j < i$. If the search is on time, we continue the search; if the search is late, then we say the remaining space is *sufficiently large* [16]. If this is the case, we may skip some unexplored space to avoid some traps, as long as we do not skip the entire T_j. To skip some unexplored space, we simply remove some open links (bottom up) on the guiding path.

At any checkpoint during the execution of the DPLL method, we can use the current guiding path to check in which subtree the current position is. Suppose we are at the end of time interval j and the current search position is still in T_1. The first three links (they must be open as we are in T_1) may or may be removed, depending on the value of j; all the other open links can be removed. If $j \geq 4$, then the first link can be skipped; if $j \geq 2$, then the second link can be skipped; if $j = 1$, then the third link can be skipped.

Example 1. Suppose the current guiding path is

$$(\langle l_1, 1 \rangle \langle l_2, 1 \rangle \langle l_3, 1 \rangle \langle l_4, 0 \rangle \langle l_5, 1 \rangle \langle l_6, 1 \rangle \langle l_7, 0 \rangle \langle l_8, 1 \rangle),$$

and we are at the end of time interval 5. The total number of open links in this case is 6 and every one can be skipped. In our implementation, a random number will be picked from $\{4, 5, 6\}$. If the chosen number is 4, then four open links will be removed and $\langle l_3, 1 \rangle$ is called the *cutoff* link which is the last open link to be removed. The guiding path after skipping will be $(\langle l_1, 1 \rangle \langle l_2, 1 \rangle \langle \overline{l_3}, 0 \rangle)$, where $\overline{l}$ is the negation of l. The search will use this path to continue the search. Note that the value of l_3 will be set to 0 when using this path.

2.3 Remembering the Skipped Search Space

Of course, it is true that each tree T_i will take different amounts of time to finish. If T_1 needs 99% of total time and the rest trees need only 1%, then we may skip too much after time interval 1 and become idle once T_2, ..., and T_8 are exhausted. To avoid the risk of being idle, we may memorize what has been skipped and this information can be stored as a guiding path.

For the previous example, the cutoff link is $\langle l_3, 1 \rangle$. To remember all the skipped open links, we just need to replace all the open links before the cutoff link, i.e., $\langle l_3, 1 \rangle$, by the corresponding closed links. That is, we need to save the following guiding path:

$$(\langle l_1, 0 \rangle \langle l_2, 0 \rangle \langle l_3, 1 \rangle \langle l_4, 0 \rangle \langle l_5, 1 \rangle \langle l_6, 1 \rangle \langle l_7, 0 \rangle \langle l_8, 1 \rangle).$$

This example illustrates how to skip a portion of the search space and how to save the skipped search space. For a more general description of the random jump strategy with guiding path, and related work, please refer to [17].

In [2], the concept of *search signature* is proposed to avoid some repeated search between restarts. At first, the search signature, which is a set of lemmas, takes more memory to store than a guiding path. Secondly, lemmas can avoid the repetition of leaf nodes but cannot eliminate the repeated visitation of internal nodes at the beginning of a search path before these lemmas become unit clauses. Because of this reason, their strategy creates bigger overhead than our strategy.

3   Conclusion

We have presented an improvement to a randomization strategy which takes the allotted run time as a parameter and checks at certain points if the remaining search space is sufficiently large comparing to the remaining run time; if yes, some space will be skipped and the skipped space is recorded as a guiding path. Like the restart strategy, it can prevent a backtrack search procedure from getting trapped in the long tails of many hard combinatorial problems and help it to find a solution quicker. Unlike the restart strategy, it never revisits any search space decided by the original search procedure. Unlike the restart strategy, it does not lose the effectiveness when working unsatisfiable problems as the overhead of the strategy is very small and is ignorable.

The motivation behind this research is to solve open quasigroup problems [14]. Without the random jump strategy, given a week of run time, SATO could not solve any of the possible exceptions in the theorems in [5,4,13,21]. Four cases were reported satisfiable in [4] and two cases were found satisfiable in [5]. In [13], four previously open cases were found satisfiabl; they are: $QG3(4^9)$, $QG3(5^9)$, $QG3(5^12)$, and $QG3(7^9)$. With the strategy, SATO solved each of them in less than a week. This clearly demonstrated the power of the new strategy. Moreover, since the random jump strategy keeps the completeness of SATO, we are able to prove that several previously unknown problems have no solutions, including $QG5(18)$ [14].

References

1. Baptista, L., and Margues-Silva, J.P., Using Randomization and Learning to Solve Hard Real-World Instances of Satisfiability, in Proceedings of the 6th International Conference on Principles and Practice of Constraint Programming (CP), September 2000.
2. Baptista, L., Lynce I., and Marques-Silva, J.P., Complete Search Restart Strategies for Satisfiability, in the IJCAI'01 Workshop on Stochastic Search Algorithms (IJCAI-SSA), August 2001.
3. Bayardo, R., Schrag, R., Using CSP look-back techniques to solve exceptionally hard SAT instances. In Proceedings of CP-96, 1996.
4. Bennett, F.E., Du, B., and Zhang, H.: Conjugate orthogonal diagonal latin squares with missing subsquares, Wal Wallis (ed.) "Designs 2002: Further Computational and Constructive Design Theory", Ch 2, Kluwer Academic Publishers, Boston, 2003.
5. Bennett, F.E., Zhang, H.: Latin squares with self-orthogonal conjugates, *Discrete Mathematics*, 284 (2004) 45–55

6. Bohm, M., Speckenmeyer, E.: A fast parallel SAT-solver – Efficient Workload Balancing, URL:http://citeseer.ist.psu.edu/51782.html, 1994.
7. Davis, M., Putnam, H. (1960) A computing procedure for quantification theory. *Journal of the ACM*, **7**, 201–215.
8. Davis, M., Logemann, G., and Loveland, D.: A machine program for theorem-proving. *Communications of the Association for Computing Machinery 5, 7* (July 1962), 394–397.
9. Feldman, Y., Dershowitz, N., Hanna, Z.: Parallel Multithreaded Satisfiability Solver: Design and Implementation. Electronic Notes in Theoretical Computer Science 128 (2005) 75-90
10. Lynce, I., Baptista, L., and Marques-Silva, J. P., Stochastic Systematic Search Algorithms for Satisfiability, in the LICS Workshop on Theory and Applications of Satisfiability Testing (LICS-SAT), June 2001.
11. Gomes, C.P., Selman, B., and Crato, C.: Heavy-tailed Distributions in Combinatorial Search. In Principles and Practices of Constraint Programming, (CP-97) Lecture Notes in Computer Science 1330, pp 121-135, Linz, Austria., 1997. Springer-Verlag.
12. Marques-Silva, J. P., and Sakallah, K. A., GRASP: A Search Algorithm for Propositional Satisfiability, in IEEE Transactions on Computers, vol. 48, no. 5, pp. 506-521, May 1999.
13. Xu, Y., Zhang, H.: Frame self-orthogonal Mendelsohn triple systems, *Acta Mathematica Sinica*, Vol.20, No.5 (2004) 913–924
14. Zhang, H.: (1997) Specifying Latin squares in propositional logic, in R. Veroff (ed.): Automated Reasoning and Its Applications, Essays in honor of Larry Wos, Chapter 6, MIT Press.
15. Zhang, H.: (1997) SATO: An efficient propositional prover, Proc. of International Conference on Automated Deduction (CADE-97). pp. 308–312, Lecture Notes in Artificial Intelligence 1104, Springer-Verlag.
16. Zhang, H.: (2002) A random jump strategy for combinatorial search. Proc. of Sixth International Symposium on Artificial Intelligence and Mathematics, Fort Lauderdale, FL, 2002.
17. Zhang, H.: (2006) A complete random jump strategy with guiding paths (full version). http://www.cs.uiowa.edu/~hzhang/crandomjump.pdf
18. Zhang, H., Bonacina, M.P., Hsiang, H.: PSATO: a distributed propositional prover and its application to quasigroup problems. *Journal of Symbolic Computation* (1996) 21, 543–560.
19. Zhang, H., Bennett, F.E.: Existence of some $(3, 2, 1)$–HCOLS and $(3, 2, 1)$–HCOLS. *J. of Combinatoric Mathematics and Combinatoric Computing*, 22 (1996) 13-22.
20. Zhang, H., Stickel, M.: Implementing the Davis-Putnam method, *J. of Automated Reasoning* 24: 277-296, 2000.
21. Zhu, L., Zhang, H.: Completing the spectrum of r-orthogonal latin squares. *Discrete Mathematics* 258 (2003)

Applications of SAT Solvers to Cryptanalysis of Hash Functions

Ilya Mironov and Lintao Zhang

Microsoft Research, Silicon Valley Campus
{mironov, lintaoz}@microsoft.com

Abstract. Several standard cryptographic hash functions were broken in 2005. Some essential building blocks of these attacks lend themselves well to automation by encoding them as CNF formulas, which are within reach of modern SAT solvers. In this paper we demonstrate effectiveness of this approach. In particular, we are able to generate full collisions for MD4 and MD5 given only the differential path and applying a (minimally modified) off-the-shelf SAT solver. To the best of our knowledge, this is the first example of a SAT-solver-aided cryptanalysis of a non-trivial cryptographic primitive. We expect SAT solvers to find new applications as a validation and testing tool of practicing cryptanalysts.

1 Introduction

Boolean Satisfiability (SAT) solvers have achieved remarkable progress in the last decade [MSS99, MMZ+01, ES03]. The record-breaking performance of the state-of-the-art SAT solvers opens new vistas for their applications beyond what conventionally has been thought feasible. Still, most of the successful real world applications of SAT solvers belong to the traditional domains of formal verification and AI. In this paper we explore applications of SAT solvers to cryptanalysis of hash functions.

Several applications of SAT solvers to cryptanalysis have been described in the literature [Mas99, MM00, FMM03, JJ05]. Their strategy can be regarded as a "head-on" approach, in the sense that they are not using any new or existing cryptanalytic methods in their attacks. Unsurprisingly, these efforts failed to produce any attacks of interest to cryptologists.

Despite the previous (arguably unsuccessful) attempts, we are convinced that SAT solvers could be of use in practical cryptanalysis. Our strategy may be described as "meet-in-the-middle": after initial, highly creative work of cryptanalysts, we are able to delegate the more laborious parts of the attack to the SAT solver.

Recently, several important cryptographic hash functions were shown to be vulnerable to collision-finding attacks [WY05, WYY05b]. The original attacks consisted of several steps each of which involves a lot of bit-tweaking and manual work. It suffices to say that the attack on the simplest function of the family, MD4, requires keeping track of as many as 122 boolean conditions.

A. Biere and C.P. Gomes (Eds.): SAT 2006, LNCS 4121, pp. 102–115, 2006.

In this paper, we show that SAT solvers can be used to automate certain elements of these attacks. In particular, we demonstrate that SAT solvers may obviate the need for compiling tables of sufficient conditions and designing clever message-modifications techniques. Our successful attacks on MD4 and MD5 suggest that SAT solvers could be a valuable addition to cryptanalysts' toolkit.

The paper is structured as follows. Section 2 is a short primer on theory and practical constructions of hash functions. Section 3 covers recent attacks on hash functions; Section 4 presents experimental results of applying SAT solvers to automation of these attacks. We conclude in Section 5.

2   Theory and Constructions of Hash Functions

Cryptographic hash functions are essential for security of many protocols. Early applications of hash functions in systems security include password tables [JKW74] and signature schemes [RSA78, Lam79]; since then virtually any cryptographic protocol uses directly or indirectly a secure hash function as a building block.

The properties required of a secure hash function differ and often depend on the protocol in question. Still, the property of being *collision-resistant* is recognized as the "gold standard" of security of hash function. The first formal definition of collision-resistant hash functions (CRHF) was given by Damgård [Dam88]. A function H is said to be collision-resistant if it is infeasible to find two different inputs x, y such that $H(x) = H(y)$. Since any compressing function has collisions, a guarantee of collision-resistance may only be computational.

A first standard hash function, MD4, was designed by Ron Rivest [Riv91]; its strengthened version MD5 followed shortly thereafter [Riv92]. A first NIST-approved hash function, SHA (Secure Hash Algorithm), adopted the general structure (and even some constants!) of MD4 [NIS93]; it was withdrawn in 1995 and replaced with a new version, dubbed SHA-1 [NIS95], that differed in one additional instruction. To avoid confusion, the original SHA is commonly referred to as SHA-0. As of 2004, two hash functions were in wide-spread (and almost exclusive) use: MD5 and SHA-1. It is fair to say that all of these functions belong to one family that shares similar design principles.

Compression function. The basic construction block of CRHFs is a collision-resistant *compression function*, which maps a fixed-length input into a shorter fixed-length output.

The heart of the construction is a *block cipher*, which is defined as a function of two inputs $E \colon \{0,1\}^k \times \{0,1\}^n \mapsto \{0,1\}^n$. Although $E(\cdot, \cdot)$ compresses its input by mapping $k + n$ bits into k bits, as is it is trivially invertible. However, the following trick, called the Davies-Meyer construction, results in a CRHF F under the assumption that E is an ideal block cipher (i.e., $E(x, \cdot)$ is an indexed collection of random permutations on $\{0,1\}^n$):

$$F(M, x) = E(x, M) \oplus M.$$

Among several methods for constructing block ciphers, the *Feistel ladder* is by far the best-known, being the method of choice for DES. The (unbalanced) Feistel ladder is a foundation of all block ciphers inside MDx and SHAx families. It is an iterative method, which consists of two separate components, a key-expansion algorithm and a collection of round functions. The ladder is parameterized by the number of rounds r and the size of the state. Assume for concreteness that the state consists of four 32-bit words (as the case of MD4 and MD5). The state goes through r rounds of transformation; let the initial, intermediate, and final states be (a_i, b_i, c_i, d_i) for $i \in \{0, \ldots, r\}$. The key-expansion algorithm K maps M to r round keys denoted $K(M) = w_0, \ldots, w_{r-1}$:

$$K \colon \{0,1\}^n \mapsto \underbrace{\{0,1\}^{32} \times \cdots \times \{0,1\}^{32}}_{r \text{ times}}.$$

Round functions $f_i \colon \{0,1\}^{128} \mapsto \{0,1\}^{32}$ for $i \in \{0, \ldots, r-1\}$ are used to update the state. MD5's transformation is one example (k_i and s_i are constants):

$$(a_{i+1}, b_{i+1}, c_{i+1}, d_{i+1}) \leftarrow$$
$$(d_i, b_i + (a_i + f(b_i, c_i, d_i) + w_i + k_i) \lll s_i, b_i, c_i).$$

Notice that the transformation is reversible if the round key w_i is known.

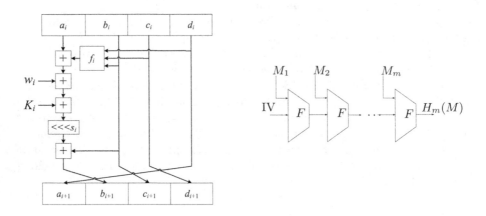

Fig. 1. One round of Feistel ladder for MD5 and the Merkle-Damgård construction

Merkle-Damgård paradigm. CRHFs are expected to take inputs of arbitrary length. A composition of fixed-length compression functions, discussed above, preserves the collision-resistance property. This method is called the Merkle-Damgård construction [Mer90, Dam90] (see Figure 1).

Generic Attacks. *Generic attacks* against hash functions are oblivious to the particulars of their constructions—they treat hash functions as black-boxes and in general provide an upper bound on security of different cryptographic properties.

We contrast two generic attacks on hash functions. Both attacks find two messages $x \neq y$ such that $H(x) = H(y)$. In the first attack the two messages are unrestricted, and thanks to the birthday paradox its generic complexity is $2^{n/2}$, where the output length of H is n bits. The goal of the second attack is to find colliding x and y such that $x = y \oplus \delta$ for some fixed δ. The generic complexity of this attack is 2^n.

Practical hash functions. Following the nomenclature developed above, designs of MD4, MD5, SHA-0, and SHA-1 hash functions follow the Merkle-Damgård paradigm, making use of a compression function built via the Davies-Meyer construction from block ciphers of the unbalanced Feistel ladder-type.

The internal state of the compression function consists of four 32-bit words (a_i, b_i, c_i, d_i) for MD4 and MD5 and five 32-bit words $(a_i, b_i, c_i, d_i, e_i)$ for SHA-0 and SHA-1. Since the size of the internal state is also the size of the output of the hash function, the output length of MD4 and MD5 is 128 bits; SHA-0 and SHA-1 produce 160-bit outputs. MD4 applies 48 rounds of the Fiestel transform, MD5 has 64, and SHA-0,1 both use 80 rounds. For details of the constructions, including the round functions, the key expansion algorithms, and the constants, omitted in the interest of brevity, we refer the reader to [MvOV96] or corresponding standards.

Commonly held belief in security of MD5 and SHA-1 had been supported by a relative absence of attacks on these and related functions. Although a collision in MD4 was discovered in 1996 [Dob96a], some weaknesses were identified in MD5 and SHA-0 [dBB94, Dob96b, BC04] and a theoretical attack was known on SHA-0 [CJ98], no collisions had been found for MD5 and SHA-1 despite more than ten years of intense scrutiny.

The year 2005 brought about a sea change in our understanding of hash functions. A new and improved attack on MD4 [WLF+05], collisions for MD5 and SHA-0 [WY05, WYY05b], and a theoretical attack on SHA-1 [WYY05a] were announced by a group of Chinese researchers led by Xiaoyun Wang in two consecutive conferences. Independently of them, attacks on SHA-0 and reduced-round SHA-1 were discovered by Biham et al. [BCJ+05].

Most of these attacks are conceptually simple but their implementations tend to be extremely laborious. Although there is considerable interest in generalizing the attacks and applying them in other contexts, the required amount of manual work may be unsurmountable. We observe that some components of the attacks may be expressed as CNF formulas and be rather efficiently solved by advanced SAT solvers.

3   Attacks on Hash Functions

In this section we develop notation and a common framework describing collision-finding attacks on hash functions.

3.1   Notation

Throughout this section and in the rest of the paper, whenever we are trying to find a collision between $M = (m_0, \ldots, m_{15})$ and $M' = (m'_0, \ldots, m'_{15})$, variables $w_i, a_i, b_i, c_i, d_i, e_i$ refer to the computation of the compression function on input M and their primed counterparts $w'_i, a'_i, b'_i, c'_i, d'_i, e'_i$ to the computation of the same function on M'.

We will be interested in two types of *differentials*: in respect to XOR and in respect to difference modulo 2^{32}. Define

$$\Delta^+ a_i = a_i - a'_i \quad (\text{mod } 2^{32}) \text{ and similarly } \Delta^+ w_i, \Delta^+ m_i, \Delta^+ b_i, \ldots [, \Delta^+ e_i]$$

and

$$\Delta^\oplus a_i = a_i \oplus a'_i \text{ and similarly } \Delta^\oplus w_i, \Delta^\oplus m_i, \Delta^\oplus b_i, \ldots, [\Delta^\oplus e_i].$$

Let

$$\Delta_i^\oplus = (\Delta^\oplus a_i, \Delta^\oplus b_i, \Delta^\oplus c_i, \Delta^\oplus d_i, [\Delta^\oplus e_i]), \text{ and similarly } \Delta_i^+.$$

The sequence $\Delta_0^\oplus, \Delta_1^\oplus, \Delta_2^\oplus, \ldots$ (resp., $\Delta_0^+, \ldots$) is a called a *differential path* in respect to the XOR differential (resp., to the difference modulo 2^{32}). In the rest of the paper, $\circ$ stands for both $+$ and $\oplus$.

Strictly speaking, the attacks due to Wang et al. fix the exact settings for most of the differing bits, subsuming both differentials. Our encodings fully use this information.

3.2   Overview of the Attacks

Conceptually, the attacks on MDx and SHA-0,1 have a lot in common. Most remarkably, the attacks solve a seemingly harder problem, i.e., finding a 512-bit message such that $H(\text{IV}, M) = H(\text{IV}, M \circ \delta)$, where H is the compression function and δ is fixed.[1] As observed earlier, the generic complexity of this attack (the one that uses the function as a black-box) is 2^n, where $n = 128$ or 160. A judicious choice of δ and a collection of clever techniques for finding M that take advantage of the weaknesses of the compression function bring the complexity of the attack to fewer than 2^{42} evaluations of the hash function.

Conceptually, the attacks consist of four distinct stages.

Stage I. Choose $\Delta^\circ m_0, \ldots, \Delta^\circ m_{15}$.

Stage II. Choose a differential path $\Delta_0^\circ, \ldots, \Delta_{r-1}^\circ$, where r is the number of rounds ($r = 48, 64$ or 80).

Stage III. Find a set of *sufficient conditions* on the message $M = (m_0, \ldots, m_{15})$ and the intermediate variables $a_i, \ldots, d_i$ that guarantee (with high probability) that the message pair $M, M' = (m_0 \circ \Delta^\circ m_0, \ldots, m_{15} \circ \Delta^\circ m_{15})$ follows the differential path $\Delta_0^\circ, \ldots, \Delta_{r-1}^\circ$.

Stage IV. Choose a message M such that all sufficient conditions hold.

---

[1] Some attacks solve the problem in two steps: they seek to find two messages M_0 and M_1, and differences δ, δ_0, and δ_1 such that $H(\text{IV}, M_0) = H(\text{IV}, M_0 \circ \delta_0) \circ \delta$ and $H(H(\text{IV}, M_0), M_1) = H(H(\text{IV}, M_0 \circ \delta_0), M_1 \circ \delta_1)$.

The attacks may seem counter-intuitive: rather than finding *any* two messages that collide under the hash function, we first severely constrain the space of possible message-pairs by fixing their difference, and by choosing the differential path we restrict the space of possible solutions even further.

There are two reasons that make this approach work. First, the differential path is carefully chosen to maximize the probability of a collision. Second, by deliberately constraining the solution space, we know some important properties of the solution, which allow us to construct one by an iterative process.

The first stage of the attack is usually done by hand or by applying some heuristics, such as trying to find a difference with low Hamming weight.

The second stage is the most creative stage of all four.There is fair amount of flexibility in it, for the curios reason that, as the round functions are non-linear, a given difference in the input may result in many possible differences in the output. Ironically, the attack turns the very foundation of security of hash functions— non-linearity of the round transformation—to its advantage. There are several constraints imposed on the differential path. First, it must be feasible. Further, it must be likely and finally it should facilitate the third stage of the attack.

The third stage is tightly coupled with the previous one. Most sufficient conditions naturally follow from the properties of the round function.

The fourth stage is computationally most intensive. Spectacular attacks due to Wang et al. would not be possible with a breakthrough in this stage. Indeed, a random pair of messages M and M' is very unlikely to follow the differential path. This is where the recent attacks depart most radically from earlier work. The idea is to start with an arbitrary message M and then carefully "massage" it into the differential path. The crucial contribution of Wang et al. was to come up with a set of tools of fixing errors in the differential path one by one.

We claim that SAT solvers might be very helpful in automating the third and the fourth stages of the attack. Since the actual attack requires a lot of iterations between the second stage and the next two, any method that would speed up testing and validation of differential paths becomes a useful cryptanalytic tool.

3.3   Attacks on MD4 and MD5

The terse exposition of the attacks due to Wang et al. was lacking some details and very short on intuition. Several papers attempted to explain, fill in omitted details, correct, improve, and automate these attacks [HPR04, Kli05, Dau05, BCH06, SO06]. In particular, we refer to Black et al. and Oswald et al. [BCH06, SO06] for intuition on the discovery process of the differential path for MD4 and MD5 (Stages I and II of our framework).

For ease of exposition, examples below are given for the attack on MD5 [WLF+05]. Attacks on MD4, SHA-0 and (reduced-round) SHA-1 are similar, with the main difference being usage of XOR differentials for the SHAx functions instead of differentials in respect to subtraction for MDx.

Stage III of the attacks produces a list of probabilistically sufficient conditions on the internal variables for the differential path to hold (of the type: lsb of c_7 is 0 and the 11th bit of a_7 is 1). There are as many as 310 of them; fortunately, most of these conditions appear in the first 25 rounds. If all conditions are met, the differential path is very likely to hold as well.

Successful completion of Stage IV of the attack is ensured by a combination of *single-message* and *multi-message modification techniques* and a probabilistic argument. More precisely, conditions in the first 16 rounds of MD5 can be fixed by going from b_{i+1}, chosen to satisfy all conditions, to m_i, as follows:

$$m_i \leftarrow ((b_{i+1} - b_i) \ggg s_i) - K_i^{(5)} - a_i - f(b_i, c_i, d_i).$$

This operation is called a single-message modification as it only affects one block of the message. Unfortunately, if we try to apply this method to correcting errors in later rounds, we are most likely to introduce new errors in earlier rounds.

Multi-message modifications do just that—they correct errors in the differential path in rounds beyond 16, by changing several blocks of the message at a time. The example given in [WY05] and discussed in more details in [BCH06] is (roughly) the following.

Suppose, we would like to force the msb of b_{17} be 1. This can be achieved by modifying $w_{16} = m_1^* \leftarrow m_1 + 2^{26}$. This modification changes the value of b_2 to b_2^* (and, since $b_2 = c_3 = d_4 = a_5$, other values as well), which is where m_1 was used for the first time. To absorb the change, we modify

$$m_2^* \leftarrow ((b_3 - b_2^*) \ggg 17) - a_2 - f(b_2^*, c_2, d_2) - K_2^{(5)},$$
$$m_3^* \leftarrow ((b_4 - b_3^*) \ggg 22) - a_3 - f(b_3, c_3^*, d_3) - K_3^{(5)},$$
$$m_4^* \leftarrow ((b_5 - b_4^*) \ggg 7) - a_4 - f(b_4, c_4, d_4^*) - K_4^{(5)},$$
$$m_5^* \leftarrow ((b_6 - b_5^*) \ggg 12) - a_5^* - f(b_5, c_5, d_5) - K_5^{(5)}.$$

As a result, none of the values computed in rounds 1–5 got changed; only the message blocks $m_1, \ldots, m_5$, intermediate variables $b_2 = c_3 = d_4 = a_5$ and $b_{17} = c_{18} = d_{19} = a_{20}$ did.

This is a representative example of a multi-message modification method, although by no means the trickiest. None of this methods expand beyond round 22, and they become progressively more complicated as more message blocks get affected.

Since as many as 37 conditions cannot be fixed using message modifications, Wang et al. fall back on the probabilistic method. Since all conditions are satisfied (heuristically) with probability 2^{-37}, it suffices to apply known message modifications to random messages 2^{37} times to have a non-trivial chance of generating a collision. Subsequent papers enriched the toolkit of message modifications, reducing the number of conditions that need to be satisfied probabilistically to around 30 [Kli05, BCH06].

4   Automation Via SAT Solvers

Arguably, the most annoying aspects of the attacks due to Wand et al. is the need to keep track of hundreds of conditions with a corresponding code to take care of them by means of message modifications. Our approach completely eliminates this difficulty, as we encode the entire differential path and leave the task of finding a message satisfying it to the SAT solver.

In this section, we report our preliminary results on using SAT solvers to implement the attacks due to Wang et al.

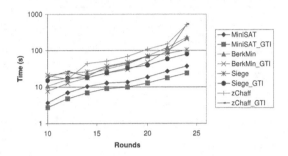

Fig. 2. Comparison of Different Solvers on MD5 Instances

To apply SAT solver to cryptanalysis of hash functions, we need to transform the hash functions' code into boolean circuits and perform CNF clausification. We experimented with several boolean circuit implementations for the adders and multiplexors, and found that different implementations of these primitives may have some tangible effect on the size of the resulting CNFs and efficiency of SAT solvers applied to these formulas. In this section, we report results on using a simple full-adder implementation and a multiplexer implemented with two AND gate and one OR gate. We found that this combination works reasonably well in our experiments. For clausification, currently we use the straightforward Tseitin transformation [Tse68] on the propositional formulas. We believe that more optimization can be obtained from careful examination of the encoding process [ES06].

Since we are dealing exclusively with satisfiable instances, we evaluated a number of stochastic and complete SAT solvers. We found that stochastic methods were not suitable for this task, which was unsurprising because the instances were highly structured with complicated interactions among the variables. For the complete solvers, we experimented with several state-of-the-art DLL search-based solvers, including the latest versions of ZChaff [MMZ+01], MiniSAT [ES03], BerkMin [GN02], and Siege [Rya04]. Each solver was tested by itself and with the SATELITE preprocessor [EB05]. Figure 2 gives the average runtime of these solvers for finding a first solution of MD5. Each data point is the average of 30 runs with randomly generated IV vectors. The SATELITE preprocessor followed by MINISAT (collectively known as SATELITEGTI) produced best results, which are reported in this section.

Wang et al.'s attacks on MD4 and MD5. The sizes of the resulting formulas appear in Table 1.

Table 1. Applying SAT solvers to MD4, MD5, and SHA-0

Hash	Total	Modifications		Formula	
	Rnd	Manual	SAT	Vars.	Cls
MD4	48	all	all	53228	221440
MD5	64	22	46	89748	375176
SHA-0	80	20	35	114809	486185

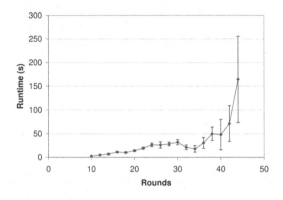

Fig. 3. Mean running time over 100 runs for MD5

Another information given in Table 1 is the number of rounds for which SATELITEGTI can find a message satisfying the differential in the median time of less than 15 min on a PC (3.2GHz PIV, 1Gb RAM) versus the number of rounds for which manual message modifications techniques are known.

As implied by Table 1, finding a collision in MD4 is easy and usually takes less than 10 minutes (the median value is approximately 500 s).

Making SATELITEGTI work for the full MD5 is less straightforward. Recall that for rounds beyond the limits of the message modifications techniques, Wang et al. use a probabilistic argument, whose running time increases exponentially with the number of sufficient conditions in the additional rounds. Although the behavior of SATELITEGTI is more complex, we discover that the best mode of operation for the solver is to follow exactly the same paradigm! The slowdown experienced by SATELITEGTI as a function of the number of rounds makes it more economical to run the solver for $n < 64$ rounds and than screen the results for solutions that hold for more rounds. The optimal value for n is 25 rounds. In other words, $n = 25$ minimizes the ratio of the number of solutions for the n-round MD5 produced in one second to the probability that a solution for n rounds will be a solution for the full MD5.

We modified SATELITEGTIto produce multiple solutions for a single SAT instance. We observe that finding subsequent solutions is much faster than finding the first solution (the runtime for finding the first solution is shown in Figure 3). In Figure 4, we plot the number of conflicts encountered between the first

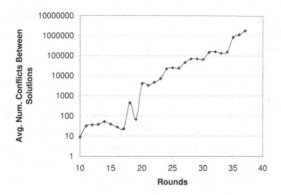

Fig. 4. Num. of Conflicts for Finding 10000 Solutions in MD5

and the 10000th solution. This graph illustrates the inverse density of solutions in the boolean space as explored by the SAT solver. For example, at round 25, on the average the solver encounters about 15 conflicts before finding a solution. We can generate around 350 solutions per second on our machine. The probability that one solution that follows the differential path for 25 rounds will lead to a collision is approximately 2^{-27}. Finding a solution for the full MD5 (first block of the attack) takes approximately 100 hours, which is comparable, albeit slower than an hour on an IBM supercomputer required by the original attack (the best known attack takes approximately 8 minutes [BCH06]). Finding solutions for the second block is significantly faster (less than an hour on our PC).

An interesting result of our experiments with SAT solvers is the importance of having a differential path encoded in the formula. Discovering a differential part is a difficult process (Stage II of the framework), which ideally should be automated. Currently, the only method of automatic enumeration of differential paths is known for MD4 [SO06], extending it to other hash functions is an open problem. Our experiments with applying SAT solvers to this task were not very encouraging. Specifically, absent restrictions imposed by the differential path the SAT solver performs much worse than in the presence of these restrictions.

Dobbertin's attack on MD4. We automated Dobbertin's attack on MD4 [Dob96a], which is considerably weaker than the more recent attacks. The reader is referred to Dobbertin's paper for details of the attack. The most difficult part of Dobbertin's method is finding a solution to a non-linear system of equations defined by rounds 12–19 of the hash function. He describes an ad hoc iterative process that starts with a random assignment and then fixes errors four bits at a time. We observe that this part may be encoded as a CNF formula *without doing any additional transformations.* Namely, we translate the rounds 12–19 of the compression function of MD4 into a CNF formula with 10364 variables and 39482 clauses. In fact, we may do it in more generality than Dobbertin, by accepting any initial value for $(a_{12}, b_{12}, c_{12}, d_{12})$, which simplifies Part 2 of his attack.

SATELITEGTI finds 2^{22} solutions to the resulting formula in less than one hour on a PC. This attack is again slower than the original attack, but has the added benefit of being substantially simpler and conceptually simpler.

Wang et al.'s attack on SHA-0. Translating the first two stages of the attack due to Wang et al. on SHA-0 into a CNF formula is analogous to the similar task for MD4 and MD5. The optimal number of rounds after which the probabilistic method outdoes the solver is 20, after which the probability that a solution results in a collision is 2^{-42}. The size of the formula for 20 rounds is 29217 variables and 117169 clauses. Based on several test runs, we estimate that generating a full collision using SATELITEGTI would require approximately 3 million CPU hours.

5   Conclusion and Directions for Future Work

In this work, we describe our initial results on using SAT solvers to automate certain components of cryptanalysis of hash functions of the MDx and SHAx families. Our implementations are considerably easier (but also slower) than the originals, since we delegate many minute details of the attacks to the SAT solver.

In particular, our method has no use for sufficient conditions and message-modification techniques, which are automatically (and implicitly) discovered by the SAT solver. We conjecture that finding a differential path, a formidable problem in itself, may be simplified if one is able to quickly test feasibility of the path without having to compile and consult tables with sufficient conditions.

Our results can be summarized as follows. We are able to launch a complete attack on MD4 by applying SATELITEGTI to the most straightforward encoding of a differential path as a CNF formula, finding a solution in less than 10 minutes. Finding a collision in MD5 follows closely the probabilistic method of Wang et al.: the SAT solver generates a lot of solutions for truncated MD5, which are then filtered in order to find a solution for the full hash function. This attack generates one collision in approximately 100 hours. A successful SAT-solver-aided attack on SHA-0 is still a theoretical possibility.

We rely in our experiments on an off-the-shelf SAT solver, SATELITEGTI, with a simple modification to allow generation of multiple solutions. To the extent of our knowledge, this is a first work that demonstrates the practicality of using SAT solvers for real cryptanalysis.

This work opens many intriguing possibilities. First, we currently use general-purpose SAT solvers. Optimizing and specializing them for cryptographic applications are interesting research directions. For example, one optimization may take advantage of the fact that many internal signals are likely to be correlated between the two input messages. Faster SAT solvers may directly translate into better cryptanalyses.

Second, the primitives studied in this work are restricted to modular addition and bit-wise boolean operations, which allow simple translation into CNF formulas. Multiplications and table lookups are widely used in construction of other cryptographic primitives. It is well known these operations pose a challenge for SAT solvers. There are some known methods for tackling such operations in a SAT solver. One approach is to use CAD tools to synthesize lookup table before translation into CNF [MM00]. Another method consists of rearranging signals during search to handle arithmetic operations more efficiently [WSK04]. The effectiveness of these approaches in cryptanalytic context is largely unexplored.

Finally, we observe that the probabilistic method of generating solutions for truncated hash function and screening them for a complete solution, a necessity in the case of Wang et al. attacks, might be avoidable in the case of a SAT solver. Indeed, there are no reasons why SAT solvers should experience a dramatic slow-down on large formulas compared to the method that uses *the same* SAT solver to produce multiple solutions for a shorter formula and test them by substitution into the longer formula. We find the current behavior counter-intuitive and amusing.

References

[BC04]   Eli Biham and Rafi Chen. Near-collisions of SHA-0. In Matthew K. Franklin, editor, *Advances in Cryptology—CRYPTO 2004*, volume 3152 of *Lecture Notes in Computer Science*, pages 290–305. Springer, 2004.

[BCH06]   John Black, Martin Cochran, and Trevor Highland. A study of the MD5 attacks: Insights and improvements. In *Fast Software Encryption 2006*, 2006.

[BCJ+05]   Eli Biham, Rafi Chen, Antoine Joux, Patrick Carribault, Christophe Lemuet, and William Jalby. Collisions of SHA-0 and reduced SHA-1. In Cramer [Cra05], pages 36–57.

[Bra90]   Gilles Brassard, editor. *Advances in Cryptology—CRYPTO '89, 9th Annual International Cryptology Conference, Santa Barbara, California, USA, August 20–24, 1989, Proceedings*, volume 435 of *Lecture Notes in Computer Science*. Springer, 1990.

[CJ98]   Florent Chabaud and Antoine Joux. Differential collisions in SHA-0. In Hugo Krawczyk, editor, *CRYPTO*, volume 1462 of *Lecture Notes in Computer Science*, pages 56–71. Springer, 1998.

[Cra05]   Ronald Cramer, editor. *Advances in Cryptology—EUROCRYPT 2005, 24th Annual International Conference on the Theory and Applications of Cryptographic Techniques, Aarhus, Denmark, May 22–26, 2005, Proceedings*, volume 3494 of *Lecture Notes in Computer Science*. Springer, 2005.

[Dam88]   Ivan Damgård. Collision free hash functions and public key signature schemes. In David Chaum and Wyn L. Price, editors, *Advances in Cryptology—EUROCRYPT '87*, volume 304 of *Lecture Notes in Computer Science*, pages 203–216. Springer, 1988.

[Dam90]   Ivan Damgård. A design principle for hash functions. In Brassard [Bra90], pages 416–427.

[Dau05]   Magnus Daum. *Cryptanalysis of Hash Functions of the MD4-Family*. PhD thesis, Ruhr-Universität Bochum, 2005.

[dBB94]   Bert den Boer and Antoon Bosselaers. Collisions for the compressin function of MD5. In Tor Helleseth, editor, *Advances in Cryptology—EUROCRYPT '93*, volume 765 of *Lecture Notes in Computer Science*, pages 293–304. Springer, 1994.

[Dob96a]   Hans Dobbertin. Cryptanalysis of MD4. In Dieter Gollmann, editor, *Fast Software Encryption '96*, volume 1039 of *Lecture Notes in Computer Science*, pages 53–69. Springer, 1996.

[Dob96b]   Hans Dobbertin. The status of MD5 after a recent attack. *CryptoBytes*, 2(2):1–6, 1996.

[EB05]   Niklas Eén and Armin Biere. Effective preprocessing in SAT through variable and clause elimination. In *Proceedings of the International Symposium on the Theory and Applications of Satisfiability Testing (SAT)*, pages 61–75, 2005.

[ES03]   Niklas Eén and Niklas Sörensson. An extensible SAT-solver. In *Proceedings of the International Symposium on the Theory and Applications of Satisfiability Testing (SAT)*, 2003.

[ES06]   Niklas Eén and Niklas Sörensson. Translating pseudo-boolean constraints into SAT. *J. Satisfiability, Boolean Modeling and Computation*, 2:1–26, 2006.

[FMM03]   Claudia Fiorini, Enrico Martinelli, and Fabio Massacci. How to fake an RSA signature by encoding modular root finding as a SAT problem. *Discrete Applied Mathematics*, 130(2):101–127, 2003.

[GN02]   Evgueni Goldberg and Yakov Novikov. BerkMin: A fast and robust Sat-solver. In *DATE*, pages 142–149, 2002.

[HPR04]   Philip Hawkes, Michael Paddon, and Gregory G. Rose. Musings on the Wang et al. MD5 collision. Cryptology ePrint Archive, Report 2004/264, 2004. http://eprint.iacr.org/.

[JJ05]   Dejan Jovanović and Predrag Janičić. Logical analysis of hash functions. In Bernhard Gramlich, editor, *Frontiers of Combining Systems (FroCoS)*, volume 3717 of *Lecture Notes in Artificial Intelligence*, pages 200–215. Springer Verlag, 2005.

[JKW74]   Arthur Evans Jr., William Kantrowitz, and Edwin Weiss. A user authentication scheme not requiring secrecy in the computer. *Commun. ACM*, 17(8):437–442, 1974.

[Kli05]   Vlastimil Klima. Finding MD5 collisions on a notebook PC using multi-message modifications. Cryptology ePrint Archive, Report 2005/102, 2005. http://eprint.iacr.org/.

[Lam79]   Leslie Lamport. Constructing digital signatures from a one-way function. Technical Report CSL-98, SRI International, October 1979.

[Mas99]   Fabio Massacci. Using Walk-SAT and Rel-SAT for cryptographic key search. In *International Joint Conference on Artificial Intelligence, IJCAI 99*, pages 290–295, 1999.

[Mer90]   Ralph C. Merkle. One way hash functions and DES. In Brassard [Bra90], pages 428–446.

[MM00]   Fabio Massacci and Laura Marraro. Logical cryptanalysis as a SAT problem. *Journal of Automated Reasoning*, 24:165–203, 2000.

[MMZ+01]   Matthew W. Moskewicz, Conor F. Madigan, Ying Zhao, Lintao Zhang, and Sharad Malik. Chaff: Engineering an efficient SAT solver. In *Proceedings of the Design Automation Conference (DAC)*, June 2001.

[MSS99]   João P. Marques-Silva and Karem A. Sakallah. GRASP—a search algorithm for propositional satisfiability. *IEEE Transactions in Computers*, 48(5):506–521, May 1999.

[MvOV96]   Alfred J. Menezes, Paul C. van Oorschot, and Scott A. Vanstone. *Handbook of Applied Cryptography*. CRC Press, 1996.

[NIS93]   NIST. Secure hash standard. FIPS PUB 180, National Institute of Standards and Technology, May 1993.

[NIS95]   NIST. Secure hash standard. FIPS PUB 180-1, National Institute of Standards and Technology, April 1995.

[Riv91]   Ronald L. Rivest. The MD4 message digest algorithm. In Alfred Menezes and Scott A. Vanstone, editors, *Advances in Cryptology—CRYPTO '90*, volume 537 of *Lecture Notes in Computer Science*, pages 303–311. Springer, 1991.

[Riv92]   Ronald L. Rivest. The MD5 message-digest algorithm. RFC 1321, The Internet Engineering Task Force, April 1992.

[RSA78]   Ronald L. Rivest, Adi Shamir, and Leonard M. Adleman. A method for obtaining digital signatures and public-key cryptosystems. *Commun. ACM*, 21(2):120–126, 1978.

[Rya04]   Lawrence Ryan. Efficient algorithms for clause-learning SAT solvers. M.Sc. Thesis, Simon Fraser University, February 2004.

[Sho05]   Victor Shoup, editor. *Advances in Cryptology—CRYPTO 2005: 25th Annual International Cryptology Conference, Santa Barbara, California, USA, August 14–18, 2005, Proceedings*, volume 3621 of *Lecture Notes in Computer Science*. Springer, 2005.

[SO06]   Martin Schläffer and Elisabeth Oswald. Searching for differential paths in MD4. In *Fast Software Encryption 2006*, 2006.

[Tse68]   Gregory Tseytin. *On the complexity of derivation in propositional calculus.*, pages 115–125. Consultant Bureau, New York-London, 1968.

[WLF+05]   Xiaoyun Wang, Xuejia Lai, Dengguo Feng, Hui Chen, and Xiuyuan Yu. Cryptanalysis of the hash functions MD4 and RIPEMD. In Cramer [Cra05], pages 1–18.

[WSK04]   Markus Wedler, Dominik Stoffel, and Wolfgang Kunz. Arithmetic reasoning in DPLL-based SAT solving. In *DATE*, pages 30–35, 2004.

[WY05]   Xiaoyun Wang and Hongbo Yu. How to break MD5 and other hash functions. In Cramer [Cra05], pages 19–35.

[WYY05a]   Xiaoyun Wang, Yiqun Lisa Yin, and Hongbo Yu. Finding collisions in the full SHA-1. In Shoup [Sho05], pages 17–36.

[WYY05b]   Xiaoyun Wang, Hongbo Yu, and Yiqun Lisa Yin. Efficient collision search attacks on SHA-0. In Shoup [Sho05], pages 1–16.

Functional Treewidth: Bounding Complexity in the Presence of Functional Dependencies

Yuliya Zabiyaka and Adnan Darwiche

Computer Science Department
University of California, Los Angeles
Los Angeles, CA 90095-1596, USA
{yuliaz, darwiche}@cs.ucla.edu

Abstract. Many reasoning problems in logic and constraint satisfaction have been shown to be exponential only in the treewidth of their interaction graph: a graph which captures the structural interactions among variables in a problem. It has long been observed in both logic and constraint satisfaction, however, that problems may be easy even when their treewidth is quite high. To bridge some of the gap between theoretical bounds and actual runtime, we propose a complexity parameter, called functional treewidth, which refines treewidth by being sensitive to non–structural aspects of a problem: functional dependencies in particular. This measure dominates treewidth and can be used to bound the size of CNF compilations, which permit a variety of queries in polytime, including clausal implication, existential quantification, and model counting. We present empirical results which show how the new measure can predict the complexity of certain benchmarks, that would have been considered quite difficult based on treewidth alone.

1 Introduction

The complexity of a number of problems in logic, constraint satisfaction, and probabilistic reasoning is bounded by the treewidth of their interaction graph [8,3,9,12]. The interaction graph is an undirected graph, with nodes representing variables in the given problem, and edges representing direct interactions between variables. For example, the interaction graph for a CNF contains an edge between two variables iff they appear in the same clause. Treewidth is a graph theoretic parameter, which measures the extent to which the graph resembles a tree [12].

Treewidth, however, appears to be too loose of a complexity bound in some cases. In particular, many problem instances that have large treewidth tend to be solvable in time and space that is much smaller than predicted by treewidth. The reason for the discrepancy between theoretical bounds and actual runtime is due to aspects of a problem structure, in particular determinism, which are not captured in the interaction graph and, hence, do not factor into the notion of treewidth. For example, for CNFs, treewidth is insensitive to the particular literals appearing in a clause, being only a function of the variables appearing in such a clause.

A. Biere and C.P. Gomes (Eds.): SAT 2006, LNCS 4121, pp. 116–129, 2006.

To bridge some of the gap between theoretical bounds based on treewidth and actual runtime, we propose in this paper a more refined parameter, which we call *functional treewidth,* that is sensitive to other aspects of a problem structure, beyond its interaction graph. In particular, functional treewidth is based on both the interaction graph and functional dependencies that are known to hold for the given problem. A functional dependency is a statement of the form $\mathbf{V} \to V$, where $\mathbf{V}$ is a set of variables and V is a single variable, indicating that each assignment of values to $\mathbf{V}$ implies a particular value for V.

Functional treewidth dominates treewidth and is therefore no easier to compute than treewidth, which is known to be NP–complete [1]. However, we show in this paper that if a CNF has functional treewidth w^f, then it has a *compilation* which is exponential only in w^f. This compilation is in the form of a deterministic, decomposable negation normal form (d-DNNF), which allows a number of queries to be answered in polytime, including clausal entailment, model counting and existential quantification [7]. In fact, we show that one of the simplest algorithms for compiling CNFs into d-DNNFs is capable of producing compilations that are only exponential in the functional treewidth. We note here that these results apply to the compilation of Bayesian networks as well, which can be reduced to the problem of compiling CNFs into d-DNNFs [5].

This paper is structured as follows. Section 2 introduces the new parameter of functional treewidth. Section 3 discusses the compilation of CNFs into d-DNNFs, showing the existence of d-DNNFs that are only exponential in functional treewidth. Section 4 presents a method for approximating functional treewidth, together with experimental results. Section 5 presents further experimental results, showing how functional treewidth can be used to bound the size of d-DNNF compilations. Finally, Section 6 closes with some concluding remarks.

2   Functional Treewidth

The treewidth of an (interaction) graph is usually defined in terms of secondary structures, such as elimination orders, jointrees[1], or dtrees, which can also be used to drive algorithms whose complexity is only exponential in treewidth. A number of these definitions are discussed in [6], with polytime transformations between these structures. We will base our treatment in this paper on dtrees, since these have been used to drive algorithms for compiling CNFs into d-DNNFs.

A *dtree* (decomposition tree) for a CNF Δ is a full binary tree whose leaves are in one–to–one correspondence with the CNF clauses; see Figure 1. We will now define the *width* of a dtree, where the treewidth of CNF Δ is the width of its best dtree (the one with smallest width). This will also correspond to the treewidth of the interaction graph for CNF Δ.

---

[1] Jointrees correspond to tree decompositions as known in the graph theoretic literature.

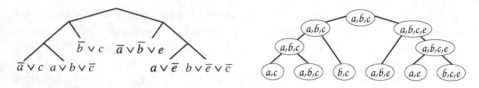

Fig. 1. A dtree (left) for the CNF equivalent to $(a \vee b \equiv c) \wedge (a \wedge b \equiv e)$ and its clusters (right). The width of this dtree is 3.

Before we define the width of a dtree, we need some additional notation. First, as is common with binary trees, we will identify a dtree with its root node. And for a dtree node T, we will use T^l and T^r to denote the left and right children of T, respectively. Moreover, for a leaf node T, the variables of T, $vars(T)$, are just the variables appearing the clause associated with T. For an internal node T, $vars(T) = vars(T^l) \cup vars(T^r)$. We will also use $vars^\uparrow(T)$ to denote $\bigcup_{T'} vars(T')$, where T' is a leaf node that is not a descendant of T.

The width of a dtree is defined in terms of the *clusters* of its nodes T, $cluster(T)$. The cluster for a leaf node T is $vars(T)$. The cluster for an internal node T is $(vars(T^l) \cap vars(T^r)) \cup (vars(T) \cap vars^\uparrow(T))$. The width of a dtree is then the size of its largest cluster minus one. Figure 1 depicts a dtree with its clusters, leading to a width of 3.

We will next define functional treewidth for a given CNF and a set of *functional dependencies* that are known to hold in the CNF. A functional dependency is a statement indicating that a variable, say, c, is functionally determined by a set of other variables, say, $\{a, b\}$. The functional dependency is denoted by $\{a, b\} \rightarrow c$ in this case. We will also find it useful to define the *closure* of a set of variables $\mathbf{V}$ under some functional dependencies $\mathbf{FD}$, denoted $\mathbf{V}+$ [11]. This set includes $\mathbf{V}$ and other variable that can be derived using the dependencies $\mathbf{FD}$. Consider the following dependencies for example:

$$\{a, b\} \rightarrow c,$$
$$\{b, c\} \rightarrow d,$$
$$\{d\} \quad \rightarrow e$$

We then have $\{a, b\}+ = \{a, b, c, d, e\}$, $\{a\}+ = \{a\}$ and $\{d\}+ = \{d, e\}$.

The basic intuition behind functional treewidth is that not all instantiations of a cluster in a dtree are indeed consistent with the given CNF, and that complexity can be linear in the number of consistent instantiations instead of all instantiations. Moreover, by reasoning about the functional dependencies that are known to hold in the CNF, one can bound the number of consistent instantiations for a given cluster. To provide such a bound, we need the notion of a (functional) implicant.

Definition 1. *Let* $\mathbf{FD}$ *be a set of functional dependencies over variables* $\mathbf{V}$, *and let* $\mathbf{X}$ *be a subset of* $\mathbf{V}$. *We will say that variables* $\mathbf{I}$ *are a* <u>*minimal implicant*</u> *for variables* $\mathbf{X}$ *under* $\mathbf{FD}$ *iff* $\mathbf{X} \subseteq \mathbf{I}+$ *and for any other set of variables* $\mathbf{J}$ *where* $\mathbf{X} \subseteq \mathbf{J}+$, *we have* $|\mathbf{I}+| \leq |\mathbf{J}+|$.

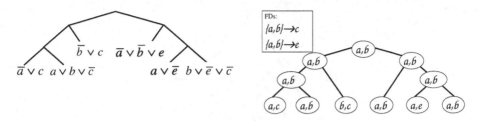

Fig. 2. Dtree (left) and functional clusters (right)

The importance of minimal implicants is this: if a cluster has a minimal implicant of size m, then it has no more than 2^m instantiations which are consistent with the given CNF. Note that the identity of the minimal implicant is not essential for this bound, only its size is.[2] Note also that this notion is different from the notion of a *key* as employed in database theory in the context of functional dependencies, where a key for $\mathbf{X}$ is an implicant for $\mathbf{X}$ that is also a subset of $\mathbf{X}$.

We are now ready to define functional treewidth.

Definition 2. *Let T be a node in a dtree for CNF Δ with functional dependencies* **FD**. *A functional cluster for node T, denoted $cluster^f(T)$, is a minimal implicant for $\overline{cluster(T)}$ under dependencies* **FD**.

Figure 2 depicts functional clusters for the dtree introduced in Figure 1.

Definition 3. *Let T be a dtree for CNF Δ with functional dependencies* **FD**. *The functional width of dtree T is the size of its maximal functional cluster minus 1. The functional treewidth of CNF Δ is the functional width of its best dtree (the one with the smallest functional cluster).*

It should be clear that functional treewidth can be no greater than treewidth, with equality in case the set of functional dependencies is empty. We will show constructively in the following section that if a CNF Δ has functional treewidth w^f, it must then have a d-DNNF compilation exponential only in w^f.

3   The Compilability of CNFs

We will consider in this section the compilability of CNFs into d-DNNFs, which is a tractable form that supports in polytime queries such as clausal entailment, model counting, and existential quantification [7]. This tractable form is also closed under conditioning (the setting of variable values), allowing an exponential number of queries to be answered each in polytime.

A d-DNNF is a rooted directed acyclic graph in which each leaf node is labeled with a literal, *true* or *false*, and each internal node is labeled with a conjunction

---

[2] If an algorithm is to take advantage of functional dependencies to improve its running time, the identity of the minimal implicant may matter then.

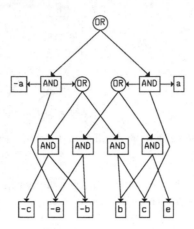

Fig. 3. A d-DNNF

Algorithm 1. cnf2ddnnf(T: dtree, α: instantiation): returns a d-DNNF.

1: $\gamma = project(\alpha, context(T))$
2: $result = CACHE_T(\gamma)$
3: **if** $result \neq NIL$ **then**
4:     return $result$
5: **if** T is a leaf **then**
6:     $result = clause2ddnnf(cnf(T), \alpha)$
7: **else**
8:     $result = \vee_\beta cnf2ddnnf(T^l, \alpha \wedge \beta) \wedge cnf2ddnnf(T^r, \alpha \wedge \beta) \wedge \beta$
9:     where β ranges over all instantiations of $cutset(T)$
10: $insert\_CACHE_T(\gamma, result)$
11: return $result$

---

or disjunction; see Figure 3. For any node N in a d-DNNF graph, $vars(N)$ denotes all propositional variables that appear in the subgraph rooted at N, and $\Delta(N)$ denotes the formula represented by N and its descendants. The nodes in a d-DNNF have the following two properties:

- **Decomposability:** $vars(N_i) \cap vars(N_j) = \emptyset$ for any two children N_i and N_j of an and-node N.
- **Determinism:** $\Delta(N_i)$ is inconsistent with $\Delta(N_j)$ for any two children N_i and N_j of an or-node N.

Algorithm 1 provides a procedure for compiling a CNF into a d-DNNF, adapted from [4]. We will explain the intuition behind the algorithm shortly, but we first point out its complexity. If the algorithm is passed a dtree with n nodes and width w, it will generate a d-DNNF for the corresponding CNF in $O(nw2^w)$ time. The algorithm must initially be called with α being the empty instantiation.

Algorithm 2. clause2ddnnf($l_1 \vee \ldots \vee l_m$: clause, α: instantiation): returns a d-DNNF for clause $(l_1 \vee \ldots \vee l_m)|\alpha$

1: **if** $m = 1$ **then**
2:     return $l_1|\alpha$
3: **else if** $\alpha \models \neg l_1$ **then**
4:     return clause2ddnnf($l_2 \vee \ldots \vee l_m, \alpha$)
5: **else if** $\alpha \models l_1$ **then**
6:     return $true$
7: **else**
8:     return $l_1 \vee (\neg l_1 \wedge clause2ddnnf(l_2 \vee \ldots \vee l_m, \alpha))$

The main technique in Algorithm 1 is that of recursive decomposition. Specifically, given a CNF Δ, we partition its clauses into Δ^l and Δ^r. If Δ^l and Δ^r share no variables, we can then compile them independently and simply conjoin the results. Suppose, however, that the two sets turn out to share a variable v. We will then use what is known as Boole's or Shannon's expansion:

$$\Delta = (v \wedge \Delta^l|v \wedge \Delta^r|v) \vee (\neg v \wedge \Delta^l|\neg v \wedge \Delta^r|\neg v),$$

where $\Delta|v$ ($\Delta|\neg v$) denotes the process of *conditioning*, which consists of replacing the occurrences of variable v by $true$ ($false$) in Δ. This recursive decomposition process is then governed by the given dtree, since each dtree node can be viewed as inducting a binary partition on the clauses below that node.

The algorithm makes use of two sets of variables at each node T. First, is $cutset(T)$ which is the set of variables that we must condition on so we can decompose the clauses below node T, $\Delta(T)$, into those below child T^l, $\Delta(T^l)$, and those below child T^r, $\Delta(T^r)$. Note that by the time we reach node T in the recursive decomposition process, the cutsets of all ancestors of node T must be instantiated. These variables are called $acutset(T)$, for ancestoral cutset of node T. Hence, $cutset(T)$ is defined as $vars(T^l) \cap vars(T^r) - acutset(T)$.

The second set of variables used at node T is $context(T) = acutset(T) \cap vars(T)$. These are variables that are guaranteed to be set when we recurse on node T, and that also appear in clauses below T. Any two recursive calls to node T which agree on the value of variables $context(T)$ must return equivalent answers. Hence, the algorithm maintains a cache at each node indexed by the instantiations of $context(T)$ to avoid recursing multiple times on the same subproblem. We note here that for an internal dtree node T, $cutset(T) \cup context(T)$ is actually the cluster of node T as defined in the previous section.

Before we present the central result in this section, we point out the following about Algorithm 1. Figure 4 depicts the d-DNNF substructure that a call $cnnf2dnnf(T, \_)$ will contribute to the final d-DNNF. In particular, for every instantiation γ of $context(T)$, Algorithm 1 will produce an OR–node, and for each instantiation β of $cutset(T)$ (under a given context instantiation γ), it will produce an AND–node. In fact, the contributed substructures can be exponential in the size of $cluster(T)$. In particular, the call will contribute $2^{|context(T)|}$

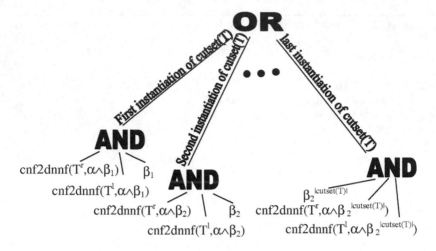

Fig. 4. The d-DNNF substructure constructed by $cnf2ddnnf(T, \alpha)$ at node T

OR–nodes, and for each such node, it will contribute $2^{|cutset(T)|}$ AND–nodes as children. Since $cutset(T)$ and $context(T)$ share no variables by definition, the size of the contributed structure is then

$$2^{|cutset(T) \cup context(T)|} = 2^{|cluster(T)|}.$$

Our central result is then as follows.

Theorem 1. *If Algorithm 1 is passed a dtree with n nodes, m variables and functional treewidth w^f, it will return a d-DNNF of size $O((n + m)2^{w^f})$.*[3]

This basically shows that if a CNF Δ has a functional treewidth of w^f, then it must have a d-DNNF compilation of size $O((n + m)2^{w^f})$. In fact, for $cluster(T)$, only instantiations which are consistent with the CNF will contribute structures to the final d-DNNF compilations, as shown in Figure 4. The proof of Theorem 1 is given in the Appendix.

4   Approximating Functional Treewidth

Determining the functional width of a dtree requires the computation of minimal implicants, which includes the computation of minimal keys as a special case (a problem known to be NP-complete [10]). We consider in this section a method for approximating functional width of a given dtree and present a number of empirical results, showing its effectiveness.

Our basic method for computing minimal implicants for $cluster(T)$ is based on an exhaustive procedure which searches for implicants of increasing sizes, up

---

[3] Alternatively, we can bound the size of produced d-DNNF by $O(nw2^{w^f})$, where w is the width of a dtree.

to size $k = |cluster(T)|$. This procedure is not practical though given the number of candidate implicants we need to consider, which is a function of both k and the number of CNF variables (from which we need to compute an implicant). We improve this procedure by not involving all CNF variables in the analysis, but only those that can be reached by traversing the functional dependencies backward, starting from variables in $cluster(T)$. We also approximate the procedure if this is not sufficient by restricting the set of variables from which the implicant is computed. In particular, we restrict our implicants to the following sets, with decreasing size: $(acutset(T) \cup vars(T))+$, $(acutset(T) \cup cluster(T))+$, and $cluster(T)$. Our method will switch from one approximation to the next if it examines more than a certain number of candidates, finally giving up and returning $cluster(T)$ as the approximation if none of the tried approximations yield a smaller set. Another approximation technique is to try to find implicants for $cutset(T)$ and $context(T)$ separately, instead of $cluster(T)$, as that provides more specific choices for reducing the set of variables from which to draw an implicant from.

To evaluate the effectiveness of proposed approximations, we experimented with many benchmarks with abundance of (easily recognizable) functional dependencies. This included various digital circuits from the LGSynth93 suite (*http://www.bdd-portal.org/benchmarks.html*), and grid CNFs from [13] which come with varying degrees of functional dependencies. Table 1 depicts results for the LGSynth93 suite, showing exponential improvements in the bounds based on (approximated) functional treewidth compared to those based on (approximated) treewidth. This basically allows us to prove the compilability of corresponding CNFs using (approximate) functional treewidth, even though the CNFs have very large (approximate) treewidths.

Grid networks were defined in [13]. They are $N \times N$ Bayesian networks with variables denoted $X_{i,j}$ for $1 \le i,j \le N$, where each variable $X_{i,j}$ has parents $X_{i-1,j}$ and $X_{i,j-1}$ when the corresponding indices are greater than zero. A fraction of the nodes, equal to d *ratio,* is determined by their parents (half of those nodes determined by both parents and another half by a single parent). Table 2 depicts results on CNF encodings of grid networks, showing how the (approximated) functional treewidth gets smaller as we increase the amount of determinism.

5   Bounding the Size of CNF Compilations

We have evaluated in the previous section the quality of our approximations for functional treewidth by comparing them to treewidth. In this section, we do another comparison with the actual size of d-DNNFs computed by cnf2ddnnf. To be able to perform this comparison, we had to restrict ourselves to problems whose (approximate) treewidth is manageable (≤ 20) since the time complexity of cnf2ddnnf is exponential in this treewidth. Hence, our results in this section are to some extent biased towards problems that are somewhat easy due to the relatively small clusters involved.

Table 1. LGsynth93 suite: n is number of clauses, m is number of variables, and a is the number of clusters for which the functional cluster was not necessarily minimal (approximated)

cnf	w	w^f	dtree nodes	m	n	a	time (min)
5xp1	27	8	1563	296	782	35	0.59
5xp1_ok	19	7	1253	255	627	9	0.23
9sym	51	14	2403	442	1202	155	1.89
9sym.scan_ok	35	9	2195	439	1098	85	1.44
9symml	31	13	1901	376	951	67	1.03
alu2	55	17	4551	833	2276	686	15.32
apex2.scan_ok	47	39	3869	784	1935	322	6.00
C1355	32	26	4775	979	2388	702	11.28
C1908	38	34	3769	770	1885	915	5.80
C2670	29	26	5985	1407	2993	415	11.19
C880	24	24	2949	633	1475	192	1.71
clip	58	12	3881	716	1941	219	6.09
clip_ok	29	14	1577	316	789	49	0.49
duke2	66	41	3743	689	1872	375	5.64
e64	83	67	4875	899	2438	633	10.89
ex4p	22	17	4497	1014	2249	197	3.74
f51m	28	12	1853	347	927	49	0.66
frg1	47	25	4541	834	2271	383	11.46
inc	20	7	1483	299	742	14	0.27
rd53	19	10	717	138	359	19	0.15
rd73	37	7	2871	536	1436	168	4.47
rd84	42	22	5331	985	2666	605	12.02
sao2	34	17	1871	346	936	155	1.18
sct	18	11	1293	266	647	45	0.32
sqrt8ml	16	10	1781	363	891	174	1.80
squar5	25	12	885	167	443	62	0.64
term1	45	25	5471	1064	2736	342	8.88
ttt2	22	15	4017	774	2009	192	4.49
vda	101	51	6691	1180	3346	772	31.86
vg2	58	33	4063	748	2032	235	4.72
x4	25	23	5919	1199	2960	281	8.71
z4ml	22	11	1581	294	791	29	0.48

For a set of circuit benchmarks from the suite LGsynth93, we extracted CNFs together with functional dependencies: for every gate constructing a functional dependency from its inputs to its output. For every CNF we constructed a dtree using hypergraph partitioning method [6], approximated its treewidth w and approximated its functional width w^f. Next, we bounded the number of edges that every node T will contribute to the d-DNNF based on on the size of $cluster^f(T)$. Using this procedure we got two bounds on the number of d-DNNF edges: a bound based on structural clusters *s-bound*, and a bound based on functional clusters *f-bound*. We then ran Algorithm 1 and calculated the true number of

Table 2. Grid networks

$N \times N$	$d\text{-}ratio$	w	w^f	dtree nodes	m	n	a	time (min)
10×10	50	15	13	455	100	228	27	0.01
10×10	75	15	9	575	100	288	49	0.01
10×10	100	15	1	627	100	314	0	0.02
14×14	50	21	18	835	196	418	37	0.02
14×14	75	21	17	1091	196	546	252	0.06
14×14	100	21	1	1295	196	648	0	0.13
18×18	50	27	26	1293	324	647	62	0.07
18×18	75	27	23	1767	324	884	408	0.19
18×18	100	27	1	2155	324	1078	0	0.59
22×22	50	33	32	2073	484	1037	138	0.25
22×22	75	33	30	2709	484	1355	885	0.63
22×22	100	33	1	3269	484	1635	0	1.97
26×26	50	39	36	2927	676	1464	229	0.60
26×26	75	39	34	3675	676	1838	848	1.74
26×26	100	40	1	4569	676	2285	89	6.73
30×30	50	46	42	3940	900	1970	319	2.20
30×30	75	43	38	5043	900	2522	1682	4.07
30×30	100	47	3	6125	900	3063	451	13.10
34×34	50	52	49	5083	1156	2542	361	3.23
34×34	75	51	47	6536	1156	3268	1950	4.68
34×34	100	52	27	7855	1156	3928	1447	48.93

edges in every d-DNNF *e-count*. Tables 3 and 4 depict the results of our experiments. To make the assessment of the quality of approximations easier, we also report s/f = *s-bound/f-bound* and f/e = *f-bound/e-count*. As f/e approaches 1, our functional treewidth bounds get closer to the true size of compilation. The point that is worth noting is that even in the cases where w and w^f are quite close, the bound provided by functional treewidth can still be much smaller than the one based on treewidth since our bounds are a function of all clusters in the dtree, not just the largest ones (captured by width).

6   Discussion

We proposed in this paper the notion of functional treewidth, as a complexity parameter for bounding the size of certain CNF compilations, which permit polytime queries, such as clausal entailment, model counting and existential quantification. We have also presented a method for approximating functional treewidth and applied it to a number of benchmarks, showing its ability to provide bounds that are exponentially better than those based on treewidth.

Our current and future work on this subject centers around three directions. First, the development of better approximation algorithms for functional

Table 3. LGsynth93 suite: A stands for approximate and E for exact value of w^f (for given dtree), n is number of clauses, m is number of variables, s-bound is edge bound based on structural clusters, f-bound is edge bound based on functional clusters, e-count is true number of edges in d-DNNF compiled by cnf2ddnnf, $s/f = s\text{-}bound/f\text{-}bound$ and $f/e = f\text{-}bound/e\text{-}count$, T_c is time to run cnf2ddnnf and T_w is time to calculate w^f

cnf	A	n	m	s-bound	f-bound	e-count	s/f	f/e	w	w^f	T_c(s)	T_w(s)
5xp1	A	627	255	7066536	26246	21427	269.24	1.22	19	7	4.92	12.30
apex7	A	1147	493	2246904	148576	35994	15.12	4.13	16	11	0.72	48.66
b1	E	66	29	5316	1266	919	4.20	1.38	8	3	0.00	0.05
b12	A	402	174	715024	62318	23887	11.47	2.61	15	11	0.84	2.52
b9	A	551	256	449264	69156	36807	6.50	1.88	15	11	0.20	6.17
bw	A	844	337	20659544	25464	19858	811.32	1.28	20	5	37.25	74.20
C17	E	30	17	900	556	355	1.62	1.57	4	3	0.01	0.02
C432	A	946	403	4446852	829182	53043	5.36	15.63	17	15	3.09	51.02
c8	A	1091	444	11843932	431312	98070	27.46	4.40	19	13	16.77	87.19
cc	A	302	140	19876	8042	5532	2.47	1.45	8	5	0.05	1.11
cht	A	1243	515	310124	44348	32945	6.99	1.35	12	7	0.20	42.72
cm138a	A	74	35	3156	1738	1268	1.82	1.37	6	4	0.00	0.06
cm150a	A	364	165	28672	8464	6458	3.39	1.31	8	6	0.03	2.55
cm151a	A	176	82	12224	4028	3074	3.03	1.31	8	5	0.01	0.48
cm152a	A	117	54	14276	5134	2287	2.78	2.24	8	7	0.01	0.19
cm162a	A	238	107	15028	6306	4034	2.38	1.56	7	7	0.01	0.80
cm163a	A	230	106	13708	5644	3873	2.43	1.46	7	6	0.00	0.75
cm42a	E	84	37	4044	1796	1381	2.25	1.30	6	4	0.00	0.08
cm82a	A	116	52	5416	2012	1528	2.69	1.32	5	3	0.00	0.17
cm85a	A	208	95	16776	6700	3395	2.50	1.97	8	8	0.01	0.64
cmb	A	214	97	65912	34124	8020	1.93	4.25	11	11	0.06	0.73
comp	A	577	258	50528	21478	10900	2.35	1.97	9	7	0.03	7.45
con1	A	93	45	6812	2678	1608	2.54	1.67	8	6	0.01	0.11
cordic	A	533	235	135768	54876	11444	2.47	4.80	12	12	0.11	6.47
count	A	641	292	39324	14702	10237	2.67	1.44	7	5	0.03	7.03
cu	A	259	110	82200	24388	10918	3.37	2.23	11	10	0.08	0.78
decod	E	98	39	8416	2580	1886	3.26	1.37	8	5	0.01	0.09
i1	A	177	95	7289	3733	2559	1.95	1.46	6	5	0.01	0.36
i2	A	1144	657	523620	440606	101623	1.19	4.34	15	15	0.45	38.05
i3	A	774	456	22565	12981	8955	1.74	1.45	5	4	0.02	14.08
i4	A	922	530	42863	25627	16000	1.67	1.60	6	6	0.03	16.81
i5	A	1538	734	171886	69408	45664	2.48	1.52	10	9	0.16	72.16
i6	A	1844	866	285256	97294	25765	2.93	3.78	14	11	0.30	112.70
i7	A	2346	1115	369436	140848	32221	2.62	4.37	13	10	0.30	267.94
inc	A	742	299	17985228	37284	27866	482.38	1.34	20	7	15.30	12.98
lal	A	643	275	7525122	2263668	68058	3.32	33.26	19	17	3.80	8.98
ldd	A	418	162	16931664	1312920	9504	12.90	138.14	20	17	0.63	3.70
majority	A	73	33	6300	2280	1709	2.76	1.33	7	5	0.00	0.11
misex1	A	320	135	136812	22928	7852	5.97	2.92	12	10	0.11	1.31
misex2	A	457	194	1153460	294684	36024	3.91	8.18	15	13	0.89	2.95

Table 4. LGsynth93 suite. Continuation of Table 3.

cnf	A	n	m	s-bound	f-bound	e-count	s/f	f/e	w	w^f	T_c(s)	T_w(s)
mux	A	501	210	464996	24372	16759	19.08	1.45	13	7	0.31	6.31
o64	A	519	325	20624	10670	7745	1.93	1.38	6	5	0.02	4.89
parity	A	257	122	10924	4000	2954	2.73	1.35	5	3	0.00	1.06
pcle	A	280	128	16100	6352	4142	2.53	1.53	6	5	0.01	1.06
pcler8	A	344	160	24560	12090	5575	2.03	2.17	7	6	0.03	1.69
pm1	A	270	121	38507	15577	13147	2.47	1.18	10	8	0.03	0.84
rd53	A	279	117	1064492	17360	5431	61.32	3.20	16	10	0.38	1.08
rd73	A	738	300	74425360	34784	26292	2139.64	1.32	22	7	88.98	27.42
sct	A	647	266	7146520	364242	60214	19.62	6.05	19	14	4.33	10.19
sqrt8	A	310	131	437540	26908	8566	16.26	3.14	15	10	0.13	1.31
sqrt8ml	A	891	363	1389624	62652	14777	22.18	4.24	16	11	1.42	40.03
squar5	A	298	122	275848	8838	6342	31.21	1.39	14	6	0.31	1.31
t481	A	409	172	45598204	7203748	98915	6.33	72.83	21	19	30.67	6.25
tcon	E	202	98	9472	3878	2969	2.44	1.31	5	3	0.00	0.41
unreg	A	568	264	56284	18402	13415	3.06	1.37	8	6	0.06	4.74
x2	A	223	93	602572	37384	8102	16.12	4.61	15	11	0.41	0.66
xor5	A	115	52	7660	2158	1332	3.55	1.62	6	5	0.00	0.19
z4ml	A	791	294	53209456	107650	38650	494.28	2.79	22	11	100.74	31.59

treewidth (there is a long tradition of such approximations for treewidth (see, e.g., [2]). Next, the development of dtree construction methods which are sensitive to functional dependencies. Note that a dtree with larger width may have a smaller functional width. This means that a method for constructing dtrees that minimizes width may miss dtrees which are optimal from a functional width viewpoint. Finally, the use of functional dependencies in improving the time and space complexity of compilation algorithms, therefore, allowing bounds on the running time based on functional treewidth.

References

1. Stefan Arnborg, D. G. Corneil, and A. Proskurowski. Complexity of finding embeddings in a k-tree. *SIAM J. Algebraic and Discrete Methods*, 8:277–284, 1987.
2. H. L. Bodlaender. A tourist guide through treewidth. *ACTA CYBERNETICA*, 11(1-2):1–22, 1993.
3. Adnan Darwiche. Decomposable negation normal form. *Journal of the ACM*, 48(4):608–647, 2001.
4. Adnan Darwiche. On the tractability of counting theory models and its application to belief revision and truth maintenance. *Journal of Applied Non-Classical Logics*, 11(1-2):11–34, 2001.
5. Adnan Darwiche. A logical approach to factoring belief networks. In *Proceedings of KR*, pages 409–420, 2002.

6. Adnan Darwiche and Mark Hopkins. Using recursive decomposition to construct elimination orders, jointrees and dtrees. In *Trends in Artificial Intelligence, Lecture notes in AI, 2143*, pages 180–191. Springer-Verlag, 2001.
7. Adnan Darwiche and Pierre Marquis. A knowledge compilation map. *Journal of Artificial Intelligence Research*, 17:229–264, 2002.
8. Rina Dechter. *Constraint Processing*. Morgan Kaufmann Publishers, Inc., San Mateo, California, 2003.
9. F. V. Jensen, S.L. Lauritzen, and K.G. Olesen. Bayesian updating in recursive graphical models by local computation. *Computational Statistics Quarterly*, 4:269–282, 1990.
10. Claudio L. Lucchesi and Sylvia L. Osborn. Candidate keys for relations. *Journal of Computer and System Sciences*, 17:270–279, 1978.
11. David Maier. *The Theory of Relational Databases*. Computer Science Press, Rockville, MD, 1983.
12. N. Robertson and P. D. Seymour. Graph minors II: Algorithmic aspects of tree-width. *J. Algorithms*, 7:309–322, 1986.
13. T. Sang, P. Beam, and H. Kautz. Solving bayesian networks by weighted model counting. In *Proceedings of AAAI*. AAAI, 2005.

Proof of Theorem 1

Without loss of generality, we will assume that the original CNF Δ is consistent, and that $cnf2ddnnf(T, \alpha)$ is initially called with $\alpha = true$.

The proof is based on a number of lemmas, which concern a recursive call $cnf2ddnnf(T, \alpha)$ to dtree node T, with $\Delta(T)$ denoting the clauses below node T:

1. Lemma 1: $cnf2ddnnf(T, \alpha)$ returns $false$ if $\Delta(T)|\alpha$ is inconsistent.
2. Lemma 2: If $\Delta \wedge \alpha$ is consistent, and $\Delta \wedge \alpha \wedge \beta$ is inconsistent for some instantiation β of $cutset(T)$, then $\Delta(T)|\alpha\beta$ is inconsistent and, moreover, either $\Delta(T^l)|\alpha\beta$ or $\Delta(T^r)|\alpha\beta$ is inconsistent.
3. Lemma 3: If $\Delta \wedge \alpha$ is inconsistent, then there is an ancestor T^a of node T for which $\Delta \wedge \alpha^a$ is consistent, where $\alpha^a = project(\alpha, acutset(T^a))$. Moreover, $\Delta \wedge \alpha^a \wedge \beta^a$ is inconsistent for instantiation $\beta^a = project(\alpha, cutset(T^a))$.

The proof is based on the observation that each disjunct on Line 8 of Algorithm 1 corresponds to an instantiation $\gamma \wedge \beta$ of $cluster(T)$ (γ is instantiation of $context(T)$ and β is instantiation of $cutset(T)$). We will prove that the disjuncts corresponding to instantiations $\gamma \wedge \beta$ of $cluster(T)$ will not be part of the returned d-DNNF if $\gamma \wedge \beta$ is inconsistent with the CNF Δ.

Consider now the call $cnf2ddnnf(T, \alpha)$, and let $\gamma = project(\alpha, context(T))$. If $\gamma \wedge \beta$ is inconsistent with Δ then $\alpha \wedge \beta$ is inconsistent with Δ as well. We have two cases:

1. $\Delta \wedge \alpha$ is consistent, but $\Delta \wedge \alpha \wedge \beta$ is inconsistent.
   Note that $context(T)$ contains all variables shared between clauses in $\Delta(T)$ and other clauses. Hence, if $\Delta \wedge \alpha \wedge \beta$ is inconsistent, then $\Delta(T) \wedge \alpha \wedge \beta$ is inconsistent. By Lemma 2, $\Delta(T)|\alpha\beta$ is inconsistent and, moreover, either

$\Delta(T^l)|\alpha\beta$ or $\Delta(T^r)|\alpha\beta$ is inconsistent. Hence, by Lemma 1, the corresponding disjunct on Line 8 of Algorithm 1 will evaluate to $false$ and will not be included in the computed d-DNNF.

2. $\Delta \wedge \alpha$ is inconsistent (and, hence, $\Delta \wedge \alpha \wedge \beta$ is inconsistent).

   By Lemma 3, there is an ancestor T^a of node T for which $\Delta \wedge \alpha^a$ is consistent, where $\alpha^a = project(\alpha, acutset(T^a))$. Moreover, $\Delta \wedge \alpha^a \wedge \beta^a$ is inconsistent for instantiation $\beta^a = project(\alpha, cutset(T^a))$. By Lemma (1), the call $cnf2ddnnf(T^a, \alpha^a)$ will return $false$ and, hence, the disjunct on Line 8 of Algorithm 1 constructed during the call $cnf2ddnnf(T, \alpha)$ will not be part of the returned d-DNNF.

We will now bound the size of returned d-DNNF, measured by the number of edges in the d-DNNF. Let cnf Δ have n_Δ clauses and m variables, then dtree T_Δ will have n_Δ leaf nodes and $n_\Delta - 1$ internal nodes. Each internal node T will contribute $\leq 2^{|cluster^f(T)|}$ disjuncts. Each disjunct will have the following edges: an edge to the parent OR–node; two edges to the solutions produced by left and right child; $|cutset(T)|$ edges to the literal nodes corresponding to the instantiation of the $cutset(T)$. Thus the total amount of edges contributed by internal nodes is (an additional factor of 2 appears when we switch from $|cluster^f(T)|$ to w^f due to -1 in the definition of w^f):

$$\sum_{\text{internal node } T} (3 + |cutset(T)|) \cdot 2^{|cluster^f(T)|} \leq 2 \cdot 2^{w^f} \cdot (3n_\Delta + m).$$

Note that $\sum_{\text{internal node } T} |cutset(T)|$ can be bounded either by the number of variables m (because all the cutsets are disjoint) or by $w \cdot n_\Delta$ (because none of the cutsets have the size greater than w).

Each consistent instantiation of the $cluster(T)$ for leaf T contributes $4 \cdot (|var(T) - acutset(T)| - 1)$ edges: each uninstantiated variable contributes at most one AND-node, one OR-node and two literal nodes (except for the last one contributing one literal node). Thus the total number of edges contributed by leaf nodes is

$$\sum_{\text{leaf node } T} 4 \cdot (|var(T) - acutset(T)| - 1) \cdot 2^{|cluster^f(T)|} \leq 4m \cdot 2 \cdot 2^{w^f}$$

The result is due to the fact that $\sum_{\text{leaf node } T}(|var(T) - acutset(T)| - 1)$ can be bounded analogous to $\sum_{\text{internal node } T} |cutset(T)|$.

Noting that the total number of nodes n in dtree T_Δ is equal to $2n_\Delta - 1$, we conclude that the number of edges in the final compilation produced by Algorithm 1 is bounded by $O((n + m)2^{w^f})$.

Encoding the Satisfiability of Modal and Description Logics into SAT: The Case Study of K(m)/$\mathcal{ALC}$

Roberto Sebastiani and Michele Vescovi

DIT, Università di Trento, Via Sommarive 14, I-38050, Povo, Trento, Italy
rseba@dit.unitn.it, vescovi@dit.unitn.it

Abstract. In the last two decades, modal and description logics have been applied to numerous areas of computer science, including artificial intelligence, formal verification, database theory, and distributed computing. For this reason, the problem of automated reasoning in modal and description logics has been throughly investigated.

In particular, many approaches have been proposed for efficiently handling the satisfiability of the core normal modal logic K_m, and of its notational variant, the description logic $\mathcal{ALC}$. Although simple in structure, $K_m/\mathcal{ALC}$ is computationally very hard to reason on, its satisfiability being PSPACE-complete.

In this paper we explore the idea of encoding $K_m/\mathcal{ALC}$-satisfiability into SAT, so that to be handled by state-of-the-art SAT tools. We propose an efficient encoding, and we test it on an extensive set of benchmarks, comparing the approach with the main state-of-the-art tools available.

Although the encoding is necessarily worst-case exponential, from our experiments we notice that, in practice, this approach can handle most or all the problems which are at the reach of the other approaches, with performances which are comparable with, or even better than, those of the current state-of-the-art tools.

1 Introduction

In the last two decades, modal and description logics have been applied to numerous areas of computer science, including artificial intelligence, formal verification, database theory, and distributed computing. For this reason, the problem of automated reasoning in modal and description logics has been throughly investigated (see, e.g., [2,9,6,10]). Many approaches have been proposed for efficiently handling the satisfiability of modal and description logics, in particular of the core normal modal logic K_m and of its notational variant, the description logic $\mathcal{ALC}$ (see, e.g., [2,10,3,4,7,8,1,12]). Notice that, although simple in structure, $K_m/\mathcal{ALC}$ is computationally very hard to reason on, as its satisfiability is PSPACE-complete [6].

In this paper we explore the idea of encoding $K_m/\mathcal{ALC}$-satisfiability into SAT, so that to be handled by state-of-the-art SAT tools. We propose an efficient encoding, with four simple variations. We test (the four variations of) it on an extensive set of benchmarks, comparing the results with those of the main state-of-the-art tools for K_m-satisfiability available. Although the encoding is necessarily worst-case exponential (unless PSPACE = NP), from our experiments we notice that, in practice, this approach can handle most or all the problems which are at the reach of the other approaches, with performances which are comparable with, or even better than, those of the current state-of-the-art tools.

A. Biere and C.P. Gomes (Eds.): SAT 2006, LNCS 4121, pp. 130–135, 2006.

For lack of space, in this short version of the paper we omit the proof of Theorem 1, the description of the empirical tests and results, and every reference to related work. All such information can be found in the extended version of the paper [13], which is publicly downloadable [1].

2 Background

We recall some basic definitions and properties of K_m. Given a non-empty set of primitive propositions $\mathcal{A} = \{A_1, A_2, \ldots\}$ and a set of m modal operators $\mathcal{B} = \{\Box_1, \ldots, \Box_m\}$, the language of K_m is the least set of formulas containing $\mathcal{A}$, closed under the set of propositional connectives $\{\neg, \wedge\}$ and the set of modal operators in $\mathcal{B}$. Notationally, we use the Greek letters $\alpha, \beta, \varphi, \psi, \nu, \pi$ to denote formulas in the language of K_m (K_m-formulas hereafter). We use the standard abbreviations, that is: "$\Diamond_r\varphi$" for "$\neg\Box_r\neg\varphi$", "$\varphi_1 \vee \varphi_2$" for "$\neg(\neg\varphi_1 \wedge \neg\varphi_2)$", "$\varphi_1 \rightarrow \varphi_2$" for "$\neg(\varphi_1 \wedge \neg\varphi_2)$", "$\varphi_1 \leftrightarrow \varphi_2$" for "$\neg(\varphi_1 \wedge \neg\varphi_2) \wedge \neg(\varphi_2 \wedge \neg\varphi_1)$", "$\top$" and "$\bot$" for the constants "true" and "false". (Hereafter formulas like $\neg\neg\psi$ are implicitly assumed to be simplified into ψ, so that, if ψ is $\neg\phi$, then by "$\neg\psi$" we mean "ϕ".) We call *depth* of φ, written *depth(φ)*, the maximum number of nested modal operators in φ. We call a *propositional atom* every primitive proposition in $\mathcal{A}$, and a *propositional literal* every propositional atom (*positive literal*) or its negation (*negative literal*).

In order to make our presentation more uniform, we adopt from [2,10] the representation of K_m-formulas from the following table:

α	α_1	α_2	β	β_1	β_2	π^r	π_0^r	ν^r	ν_0^r
$(\varphi_1 \wedge \varphi_2)$	φ_1	φ_2	$(\varphi_1 \vee \varphi_2)$	φ_1	φ_2	$\Diamond_r\varphi_1$	φ_1	$\Box_r\varphi_1$	φ_1
$\neg(\varphi_1 \vee \varphi_2)$	$\neg\varphi_1$	$\neg\varphi_2$	$\neg(\varphi_1 \wedge \varphi_2)$	$\neg\varphi_1$	$\neg\varphi_2$	$\neg\Box_r\varphi_1$	$\neg\varphi_1$	$\neg\Diamond_r\varphi_1$	$\neg\varphi_1$
$\neg(\varphi_1 \rightarrow \varphi_2)$	φ_1	$\neg\varphi_2$	$(\varphi_1 \rightarrow \varphi_2)$	$\neg\varphi_1$	φ_2				

in which non-literal K_m-formulas are grouped into four categories: α's (conjunctive), β's (disjunctive), π's (existential), ν's (universal).

A *Kripke structure* for K_m is a tuple $\mathcal{M} = \langle \mathcal{U}, \mathcal{L}, \mathcal{R}_1, \ldots, \mathcal{R}_m \rangle$, where $\mathcal{U}$ is a set of states, $\mathcal{L}$ is a function $\mathcal{L} : \mathcal{A} \times \mathcal{U} \longmapsto \{True, False\}$, and each $\mathcal{R}_r$ is a binary relation on the states of $\mathcal{U}$. With an abuse of notation we write "$u \in \mathcal{M}$" instead of "$u \in \mathcal{U}$". We call a *situation* any pair $\mathcal{M}, u$, $\mathcal{M}$ being a Kripke structure and $u \in \mathcal{M}$. The binary relation $\models$ between a modal formula φ and a situation $\mathcal{M}, u$ is defined as follows:

$$\mathcal{M}, u \models A_i, A_i \in \mathcal{A} \iff \mathcal{L}(A_i, u) = True;$$
$$\mathcal{M}, u \models \neg A_i, A_i \in \mathcal{A} \iff \mathcal{L}(A_i, u) = False;$$
$$\mathcal{M}, u \models \alpha \iff \mathcal{M}, u \models \alpha_1 \text{ and } \mathcal{M}, u \models \alpha_2;$$
$$\mathcal{M}, u \models \beta \iff \mathcal{M}, u \models \beta_1 \text{ or } \mathcal{M}, u \models \beta_2;$$
$$\mathcal{M}, u \models \pi^r \iff \mathcal{M}, w \models \pi_0^r \text{ for some } w \in \mathcal{U} \text{ s.t. } \mathcal{R}_r(u, w) \text{ holds in } \mathcal{M};$$
$$\mathcal{M}, u \models \nu^r \iff \mathcal{M}, w \models \nu_0^r \text{ for every } w \in \mathcal{U} \text{ s.t. } \mathcal{R}_r(u, w) \text{ holds in } \mathcal{M}.$$

"$\mathcal{M}, u \models \varphi$" should be read as "$\mathcal{M}, u$ satisfy φ in K_m" (alternatively, "$\mathcal{M}, u$ K_m-satisfies φ"). We say that a K_m-formula φ is satisfiable in K_m (K_m-satisfiable from now on) if

---

[1] Available at http://www.dit.unitn.it/~rseba/sat06/extended.ps

and only if there exist $\mathcal{M}$ and $u \in M$ s.t. $\mathcal{M}, u \models \varphi$. (When this causes no ambiguity, we sometimes drop the prefix "K_m-".) We say that w is a *successor* of u through $\mathcal{R}_r$ iff $\mathcal{R}_r(u, w)$ holds in $\mathcal{M}$.

The problem of determining the K_m-satisfiability of a K_m-formula φ is decidable and PSPACE-complete [9,6], even restricting the language to a single boolean atom (i.e., $\mathcal{A} = \{A_1\}$) [5]; if we impose a bound on the modal depth of the K_m-formulas, the problem reduces to NP-complete [5]. For a more detailed description on K_m— including, e.g., axiomatic characterization, decidability and complexity results — see [6,5].

A K_m-formula is said to be in *Negative Normal Form (NNF)* if it is written in terms of the symbols $\Box_r, \Diamond_r, \wedge, \vee$ and propositional literals $A_i, \neg A_i$ (i.e., if all negations occur only before propositional atoms in $\mathcal{A}$). Every K_m-formula φ can be converted into an equivalent one $NNF(\varphi)$ by recursively applying the rewriting rules: $\neg \Box_r \varphi \Longrightarrow \Diamond_r \neg \varphi$, $\neg \Diamond_r \varphi \Longrightarrow \Box_r \neg \varphi$, $\neg(\varphi_1 \wedge \varphi_2) \Longrightarrow (\neg \varphi_1 \vee \neg \varphi_2)$, $\neg(\varphi_1 \vee \varphi_2) \Longrightarrow (\neg \varphi_1 \wedge \neg \varphi_2)$, $\neg \neg \varphi \Longrightarrow \varphi$.

A K_m-formula is said to be in *Box Normal Form (BNF)* [11,12] if it is written in terms of the symbols $\Box_r, \neg \Box_r, \wedge, \vee$, and propositional literals $A_i, \neg A_i$ (i.e., if no diamonds are there, and all negations occurs only before boxes or before propositional atoms in $\mathcal{A}$). Every K_m-formula φ can be converted into an equivalent one $BNF(\varphi)$ by recursively applying the rewriting rules: $\Diamond_r \varphi \Longrightarrow \neg \Box_r \neg \varphi$, $\neg(\varphi_1 \wedge \varphi_2) \Longrightarrow (\neg \varphi_1 \vee \neg \varphi_2)$, $\neg(\varphi_1 \vee \varphi_2) \Longrightarrow (\neg \varphi_1 \wedge \neg \varphi_2)$, $\neg \neg \varphi \Longrightarrow \varphi$.

3   The Encoding

We borrow some notation from the *Single Step Tableau (SST)* framework [10]. We represent univocally states in $\mathcal{M}$ as labels σ, represented as non empty sequences of integers $1.n_1^{r_1}.n_2^{r_2}. \ldots .n_k^{r_k}$, s.t. the label 1 represents the root state, and $\sigma.n^r$ represents the n-th successor of σ through the relation $\mathcal{R}_r$.

Notationally, we often write "$(\bigwedge_i l_i) \to \bigvee_j l_j$" for the clause "$\bigvee_j \neg l_i \vee \bigvee_j l_j$", and "$(\bigwedge_i l_i) \to (\bigwedge_j l_j)$" for the conjunction of clauses "$\bigwedge_j (\bigvee_i \neg l_i \vee l_j)$".

3.1   The Basic Encoding

Let $A_{[,]}$ be an *injective* function which maps a pair $\langle \sigma, \psi \rangle$, s.t. σ is a state label and ψ is a K_m-formula which is not in the form $\neg \phi$, into a boolean variable $A_{[\sigma, \psi]}$. Let $L_{[\sigma, \psi]}$ denote $\neg A_{[\sigma, \phi]}$ if ψ is in the form $\neg \phi$, $A_{[\sigma, \psi]}$ otherwise. Given a K_m-formula φ, the encoder $K_m 2SAT$ builds a boolean CNF formula as follows:

$$K_m 2SAT(\varphi) := A_{[1, \varphi]} \wedge Def(1, \varphi), \tag{1}$$

$$Def(\sigma, A_i), := \top \tag{2}$$

$$Def(\sigma, \neg A_i) := \top \tag{3}$$

$$Def(\sigma, \alpha) := (L_{[\sigma, \alpha]} \to (L_{[\sigma, \alpha_1]} \wedge L_{[\sigma, \alpha_2]})) \wedge Def(\sigma, \alpha_1) \wedge Def(\sigma, \alpha_2) \tag{4}$$

$$Def(\sigma, \beta) := (L_{[\sigma, \beta]} \to (L_{[\sigma, \beta_1]} \vee L_{[\sigma, \beta_2]})) \wedge Def(\sigma, \beta_1) \wedge Def(\sigma, \beta_2) \tag{5}$$

$$Def(\sigma, \pi^{r,j}) := (L_{[\sigma, \pi^{r,j}]} \to L_{[\sigma.j, \pi_0^{r,j}]}) \wedge Def(\sigma.j, \pi_0^{r,j}) \tag{6}$$

$$Def(\sigma, \nu^r) := \bigwedge_{\langle \sigma : \pi^{r,i} \rangle} ((L_{[\sigma, \nu^r]} \wedge L_{[\sigma, \pi^{r,i}]}) \to L_{[\sigma.i, \nu_0^r]}) \wedge \bigwedge_{\langle \sigma : \pi^{r,i} \rangle} Def(\sigma.i, \nu_0^r). \tag{7}$$

Here by "$\langle \sigma : \pi^{r,i} \rangle$" we mean that $\pi^{r,i}$ is the j-th dinstinct π^r formula labeled by σ.

We assume that the K_m-formulas are represented as DAGs, so that to avoid the expansion of the same $Def(\sigma, \psi)$ more than once. Moreover, following [10], we assume that, for each σ, the $Def(\sigma, \psi)$'s are expanded in the order: α, β, π, ν. Thus, each $Def(\sigma, \nu^r)$ is expanded after the expansion of all $Def(\sigma, \pi^{r,i})$'s, so that $Def(\sigma, \nu^r)$ will generate one clause $((L_{[\sigma, \pi^{r,i}]} \wedge L_{[\sigma, \Box_r \nu_0^r]}) \rightarrow L_{[\sigma.i, \nu_0^r]})$ and one novel definition $Def(\sigma.i, \nu_0^r)$ for each $Def(\sigma, \pi^{r,i})$ expanded.

Theorem 1. *A K_m-formula φ is K_m-satisfiable if and only if the corresponding boolean formula $K_m2SAT(\varphi)$ is satisfiable.*

Notice that, due to (7), the number of variables and clauses in $K_m2SAT(\varphi)$ may grow exponentially with $depth(\varphi)$. This is in accordance to what stated in [5].

3.2 Variants

Before the encoding, some potentially useful preprocessing can be performed.

First, the input K_m-formulas can be converted into NNF (like, e.g., in [10]) or into BNF (like, e.g., in [3,11]). One potential advantage of the latter is that, when one $\Box_r \psi$ occurs both positively and negatively (like, e.g., in $(\Box_r \psi \vee ...) \wedge (\neg \Box_r \psi \vee ...) \wedge ...)$, then both occurrences of $\Box_r \psi$ are labeled by the same boolean atom $A_{[\sigma, \Box_r \psi]}$, and hence they are always assigned the same truth value by DPLL; with NNF, instead, the negative occurrence $\neg \Box_r \psi$ is rewritten into $\Diamond_r \neg \psi$, so that two distinct boolean atoms $A_{[\sigma, \Box_r \psi]}$ and $A_{[\sigma, \Diamond_r \neg \psi]}$ are generated; DPLL can assign them the same truth value, creating a hidden conflict which may require some extra boolean search to reveal.

Example 1 (NNF).
Let φ_{nnf} be $(\Diamond A_1 \vee \Diamond(A_2 \vee A_3)) \wedge \Box \neg A_1 \wedge \Box \neg A_2 \wedge \Box \neg A_3$.[2] It is easy to see that φ_{nnf} is K_1-unsatisfiable. $K_m2SAT(\varphi_{nnf})$ is:

1. $\quad A_{[1, \varphi_{nnf}]}$
2. $\wedge \; (A_{[1, \varphi_{nnf}]} \rightarrow (A_{[1, \Diamond A_1 \vee \Diamond(A_2 \vee A_3)]} \wedge A_{[1, \Box \neg A_1]} \wedge A_{[1, \Box \neg A_2]} \wedge A_{[1, \Box \neg A_3]}))$
3. $\wedge \; (A_{[1, \Diamond A_1 \vee \Diamond(A_2 \vee A_3)]} \rightarrow (A_{[1, \Diamond A_1]} \vee A_{[1, \Diamond(A_2 \vee A_3)]}))$
4. $\wedge \; (A_{[1, \Diamond A_1]} \rightarrow A_{[1.1, A_1]})$
5. $\wedge \; (A_{[1, \Diamond(A_2 \vee A_3)]} \rightarrow A_{[1.2, A_2 \vee A_3]})$
6. $\wedge \; ((A_{[1, \Box \neg A_1]} \wedge A_{[1, \Diamond A_1]}) \rightarrow \neg A_{[1.1, A_1]})$
7. $\wedge \; ((A_{[1, \Box \neg A_2]} \wedge A_{[1, \Diamond A_1]}) \rightarrow \neg A_{[1.1, A_2]})$
8. $\wedge \; ((A_{[1, \Box \neg A_3]} \wedge A_{[1, \Diamond A_1]}) \rightarrow \neg A_{[1.1, A_3]})$
9. $\wedge \; ((A_{[1, \Box \neg A_1]} \wedge A_{[1, \Diamond(A_2 \vee A_3)]}) \rightarrow \neg A_{[1.2, A_1]})$
10. $\wedge \; ((A_{[1, \Box \neg A_2]} \wedge A_{[1, \Diamond(A_2 \vee A_3)]}) \rightarrow \neg A_{[1.2, A_2]})$
11. $\wedge \; ((A_{[1, \Box \neg A_3]} \wedge A_{[1, \Diamond(A_2 \vee A_3)]}) \rightarrow \neg A_{[1.2, A_3]})$
12. $\wedge \; (A_{[1.2, A_2 \vee A_3]} \rightarrow (A_{[1.2, A_2]} \vee A_{[1.2, A_3]}))$

After a run of BCP, 3. reduces to the implicate disjunction. If the first element $A_{[1, \Diamond A_1]}$ is assigned, then by BCP we have a conflict on 4.,6. If the second element $A_{[1, \Diamond(A_2 \vee A_3)]}$ is assigned, then by BCP we have a conflict on 12. Thus $K_m2SAT(\varphi_{nnf})$ is unsatisfiable. $\Diamond$

---

[2] For K_1 formulas, we omit the box and diamond indexes.

Example 2 (BNF).

Let $\varphi_{bnf} = (\neg\Box\neg A_1 \vee \neg\Box(\neg A_2 \wedge \neg A_3)) \wedge \Box\neg A_1 \wedge \Box\neg A_2 \wedge \Box\neg A_3$. It is easy to see that φ_{bnf} is K_1-unsatisfiable. $K_m2SAT(\varphi_{bnf})$ is:

1. $\quad A_{[1, \varphi_{bnf}]}$
2. $\wedge \quad (A_{[1, \varphi_{bnf}]} \rightarrow (A_{[1, (\neg\Box\neg A_1 \vee \neg\Box(\neg A_2 \wedge \neg A_3))]} \wedge A_{[1, \Box\neg A_1]} \wedge A_{[1, \Box\neg A_2]} \wedge A_{[1, \Box\neg A_3]}))$
3. $\wedge \quad (A_{[1, (\neg\Box\neg A_1 \vee \neg\Box(\neg A_2 \wedge \neg A_3))]} \rightarrow (\neg A_{[1, \Box\neg A_1]} \vee \neg A_{[1, \Box(\neg A_2 \wedge \neg A_3)]}))$
4. $\wedge \quad (\neg A_{[1, \Box\neg A_1]} \rightarrow A_{[1.1, A_1]})$
5. $\wedge \quad (\neg A_{[1, \Box(\neg A_2 \wedge \neg A_3)]} \rightarrow \neg A_{[1.2, (\neg A_2 \wedge \neg A_3)]})$
6. $\wedge \quad ((A_{[1, \Box\neg A_1]} \wedge \neg A_{[1, \Box\neg A_1]}) \rightarrow \neg A_{[1.1, A_1]})$
7. $\wedge \quad ((A_{[1, \Box\neg A_2]} \wedge \neg A_{[1, \Box\neg A_1]}) \rightarrow \neg A_{[1.1, A_2]})$
8. $\wedge \quad ((A_{[1, \Box\neg A_3]} \wedge \neg A_{[1, \Box\neg A_1]}) \rightarrow \neg A_{[1.1, A_3]})$
9. $\wedge \quad ((A_{[1, \Box\neg A_1]} \wedge \neg A_{[1, \Box(\neg A_2 \wedge \neg A_3)]}) \rightarrow \neg A_{[1.2, A_1]})$
10. $\wedge \quad ((A_{[1, \Box\neg A_2]} \wedge \neg A_{[1, \Box(\neg A_2 \wedge \neg A_3)]}) \rightarrow \neg A_{[1.2, A_2]})$
11. $\wedge \quad ((A_{[1, \Box\neg A_3]} \wedge \neg A_{[1, \Box(\neg A_2 \wedge \neg A_3)]}) \rightarrow \neg A_{[1.2, A_3]})$
12. $\wedge \quad (\neg A_{[1.2, (\neg A_2 \wedge \neg A_3)]} \rightarrow (A_{[1.2, A_2]} \vee A_{[1.2, A_3]}))$

Unlike with NNF, $K_m2SAT(\varphi_{bnf})$ is found unsatisfiable directly by BCP. Notice that the unit-propagation of $A_{[1, \Box\neg A_1]}$ from 2. causes $\neg A_{[1, \Box\neg A_1]}$ in 3. to be false, so that one of the two (unsatisfiable) branches induced by the disjunction is cut a priori. With NNF, the corresponding atoms $A_{[1, \Box\neg A_1]}$ and $A_{[1, \Diamond A_1]}$ are not recognized to be one the negation of the other, s.t. DPLL may need exploring one boolean branch more.   ◇

Second, the (NNF or BNF) K_m-formula can also be rewritten by recursively applying the validity-preserving "box/diamond lifting rules":

$$(\Box_r\varphi_1 \wedge \Box_r\varphi_2) \Longrightarrow \Box_r(\varphi_1 \wedge \varphi_2), \quad (\Diamond_r\varphi_1 \vee \Diamond_r\varphi_2) \Longrightarrow \Diamond_r(\varphi_1 \vee \varphi_2). \qquad (8)$$

This has the potential benefit of reducing the number of $\pi^{r,i}$ formulas, and hence the number of labels $\sigma.i$ to take into account in the expansion of the $Def(\sigma, v^r)$'s.

Example 3 (BNF with LIFT).

Let $\varphi_{bnflift} = \neg\Box(\neg A_1 \wedge \neg A_2 \wedge \neg A_3) \wedge \Box(\neg A_1 \wedge \neg A_2 \wedge \neg A_3)$. It is easy to see that $\varphi_{bnflift}$ is K_1-unsatisfiable. $K_m2SAT(\varphi_{bnflift})$ is:

1. $\quad A_{[1, \varphi_{bnflift}]}$
2. $\wedge \quad (A_{[1, \varphi_{bnflift}]} \rightarrow (\neg A_{[1, \Box(\neg A_1 \wedge \neg A_2 \wedge \neg A_3)]} \wedge A_{[1, \Box(\neg A_1 \wedge \neg A_2 \wedge \neg A_3)]}))$
3. $\wedge \quad (\neg A_{[1, \Box(\neg A_1 \wedge \neg A_2 \wedge \neg A_3)]} \rightarrow \neg A_{[1.1, (\neg A_1 \wedge \neg A_2 \wedge \neg A_3)]})$
4. $\wedge \quad (\neg A_{[1.1, (\neg A_1 \wedge \neg A_2 \wedge \neg A_3)]} \rightarrow (A_{[1.1, A_1]} \vee A_{[1.1, A_2]} \vee A_{[1.1, A_3]}))$

$K_m2SAT(\varphi_{bnflift})$ is found unsatisfiable by BCP.   ◇

One potential drawback of applying the lifting rules is that, by collapsing $(\Box_r\varphi_1 \wedge \Box_r\varphi_2)$ into $\Box_r(\varphi_1 \wedge \varphi_2)$ and $(\Diamond_r\varphi_1 \vee \Diamond_r\varphi_2)$ into $\Diamond_r(\varphi_1 \vee \varphi_2)$, the possibility of sharing box/diamond subformulas in the DAG representation of φ is reduced.

4   Conclusions and Future Work

In this paper (see also the extended version) we have explored the idea of encoding $K_m/\mathcal{ALC}$-satisfiability into SAT, so that to be handled by state-of-the-art SAT tools. We have showed that, despite the intrinsic risk of blowup in the size of the encoded formulas, the performances of this approach are comparable with those of current state-of-the-art tools on a rather extensive variety of empirical tests. (Notice that, as a byproduct of this work, the encoding of hard K_m-formulas could be used as benchmarks for SAT solvers.)

We see many possible direction to explore in order to enhance and extend this approach. First, our current implementation of the encoder is very straightforward, and optimizations for making the formula more compact can be introduced. Second, techniques implemented in other approaches (e.g., the pure literal optimization of [12]) could be imported. Third, hybrid approaches between K_m2SAT and KSAT-style tools could be investigated.

Another important open research line is to explore encodings for other modal and description logics. Whilst for logics like T_m the extension should be straightforward, logics like $S4_m$, or more elaborated description logics than $\mathcal{ALC}$, should be challenging.

References

1. S. Brand, R. Gennari, and M. de Rijke. Constraint Programming for Modelling and Solving Modal Satisfability. In *Proc. CP 2003*, volume 3010 of *LNAI*. Springer, 2003.
2. M. Fitting. *Proof Methods for Modal and Intuitionistic Logics*. D. Reidel Publishg, 1983.
3. F. Giunchiglia and R. Sebastiani. Building decision procedures for modal logics from propositional decision procedures - the case study of modal K. In *Proc. CADE'13*. Springer, 1996.
4. F. Giunchiglia and R. Sebastiani. Building decision procedures for modal logics from propositional decision procedures - the case study of modal K(m). *Information and Computation*, 162(1/2), 2000.
5. J. Y. Halpern. The effect of bounding the number of primitive propositions and the depth of nesting on the complexity of modal logic. *Artificial Intelligence*, 75(3):361–372, 1995.
6. J.Y. Halpern and Y. Moses. A guide to the completeness and complexity for modal logics of knowledge and belief. *Artificial Intelligence*, 54(3):319–379, 1992.
7. I. Horrocks and P. F. Patel-Schneider. Optimizing Description Logic Subsumption. *Journal of Logic and Computation*, 9(3):267–293, 1999.
8. U. Hustadt, R. A. Schmidt, and C. Weidenbach. MSPASS: Subsumption Testing with SPASS. In *Proc. DL'99*, pages 136–137, 1999.
9. R. Ladner. The computational complexity of provability in systems of modal propositional logic. *SIAM J. Comp.*, 6(3):467–480, 1977.
10. F. Massacci. Single Step Tableaux for modal logics: methodology, computations, algorithms. *Journal of Automated Reasoning*, Vol. 24(3), 2000.
11. G. Pan, U. Sattler, and M. Y. Vardi. BDD-Based Decision Procedures for K. In *Proc. CADE*, LNAI. Springer, 2002.
12. G. Pan and M. Y. Vardi. Optimizing a BDD-based modal solver. In *Proc. CADE*, LNAI. Springer, 2003.
13. R. Sebastiani and M. Vescovi. Encoding the satisfiability of modal and description logics into SAT: the case study of K(m)/$\mathcal{ALC}$. Extended version. Technical Report DIT-06-033, DIT, University of Trento., May 2006. Available at http://www.dit.unitn.it/~rseba/sat06/extended.ps.

SAT in Bioinformatics:
Making the Case with Haplotype Inference

Inês Lynce[1] and João Marques-Silva[2]

[1] IST/INESC-ID, Technical University of Lisbon, Portugal
ines@sat.inesc-id.pt
[2] School of Electronics and Computer Science, University of Southampton, UK
jpms@ecs.soton.ac.uk

Abstract. Mutation in DNA is the principal cause for differences among
human beings, and Single Nucleotide Polymorphisms (SNPs) are the
most common mutations. Hence, a fundamental task is to complete a
map of haplotypes (which identify SNPs) in the human population. Associated with this effort, a key computational problem is the inference
of haplotype data from genotype data, since in practice genotype data
rather than haplotype data is usually obtained. Recent work has shown
that a SAT-based approach is by far the most efficient solution to the
problem of haplotype inference by pure parsimony (HIPP), being several
orders of magnitude faster than existing integer linear programming and
branch and bound solutions. This paper proposes a number of key optimizations to the the original SAT-based model. The new version of the
model can be orders of magnitude faster than the original SAT-based
HIPP model, particularly on biological test data.

1 Introduction

Over the last few years, an emphasis in human genomics has been on identifying genetic variations among different people. This allows to systematically test
common genetic variants for their role in disease; such variants explain much of
the genetic diversity in our species. A particular focus has been put on the identification of Single Nucleotide Polymorphisms (SNPs), point mutations found
with only two possible values in the population, and tracking their inheritance.
However, this process is in practice very difficult due to technological limitations.
Instead, researchers can only identify whether the individual is heterozygotic at
that position, i.e. whether the values inherited from both parents are different.
This process of going from genotypes (which include the ambiguity at heterozygous positions) to haplotypes (where we know from which parent each SNP is
inherited) is called *haplotype inference.*

A well-known approach to the haplotype inference problem is called Haplotype Inference by Pure Parsimony (HIPP). The problem of finding such solutions is APX-hard (and, therefore, NP-hard) [7]. Current methods for solving the
HIPP problem utilize Integer Linear Programming (ILP) [5,1,2] and branch and
bound algorithms [10]. Recent work [8] has proposed the utilization of SAT for

A. Biere and C.P. Gomes (Eds.): SAT 2006, LNCS 4121, pp. 136–141, 2006.

the HIPP problem. Preliminary results are significant: on existing well-known problem instances, the SAT-based HIPP solution (SHIPs) is by far the most efficient approach to the HIPP problem, being orders of magnitude faster than *any* other alternative exact approach for the HIPP problem. Nevertheless, additional testing revealed that the performance of SHIPs can deteriorate for larger problem instances. This motivated the development of a number of optimizations to the basic SHIPs model, which are described in this paper. The results of the improved model are again very significant: the improved model can be *orders of magnitude faster* on biological test data than the basic model.

2   Haplotype Inference

A *haplotype* is the genetic constitution of an individual chromosome. The underlying data that forms a haplotype can be the full DNA sequence in the region, or more commonly the SNPs in that region. Diploid organisms pair homologous chromosomes, and thus contain two haplotypes, one inherited from each parent. The *genotype* describes the conflated data of the two haplotypes. In other words, an *explanation* for a genotype is a pair of haplotypes. Conversely, this pair of haplotypes explains the genotype. If for a given site both copies of the haplotype have the same value, then the genotype is said to be *homozygous* at that site; otherwise is said to be *heterozygous*.

Given a set $\mathcal{G}$ of n genotypes, each of length m, the haplotype inference problem consists in finding a set $\mathcal{H}$ of $2 \cdot n$ haplotypes, such that for each genotype $g_i \in G$ there is at least one pair of haplotypes (h_j, h_k), with h_j and $h_k \in \mathcal{H}$ such that the pair (h_j, h_k) explains g_i. The variable n denotes the number of individuals in the sample, and m denotes the number of SNP sites. g_i denotes a specific genotype, with $1 \leq i \leq n$. (Furthermore, g_{ij} denotes a specific site j in genotype g_i, with $1 \leq j \leq m$.) Without loss of generality, we may assume that the values of a SNP are always 0 or 1. Value 0 represents the wild type and value 1 represents the mutant. A haplotype is then a string over the alphabet {0,1}. Moreover, genotypes may be represented by extending the alphabet used for representing haplotypes to {0,1,2}. Homozygous sites are represented by values 0 or 1, depending on whether both haplotypes have value 0 or 1 at that site, respectively. Heterozygous sites are represented by value 2.

One of the approaches to the haplotype inference problem is called Haplotype Inference by Pure Parsimony (HIPP). A solution to this problem minimizes the total number of distinct haplotypes used. Experimental results provide support for this method: the number of haplotypes in a large population is typically very small, although genotypes exhibit a great diversity.

3   SAT-Based Haplotype Inference

This section summarizes the model proposed in [8], where a more detailed description of the model (and associated optimizations) can be found. The SAT-based formulation models whether there exists a set $\mathcal{H}$ of haplotypes, with

$r = |\mathcal{H}|$ haplotypes, such that each genotype $g_i \in \mathcal{G}$ is explained by a pair of haplotypes in $\mathcal{H}$. The SAT-based algorithm considers increasing sizes for $\mathcal{H}$, from a lower bound lb to an upper bound ub. Trivial lower and upper bounds are, respectively, 1 and $2 \cdot n$. The algorithm terminates for a size of $\mathcal{H}$ for which there exists $r = |\mathcal{H}|$ haplotypes such that every genotype in $\mathcal{G}$ is explained by a pair of haplotypes in $\mathcal{H}$. In what follows we assume n genotypes each with m sites. The same indexes will be used throughout: i ranges over the genotypes and j over the sites, with $1 \leq i \leq n$ and $1 \leq j \leq m$. In addition r candidate haplotypes are considered, each with m sites. An additional index k is associated with haplotypes, $1 \leq k \leq r$. As a result, $h_{kj} \in \{0,1\}$ denotes the jth site of haplotype k. Moreover, a haplotype h_k, is viewed as a m-bit word, $h_{k1} \ldots h_{km}$. A valuation $v : \{h_{k1}, \ldots, h_{km}\} \rightarrow \{0,1\}$ to the bits of h_k is denoted by h_k^v. Observe that valuations can be extended to other sets of variables.

For a given value of r, the model considers r haplotypes and seeks to associate two haplotypes (which can possibly represent the same haplotype) with each genotype g_i, $1 \leq i \leq n$. As a result, for each genotype g_i, the model uses *selector* variables for selecting which haplotypes are used for explaining g_i. Since the genotype is to be explained by *two* haplotypes, the model uses two sets, a and b, of r selector variables, respectively s_{ki}^a and s_{ki}^b, with $k = 1, \ldots, r$. Hence, genotype g_i is explained by haplotypes h_{k_1} and h_{k_2} if $s_{k_1 i}^a = 1$ and $s_{k_2 i}^b = 1$. Clearly, g_i is also explained by the same haplotypes if $s_{k_2 i}^a = 1$ and $s_{k_1 i}^b = 1$.

If a site g_{ij} of a genotype g_i is either 0 or 1, then this is the value required at this site and so this information is used by the model. If a site g_{ij} is 0, then the model requires, for $k = 1, \ldots, r$, $(\neg h_{kj} \vee \neg s_{ki}^a) \wedge (\neg h_{kj} \vee \neg s_{ki}^b)$. If a site g_{ij} is 1, then the model requires, for $k = 1, \ldots, r$, $(h_{kj} \vee \neg s_{ki}^a) \wedge (h_{kj} \vee \neg s_{ki}^b)$. Otherwise, one requires that the haplotypes explaining the genotype g_i have opposing values at site i. This is done by creating two variables, $g_{ij}^a \in \{0,1\}$ and $g_{ij}^b \in \{0,1\}$, such that $g_{ij}^a \neq g_{ij}^b$. In CNF, the model requires two clauses, $(g_{ij}^a \vee g_{ij}^b) \wedge (\neg g_{ij}^a \vee \neg g_{ij}^b)$. In addition, the model requires, for $k = 1, \ldots, r$, $(h_{kj} \vee \neg g_{ij}^a \vee \neg s_{ki}^a) \wedge (\neg h_{kj} \vee g_{ij}^a \vee \neg s_{ki}^a) \wedge (h_{kj} \vee \neg g_{ij}^b \vee \neg s_{ki}^b) \wedge (\neg h_{kj} \vee g_{ij}^b \vee \neg s_{ki}^b)$. Clearly, for each i, and for a or b, it is necessary that exactly one haplotype is used, and so exactly one selector variable be assigned value 1. This can be captured with cardinality constraints, $(\sum_{k=1}^{r} s_{ki}^a = 1) \wedge (\sum_{k=1}^{r} s_{ki}^b = 1)$. Since the proposed model is purely SAT-based, a simple alternative solution is used, which consists of the CNF representation of a simplified adder circuit [8].

The model described above is not effective in practice. Hence a number of improvements have been added to the basic model. One technique, common to other approaches to the HIPP problem, is the utilization of structural simplifications techniques, for reducing the number of genotypes and sites [2,8]. Another technique is the utilization of lower bound estimates, which reduce the number of iterations of the algorithm, but also effectively prune the search space. Finally, one additional key technique for pruning the search space is motivated by observing the existence of symmetry in the problem formulation. Consider two haplotypes h_{k_1} and h_{k_2}, and the selector variables $s_{k_1 i}^a$, $s_{k_2 i}^a$, $s_{k_1 i}^b$ and $s_{k_2 i}^b$. Furthermore, consider Boolean valuations v_x and v_y to the sites of haplotypes h_{k_1}

and h_{k_2}. Then, $h_{k_1}^{v_x}$ and $h_{k_2}^{v_y}$, with $s_{k_1i}^a s_{k_2i}^a s_{k_1i}^b s_{k_2i}^b = 1001$, corresponds to $h_{k_1}^{v_y}$ and $h_{k_2}^{v_x}$, with $s_{k_1i}^a s_{k_2i}^a s_{k_1i}^b s_{k_2i}^b = 0110$, and one of the assignments can be eliminated. To remedy this, one possibility is to enforce an ordering of the Boolean valuations to the haplotypes [1]. Hence, for any valuation v to the problem variables we require $h_1^v < h_2^v < \ldots < h_r^v$ (see [8] for further details).

4   Improvements to SAT-Based Haplotype Inference

Motivated by an effort to apply the SHIPs model to biological test data, we were able to identify a number of additional improvements to the basic model. For difficult problem instances, the run time is very sensitive to the number of g variables used. The basic model creates two variables for each heterozygous site. One simple optimization is to replace each pair of g variables associated with a heterozygous site, g_{ij}^a and g_{ij}^b, by a single variable t_{ij}. Consequently, the new set of constraints becomes, $(h_{kj} \vee \neg t_{ij} \vee \neg s_{ki}^a) \wedge (\neg h_{kj} \vee t_{ij} \vee \neg s_{ki}^a) \wedge (h_{kj} \vee t_{ij} \vee \neg s_{ki}^b) \wedge (\neg h_{kj} \vee \neg t_{ij} \vee \neg s_{ki}^b)$. Hence, if selector variable s_{ki}^a is activated (i.e. assumes value 1), then h_{kj} is equal to t_{ij}. In contrast, if selector variable s_{ki}^b is activated, then h_{kj} is the complement of t_{ij}. Observe that, since the genotype has at least one heterozygous site, then it must be explained by *two different* haplotypes, and so s_{ki}^a and s_{ki}^b cannot be simultaneously activated.

The basic model utilizes lower bounds, which are obtained by identifying incompatibility relations among genotypes. These incompatibility relations find other applications. Consider two *incompatible* genotypes, g_{i_1} and g_{i_2}, and a candidate haplotype h_k. Hence, if either $s_{ki_1}^a$ or $s_{ki_1}^b$ is activated, and so h_k is used for explaining genotype g_{i_1}, then haplotype h_k *cannot* be used for explaining g_{i_2}; hence both $s_{ki_2}^a$ and $s_{ki_2}^b$ *must not* be activated. The implementation of this condition is achieved by adding the following clauses for each pair of incompatible genotypes g_{i_1} and g_{i_2} and for each candidate haplotype h_k, $(\neg s_{ki_1}^a \vee \neg s_{ki_2}^a) \wedge (\neg s_{ki_1}^a \vee \neg s_{ki_2}^b) \wedge (\neg s_{ki_1}^b \vee \neg s_{ki_2}^a) \wedge (\neg s_{ki_1}^b \vee \neg s_{ki_2}^b)$.

One of the key techniques proposed in the basic model is the utilization of the sorting condition over the haplotypes, as an effective symmetry breaking technique. Additional symmetry breaking conditions are possible. Observe that the model consists of selecting a candidate haplotype for the a representative and another haplotype for the b representative, such that each genotype is explained by the a and b representatives. Given a set of r candidate haplotypes, let h_{k_1} and h_{k_2}, with $k_1, k_2 \leq r$, be two haplotypes which explain a genotype g_i. This means that g_i can be explained by the assignments $s_{k_1i}^a s_{k_2i}^a s_{k_1i}^b s_{k_2i}^b = 1001$, but also by the assignments $s_{k_1i}^a s_{k_2i}^a s_{k_1i}^b s_{k_2i}^b = 0110$. This symmetry can be eliminated by requiring that only one arrangement of the s variables can be used to explain each genotype g_i. One solution is to require that the haplotype selected by the s_{ki}^a variables always has an index *smaller* than the haplotype selected by the s_{ki}^b variables. This requirement is captured by the conditions $\left(s_{k_1i}^a \rightarrow \bigwedge_{k_2=1}^{k_1-1} \neg s_{k_2i}^b \right)$

---

[1] See for example [4] for a survey of work on the utilization of lexicographic orderings for symmetry breaking.

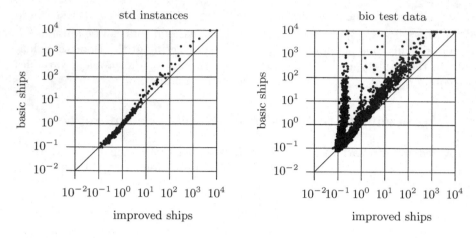

Fig. 1. Comparison of the basic and improved SHIPs models (run times)

and $\left(s_{k_2 i}^b \rightarrow \bigwedge_{k_1=k_2+1}^r \neg s_{k_1 i}^a\right)$. Clearly, each condition above can be represent by a single clause. Moreover, observe that for genotypes with heterozygous sites, the upper limit of the first constraint can be set to $k_1 - 1$ and the lower limit of the second condition can be set to $k_2 + 1$.

5   Experimental Results

The models described in the previous section, referred to as SHIPs (Sat-based Haplotype Inference by Pure Parsimony), have been implemented as a Perl script, which iteratively generates CNF formulas to be given to a SAT solver. Currently, MiniSAT [3] is used.

With the purpose of comparing the basic and the improved versions of SHIPs, two sets of problem instances are considered. The first set of instances were generated using Hudson's program ms [6] (denoted std instances). The second set of instances are the instances currently available from biological test data (denoted bio test data) (e.g. from [9]).

The results are shown in Figure 1. Each plot compares the CPU time required by both the basic and the improved SHIPs models for solving each problem instance. The limit CPU time was set to 10000s using a 1.9 GHz AMD Athlon XP with 1GB of RAM running RedHat Linux. For the std instances the results are clear. The improved model is consistently faster than the basic model, especially for the most difficult problem instances. For problem instances requiring more than 10 CPU seconds, and with a single exception, the improved model is always faster than the basic model. For most of these instances, and by noting that the plot uses a log scale, we can conclude that the speedups range from a factor of 2 to a factor of 10. For the bio test data instances the performance differences become even more clear. The improved model significantly outperforms the basic model. Observe that the speedups introduced by the improved model can exceed 4 orders of magnitude.

6   Conclusions and Future Work

This paper provides further evidence that haplotype inference is a new very promising application area for SAT. The results in this paper and in [8] provide unquestionable evidence that the utilization of SAT yields the most efficient approach to the problem of haplotype inference by pure parsimony. Indeed, the SAT-based approach is the *only* approach currently capable of solving a large number of practical instances. Moreover, the optimizations proposed in this paper are shown to be essential for solving challenging problem instances from biological test data. Despite the promising results, several challenges remain. Additional biological test data may yield new challenging problem instances, which may motivate additional optimizations to the SAT-based approach.

Acknowledgments. The authors thank Arlindo Oliveira for having pointed out the haplotype inference by pure parsimony problem. This work is partially supported by FCT under research project POSC/EIA/61852/2004.

References

1. D. Brown and I. Harrower. A new integer programming formulation for the pure parsimony problem in haplotype analysis. In *Workshop on Algorithms in Bioinformatics (WABI'04)*, 2004.
2. D. Brown and I. Harrower. Integer programming approaches to haplotype inference by pure parsimony. *IEEE/ACM Transactions on Computational Biology and Bioinformatics*, 3(2):141–154, April-June 2006.
3. N. Eén and N. Sörensson. An extensible SAT-solver. In *International Conference on Theory and Applications of Satisfiability Testing (SAT)*, pages 502–518, 2003.
4. A. Frisch, B. Hnich, Z. Kiziltan, I. Miguel, and T. Walsh. Global constraints for lexicographic orderings. In *International Conference on Principles and Practice of Constraint Programming (CP)*, 2002.
5. D. Gusfield. Haplotype inference by pure parsimony. In *14th Annual Symposium on Combinatorial Pattern Matching (CPM'03)*, pages 144–155, 2003.
6. R. R. Hudson. Generating samples under a wright-fisher neutral model of genetic variation. *Bioinformatics*, 18(2):337–338, February 2002.
7. G. Lancia, C. M. Pinotti, and R. Rizzi. Haplotyping populations by pure parsimony: complexity of exact and approximation algorithms. *INFORMS Journal on Computing*, 16(4):348–359, 2004.
8. I. Lynce and J. Marques-Silva. Efficient haplotype inference with Boolean satisfiability. In *National Conference on Artificial Intelligence (AAAI)*, July 2006.
9. M. J. Rieder, S. T. Taylor, A. G. Clark, and D. A. Nickerson. Sequence variation in the human angiotensin converting enzyme. *Nature Genetics*, 22:481–494, 2001.
10. L. Wang and Y. Xu. Haplotype inference by maximum parsimony. *Bioinformatics*, 19(14):1773–1780, 2003.

Lemma Learning in SMT on Linear Constraints

Yinlei Yu and Sharad Malik

Department of Electrical Engineering
Princeton University
Princeton, NJ 08544, USA
{yyu, sharad}@princeton.edu

Abstract. The past decade has seen great improvement in Boolean Satisfiability(SAT) solvers. SAT solving is now widely used in different areas, including electronic design automation, software verification and artificial intelligence. However, many applications have non-Boolean constraints, such as linear relations and uninterpreted functions. Converting such constraints into SAT is very hard and sometimes impossible. This has given rise to a recent surge of interest in Satisfiability Modulo Theories (SMT). SMT incorporates predicates in other theories such as linear real arithmetic, into a Boolean formula. Solving an SMT problem entails either finding an assignment for all Boolean and theory specific variables in the formula that evaluates the formula to TRUE or proving that such an assignment does not exist. To solve such an SMT instance, a solver typically combines SAT and theory-specific solving under the Nelson-Oppen procedure framework. Fast SAT and theory-specific solvers and good integration of the two are required for efficient SMT solving.

Efficient learning contributes greatly to the success of the recent SAT solvers. However, the learning technique in SMT is limited in the current literature. In this paper, we propose methods of efficient lemma learning on SMT problems with linear real/integer arithmetic constraints. We describe a static learning technique that analyzes the relationship of the linear constraints. We also discuss a conflict driven learning technique derived from infeasible sets of linear real/integer constraints. The two learning techniques can be expanded to many other theories. Our experimental results show that lemma learning can significantly improve the speed of SMT solvers.

1 Introduction

Boolean Satisfiability(SAT) has been widely used in verification, artificial intelligence and many other areas. However, in many cases, constraints and variables are not necessarily Boolean; some constraints may be linear relationships among integer or real variables. Many verification systems have been designed to address such problems. Some systems, such as Alloy [1], choose to convert the integer constraints into Boolean formulae by encoding the arithmetic constraints and bit-blasting the variables with a limited range (for instance, it converts an 32-bit integer variable into 32 Boolean variables); this results in a huge CNF formula that is hard for the current state-of-art SAT solvers.

Satisfiability Modulo Theory (SMT) is an extension of SAT addressing the problem of different types of constraints. An SMT problem determines the satisfiability of a

A. Biere and C.P. Gomes (Eds.): SAT 2006, LNCS 4121, pp. 142–155, 2006.

Boolean formula with predicates on a decidable theory. It is particularly useful in situations where linear constraints and other types of constraints are required, such as timing verification, software verification[2] and artificial intelligence.

For satisfiability problems like SAT and SMT, *lemma learning* or just *learning* means extracting new constraints from previous search process and adding them into the problem to reduce search space. It has been shown to be very effective in SAT[3][4][5]. In this paper, we show that SMT solvers can benefit significantly from two types of learning processes, static and dynamic.

2   Basic Definitions and SMT Algorithm

In this section, we review the basic definitions of Satisfiability Modulo Theory and basic algorithms for solving SMT problems.

2.1   Basic Definitions

SMT solving is a procedure of finding an satisfying assignment for a quantifier-free formula F with predicates on a certain background theory T, or showing that such assignment does not exist. A satisfying assignment M is an assignment on all variables (may also be referred to as a *model*), such that the formula evaluates to TRUE under a given background theory T. We can say M entails F under theory T, or expressed as $M \models_T F$. Theory T must be decidable, i.e., given a conjunction of constraints in T, there must exist a procedure of finite steps that can test the existence of a satisfying assignment for these constraints. Otherwise, the underlying SMT problem would become undecidable – a solver for such SMT problems would be impossible.

Typical theories used in SMT are Linear Real Arithmetic (LRA), Linear Integer Arithmetic(LIA), Real/Interger Difference Arithmetic(RDA/IDA), List theory (L) and Equality and Uninterpreted Functions (EUF). A simple example of an SMT instance on LRA is:

Example 1.

$$((b_1 \vee (x_1 + x_2 \leq 5)) \wedge (\neg b_1 \vee (x_1 + x_2 \geq 7)) \wedge (x_1 > 10) \tag{1}$$

Where $b_1 \in \{T, F\}$ and $x_1, x_2 \in \mathbb{R}$. Assignment $\{b_1 = F, x_1 = 11, x_2 = -6\}$ makes the formula evaluate to TRUE. Therefore this SMT instance is satisfiable.

In this paper, a *theory-specific predicate* is an expression in theory T, such that it evaluates to a Boolean value (T or F) under T when a T-value is assigned to each of the variables in the expression. The predicate is atomic in the Boolean formula, containing no Boolean operators AND($\wedge$), OR($\vee$) or NOT($\neg$). A *Boolean variable* is a variable that can only take Boolean constant T or F as its value. A Satisfiability Modulo Theory (SMT) instance on a theory T is defined recursively as following:

1. A predicate on theory T or a Boolean variable is an SMT instance.
2. The logic AND($\wedge$), OR($\vee$) of two SMT instances or the logic NOT($\neg$) of an SMT instance is an SMT instance.

A *literal* is an occurrence of a Boolean variable or its inverse. It is either a part of a Boolean formula, or represents an assignment on this Boolean variable such that the literal evaluates to true. A *clause* is the disjunction(OR) of a set of literals. A Boolean formula in *Conjunctive Normal Form*(CNF) is a conjunction of clauses. *Adding a clause* to a CNF means conjuncting(AND) it with a *clause*. A SAT solver is capable of checking the satisfiability of a Boolean formula in CNF. It is trivial to convert a non-CNF Boolean formula into an equivalent formula in CNF by introducing additional Boolean variables to represent interim terms. SMT solvers typically call SAT internally and use CNF function representations.

2.2 Basic SMT Algorithm

The basic SMT algorithm uses the DPLL[6][7] algorithm for Boolean satisfiability and employs the Nelson-Oppen[8] procedure to combine theory-specific solvers with DPLL.

In the basic algorithm, given an SMT formula F, every theory-specific predicate p is abstracted and represented by a Boolean variable b_p. b_p is required to have the same Boolean value as the predicate p in any assignment. All occurrences of p in F is replaced by b_p. The resulting formula B is a Boolean formula. Some additional Boolean variables may be introduced to convert a general Boolean formula into CNF.

The Boolean variables that represent theory-specific predicates themselves are also referred to as predicate variables. An assignment on a predicate variable b_p means the underlying theory-specific predicate must evaluate to T if $b_p = $ T and must evaluate to F if otherwise. A *predicate literal* is a literal with a predicate variable as its Boolean variable.

For linear constraints on real/integer variables, we may further assume the comparison predicates of *bounds* are only $\leq, =$ and $\geq$ and those of *constraints* are only $\leq$ or $=$. As we can choose the inverse form of a variable for both *bounds* and *constraints* and negate both sides for *constraints*, these assumptions do not limit generality.

In Example 1, the Boolean formula B after abstraction becomes:

$$B \equiv ((b_1 \vee b_2) \wedge (\neg b_1 \vee b_3) \wedge b_4 \qquad (2)$$

while the theory specific predicates are:

$$b_2 : (x_1 + x_2 \leq 5)$$
$$b_3 : (x_1 + x_2 \geq 7)$$
$$b_4 : (x_1 > 10)$$

SAT solvers that can enumerate all solutions are used to find satisfying assignments for the resulting Boolean formula in CNF.

A satisfying assignment for (2) is $b_1 = $ F, $b_2 = $ T, $b_3 = $ F and $b_4 = $ T.

Upon finding a satisfying assignment A, each theory-specific predicate p is required to evaluate to the Boolean value assigned to its corresponding predicate variable b_p. Thus these predicates become constraints on the particular theory. The consistency of these constraints are tested with theory-specific solvers $Sover_T$. For instance, for linear real constraints, we may linear programming solvers such as `Cassowary`[9]. If the theory-specific solver finds a consistent assignment for these constraints, the SMT instance is satisfiable; the satisfying assignment is the combination of the theory-specific assignment and assignment on Boolean variables. An inconsistent verdict from $Sovler_T$ requires the SAT solver to find another satisfying assignment until no more such assignments exist. If no satisfying assignment for the Boolean part corresponds to a consistent theory specific constraint set, the SMT instance is *unsatisfiable*, or *invalid*. Fig. 1 shows the basic SMT algorithm. Here $AtomMap$ is the map of theory-specific atoms to Boolean variables generated by the abstraction process SMT_Atomize. Find-NextSAT is a call to a SAT solver. Calling it consecutively will obtain different satisfying assignments until all possible assignments are returned, in which case FindNextSAT will return empty. $Solver_T$ tests the consistency of theory-specific constraints and returns `TRUE` for consistent and `FALSE` otherwise.

```
Function SMT_Solve (F : SMT Instance)
{B, AtomMap} ← SMT_Atomize(F)
loop
    A ← FindNextSAT(B)
    if A does not exist then
        return UNSATISFIABLE
    else
        P ← AtomMap(A)
        if Solver_T(P) = SATISFIABLE then
            return SATISFIABLE
        end if
    end if
end loop
```

Fig. 1. Basic SMT algorithm

2.3 SMT with Linear Constraints

In this paper, we focus on SMT instances with linear predicates on real/integer variables. Without loss of generality, a predicate on a linear constraint is in the form of $\sum_{i=1}^{k} a_i x_i \sim b$, in which $\sim \in \{=, \leq, \geq\}$ ($\sim \in \{=, \leq\}$ if $k \geq 2$).

To facilitate discussion in following sections, we categorize linear predicates into two types:

1. *Bound predicates* are predicates with only one real/integer variable involved, in the form of $x \sim c$.
2. *Constraint predicates* are predicates that compares the linear combination of at least two real variables with a constant, in the form of $\sum_{i=1}^{k} a_i x_i \sim c$, $k \geq 2$.

The constraints in the first category correspond to the column constraints in linear program solving and the ones in the second category correspond to the row constraints.

The basic solving algorithm for SMT with linear constraints is the same as in Figure 1. The TheorySolve function can use any linear programming solver or integer linear programming solver, such as CPLEX[10], Xpress-MP[11] and Mosek[12].

2.4 Previous Works on SMT

In the recent years, a number of good SMT solvers are developed, such as ario[13], MathSAT[14], CVCLite[15], DPLL(T)[16], Simplify[2], ICS, simplics and Yices[17]. All of them are based on the Nelson-Oppen algorithm, but with different techniques to improve the solving speed. In particular, ario incorporated UTVPI (*unit two variable per inequality*) that can propagate constraints with only two unit coefficient variables efficiently. DPLL(T) derives implications in the theory domain extensively and shows the effectiveness of learning in integer difference logic. MathSAT uses layers of solvers so that some conflicts that only involve simpler theories may be detected much more efficiently, such as the conflicts on linear difference constraints for an SMT with general linear arithmetic constraints. Some of these solvers also use "early" theory-specific inference procedure. Such procedure will not wait till the Boolean variables are assigned; rather, it checks on-the-fly the consistency of the underlying constraints based on a partial Boolean assignment.

3   Learning in SMT

The basic SMT algorithm suffers from an overly relaxed Boolean formula. In the conversion from SMT to SAT, many predicate variables are not well constrained as they typically have few occurrences in the converted Boolean formula. This leads to a huge number of satisfying assignments; most of which are conflicting when they are mapped back as theory specific constraints. In this section, we address this issue by proposing to add clauses to reflect the relationship of underlying theories.

Boolean Learning elevates theory specific constraints into the Boolean domain. If a set of constraints in an SMT problem P are inconsistent, and they correspond to a set of literals $\{l_1, \ldots, l_k\}$, it means these set of literals cannot be satisfied simultaneously. This can be represented as a clause of negated literals of all predicates involved as $(l'_1 + \cdots + l'_k)$. This clause is implied by the theory specific predicates; adding it (ANDing it to the original Boolean formula, in fact) will not change the satisfiability of the SMT problem, but will prune the search space in Boolean search. $P \models_T (l'_1 + \cdots + l'_k)$. This technique is *Theory specific Learning*; it is referred to as *learning* heretofore in this paper for simplicity.

To illustrate *learning*, we reuse the example from last section. It is clear that b_2 and b_3 can never be true simultaneously, as $x_1 + x_2$ cannot be ≤ 5 and ≥ 7 at the same time. A Boolean clause $(b'_2 + b'_3)$ can be added to avoid such local constraint conflicts. In this section, we mainly discuss learning methods for SMT with linear constraints.

3.1   Static Learning for SMT with Linear Constraints

A traditional SAT solver has no knowledge to avoid conflicting assignments on theory-specific predicates. If learning is not involved, the SAT solver will enumerate all the possible satisfying assignments (or partial assignments for solvers that justify theory-specific predicates whenever their corresponding predicates variables are assigned) Because of the relaxed nature of the related Boolean formula, there may be huge number of satisfying assignments. Learning can be used to mitigate the problem.

To address the problem, some previous work has proposed to use theory based implications [14]. However, implication based on Boolean clauses are much more efficient due to the effectiveness of fast Boolean Constraint Propagation(BCP) algorithms like two-literal watching[4]. We propose a preprocessing procedure that directly elevate most of *local* theory-specific inconsistencies into Boolean domain: A clause is added to the CNF when a possible conflict exists among a set of predicates. In the previous example, the clause $(b'_2 + b'_3)$ will be added to the CNF in this preprocessing step.

Deriving clauses from simple linear constraints. Many simple constraints can be derived from analyzing the relationship of constraints. In a typical SMT problem with linear constraints, a majority of constraints has only one or two variables. Table 1 shows the distribution of number of variables in a linear predicate. In this table, each row represent a group of benchmarks from SMTLIB's QF_LRA and QF_LIA groups. The number in each column represents the average percentage of predicates with a specific number of variables (the percentage in each instance, unweighed average).

Table 1. Distribution of # of variables in linear predicates

Name	1	2	3	4	5	6+
Carpark2	49.06	45.99	4.94	0	0	0
gasburner-prop3	65.09	21.70	13.21	0	0	0
pursuit-safety	61.42	27.50	11.08	0	0	0
scheduler	0.83	99.17	0	0	0	0
TM	75.88	17.99	4.19	.40	.14	1.40
tgc_io	77.56	17.50	4.93	0	0	0
windowreal	83.17	13.49	3.43	0	0	0
ckt	95.09	.43	.30	.30	.30	3.56
FISCHER	0	97.23	0	2.77	0	0
MULTIPLIER	84.87	12.37	.44	.33	.27	1.70
wisa	.10	92.33	7.32	.13	.10	0

The simpler constraints enable us to exploit their relationship and derive clauses. Here we show three different types of simple constraint relationships and the possible implications they may generate.

Bound predicates on the same variable
Consider two bound predicates p_1 and p_2 on the same variable x with bounds b_1 and b_2 respectively. If both comparison signs are $\geq$ and $b_1 < b_2$, then the assignment $p_1 = F$ and $p_2 = T$ would require $x < b_1$ and $x \geq b_2$, which conflicts with the fact $b_1 < b_2$. This can be represented as a clause $(p_1 + p'_2)$. Similar conflict clauses can be added for pairs of different bounds. The detailed list of derived clauses is shown in Table 2. The clause may also be understood as the fact of $p_2 \rightarrow p_1$, which is equivalent to the new clause $(p_1 + p'_2)$. The new clause for each situation is listed in Table 2. Without loss of generality, we assume that $b_1 \leq b_2$ and the predicate signs $\sim \in \{\leq, =, \geq\}$. We may

Table 2. The implication table for a pair of bounds on the same real variable

p_1	p_2	$b_1 \sim b_2$	new clause
$\leq, =$	$\leq$	$b_1 < b_2$	$(p_1' + p_2)$
$\leq, =$	$=, \geq$	$b_1 < b_2$	$(p_1 + p_2)$
$\geq$	$\leq$	$b_1 < b_2$	$(p_1' + p_2')$
$\geq$	$=, \geq$	$b_1 < b_2$	$(p_1 + p_2')$
$\leq$	$\geq$	$b_1 = b_2$	$(p_1 + p_2)$
$\leq, \geq$	$=$	$b_1 = b_2$	$(p_1 + p_2')$
$\leq$	$\leq$	$b_1 = b_2$	$(p_1 + p_2'), (p_1' + p_2)$
$=$	$=$	$b_1 = b_2$	$(p_1 + p_2'), (p_1' + p_2)$
$\geq$	$\geq$	$b_1 = b_2$	$(p_1 + p_2'), (p_1' + p_2)$

swap p_1 and p_2 if $b_1 > b_2$ and/or take the inverse of the predicate if the predicate signs are $<$ or $>$.

In fact not all such predicates are necessary. Inequality predicates ($\leq, \geq$) are transitive, so only adjacent inequality predicates on the same variable need to add Boolean predicates according to Table 2; for equality predicates, it only need to add clauses for other equality predicates and *closest* inequality predicates.

For predicates on integer variables, stronger clauses are possible. For example, two inequality predicates with their non-integer bounds in a same open range $(k, k + 1)$, $k \in \mathbb{Z}$ are equivalent (or inversely equivalent if on is $\leq$ and the other is $\geq$). The fact that $(x > k) \equiv (x \geq k+1)$ if $x, k \in \mathbb{Z}$ and $(x > k) \equiv (x \geq \lceil k \rceil)$ if $x \in \mathbb{Z}, k \notin \mathbb{Z}$ can be also used to generate stronger clauses.

Bound predicates on variables used in constraint predicates

Consider a linear constraint predicate p, if every real variables x_i in p has one or more bound predicates associated with it, we may derived new bound predicates out of them: Consider constraint predicate

$$p : (a_1 x_1 + \ldots + a_k x_k \sim b)$$

Suppose each of $k - 1$ variables $x_1, \ldots, x_{k-1}$ has a bound predicate

$$b_i : (x_i \sim x_{i0}), (i = 1, \ldots, k - 1)$$

A new predicate for x_k can be generated as

$$\left(x_k \sim \frac{b - \sum_{i=1}^{k-1} a_i x_{i0}}{a_k} \right)$$

The '$\sim$' is '$=$' if all predicate signs of $b_1, \ldots, b_{k-1}$ and p are '$=$'; if otherwise, '$\sim$' can be '$\leq$' or '$\geq$' depending on the simple inequality relationship.

For example, given predicates $p : 3x_1 + x_2 + x_3 \leq 12$, $q_1 : x_1 \leq 3$ and $q_2 : x_2 \leq 2$. We may add a new predicate $q_3 : x_2 \geq 1$ that subject to: $(p \wedge q_1 \wedge q_2 \rightarrow q_3)$, which is a clause $(p' + q_1' + q_2' + q_3)$.

Two constraint predicates share the same set of variables
If two constraint predicates share the same set of real variables, We may extract new predicates from them. Here we illustrate the method on two constraints on same two real variables. To facilitate discussion, we denote the two constraints as $p_1 : a_{11}x_1 + a_{12}x_2 \sim b_1$ and $p_2 : a_{21}x_1 + a_{22}x_2 \sim b_2$, in which $\sim$ can be either $\leq$ or $=$.

- If the two lines represented by the two constraints are parallel, i.e., $a_{12}/a_{11} = a_{22}/a_{21}$, then their corresponding predicates will not allow certain combinations. For instance, if one predicate p_1 is $(x_1 + 2x_2 \leq 5)$ while the other predicate p_2 is $(-2x_1 - 4x_2 \leq 8)$, p_1 and p_2 being FALSE is impossible, as $x_1 + 2x_2$ cannot be greater than 5 and less than -4 simultaneously. A clause $(p_1 + p_2)$ is added to the clause database accordingly. The detailed implication step is omitted here as it is easy to infer and similar to the implication table for a pair of bounds.
- If the two lines are not parallel, they must cross at a certain point (x_{10}, x_{20}), in which

$$x_{10} = \frac{a_{22}b_1 - a_{21}b_2}{a_{11}a_{22} - a_{21}a_{12}}, x_{20} = \frac{a_{11}b_2 - a_{21}b_1}{a_{11}a_{22} - a_{21}a_{12}}$$

In this case, we may derive additional predicates of $q_1 : (x_1 \sim x_{10})$, $q_2 : (x_2 \sim x_{20})$ based on basic arithmetic inequality rules. An assignment of p_1 and p_2 may imply q_1, q_2 and/or their inverse depending on the signs of the coefficients a_{ij} and b_i.
An example of such new predicates is: $p_1 : x_1 + x_2 \leq 3$, $p_2 : x_1 - x_2 \leq 5$. Their crossing point is $(4, -1)$. This entails the following implication relationships:

$$p_1 \wedge p_2 \rightarrow x_1 \leq 4$$

$$p_1 \wedge p_2' \rightarrow x_2 < -1$$

$$p_1' \wedge p_2 \rightarrow x_2 > -1$$

$$p_1' \wedge p_2' \rightarrow x_1 > 4$$

The detailed implication table for all conditions is quite long but nevertheless can be implemented efficiently. As all the predicate relationships can be easily derived by arithmetic inequality rules similar to the above.
For constraints with n variables, we may add $n - 1$ variable predicates similarly as what we describe above.

Note that adding these predicates does not generate more implications for existing Boolean variables in most cases. However, such new predicates, especially if they are bound predicates, may be used to generate static learned clauses (such as what is described in "Bound predicates on the same variable"). For example, a bound predicates $b_n : x_1 \leq 0$ can be generated by pairs of two-variable predicates $b_1 : x_1 + x_2 \leq 0$ and $b_2 : x_1 - x_2 \leq 0$ with a clause $(b_1' + b_2' + b_n)$. If we have another bound predicate $b : x_1 \leq 1$, we may derive another clause $(b + b_n')$ and obtain an implication chain of $b_1 \wedge b_2 \rightarrow b_n \rightarrow b$ which is not otherwise available.

3.2 Conflict Driven Learning

The constraints derived from static learning we described in the previous subsection is not complete – satisfying all additional constraints does not guarantee the underlying LP/ILP feasible; though such assignments are more likely to have an feasible underlying LP/ILP, but linear constraint solving is still needed to check the actual feasibility.

Conflict driven learning is widely used in SAT solvers to avoid searching previously visited Boolean sub-spaces. It has led to significant improvement on SAT solvers. For a problem with difference logic, its underlying theory-specific problem is a network flow; a conflict means a negative cycle in the network flow graph; a conflict clause based on the Boolean assignments on the cycle can be added to the clause database, as implemented in many solvers such as MathSAT[14].

For problems with linear real/integer constraints, a naïve conflict driven learning method is just adding the inverse of all literals that determine theory-specific predicates.

A smaller conflict clause can prune more search space than larger ones. To generate a smaller conflict clause from a set of infeasible constraints, we should reduce the number of constraints while maintaining its infeasibility. *Irreducible Infeasible Set* (IIS)[18][19] of the corresponding LP/ILP can be used. An IIS is an infeasible subset of constraints in an infeasible linear program, such that removing a single constraint from IIS will make the subset feasible. The size of an IIS is typically much smaller than that of an original linear program. IIS-finding capability is provided by most linear programming solvers, such as CPLEX, Xpress-MP, Mosek and MINOS(IIS). The disjunction of the inverse of all predicate literals that are involved in the IIS can be added to the original Boolean formula as a conflict clause. In fact, in our experiments, the average size of conflict clause derived from IIS is only 3.66% of the size of naïve conflict clause for benchmark group sal and 20.03% for benchmark group TM in SMTLIB.

To illustrate the difference of the naïve method and conflict driven learning, we may revisit Example 1. If we obtained a satisfying assignment for the Boolean formula 1: $b_1 = \text{T}$, $b_2 = \text{T}$, $b_3 = \text{T}$ and $b_4 = \text{T}$. This assignment will lead to a set of inconsistent linear constraints: $b_2 = \text{T}$ requires $(x_1 + x_2 \leq 5)$ while $b_3 = \text{T}$ requires $(x_1 + x_2 \geq 7)$. The naïve method requires inverting *all* predicate assignments to form a conflict clause; in this case the clause is $(b'_2 + b'_3 + b'_4)$. Obviously, only $b_2 = \text{T}$ and $b_3 = \text{T}$ contribute to the inconsistency while the assignment on b_4 is not needed. Therefore the clause generated from conflict driven learning is $(b'_2 + b'_3)$, which is shorter than the one from naïve learning.

The SMT algorithm that incorporates both types of learning is shown in Fig. 2.

4   Implementation and Experimental Results

4.1   Implementation

To evaluate the methodology described above, we implemented a test SMT solver on linear constraints based on zChaff[4] revision 2004.11.15 as the propositional solver and CPLEX[10] version 10.0 as the LP/ILP solver. zChaff is modified to accommodate incremental clause addition and deletion. Our SMT solver incorporated the static clause and predicate generation as described in Section 3.1 and conflict

Function SMT_Solve_With_Learning (F : SMT Instance)
$\{B, AtomMap\} \leftarrow$ SMT_Atomize(F)
$B \leftarrow B \wedge$ SMT_Static_Learn($AtomMap$)
loop
   $A \leftarrow$ SolveSAT(B)
   if A does not exist **then**
     return UNSATISFIABLE
   else
     $P \leftarrow AtomMap(A)$
     if TheorySolve(P) $=$ SATISFIABLE **then**
       return SATISFIABLE
     else
       $C \leftarrow$ FindConflictSet(P)
       $B \leftarrow B \wedge C$
     end if
   end if
end loop

Fig. 2. SMT algorithm with learning

driven learning as described in Section 3.2. Among the methods described in Section 3.1, clause generation for adjacent bound predicate pairs was implemented; for the constraint-bound predicate pairs, we generate new bound predicates. For two variable predicates with same real variables, we generate predicates on one variable accordingly. The preprocessing step ignores predicates and bounds with variable coefficients (coefficients that may change depending on the Boolean assignments, for instance $3 * x + (a?5:3) * y \leq 4$, in which $x, y \in \mathbb{R}$ and $a \in \{T, F\}$), because the implication relationship is uncertain when predicates with variable coefficients are involved. As a proof-of-concept implementation, we did not include certain advanced features like dynamic theory specific implications and optimizations for difference constraints so that we can separate the benefit of learning from other techniques.

The SMT solver reads in the SMT formula in SMT-LIB format[20]. The Boolean formula is represented as a binary and-inverter graph(BAIG) internally[21]. Light-weight logic simplification is applied on the Boolean formula, including structural hashing and two-level optimizations. The linear part of the formulas are hashed as well; identical formulas are merged. The simplified Boolean formula is converted into Conjunctive Normal Form by introducing interim Boolean variables. A Boolean variable is assigned on each linear predicate. For conflict driven learning, our SMT solver uses the IIS procedure in CPLEX to find the conflict sets. The program is written in C++, it is compiled by g++ 4.0.1 with option -march=pentium4 -O3 and run on Fedora Core 4.0 Linux system.

4.2 Experimental Results

The effectiveness of learning. We report the experimental results here. The first experiment is to evaluate the effectiveness of learning. We implemented the basic algorithm and compared the results with and without static and conflict driven learning. The benchmarks we used are from SMTLIB[20]. We tested the sal and TM groups from linear real arithmetic section (QF_LRA) of the benchmark suite as well as wisa group from linear integer arithmetic section(QF_LIA). The test cases in these groups have significant number of predicates that are not just difference predicates. This will give a fair evaluation for solvers that are not optimized for difference arithmetic. The results of a random selection in each benchmark group is listed, as we have more than 100 benchmark cases and the omitted ones shows the same data trend. The test results are shown in Table 3. The time limit for each single instance is 300 seconds. All the test cases are run on a Linux machine with a 2.8GHz Intel Pentium 4 CPU (1MB L2 cache) and 1GB memory.

The columns of Table 3 are as following: In Table 3(a), T_x means the CPU time for solving a particular instance (in seconds). "TIME" in a T_x column means the instance cannot be solved in the given time limit. L_x means the number of LP solver calls. The subscript possibilities of x are $Base$, for SMT solver with naïve learning (adding a conflict clause that disjuncts the inversion of all assignments on predicates), SL for static learning only, CL for conflict driven learning by IIS only and AL for both static and conflict driven learning enabled. Table 3(b) shows how many instances each of the four cases can solve for each group of SMT instances.

Table 3. SMT solver with and without learning

Name	T_{Base}	L_{Base}	T_{SL}	L_{SL}	T_{CL}	L_{CL}	T_{AL}	L_{AL}
Carpark2-t1-3	TIME	-	0.15	0	0.21	35	0.17	0
gasburner-prop3-10	78.51	68,222	16.13	42,875	0.13	43	0.08	5
pursuit-safety-4	102.32	69,615	10.14	31,752	0.06	14	0.04	5
pursuit-safety-20	TIME	-	TIME	-	2.67	686	1.21	164
tgc_io_safe-12	TIME	-	TIME	-	0.95	322	0.57	31

(a) Individual instances

Group Name	# of Inst	BASE	Stat. Learn	Conf. Learn	All Learn
Carpark2	12	10	12	12	12
gasburner-prop3	20	10	18	20	20
pursuit-safety	20	4	10	20	20
tgc_io-nosafe	7	4	7	7	7
tgc_io-safe	20	4	13	20	20
windowreal-no_t_deadlock	20	20	20	20	20
wisa	5	0	0	4	4

(b) Instance groups

The test results show that the conflict driven learning can significantly improve the solving speed of SMT solvers. Adding static constraints can further improve the solving performance by reducing solving time and number of LP solver calls. It can even completely avoid any linear program solving in the test case Carpark2-t1-3. However, although only adding static constraints can significantly improve the performance of the basic SMT solver, it is not as good as conflict driven learning by itself. It is not surprising as conflict driven learning is capable of deriving all the implications that can be found by static learning, although not as efficiently. The conflict between bound constraints can always be detected when transforming a satisfying assignment into a linear program; all the other types of statically learned clauses can also be found by conflict driven learning as CPLEX can find minimum conflicting constraints sets. However, it is not as efficient because such conflicts are found in the linear programming phase, which requires a full assignment (some solvers like MathSAT can check some types of conflict dynamically, which make it more effective).

Performance test. The second experiment compares our SMT solver with a few state-of-art SMT solvers available. We chose MathSAT 3.3.1[14], Yices 0.1[17], Ario 1.1[13] and Simplics 1.1[22] to compare with our SMT solver with learning. All these versions are the latest from their respective websites. Among them Simplics is the best in thelinear real arithmetic part of SMT-COMP'05[23]. Each row is a group of test benches.

The first column in Table 4 is the name of a group of benchmarks and the second is the number of instances in each group. The next five columns are the result from MathSAT, Yices, ario, Simplics and our SMT solver respectively. The number outside the parenthesis is the number of instances the solver can solve in 15 minutes (900 seconds). The number inside is the amount of time the solver used to solve all the SMT instance in this group. TIME_OUT or MEM_OUT is charged 900 seconds as well.

Table 4. SMT solver performance comparisons

Name	# of Inst.	MathSAT	Yices	ario	simplics	SMT/WL
TM	25	19(6969.61)	17(8402.9)	12(12624.7)	20(5585.73)	17(8897.2)
Carpark2	12	12(1.62)	12(0.62)	12(0.77)	12(0.4)	12(2.46)
gasburner-prop3	20	20(41.18)	20(0.81)	20(1.7)	20(0.55)	20(1.35)
pursuit-safety	20	20(523.5)	20(7.15)	13(7095.46)	20(5.7)	20(10.88)
tgc_io_nosafe	7	7(0.76)	7(0.17)	7(0.19)	7(0.21)	7(1.00)
tgc_io_safe	20	20(157.60)	20(6.99)	20(37.85)	20(4.81)	20(11.82)
windowreal-no_t_deadlock	20	20(14.09)	20(1.03)	20(3.03)	20(1.83)	20(9.84)
windowreal-safe	4	4(0.27)	4(0.09)	4(0.07)	4(0.09)	4(0.45)
windowreal-safe2	4	4(0.26)	4(0.11)	4(0.08)	4(0.04)	4(0.4)
scheduler_abz5	5	5(103.8)	5(20.7)	5(117.03)	5(199.12)	5(896.12)
wisa	5	2(3643.19)	4(1014.17)	4(961.56)	5(317.55)	4(1281.52)

From Table 4, we can see the performance of our solver is comparable with the state-of-art SMT solvers available in most cases. This is impressive as we use only the two learning schemes proposed in this paper compared to optimizations such as eager theory propagation, UTVPI, network optimization, layered solver structure, and so on in current state-of-art SMT solvers. The performance on the scheduling_abz5 group is not as good because each instance in this group contains a large number of difference constraints. Our solver is not optimized for this network type of constraint while other solvers like MathSAT optimize for difference constraints. In the TM benchmarks, learning is not as effective as the conflict clause obtained from IIS is much larger than other cases.

5   Conclusions and Further Work

This paper proposes static and dynamic learning for linear real constraints. The overall goal is to have a much tighter coupling of the Boolean and real/integer reasoning engines. With static learning, we have shown how relatively simple relationships between the predicates may be statically captured in the Boolean part and dynamically exploited. Using efficient Boolean Constraint Propagation, this basic framework can serve as the basis for adding additional clauses using more sophiscated reason. The important benefit of such learning is reduction in the number of calls to the LP/ILP solvers.

With conflict driven learning in the LP/ILP solvers, we have successfully used the notion of the Irreducible Infeasible Set(IIS) (also known as "conflict" in CPLEX) to identify the core of the infeasibility and exclude this from future search. This has demonstrated clear benefit in the solver.

As for future work, the methodology can be applied on mixed integer linear constraints and as well as other types of constraints. More sophiscated static and dynamic learning methods may be applied; the algorithms for special cases like difference constraints may be incorporated into this framework.

References

1. Sofware Design Group, MIT: The Alloy Analyzer. (2006) http://alloy.mit.edu/ as retrieved on May 12, 2006.
2. Detlefs, D., Nelson, G., Saxe, J.B.: Simplify: a theorem prover for program checking. Journal of the ACM **52**(3) (2005) 365–473
3. Marques-Silva, J.P., Sakallah, K.A.: GRASP - a search algorithm for propositional satisfiability. IEEE Transactions on Computers **48**(5) (1999) 506–521
4. Moskewicz, M.W., Madigan, C.F., Zhao, Y., Zhang, L., Malik, S.: Chaff: Engineering an efficient SAT solver. In: Proc. DAC. (2001) 530–535
5. Zhang, L., Madigan, C.F., Moskewicz, M.W., Malik, S.: Efficient conflict driven learning in a Boolean satisfiability solver. In: Proc. ICCAD. (2001)
6. Davis, M., Putnam, H.: A computing procedure for quantification theory. Journal of the ACM **7**(3) (1960) 201–215
7. Davis, M., Logemann, G., Loveland, D.: A machine program for theorem proving. Communications of the ACM **5**(7) (1962) 394 – 397
8. Nelson, G., Oppen, D.C.: Simplification by cooperating decision procedures. ACM Transactions on Programming Languages and Systems **1** (1979) 245–257
9. Borning, A., Marriott, K., Stuckey, P., Xiao, Y.: Solving linear arithmetic constraints for user interface applications. In: Proc. ACM Symp. on User Interface Software and Technology. (1996) 87–96
10. ILog S. A.: CPLEX Manual, Version 10.0. (2006)
11. Dash optimization Inc.: Xpress-MP optimization software (2006) http://www.dashoptimization.com/.
12. Mosek ApS: MOSEK optimization software (2006) http://www.mosek.com/.
13. Sheini, H.M., Sakallah, K.A.: A scalable method for solving satisfiability of integer linear arithmetic logic. In: Proc. SAT. Volume 3569 of LNCS. (2005) 241–256
14. Bozzano, M., Bruttomesso, R., Cimatti, A., Junttila, T., van Rossum, P., Schulz, S., Sebastiani, R.: An incremental and layered procedure for the satisfiability of linear arithmetic logic. In: Proc. TACAS. Volume 3440 of LNCS. (2005) 317–333
15. Barrett, C., Dill, D.L., Stump, A.: Checking satisfiability of first-order formulas by incremental translation into SAT. In: Proc. CAV. Volume 2404 of LNCS. (2002)

16. Nieuwenhuis, R., Oliveras, A.: DPLL(T) with Exhaustive Theory Propagation and its Application to Difference Logic. In: Proc. CAV. Volume 3576 of LNCS. (2005) 321–334
17. Rueß, H., Shankar, N.: Solving linear arithmetic constraints. Technical report, SRI International (2004)
18. Chinneck, J.W.: MINOS(IIS): Infeasibility analysis using MINOS. Computers and Operations Research **21**(1) (1994) 1–9
19. Guieu, O., Chinneck, J.W.: Analyzing infeasible mixed-integer and integer linear programs. INFORMS J. of Computing **11**(1) (1999) 63–77
20. Ranise, S., Tinelli, C.: The SMT-LIB standard: Version 1.1 (2005) http://combination. cs.uiowa.edu/smtlib/papers/format-v1.1-r05.04.12.pdf.
21. Bjesse, P., Borälv, A.: DAG-aware circuit compression for formal verification. In: Proc. ICCAD. (2004) 42–49
22. Dutertre, B., de Moura, L.: Simplics: Tool description (2005) `http://fm.csl.sri. com/simplics/description.pdf` as retrieved on May 12, 2006.
23. Barrett, C., de Moura, L., Stump, A.: SMT-COMP: Satisfiability modulo theories competition. In: Proc. CAV. Volume 3576 of LNCS. (2005) 20–23

On SAT Modulo Theories and Optimization Problems

Robert Nieuwenhuis and Albert Oliveras[*]

Abstract. Solvers for *SAT Modulo Theories (SMT)* can nowadays handle large industrial (e.g., formal hardware and software verification) problems over theories such as the integers, arrays, or equality. Here we show that SMT approaches can also efficiently solve problems that, at first sight, do not have a typical SMT flavor. In particular, here we deal with SAT and SMT problems where models M are sought such that a given cost function $f(M)$ is minimized.

For this purpose, we introduce a variant of SMT where the theory T becomes progressively stronger, and prove it correct using the Abstract DPLL Modulo Theories framework. We discuss two different examples of applications of this SMT variant: weighted Max-SAT and weighted Max-SMT. We show how, with relatively little effort, one can obtain a competitive system that, in the case of weighted Max-SMT in the theory of Difference Logic, can even handle well-known hard radio frequency assignment problems without any tailored heuristics. These results seem to indicate that Max-SAT/SMT techniques can already be used for realistic applications.

1 Introduction

The Davis-Putnam-Logemann-Loveland (DPLL) procedure for propositional SAT [DP60, DLL62] has also been adapted for handling problems in more expressive logics, and, in particular, for the *SAT Modulo Theories (SMT)* problem: deciding the satisfiability of ground first-order formulas with respect to background theories such as the integer or real numbers, or arrays. SMT problems frequently arise in formal hardware and software verification applications, where typical formulas consist of very large sets of clauses like:

$$p \quad \vee \quad \neg q \quad \vee \quad a = b - c \quad \vee \quad read(v,\, b{+}c) = d \quad \vee \quad a{-}c \leq 7$$

with propositional atoms as well as atoms over (combined) theories like the integers, the reals, or arrays. SMT has become a very active area of research, and efficient SMT solvers exist that can handle (combinations of) many such theories (see also the SMT problem library [TR05] and the SMT Competition [BdMS05]). Currently most SMT solvers follow the so-called *lazy* approach to SMT, combining (i) *theory solvers* to process *conjunctions* of literals over the given theory T, with (ii) DPLL-based *engines* for dealing with the boolean structure of the formulas.

[*] Technical Univ. of Catalonia, Barcelona, `www.lsi.upc.edu/~roberto|~oliveras`. Partially supported by Spanish Ministry of Educ. and Science through the LogicTools project (TIN2004-03382, both authors), and FPU grant AP2002-3533 (Oliveras).

A. Biere and C.P. Gomes (Eds.): SAT 2006, LNCS 4121, pp. 156–169, 2006.

$DPLL(T)$ is a general SMT architecture for the lazy approach [GHN+04]. It consists of a $DPLL(X)$ engine, whose parameter X can be instantiated with a T-solver $Solver_T$, thus producing a $DPLL(T)$ system. The $DPLL(X)$ engine always considers the problem as a purely propositional one. For example, if the theory T is the integers, at some point $DPLL(X)$ might consider a partial assignment containing, among many others, the three literals $a \geq b + 3$, $b - 2 \geq c$, and $a \not> c$ without noticing its T-inconsistency, because it just considers such literals as propositional (syntactic) objects. But $Solver_T$ continuously analyzes the partial model that $DPLL(X)$ is building (a conjunction of literals). It can warn $DPLL(X)$ about this T-inconsistency, and generate a clause, called a *theory lemma*, $a \not\geq b + 3 \ \lor \ b - 2 \not\geq c \ \lor \ a > c$ that can be used by $DPLL(X)$ for backjumping. $Solver_T$ sometimes also does *theory propagation*: as soon as, e.g., $a \geq b + 3$ and $b - 2 \geq c$ become true, it can notify $DPLL(X)$ about T-consequences like $a > c$ that occur in the input formula. The modular $DPLL(T)$ architecture is flexible, and, compared with other SMT techniques, $DPLL(T)$ is also very efficient and has good scaling properties: the BarcelogicTools implementation of $DPLL(T)$ won all the four divisions it entered at the 2005 SMT Competition [BdMS05].

The aim of this paper is to show that SMT techniques such as $DPLL(T)$ can be easily adapted to efficiently solve problems that, at first sight, do not have a typical SMT flavor. In particular, here we deal with SAT and SMT problems where models M are sought such that a given cost function $f(M)$ is minimized.

For this purpose, in Section 2 we introduce a variant of SMT where the (first-order) theory T becomes *progressively stronger*, that is, T may be periodically replaced by $T \land T'$ for some first-order theory T'. Then, after mentioning some applications to optimization and other problems, we prove this variant correct by extending *Abstract DPLL Modulo Theories*, a uniform, declarative framework introduced in [NOT05] for modeling and reasoning about lazy SMT procedures.

In Section 3 we apply this SMT variant in a branch-and-bound setting, where the theory T "knows", possibly among the information about other theories, the cost function f and its current best bound. Each time a better bound is found, the SMT procedure continues with a theory that has become stronger, in the sense that models with a cost higher than this new bound now become T-inconsistent.

We then show how to deal in this framework with the *exact weighted Max-SAT* problem: given a set of pairs $\{(C_1, w_1) \ldots, (C_m, w_m)\}$ where each C_i is a propositional clause and w_i is its *weight* (a positive natural or real number), find a propositional assignment M that minimizes the sum of the weights of clauses that are false in M.

In Section 4 we report experimental results on an implementation of $DPLL(T)$ for Max-SAT and also explain how specialized propagation rules for Max-SAT in the style of [LH05] can be easily and flexibly incorporated. For instance, when two pairs of the form (l, w_1) and $(\neg l, w_2)$ appear, one can propagate $min\{w_1, w_2\}$.

An interesting aspect of this approach is that DPLL(T) allows one to obtain with relatively little effort Max-SAT implementations that are competitive w.r.t. state-of-the-art systems. We give experimental results that show this, for settings with and without the additional propagation rules.

In Section 5 we show how this approach can be smoothly extended to Max-SMT. As an example, we deal with *weighted Max-SMT* modulo the theory of Integer Difference Logic, a fragment of integer linear arithmetic. In this case, formulas are built over propositional atoms, as well as (ground) atoms of the form $a - b \leq k$, where a and b are (Skolemized) integer variables and k is an integer.

This logic is used in the context of hardware and software verification; for instance, some properties of timed automata are naturally expressed in it. But again, also problems that do not look a priori like typical SMT problems can be handled very efficiently with it, and also optimization problems can be solved using our approach.

We give experimental results on the well-known hard CELAR radio frequency assignment problems [CdGL+99]. In these problems, integer variables must belong to certain intervals, and constraints express minimal distances between variables, all of which can be very nicely modeled in Difference Logic.

From our BarcelogicTools DPLL(T) implementation we have obtained, with very little effort, our first Max-SMT system. In spite of its unlabored development, and of its single standard SMT decision heuristic, our experiments reveal that it can already handle these CELAR problems that, according to our experiments, appear to be far beyond the capabilities of systems dealing with translations into, e.g., Weighted Max-SAT, pseudo-Boolean, or Integer Linear Programming Problems. On the CELAR problems, this implementation importantly outperforms the best-known weighted CSP solver Toolbar [dGHZL05] in its default settings, and is still close or superior to Toolbar with its best (according to its authors) branching heuristic for these problems.

2   SMT with Progressively Stronger Theories

Abstract DPLL Modulo Theories [NOT05] is a framework for modeling and reasoning about DPLL-based SAT and SMT systems in terms of simple transition rules and rule application strategies. The framework eases the understanding and the comparison of different approaches as well as the proving of their correctness. In this section, we briefly describe the framework (see [NOT05] for details) and then extend it to accommodate *progressively stronger* theories, that is, the theory T may be periodically replaced by $T \wedge T'$ for some first-order theory T', and prove the correctness of this extension.

2.1   Potential Applications

Such a SAT or SMT procedure where the theory becomes progressively stronger has applications in the context of branch-and-bound-like applications, where a single model is sought that minimizes a given cost function. The weighted Max-SAT and Max-SMT problems addressed in the next sections are just a particular case of this; one could also handle problems like Max-Ones, or even problems with non-linear cost functions.

But, apart from optimization problems, it can also be useful for problems where, given a set of pairwise T-incompatible (first-order) properties $P_1 \ldots P_n$, different models $M_1 \ldots M_n$ are sought such that each M_i satisfies the corresponding P_i. Initially, one can add $P_1 \vee \ldots \vee P_n$ to the theory. Then, each time such an M_i is found, before backjumping to continue the search one can strengthen the theory T, replacing it by $T \wedge \neg P_i$, which can help pruning the search of models for the remaining P_j's.

A more concrete problem of this kind is, for instance, a company where every month the best employee is rewarded with a favorable working schedule for the following month. Hence the company needs to prepare in advance n schedules with properties $P_1 \ldots P_n$, where each P_i expresses that employee i is the one that works less hours, (or works least night shifts, or gets most money, etc).

Another completely different application is automatic classification of finite algebras [CMSM04], where one may be searching for, say, finite groups satisfying a set $P_1 \ldots P_n$ of different properties, one group for each P_i.

We stress that this is particularly useful if the properties $P_1 \ldots P_n$ cannot (efficiently) be expressed at the level of the SMT formula itself, and if $Solver_T$ can adequately handle their negations.

2.2   Abstract DPLL Modulo Theories

As usual in SMT, given a background theory T (a set of closed first-order formulas), we will only consider the SMT problem for *ground* (and hence quantifier-free) CNF formulas F. Such formulas may contain *free* constants, i.e., constant symbols not in the signature of T, which, as far as satisfiability is concerned, can be equivalently seen as existential variables. Other than free constants, all other predicate and function symbols in the formulas will instead come from the signature of T. From now on, we will assume that all formulas satisfy these restrictions.

The formalism we describe is based on a set of *states* together with a binary relation $\Longrightarrow$ (called the *transition relation*) over these states, defined by means of *transition rules*. Starting with a state containing an input formula F, one can use the rules to generate a finite sequence of states, where the final state indicates, for a certain theory T, whether or not F is T-consistent.

A *state* is either the distinguished state $T \parallel fail$ (denoting T-unsatisfiability) or a triple of the form $T \parallel M \parallel F$, where T is a theory, M is a sequence of literals, and F is a formula in conjunctive normal form (CNF), i.e., a finite set of disjunctions of literals. We additionally require that M never contains both a literal and its negation and that each literal in M is annotated as either a *decision* literal (indicated by l^d) or not. Frequently, we will refer to M as a *partial assignment* or consider M just as a set or conjunction of literals, ignoring both the annotations and the order of its elements. We use $\emptyset$ to denote the empty sequence.

In what follows, a possibly subscripted or primed lowercase l *always* denotes a literal. Similarly T and T' always denote theories, C and D always denote clauses (disjunctions of literals), F and G denote conjunctions of clauses, and M and N denote assignments.

We write $M \models F$ to indicate that M propositionally satisfies F. If C is a clause $l_1 \vee \ldots \vee l_n$, we sometimes write $\neg C$ to denote the formula $\neg l_1 \wedge \ldots \wedge \neg l_n$. We say that C is *conflicting* in a state $T \parallel M \parallel F, C$ if $M \models \neg C$.

A formula F is called *T-(in)consistent* if $F \wedge T$ is (un)satisfiable in the first-order sense. We say that *M is a T-model of F* if $M \models F$ and M, seen as a conjunction of literals, is T-consistent. It is not difficult to see that F is T-consistent if, and only if, it has a T-model. If F and G are formulas, then F *entails G in T*, written $F \models_T G$, if $F \wedge \neg G$ is T-inconsistent. If $F \models_T G$ and $G \models_T F$, we say that F and G are *T-equivalent*.

We start presenting a small variant, to accomodate the presence of the theory in the states, of the transition system first presented in [NOT05]:

Definition 1. *The* Basic DPLL Modulo Theories system *consists of the following five rules:*

UnitPropagate :

$$T \parallel M \parallel F, C \vee l \implies T \parallel M\, l \parallel F, C \vee l \quad \text{if} \begin{cases} M \models \neg C \\ l \text{ is undefined in } M \end{cases}$$

Decide :

$$T \parallel M \parallel F \implies T \parallel M\, l^{\mathsf{d}} \parallel F \quad \text{if} \begin{cases} l \text{ or } \neg l \text{ occurs in a clause of } F \\ l \text{ is undefined in } M \end{cases}$$

Fail :

$$T \parallel M \parallel F, C \implies T \parallel fail \quad \text{if} \begin{cases} M \models \neg C \\ M \text{ contains no decision literals} \end{cases}$$

Backjump :

$$T \parallel M\, l^{\mathsf{d}}\, N \parallel F, C \implies T \parallel M\, l' \parallel F, C \quad \text{if} \begin{cases} M\, l^{\mathsf{d}}\, N \models \neg C, \text{ and there is} \\ \text{some clause } C' \vee l' \text{ such that:} \\ F, C \models_T C' \vee l' \text{ and } M \models \neg C', \\ l' \text{ is undefined in } M, \text{ and} \\ l' \text{ or } \neg l' \text{ is in } F \text{ or in } M\, l^{\mathsf{d}}\, N \end{cases}$$

Theory Propagate :

$$T \parallel M \parallel F \implies T \parallel M\, l \parallel F \quad \text{if} \begin{cases} M \models_T l \\ l \text{ or } \neg l \text{ occurs in } F \\ l \text{ is undefined in } M \end{cases}$$

Definition 2. *The* Strengthening DPLL Modulo Theories system *consists of the five Basic DPLL Modulo Theories rules, and the following four rules:*

Restart :

$$T \parallel M \parallel F \qquad \Longrightarrow T \parallel \emptyset \parallel F$$

Theory Learn :

$$T \parallel M \parallel F \qquad \Longrightarrow T \parallel M \parallel F, C \qquad \text{if} \begin{cases} each\ atom\ of\ C\ is\ in\ F\ or\ in\ M \\ F \models_T C \end{cases}$$

Theory Forget :

$$T \parallel M \parallel F, C \qquad \Longrightarrow T \parallel M \parallel F \qquad \text{if} \{\, F \models_T C$$

Theory Strengthen :

$$T \parallel M \parallel F \qquad \Longrightarrow T \wedge T' \parallel M \parallel F$$

We denote the transition relation defined by all nine rules by $\Longrightarrow_S$.

For a transition relation $\Longrightarrow$, we denote by $\Longrightarrow^*$ the reflexive-transitive closure of $\Longrightarrow$. We call any sequence of the form $S_0 \Longrightarrow S_1$, $S_1 \Longrightarrow S_2$, ... a *derivation*, and denote it by $S_0 \Longrightarrow S_1 \Longrightarrow S_2 \Longrightarrow \ldots$. We call any subsequence of a derivation a *subderivation*. If $S \Longrightarrow S'$ we say that there is a *transition* from S to S'. A state S is *final* with respect to $\Longrightarrow$ if there are no transitions from S.

2.3   Correctness of Strengthening Abstract DPLL Modulo Theories

A decision procedure for SMT can be obtained by generating a derivation using $\Longrightarrow_S$ with a particular strategy. The relevant derivations are those that start with a state of the form $T_0 \parallel \emptyset \parallel F_0$, where F_0 is the initial formula. The aim of a derivation is to compute a state S such that: (i) S is final with respect to the five rules of Basic DPLL Modulo Theories and (ii) if S is of the form $T \parallel M \parallel F$ then M is T-consistent. We start by stating some invariants.

Lemma 3. *If* $T_0 \parallel \emptyset \parallel F_0 \Longrightarrow_S^* T \parallel M \parallel F$ *then the following hold.*

1. *All the atoms in M and all the atoms in F are atoms of F_0.*
2. *M contains no literal more than once and is indeed an assignment, i.e., it contains no pair of literals of the form l and $\neg l$.*
3. *F and F_0 are T-equivalent.*
4. *If M is of the form $M_0\, l_1\, M_1\, \ldots\, l_n\, M_n$, where $l_1, \ldots, l_n$ are all the decision literals of M, then $F_0, l_1, \ldots, l_i \models_T M_i$ for all i in $0 \ldots n$.*

The following termination result says that derivations are finite provided some standard conditions are fulfilled (e.g., Restart is applied with increasing periodicity), and that the theory is not strengthened infinitely many times, which is indeed the case in branch and bound and the other mentioned applications. The proof is a simple extension of the one for the standard conditions [NOT05]. This is also the case for the other proofs we omit here for space reasons.

Theorem 4 (Termination of $\Longrightarrow_S$). *Every derivation Der of the form* $T \parallel \emptyset \parallel F = S_0 \Longrightarrow_S S_1 \Longrightarrow_S \ldots$ *is finite if the following conditions hold:*

1. *Der has no infinite subderivations consisting of only* Theory Learn *and* Theory Forget *steps.*
2. Theory Strengthen *is applied only finitely many times.*
3. *For every subderivation of Der of the form*
   $S_{i-1} \Longrightarrow_S S_i \Longrightarrow_S \ldots \Longrightarrow_S S_j \Longrightarrow_S \ldots \Longrightarrow_S S_k$ *where the only three* Restart *steps are the ones producing* S_i, S_j, *and* S_k, *either:*
   - *there are more Basic DPLL Modulo Theories steps in* $S_j \Longrightarrow_S \ldots \Longrightarrow_S S_k$ *than in* $S_i \Longrightarrow_S \ldots \Longrightarrow_S S_j$, *or*
   - *in* $S_j \Longrightarrow_S \ldots \Longrightarrow_S S_k$ *a new clause is learned that is not forgotten in Der.*

Lemma 5. *If* $T_0 \parallel \emptyset \parallel F_0 \Longrightarrow_S^* T \parallel M \parallel F$ *and there is some conflicting clause in* $T \parallel M \parallel F$, *i.e.,* $M \models \neg C$ *for some clause* C *in* F, *then either* Fail *or* Backjump *applies to* $T \parallel M \parallel F$.

Property 6. If $T_0 \parallel \emptyset \parallel F_0 \Longrightarrow_S^* T \parallel M \parallel F$ and M is T-inconsistent, then either there is a conflicting clause in $T \parallel M \parallel F$, or else Theory Learn applies to $T \parallel M \parallel F$, generating a conflicting clause.

Even if it is very easy to generate non-terminating derivations for $\Longrightarrow_S$, Theorem 4 defines a very general strategy for avoiding that.

Lemma 5 and Property 6 show that, for a state of the form $T \parallel M \parallel F$, if there is some literal of F undefined in M, or there is some conflicting clause, or M is T-inconsistent, then a rule of Basic DPLL Modulo Theories is always applicable, possibly after a single Theory Learn step. Together with Theorem 4 (Termination), this shows how to compute a state to which the following main theorem is applicable.

Theorem 7. *Let Der be a derivation* $T_0 \parallel \emptyset \parallel F_0 \Longrightarrow_S^* S$, *where* S *is (i) final with respect to Basic DPLL Modulo Theories, and (ii) if* S *is of the form* $T \parallel M \parallel F$ *then* M *is* T-consistent. *Then*

1. S *is* $T \parallel$ fail *if, and only if,* F *is* T-inconsistent.
2. *If* S *is of the form* $T \parallel M \parallel F$ *then* M *is a* T-model of F.

These results are easy to apply. For example, in the context of branch and bound, each time a final state $T \parallel M \parallel F$ is obtained (final in the sense of conditions (i) and (ii) of Theorem 7), M is the current best model found. After that, one can apply Theory Strengthen to decrease the current upper bound and make M inconsistent with the strengthened theory that says that an M with smaller cost is needed (see the next section). By property 6, this will trigger further rule applications. When no smaller cost solution exists, the theorem implies that $T \parallel$ fail will be eventually obtained.

Similarly, when different models $M_1 \ldots M_n$ are sought satisfying properties $P_1 \ldots P_n$, one can initially add $P_1 \vee \ldots \vee P_n$ to the theory. Then, each time one

M_i is found for a P_i, the theory is strengthened by adding $\neg P_i$ to it, and again by property 6 the derivation continues. Once all M_i have been found, instead of adding the last $\neg P_i$ (which would make the theory inconsistent), the process is stopped.

3   Expressing Max-SAT and Max-SMT in This Framework

Here we apply this SMT variant in a branch-and-bound setting, where, given a cost function f, a model M is sought with minimum $f(M)$. In this case, the progressively stronger theory T "knows", possibly among the information about other theories, the cost function f and its current best upper bound.

In particular, here we consider the *exact weighted Max-SAT or Max-SMT* problem: given a set of pairs $\{(C_1, w_1) \ldots, (C_m, w_m)\}$ where each C_i is a (propositional or SMT) clause and w_i is its *weight* (a positive natural or real number), find an assignment M (consistent with the initial background theory T) that minimizes the sum of the weights of clauses that are false in M.

We use the well-known encoding where each weighted clause (C_i, w_i) gets a distinct additional positive literal p_i, i.e., it becomes $C_i \vee p_i$, where p_i is a fresh propositional symbol.

Given this encoding, apart from the initial background theory T (which is empty in the propositional case), the theory consists of the integers plus

$$p_1 \rightarrow (k_1 = w_1) \qquad \neg p_1 \rightarrow (k_1 = 0)$$
$$\ldots \qquad\qquad \ldots \qquad\qquad k_1 + \cdots + k_m \leq B$$
$$p_m \rightarrow (k_m = w_m) \qquad \neg p_m \rightarrow (k_m = 0)$$

In addition, we will have an initial cost bound B_0, and the relation $B < B_0$ will also be part of the theory. Then, each time the theory is strengthened with a new upper bound B_i, the relation $B < B_i$ is added.

Note that one can also express that a certain (disjoint) subset of the clauses must be true with a single common weight w_i, by simply adding the same p_i to all clauses in the subset.

Also note that initially each clause contains at most one weight literal p_i, but during the search these literals receive the same treatment as any other literal. Hence, due to conflict-driven learning, clauses with many (positive and negative) occurrences of such weight literals appear. The truth value of such weight literals can be set by theory propagation, since $Solver_T$ may communicate DPLL(X) that a certain p_i must be false in order not to exceed the current best upper bound for the function cost f.

4   Experiments with Max-SAT and Further Pruning Rules

In this section we give experimental results for propositional Max-SAT, showing that a competitive (wrt. pseudo-Boolean solvers) DPLL(T) system can be obtained with relatively little effort. Moreover, we also discuss how specialized propagation rules can be incorporated into a DPLL(T) implementation.

4.1   Comparison with Other Approaches

Existing specific algorithms for Max-SAT (e.g., [LMP05, LH05, XZ05]) have been mainly designed to attack relatively small but challenging problems. To solve these problems, they use several pruning techniques that detect when a certain partial assignment cannot be extended to a complete assignment that improves the current upper bound. However, on larger examples such as most of the ones analyzed here, these techniques (and their implementations) become extremely time and memory consuming, and hence here we only give experimental results for one of them, namely Toolbar [LH05].

Another possibility for attacking Max-SAT problems is to use pseudo-Boolean solvers [ARMS02, SS06, ES06]. The encoding for Max-SAT presented in Section 3 can be easily adapted to convert the clauses, including the additional weight literals, into pseudo-Boolean constraints. Then, the objective function the pseudo-Boolean solver has to minimize, subject to the pseudo-Boolean constraints, is $w_1 * p_1 + \ldots + w_m * p_m$. Since these solvers are designed to deal with large input pseudo-Boolean problems, they do not incorporate any ad-hoc technique for Max-SAT, which makes them inefficient on the previously mentioned small challenging problems, but competitive on problems whose difficulty is essentially due to its size.

As we will see in the experiments below, the DPLL(T) system we propose here is competitive with the pseudo-Boolean solvers (in fact, it is usually faster), but in addition, due to its modular architecture, it is easy to develop. Once a DPLL(T) system for SMT has been constructed, almost no additional work has to be done to convert it into a tool for Max-SAT. The DPLL(X) engine already incorporates all the necessary machinery and the only thing needed is to implement $Solver_T$ for the theory described in Section 3, something doable in less than 200 lines of C code. It can also easily be adapted in order to incorporate additional pruning rules.

4.2   Additional Pruning Rules

The resulting DPLL(T) system can be further improved by providing it with specialized deduction rules for Max-SAT.

Example 8. If, due to conflict-driven lemma learning, DPLL(X) learns a clause consisting only of positive-weight literals $p_1 \vee \ldots \vee p_n$, each p_i with its associated weight w_i, one can immediately add $min\{w_1 \ldots w_n\}$ to the current cost of the assignment.

Some more complicated resolution-like rules were studied and shown to be very effective in [LH05]. Hence, we have investigated up to what extent such specialized resolution rules can be incorporated into our architecture.

Example 9. Assume our input formula contains, among many others, the binary clauses $l \vee p_1$ and $\neg l \vee p_2$, where each p_i is a literal of weight w_i. Since any total assignment will contain either l or $\neg l$, it is easy to observe that any model will have a cost of at least $min\{w_1, w_2\}$.

All the other rules presented in [LH05] are similar forms of resolution. In order to implement them, one should detect binary or ternary clauses to which resolution is applicable, and for greater effectivity, this should be done on the fly, not only as a preprocessing step. This is quite expensive in general for DPLL-based systems using the two-watched literal scheme [MMZ+01]. In this scheme, one can detect newly generated unit clauses by only watching two literals in each clause, but the detection of binary clauses would require to watch three literals per clause. However, special situations like the one in Example 9 are still detectable with only two watched literals per clause if one watches a positive literal weight p only if there is no other possibility. This restriction ensures that, as soon as such a literal p becomes watched, we have found a binary clause of the form $l \vee p$.

With this small modification, DPLL(X) can efficiently detect the presence of binary clauses of the form $l \vee p_1$ and $\neg l \vee p_2$ and then notify to $Solver_T$ an increment in the cost of the current assignment of $m=min\{w_1, w_2\}$, thus allowing $Solver_T$ to further prune partial assignments that have no possibility to improve the current upper bound. Since part of the weights w_1 and w_2 has already been amortized in the cost of the assignment, $Solver_T$ also has to be notified that the weights of p_1 and p_2 now become $w_1 - m$ and $w_2 - m$, respectively.

4.3   Experimental Evaluation

Experiments have been done on several well-known already existing benchmark families. The DIMACS suite consists of unsatisfiable propositional formulas with a weight of 1 for each clause, similarly to the Weighted DIMACS family, with random weights between 1 and 1000 for each clause [dGLMS03]. Finally, the Quasi-group instances[1] encode quasi-group completion problems in which the clauses enforcing the quasi-group structure and that some cells must contain a given element have been given a certain weight.

We compare with three other systems: Toolbar [dGHZL05], a weighted-CSP solver which incorporates specialized algorithms and data structures for the Max-SAT problem; Pueblo [SS06], a pseudo-Boolean solver implementing a branch-and-bound approach to minimize a given goal function; and Minisat+ [ES06], a pseudo-Boolean solver based on translations to propositional satisfiability. This is by no means an exhaustive comparison with all available tools. We chose three tools that —we believe— represent the state of the art in these three different approaches, and that can handle problems of a reasonable size.

We ran our system in two settings: the basic one (General DPLL(T) in the table) and one implementing the specialized deduction rule mentioned in Example 9 (Special DPLL(T)). In none of them specialized heuristics were developed.

---

[1] We thank Felip Manyà for providing us with them.

We used our standard branching heuristic for solving general SMT problems, an extension of VSIDS [MMZ+01].

Results are in seconds and aggregated per family of benchmarks. Each benchmark was run on a 2GHz 512MB Pentium 4 for 10 minutes, i.e. 600 seconds. An annotation of the form $(n\ t)$ indicates that the system timed out in n benchmarks. Each timeout is counted as 600s in the table.

Benchmark family	#pblms.	Toolbar	Minisat+	Pueblo	General DPLL(T)	Special. DPLL(T)
DIMACS:						
aim	24	(15 t) 9397	**0.5**	**0.5**	**0.5**	**0.5**
bridge fault	4	(4 t) 2400	220	73	**0.3**	0.6
dubois	13	(9 t) 6442	**0.4**	1	0.9	0.97
hole	5	(1 t) 795	102	**57**	470	(1 t) 605
jnh	30	**587**	(2 t) 5433	(3 t) 3798	(1 t)1485	948
pret	8	(4 t) 2782	0.38	0.4	0.28	**0.27**
ssa	4	(4 t) 2400	(1 t) 626	(1 t) 601	(1t)601	(1 t) **601**
Weighted DIMACS:						
wjnh	30	105	(19 t) 14025	1415	**42.6**	54.3
Quasi-groups:						
Size 6	100	(5 t) 4194	16	2.3	**1.2**	3.5
Size 7	100	(69 t) 46202	178	6.4	**2.9**	13
Size 8	100	(100 t) 60000	582	17	**9.1**	41
Size 9	100	(100 t) 60000	1599	47	**29**	157
Size 10	100	(100 t) 60000	4783	203	**151**	668

These experiments only have a relative value, given the fact that tools such as Minisat+ or Pueblo are designed to attack the broader class of pseudo-boolean optimization problems, and also because Toolbar's specialized Max-SAT algorithms are more tailored towards small but challenging Max-SAT problems.

One can also observe from the results that the integration of specialized deduction rules was not always successful. In some benchmarks, like Quasi-groups, the rule we implemented did never apply. In this case one pays a non-negligible overhead (the system gets as much as 5 times slower) without obtaining any gain. On the other hand, in the benchmarks where it was more productive (e.g. in the jnh family, where on average at least one rule application was possible out of every 10 decisions) the overhead was compensated by a reduction in the search space. We believe it is still unclear whether to pursue the integration of such rules is worthwhile. A more careful analysis should be made on more realistic benchmarks, coming from real applications where weights are not assigned at random.

5   Max-SMT: The Example of Difference Logic

We now show how this approach can be smoothly extended to Max-SMT. We focus here on the case of *weighted Max-SMT* modulo the theory of Integer Difference Logic, a fragment of integer linear arithmetic, but our approach is not limited to this theory.

In Integer Difference Logic, formulas are built over propositional atoms, as well as (ground) atoms of the form $a - b \leq k$, where a and b are (Skolemized) integer variables and k is an integer[2]. This logic is used in the context of hardware and software verification; for instance, some properties of timed automata are naturally expressed in it. But here we will show how a problem that a priory doesn't look well-suited for Difference Logic can be encoded in it, and how also optimization problems can be solved using our approach.

5.1   Encoding the CELAR Problems in Integer Difference Logic

The well-known CELAR Radio Link Frequency Assignment problems [CdGL+99] consist of, given a set of radio links and a set of radio frequencies, assigning a frequency to each radio link. For some pairs of radio links, their frequencies must be at a certain exact distance, and for others, they must be at least at a certain distance. The latter constraints are soft, i.e., each one of them has an associated cost if it is not satisfied, and the solution with minimum total cost has to be found. Here we shortly explain how we encode these problems as a Max-SMT problem for Integer Difference Logic.

Each radio link l_i has a finite set of available frequencies D_i that can always be seen as the disjoint union of four sets. For example:

$$\{2 + 14k \mid 1 \leq k \leq 11\} \qquad \{2 + 14k \mid 18 \leq k \leq 28\}$$
$$\{8 + 14k \mid 29 \leq k \leq 39\} \qquad \{8 + 14k \mid 46 \leq k \leq 56\}$$

This observation is crucial to express in a compact manner that the frequency f_i for the radio link l_i has to be in D_i. For that purpose, we will encode the value of f_i using two variables: a propositional variable t_i expressing whether f_i is 2 modulo 14 or not, and an integer variable m_i, representing $f_i \bmod 14$. With these additional variables we can express that f_i is in D_i with the formulas:

$$t_i \rightarrow (1 \leq m_i \leq 11 \ \lor \ 18 \leq m_i \leq 28)$$
$$\neg t_i \rightarrow (29 \leq m_i \leq 39 \ \lor \ 46 \leq m_i \leq 56)$$

It now remains to encode distance constraints of the form $|f_i - f_j| > k$, with their costs w_{ij}. Our encoding of f_i and f_j using the auxiliary variables makes it natural to reason by cases, depending on whether f_i and f_j are 2 modulo 14 or not. For example if f_i is 2 modulo 14 and f_j is not (and hence is 8 modulo 14), then $|f_i - f_j| > k$ is equivalent to $|2 + 14m_i - 8 - 14m_j| > k$. After the necessary manipulations, the corresponding Difference Logic clause is:

$$(t_i \land \neg t_j) \rightarrow \left(m_i - m_j \geq \left\lfloor \frac{k+6}{14} \right\rfloor + 1 \ \lor \ m_i - m_j \leq \left\lceil \frac{-k+6}{14} \right\rceil - 1 \right)$$

The exact distance constraints are encoded similarly. Costs are expressed by additional weight literals as explained in Section 3.

---

[2] Atoms of the form $a \leq k$ are also allowed because one can use an auxiliary integer variable z_0, and consider the inequality $a - z_0 \leq k$ instead. It is not difficult to see that this transformation preserves T-satisfiability. See [NO05] for details on our DPLL(T) system for Difference Logic.

5.2   Experimental Evaluation

As before, from our BarcelogicTools DPLL(T) implementation we obtained, with little effort, our first Max-SMT system.

We compared our system with the best-known weighted CSP solver Toolbar [dGHZL05] in three different settings. The first one used a static branching heuristic, the second one was the default setting for Toolbar and the third one, the best possible choice according to the authors, used a Jeroslow-like branching heuristic. We used the same machine as before and again results are in seconds.

Benchmark	Toolbar			DPLL(T)
name	Static	Default	Jeroslow	
SUBCELAR_6_0	**0.2**	0.4	**0.2**	5
SUBCELAR_6_1	85	252	**65**	90
SUBCELAR_6_2	127	982	**25**	132
SUBCELAR_6_3	7249	2169	**355**	636
SUBCELAR_6_4	7021	> 5 hours	1942	**1417**

The choice of different branching heuristics leads to dramatic changes in Toolbar's runtime. Hence, we believe that specialized heuristics for DPLL(T) could still improve its performance. From these limited results, one cannot infer that DPLL(T) has better scaling properties than Toolbar in its best setting, although it is 20 times slower on the smallest problem, but faster on the largest one. In any case, we believe these results indicate that SMT tools can be already used for efficiently solving industrial optimization problems.

We also carefully translated these problems into weighted Max-SAT and pseudo-Boolean problems. Somewhat to our surprise, the tools mentioned in Section 4 needed around 30 seconds on the smallest of these problems, and did not terminate in a day on the second smallest one. These problems are also not known to be tractable by means of translations into pure Integer Linear Programming, in spite of attempts using the best ILP solvers (Javier Larrosa, private communication).

6   Conclusions and Further Work

By developing DPLL(T) techniques for weighted Max-SAT, we have shown that DPLL(T) can be very competitive for problems that do not look a priori like typical SMT problems. Since this was achieved with relatively little effort, we see this as an indication of the quality, in terms of efficiency and flexibility, of our approach.

The success of our Max-SMT implementation on the CELAR benchmarks reveals that realistic problems can be modeled as Max-SMT problems and solved with small variants of SMT solvers. Effectivity of SMT solvers for that purpose has also been recently shown in [SPSP05], using SMT over another fragment of linear arithmetic, for solving soft temporal constraints, where extensive experiments were done on random problems.

Future work concerns other problems that are not typical SMT-like. For example, we are currently investigating the use of DPLL(T) for expressing finite-domain constraints, and in particular global constraints from the constraint programming world, such as `alldifferent`.

References

[ARMS02]   F. A. Aloul, A. Ramani, I. L. Markov, and K. A. Sakallah. PBS: A backtrack-search pseudo-boolean solver and optimizer. In *SAT'02*, LNCS, pp. 346–353.

[BdMS05]   C. Barrett, L. de Moura, and A. Stump. SMT-COMP: Satisfiability Modulo Theories Competition. In *CAV'05*, LNCS 3576, pp. 20–23. http://www.csl.sri.com/users/demoura/smt-comp/

[CdGL+99] B. Cabon, S. de Givry, L. Lobjois, T. Schiex, and J. P. Warners. Radio link frequency assignment. *Constraints*, 4(1):79–89, 1999.

[CMSM04]   S. Colton, A. Meier, V. Sorge, and R. McCasland. Automatic generation of classification theorems for finite algebras. In *IJCAR'04*, LNCS 3097, pp. 400–414.

[dGHZL05] S. de Givry, F. Heras, M. Zytnicki, and J. Larrosa. Existential arc consistency: Getting closer to full arc consistency in weighted CSPs. In *IJCAI'05*, pp. 84–89, 2005.

[dGLMS03] Simon de Givry, Javier Larrosa, Pedro Meseguer, and Thomas Schiex. Solving max-sat as weighted csp. In *CP'03*, LNCS 2833, 363–376, 2003.

[DLL62]   M. Davis, G. Logemann, and D. Loveland. A machine program for theorem-proving. *Comm. of the ACM*, 5(7):394–397, 1962.

[DP60]   M. Davis and H. Putnam. A computing procedure for quantification theory. *Journal of the ACM*, 7:201–215, 1960.

[ES06]   N. Eén and N. Sörensson. Translating pseudo-boolean constraints into SAT. *Journal on Satisfiability, Boolean Modeling and Computation*, 2:1–26, 2006.

[GHN+04]   H. Ganzinger, G. Hagen, R. Nieuwenhuis, A. Oliveras, and C. Tinelli. DPLL(T): Fast Decision Procedures. In *CAV'04*, LNCS 3114, 175–188.

[LH05]   J. Larrosa and F. Heras. Resolution in Max-SAT and its relation to local consistency in weighted CSPs. In *IJCAI'05*, pp. 193–198, 2005.

[LMP05]   C. Li, F. Manyà, and J. Planes. Solving Over-Constrained Problems with SAT. In *CP'05*, LNCS 3709, pp. 403–414.

[MMZ+01]   M. W. Moskewicz, C. F. Madigan, Y. Zhao, L. Zhang, and S. Malik. Chaff: Engineering an Efficient SAT Solver. In *DAC'01*, 2001.

[NO05]   R. Nieuwenhuis and A. Oliveras. DPLL(T) with Exhaustive Theory Propagation and its Application to Difference Logic. In *CAV'05* LNCS 3576, pp. 321–334.

[NOT05]   R. Nieuwenhuis, A. Oliveras, and C. Tinelli. Abstract DPLL and Abstract DPLL Modulo Theories. In *LPAR'04*, LNAI 3452, pp. 36–50.

[SPSP05]   H. Sheini, B. Peintner, K. Sakallah, and M. Pollack. On solving soft temporal constraints using SAT techniques. In *CP'05*, LNCS 3709, pp. 607–621.

[SS06]   H. M. Sheini and K. A. Sakallah. Pueblo: A hybrid pseudo-boolean SAT solver. *J. Satisfiability, Boolean Modeling and Comp.*, 2:165–189, 2006.

[TR05]   C. Tinelli and S. Ranise. SMT-LIB: The Satisfiability Modulo Theories Library, July 2005. http://goedel.cs.uiowa.edu/smtlib/.

[XZ05]   Z. Xing and W. Zhang. Maxsolver: an efficient exact algorithm for (weighted) maximum satisfiability. *Artif. Intell.*, 164(1-2):47–80, 2005.

Fast and Flexible Difference Constraint Propagation for DPLL(T)

Scott Cotton and Oded Maler

Verimag, Centre Équation
2, Avenue de Vignate
38610 Gières, France
{Scott.Cotton, Oded.Maler}@imag.fr

Abstract. In the context of DPLL(T), theory propagation is the process of dynamically selecting consequences of a conjunction of constraints from a given set of candidate constraints. We present improvements to a fast theory propagation procedure for *difference constraints* of the form $x - y \leq c$. These improvements are demonstrated experimentally.

1 Introduction

In this paper, theory propagation refers to the process of of dynamically selecting consequences of a conjunction of constraints from a given set of constraints whose truth values are not yet determined. The problem is central to an emerging method, known as DPLL(T) [10,17,3,18] for determining the satisfiability of arbitrary Boolean combinations of constraints. We present improvements to theory propagation procedure for *difference logic* whose atomic constraints are of the form $x - y \leq c$. Our contribution is summarized below:

1. We introduce flexibility in the invocation times and scope of constraint propagation in the DPLL(T) framework. This feature is theory independent and is described in more detail in [4].
2. We identify conditions for early termination of single source shortest path (SSSP) based incremental propagation. These conditions allow us to ignore nodes in the constraint graph which are not affected by the assignment of the new constraint.
3. We implement a fast consistency check algorithm for difference constraints based on an algorithm presented in [9]. As a side effect, the performance of subsequent theory propagation is improved. We present an adaptation of Goldberg's smart-queue algorithm [11] for the theory propagation itself.
4. We show that incremental complete difference constraint propagation can be achieved in $\mathcal{O}(m + n \log n + |U|)$ time, where m is the number of constraints whose consequences are to be found, n the number of variables in those constraints, and U the set of constraints which are candidates for being deduced. This is a major improvement over the $O(mn)$ complexity of [17].

A. Biere and C.P. Gomes (Eds.): SAT 2006, LNCS 4121, pp. 170–183, 2006.

The rest of this paper is organized as follows. In Section 2 we describe the context of theory propagation in DPLL(T) and review difference constraint propagation in particular. In Section 3, we present a simplified consistency checker which also speeds up subsequent propagation. Our propagation algorithm including the early termination feature is presented in Section 4; experimental evidence is presented in Section 5; and we conclude in Section 6.

2   Background

Propagation in DPLL(T). The Davis-Putnam-Loveland-Logemann (DPLL) satisfiability solver is concerned with propositional logic [6,5]. Its input formula φ is assumed to be in conjunctive normal form. Given such a formula, a DPLL solver will search the space of truth assignments to the variables by incrementally building up partial assignments and backtracking whenever a partial assignment falsifies a clause. Assignments are extended both automatically by *unit propagation* and by *guessing* truth values. Unit propagation is realized by keeping track of the effect of partial assignments on the clauses in the input formula. For example, the clause $x \vee \neg y$ is solved under an assignment including $y \mapsto \texttt{false}$ and is reduced to the clause x under an assignment which includes $y \mapsto \texttt{true}$ but which contains no mapping for x. Whenever a clause is reduced to contain a single literal, *i.e.* it is in the form v or the form $\neg v$, the DPLL engine extends the partial assignment to include the truth value of v which solves the clause. If a satisfying assignment is found, the formula is satisfiable. If the procedure exhausts the (guessed) assignment space without finding a satisfying assignment, the formula is not satisfiable.

DPLL(T) is a DPLL-based decision procedure for satisfiability modulo theories. The DPLL(T) framework extends a DPLL solver to the case of a Boolean combination of constraints which are interpreted with respect to some background theory T. An external theory-specific solver called Solver$_T$ is invoked at each assignment extension in the DPLL procedure. Solver$_T$ is responsible for checking the consistency of the assignment with respect to the theory T. If an assignment is inconsistent, the DPLL procedure backtracks just as it would if an empty clause were detected. In addition, a central feature of DPLL(T) is that Solver$_T$ may also find T-consequences of the assignment and communicate them to the DPLL engine. This latter activity, called *theory propagation*, is intended to help guide the DPLL search so that the search is more informed with respect to the underlying theory. Theory propagation is thus *interleaved* with unit propagation and moreover the two types of propagation feedback into one another. This quality gives the resulting system a strong potential for reducing the guessing space of the DPLL search. Consequently, DPLL(T) is an effective framework for satisfiability modulo theories [10,3,17].

Flexible Propagation. We define an interface to Solver$_T$ which allow it to interact with the DPLL engine along *any such interleaving* of unit propagation and theory propagation. Below are three methods to be implemented by Solver$_T$, which can be called in any sequence:

SetTrue. This method is called with a constraint c every time the DPLL engine extends the partial assignment A with c. Solver$_T$ is expected to indicate whether or not $A \cup \{c\}$ is T-consistent. If consistent, an empty set is returned. Otherwise, an inconsistent subset of $A \cup \{c\}$ is returned.

TheoryProp. This method returns a set C of T-consequences of the current assignment A. A set of reasons $R_c \subseteq A$ is associated with each $c \in C$, satisfying $R_c \models c$. Unlike the original DPLL(T) of [10], the method is entirely decoupled from SetTrue.

Backtrack. This method simply indicates which assigned constraints become unassigned as a result of backtracking.

The additional flexibility of the timing of occurences of calls to SetTrue in relation to occurences of calls to TheoryProp allows the system to propagate constraints either *eagerly* or *lazily*. Eager propagation follows every call to SetTrue with a call to TheoryProp. Lazy propagation calls TheoryProp only after a sequence of calls to SetTrue, in particular when unit propagation is not possible.

Whatever the sequence of calls, it is often convenient to identify the source of an assignment in the method SetTrue. At the same time, it is the DPLL engine which calls these methods and we would like to minimize its responsibilities to facilitate using off-the-shelf DPLL solvers along with the host of effective optimizations associated with them. Hence, we do not require that the DPLL engine keep track of the source of every assignment. Rather we allow it to treat constraints more or less the same way it treats propositional literals, and put the burden on Solver$_T$ instead.

Towards this end, we have Solver$_T$ associate an annotation α_c with each constraint (or its negation) which appears in a problem. In addition to tracking the origin of constraints passed to SetTrue, the annotation is used to keep track of a set of assigned constraints whose consequences have been found. The annotation can take any value from $\{\Pi, \Sigma, \Delta, \Lambda\}$ with the following intended meanings.

Π: Constraints whose consequences have been found (propagated constraints).
Δ: Constraints which have been identified as consequences of constraints labelled Π.
Σ: Constraints which have been assigned, but whose consequences have not been found yet.
Λ: Unassigned constraints.

For convenience, we use the labels Π, Δ, Σ and Λ interchangeably with the set of constraints which have the respective label.

It is fairly straightforward to maintain labels with these properties via the methods SetTrue, TheoryProp, and Backtrack. Whenever a constraint is passed to SetTrue which is labelled Λ we label it Σ and perform a consistency check. Whenever TheoryProp is called, constraints labelled Σ are labelled Π one at a time. After each such relabelling, all the constraints in $\Sigma \cup \Lambda$ which are consequences of constraints labelled Π are re-labelled Δ. Whenever backtracking occurs, all constraints which become unassigned together with all consequences which are not yet assigned are labelled Λ.

As explained at length in [4] for the more general context of theory decomposition, such constraint labels provide for a more flexible form of theory propagation which, in particular, exempts constraints labelled Δ from consistency checks and from participating as antecedents in theory propagation. This feature reduces the cost of theory propagation without changing its outcome, and is independent of the theory and propagation method used.

2.1   Difference Constraints and Graphs

Difference constraints can express naturally a variety of timing-related problems including schedulability, circuit timing analysis, and bounded model checking of timed automata [16,3]. In addition, difference constraints can be used as an abstraction for general linear constraints and many problems involving general linear constraints are dominated by difference constraints. Difference constraints are also much more easily decided than general linear constraints, in particular using the following convenient graphical representation.

Definition 1 (Constraint graph). *Let S be a set of difference constraints and let G be the graph comprised of one weighted edge $x \xrightarrow{c} y$ for every constraint $x - y \leq c$ in S. We call G the constraint graph of S.*

The constraint graph may be readily used for consistency checking and constraint propagation, as is indicated in the following well-known theorem.

Theorem 1. *Let Γ be a conjunction of difference constraints, and let G be the constraint graph of Γ. Then Γ is satisfiable if and only if there is no negative cycle in G. Moreover, if Γ is satisfiable, then $\Gamma \models x - y \leq c$ if and only if y is reachable from x in G and $c \geq d_{xy}$ where d_{xy} is the length of a shortest path from x to y in G.*

As the semantics of conjunctions of difference constraints are so well characterized by constraint graphs, we refer to sets of difference constraints interchangeably with the corresponding constraint graph. In this way, we also further abuse the notation associated with constraint labels introduced in section 2. In particular, the labels Π, Σ, Δ, and Λ are used not only to refer to the set of difference constraints with the respective label, but also to the constraint graph induced by that set of constraints. We also often refer to a difference constraint $x - y \leq c$ by an edge $x \xrightarrow{c} y$ in a constraint graph and *vice versa*.

3   Consistency Checks

In accordance with Theorem 1, one way to show that a set of difference constraints Γ is consistent is to show that Γ's constraint graph G contains no negative cycle. This in turn can be accomplished by establishing a *valid potential function*, which is a function π on the vertices of a graph satisfying $\pi(x) + c - \pi(y) \geq 0$ for every edge $x \xrightarrow{c} y$ in G. A valid potential function may readily be used to establish lower bounds on shortests path lengths between arbitrary vertices (v_1, v_n):

$$\Sigma_{i=1}^{n-1} \pi(v_i) + c_i - \pi(v_{i+1}) \geq 0$$
$$\pi(v_1) - \pi(v_n) + \Sigma_{i=1}^{n-1} c_i \geq 0$$
$$\Sigma_{i=1}^{n-1} c_i \geq \pi(v_n) - \pi(v_1)$$

If one considers the case that $v_1 = v_n$, it follows immediately that the existence of a valid potential function guarantees that G contains no negative cycles. In addition, a valid potential function for a constraint graph G defines a satisfying assignment for the set Γ of difference constraints used to form G. In particular, if π is a valid potential function for G, then the function $v \mapsto -\pi(v)$ is a satisfying assignment for Γ.

In the DPLL(T) framework, consistency checks occur during calls to SetTrue, when a constraint $u \xrightarrow{d} v$ is added to the set of assigned constraints. If the constraint is labelled Δ, then there is no reason to perform a consistency check. Otherwise, the constraint is labelled Σ. In this latter case, SetTrue must perform a consistency check on the set $\Pi \cup \Sigma$. To solve this problem, we make use of an incremental consistency checking algorithm based largely on an incremental shortests paths and negative cycle detection algorithm due to Frigioni[1] et al [9]. Before detailing this algorithm, we first formally state the incremental consistency checking problem in terms of constraint graphs and potential functions:

Definition 2 (Incremental Consistency Checking). *Given a directed graph* G *with weighted edges, a potential function* π *satisfying* $\pi(x) + c - \pi(y) \geq 0$ *for every edge* $x \xrightarrow{c} y$, *and an edge* $u \xrightarrow{d} v$ *not in* G, *find a potential function* π' *for the graph* $G' = G \cup \{u \xrightarrow{d} v\}$ *if one exists.*

The complete algorithm for this problem is given in pseudocode in Figure 1. The algorithm maintains a function γ on vertices which holds a conservative estimate on how much the potential function must change if the set of constraints is consistent. The function γ is refined by scanning outgoing edges from vertices for which the value of π' is known.

3.1   Proof of Correctness and Run Time

Lemma 1. *The value* $\min(\gamma)$ *is non-decreasing throughout the procedure.*

Proof. Whenever the algorithm updates $\gamma(z)$ to $\gamma'(z) \neq \gamma(z)$ for some vertex z, it does so either with the value 0, or with the value $\pi'(s) + c - \pi(t)$ for some

---

[1] The algorithm and its presentation here are much simpler primarily because Frigioni et al. maintain extra information in order to solve the fully dynamic shortests paths problem, whereas this context only demands incremental consistency checks. In particular, we do not compute single source shortests paths, but rather simply use a potential function which reduces the graph size.

$$\gamma(v) \leftarrow \pi(u) + d - \pi(v)$$
$$\gamma(w) \leftarrow 0 \text{ for all } w \neq v$$
while $\min(\gamma) < 0 \wedge \gamma(u) = 0$
    $s \leftarrow \text{argmin}(\gamma)$
    $\pi'(s) \leftarrow \pi(s) + \gamma(s)$
    $\gamma(s) \leftarrow 0$
    for $s \xrightarrow{c} t \in G$ **do**
        if $\pi'(t) = \pi(t)$ **then**
            $\gamma(t) \leftarrow \min\{\gamma(t), \pi'(s) + c - \pi(t)\}$

Fig. 1. Incremental consistency checking algorithm, invoked by `SetTrue` for a constraint $u \xrightarrow{d} v$ labelled Λ. If the outer loop terminates because $\gamma(u) < 0$, then the set of difference constraints is not consistent. Otherwise, once the outer loop terminates, π' is a valid potential function and $-\pi'$ defines a satisfying assignment for the set of difference constraints.

edge $s \xrightarrow{c} t$ in G such that $t = z$. In the former case, we know $\gamma(z) < 0$ by the termination condition, and in the latter we have $\gamma'(z) = \pi'(s) + c - \pi(t) = (\pi(s) + c - \pi(t)) + \gamma(s) \geq \gamma(s)$, since $\pi(s) + c - \pi(t) \geq 0$. $\qquad \square$

Lemma 2. *Assume the algorithm is at the beginning of the outer loop. Let z be any vertex such that $\gamma(z) < 0$. Then there is a path from u to z with length $L(z) = \pi(z) + \gamma(z) - \pi(u)$.*

Proof. (sketch) By induction on the number of times the outermost loop is executed. $\qquad \square$

Theorem 2. *The algorithm correctly identifies whether or not G' contains a negative cycle. Moreover, when there is no negative cycle the algorithm establishes a valid potential function for G'.*

Proof. We consider the various cases related to termination.

- **Case 1.** $\gamma(u) < 0$. From this it follows that $L(u) < 0$ and so there is a negative cycle. In this case, since the DPLL engine will backtrack, the original potential function π is kept and π' is discarded.
- **Case 2.** $\min(\gamma) = 0$ and $\gamma(u) = 0$ throughout.
  In this case we claim π' is a valid potential function. Let γ_i be the value of γ at the beginning of the ith iteration of the outer loop. We bserve that $\forall v \,.\, \pi'(v) \leq \pi(v)$ and consider the following cases.
  - For each vertex v such that $\pi'(v) < \pi(v)$, $\pi'(v) = \pi(v) + \gamma_i(v)$ for some refinement step i. Then for every edge $v \xrightarrow{c} w \in G$, we have that $\gamma_{i+1}(w) \leq \pi'(v) + c - \pi(w)$ and so $\pi'(w) \leq \pi(w) + \gamma_{i+1}(w) \leq \pi(w) + \pi'(v) + c - \pi(w) = \pi'(v) + c$. Hence $\pi'(v) + c - \pi'(w) \geq 0$.
  - For each vertex v such that $\pi'(v) = \pi(v)$, we have $\pi'(v) + c - \pi'(w) = \pi(v) + c - \pi'(w) \geq \pi(v) + c - \pi(w) \geq 0$ for every $v \xrightarrow{c} w \in G$
  We conclude π' is a valid potential function with respect to all edges in G'.

In all cases, the algorithm either identifies the presence of a negative cycle, or it establishes a valid potential function π'. As noted above, a valid potential function precludes the existence of a negative cycle. $\quad\square$

Theorem 3. *The algorithm runs in time[2] $\mathcal{O}(m + n \log n)$.*

Proof. The algorithm scans every vertex once. If a Fibonacci heap is used to find $\operatorname{argmin}(\gamma)$ at each step, and for decreases in γ values, then the run time is $\mathcal{O}(m + n \log n)$. $\quad\square$

3.2 Experiences and Variations

For simplicity, we did not detail how to identify a negative cycle if the set of constraints is inconsistent. A negative cycle is a minimal inconsistent set of constraints and is returned by `SetTrue` in the case of inconsistency. Roughly speaking, this can be accomplished by keeping track of the last edge $x \overset{c}{\to} y$ along which $\gamma(y)$ was refined for every vertex. Then every vertex in the negative cycle will have such an associated edge, those edges will form the negative cycle and may easily recovered.

In practice we found that the algorithm is much faster if we maintain for each vertex v a bit indicating whether or not its new potential $\pi'(v)$ has been found. With this information at hand, it is straightforward to update a single potential function rather than keeping two. In addition, this information can readily be used to skip the $\mathcal{O}(n)$ initialization of γ and to keep only vertices v with $\gamma(v) < 0$ in the priority queue. We found that the algorithm ran faster with a binary priority queue than with a Fibonacci heap, and also a bit faster when making use of Tarjan's subtree-enumeration trick [20,1] for SSSP algorithms. Profiling benchmark problems each of which invokes hundreds of thousands of consistency checks indicated that this procedure was far less expensive than constraint propagation in the DPLL(T) context, although the two have similar time complexity.

4   Propagation

The method `TheoryProp` described in Section 2 is responsible for constraint propagation. The procedure's task is to find a set of consequences C of the current assignment A, and a set of reasons R_c for each consequence $c \in C$. For difference constraints, by Theorem 1, this amounts to computing shortests paths in a constraint graph.

We present a complete incremental method for difference constraint propagation which makes use of the constraint labels Π, Σ, Δ, and Λ. The constraint propagation is divided into incremental steps, each of which selects a constraint c labelled Σ, relabels c with Π, and then finds the consequences of those constraints labelled Π from the set $\Sigma \cup \Lambda$, labelling them Δ. A single step may

---

[2] Whenever stating the complexity of graph algorithms, we use n for the number of vertices in the graph and m for the number of edges.

or may not find unassigned consequences. On every call to `TheoryProp`, these incremental steps occur until either there are no constraints labelled Σ, or some unassigned consequences are found. Any unassigned consequences are returned to the DPLL(X) engine for assignment and further unit propagation. We state the problem of a single incremental step in terms of constraint graphs and shortests paths below.

Definition 3 (Incremental complete difference constraint propagation). *Let G, H be two edge disjoint constraint graphs, and let $x \xrightarrow{c} y \in H$ be a distinguished edge. Suppose that for every edge $u \xrightarrow{d} v \in H$, the length of a shortest path from u to v in G exceeds d. Let $G' = G \cup \{x \xrightarrow{c} y\}$ and $H' = H \setminus \{x \xrightarrow{c} y\}$. Find the set of all edges $u \xrightarrow{d} v$ in H' such that the length of a shortest path from u to v in G' does not exceed d.*

The preconditions relating the graphs G and H are a result of labelling and complete propagation. If all consequences of G are found and removed from H prior to every step, then no consequnces of G are found in H and so the length of a shortest path from x to y in G exceeds the weight of any edge $u \xrightarrow{d} v \in H$.

As presented by Nieuwenhuis *et al.* [17], this problem may be reduced to solving two SSSP problems. First, for the graph G', the SSSP weights $\delta_y^{\rightarrow}$ from y are computed and then SSSP weights $\delta_x^{\leftarrow}$ to x are computed, the latter being accomplished simply by computing $\delta_x^{\rightarrow}$ in the reverse graph. Then for any constraint $u \xrightarrow{d} v \in H'$, the weight of the shortests path from u to v passing through $x \xrightarrow{c} y$ in G' is given in constant time by $\delta_x^{\leftarrow}(u) + c + \delta_y^{\rightarrow}(v)$. In accordance with Theorem 1, the weight of this path determines whether or not the constraint $u \xrightarrow{d} v$ is implied, in particular by the condition $\delta_x^{\leftarrow}(u) + c + \delta_y^{\rightarrow}(v) \le d$. It then suffices to check every constraint in H' in this fashion. We now present several improvements to this methodology.

4.1   Completeness, Candidate Pruning, and Early Termination

A slight reformulation of the method above allows for *early termination* of the SSSP computations under certain conditions. That is, nodes for which it becomes clear that their minimal distance will not be improved due to the insertion of $x \xrightarrow{c} y$ to the constraint graph will not be explored. We introduce the idea of *relevancy* below to formalize how we can identify such vertices and give an example in Figure 2. For a new edge $x \xrightarrow{c} y$, relevancy is based on shortest path distances $\delta_x^{\rightarrow}$ (from x) and $\delta_y^{\leftarrow}$ (to y), in contrast to the formulation above. Under this new formulation, if the shortest path from u to v passes through $x \xrightarrow{c} y$, then the path length is $\delta_y^{\leftarrow}(u) + \delta_x^{\rightarrow}(v) - c$.

Definition 4 (δ-relevancy with respect to $x \xrightarrow{c} y$). *A vertex z is $\delta_x^{\rightarrow}$-relevant if every shortest path from x to z passes through $x \xrightarrow{c} y$; similarly, a vertex z is $\delta_y^{\leftarrow}$-relevant if every shortest path from z to y passes through $x \xrightarrow{c} y$. A constraint $u \xrightarrow{d} v$ is δ-relevant if both u is $\delta_y^{\leftarrow}$-relevant and v is $\delta_x^{\rightarrow}$-relevant. A set C of constraints is δ-relevant if every $u \xrightarrow{d} v \in C$ is δ-relevant.*

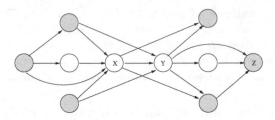

Fig. 2. An example graph showing δ-relevant vertices with respect to the edge (x, y). For simplicity, all edges are assumed to have weight 1. The relevant vertices are white and the irrelevant vertices are shaded. As an example, the vertex z is not $\delta_x^{\rightarrow}$-relevant because there is a shortest path from x to z which does not pass through y. As a result, any constraint $u \xrightarrow{d} z \in H'$ is not member of the incremental complete propagation solution set.

Lemma 3. *The solution set for complete incremental difference constraint propagation is δ-relevant.*

Proof. Let $x \xrightarrow{c} y$ be the new edge in G, and suppose for a contradiction that some constraint $u \xrightarrow{d} v \in H'$ in the solution set is not δ-relevant. Then the length of a shortest path from u to v in G' does not exceed d. By definition of δ-relevancy, some path p from u to v in G' which does not pass through $x \xrightarrow{c} y$ is at least as short as the shortest path from u to v passing through $x \xrightarrow{c} y$. Observe that p is a path in G. By the problem definition, $u \xrightarrow{d} v \notin H$ and $H' \subset H$. Hence $u \xrightarrow{d} v \notin H'$, a contradiction.   □

Corollary 1 (Early Termination). *It suffices to check every δ-relevant member of H' for membership in the solution set. As a result, each SSSP algorithm computing $\delta \in \{\delta_x^{\rightarrow}, \delta_y^{\leftarrow}\}$ need only compute correct shortests path distances for δ-relevant vertices.*

Early termination is fairly easy to implement with most SSSP algorithms in the incremental constraint propagation context. First, for $\delta \in \{\delta_x^{\rightarrow}, \delta_y^{\leftarrow}\}$, we maintain a label for each vertex indicating whether or not it is δ-relevant. We then define an order $\prec$ over shortest path distances of vertices in a way that favors irrelevancy:

$$\delta(u) \prec \delta(v) \iff \delta(u) < \delta(v) \text{ or } \begin{cases} \delta(u) = \delta(v) \\ u \text{ is } \delta\text{-irrelevant} \\ v \text{ is } \delta\text{-relevant} \end{cases}$$

Since the new constraint $x \xrightarrow{c} y$ is a unique shortest path, we initially give y the label $\delta_x^{\rightarrow}$-relevant and x the label $\delta_y^{\leftarrow}$-relevant. During the SSSP computation of δ, whenever an edge $u \xrightarrow{d} v$ is found such that $\delta(u) \prec \delta(v) + d$, the distance to v is updated and we propagate u's δ-relevancy label to v. If at any point in time all such edges are not δ-relevant, then the algorithm may terminate.

To facilitate checking only δ-relevant constraints in H', one may adopt a trick described in [17] for checking only a reachable subset of H'. In particular, one may maintain the constraint graph H' in an adjacency list form which allows iteration over incoming and outgoing edges for each vertex as well as finding the in- and out-degree of each vertex. If the sets of $\delta_x^{\rightarrow}$-relevant and $\delta_y^{\leftarrow}$-relevant vertices are maintained during the SSSP algorithm, the smaller of these two sets, measured by total in- or out-degree in H', may be used to iterate over a subset of constraints in H' which need to be checked.

4.2 Choice of Shortests Path Algorithm

There are many shortest path algorithms, and it is natural to ask which one is best suited to this context. An important observation is that whenever shortests paths $\delta_x^{\rightarrow}$ or $\delta_y^{\leftarrow}$ are computed, the graph G has been subject to a consistency check. Consistency checks establish a potential function π which can be used to speed up the shortests path computations a great deal. In particular, as was first observed by Johnson [13], we can use $\pi(x) + c - \pi(y)$ as an alternate, non-negative edge weight for each edge $x \overset{c}{\rightarrow} y$. This weight is called the *reduced cost* of the edge. The weight w of path p from a to b under reduced costs is non-negative and the original weight of p, that is, without using reduced costs, is easily retrieved as $w + \pi(b) - \pi(a)$. Our implementation of constraint propagation exploits this property by using an algorithm for shortests paths on graphs with *non-negative edge weights*. The most common such algorithm is Dijkstra's [7], which runs in $\mathcal{O}(m + n \log n)$ time. This is an improvement over algorithms which allow arbitrary edge weights, the best of which run in $\mathcal{O}(mn)$ time [2]. A direct result follows.

Theorem 4. *Complete incremental difference constraint propagation can be accomplished in $\mathcal{O}(m + n \log n + |H'|)$ time where m is the number of assigned constraints, n the number of variables occuring in assigned constraints, and H' is the set of unassigned constraints.*

Proof. The worst case execution time of finding all consequences over a sequence of calls to SetTrue and TheoryProp, is $\mathcal{O}(m + n \log n + |H'|)$ per call to TheoryProp and $\mathcal{O}(m + n \log n)$ per call to SetTrue. Thus if every call to SetTrue is followed by a call to TheoryProp, then the combined time for both calls is $\mathcal{O}(m + n \log n + |H'|)$. □

4.3 Adaption of a Fast SSSP Algorithm

In order to fully exploit the use of the potential function in constraint propagation, we make use of a state-of-the-art SSSP algorithm for a graph with non-negative edge weights. In particular, we implemented (our own interpretation of) Goldberg's smart-queue algorithm [11]. The application of this algorithm to difference constraint propagation context is non-trivial because it makes use of a heuristic requiring that we keep track of some information for each vertex as the graph Π and its potential function changes. Even in the face of the extra book-keeping the algorithm turns out to run significantly faster than standard implementations of Dijkstra's algorithm with a Fibonacci heap or a binary priority queue.

The smart-queue algorithm is a priority queue based SSSP algorithm for a graph with non-negative edge weights which maintains a priority queue on vertices. Each vertex is prioritized according to the shortest known path from the source to it. The smart-queue algorithm also makes use of the *caliber heuristic* which maintains for each vertex the minimum weight of any edge leading to it. This weight is called the *caliber* of the vertex. After removing the minimum element of distance d from the priority queue, d serves as a lower bound on the distance to all remaining vertices. When scanning a vertex, we know the lower bound d, and if we come accross a vertex v with caliber c_v and tentative distance d_v, we know that the distance d_v is exact if $d + c_v \geq d_v$. Vertices whose distance is known to be exact are not put in the priority queue, and may be removed from the priority queue prematurely if they are already there. The algorithm scans exact vertices greedily in depth first order. When no exact vertices are known it backs off to use the priority queue to determine a new lower bound. The priority queue is based on lazy radix sort, and allows for constant time removal of vertices. For full details, the reader is referred to [11].

In this context, the caliber of a vertex may change whenever either the graph Π or its potential function changes. This in turn requires that the graph Π be calibrated before each call to `TheoryProp`. Calibration may be accomplished with linear cost simply by traversing the graph Π once prior to each such call. However, we found that if, between calls to `TheoryProp`, we keep track of those vertices whose potential changes as well as those vertices which have had an edge removed during backtracking, then we can reduced the cost of recalibration. In particular, the recalibration associated with each call to `TheoryProp` can then be restricted to the subgraph which is reachable in one step from any such affected vertex.

5 Experiments

We present various comparisons between different methods for difference constraint propagation. With one exception, the different methods are implemented in the same basic system: a Java implementation of DPLL(T) which we call Jat. The underlying DPLL solver is fairly standard with two literal watching [14], 1UIP clause learning and VSID+stack heuristics [12] as in the current version of ZChaff [15], and conflict clause minimization as in MiniSat [8]. Within this fixed framework, we present a comparison of reachability-based and relevancy-based early termination in Figure 3 as well as a comparison of lazy and eager strategies in Figure 4. These comparisons are performed on scheduling problems encoded as difference logic satisfiability problems on a 2.4GHz intel based box runnning linux. The scheduling problems are taken from standard benchmarks, predominately from the SMT-LIB QF_RDL section [19]. In Figure 5, we also present a comparison of our best configuration, implemented in Java, against BarceLogicTools (BCLT) which is implemented in C and which, in 2005, won the SMT competition for difference logic.

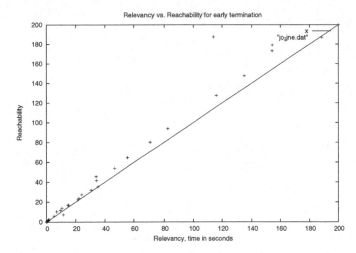

Fig. 3. A comparison of relevancy based early termination and reachability based early termination. The relevancy based early termination is consistently faster and the speed difference is roughly proportional to the difficulty of the underlying problem.

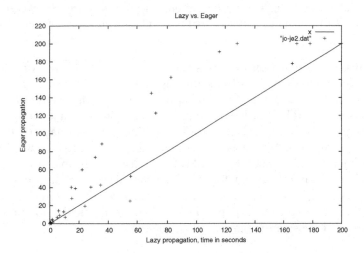

Fig. 4. A comparison of lazizess and eagerness in theory propagation for difference logic. Both lazy and eager implementations use relevancy based early termination and the same underlying SSSP algorithm. The lazy strategy is in general significantly faster than the eager strategy. This difference arises because the eager strategy performs constraint propagation more frequently.

Job-shop scheduling problems encoded as difference logic satisfiability problems, like the ones used in our experiments, are strongly numerically constrained and weakly propositionally constrained. These problems are hence a relatively pure measure of the efficiency of difference constraint propagation. While it

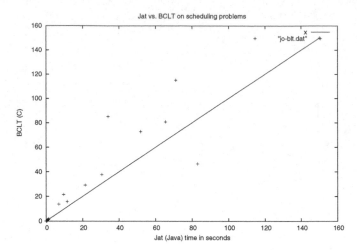

Fig. 5. Jat with lazy propagation and relevancy based early termination compared with BarceLogicTools [17] on job-shop scheduling problems. The Jat propagation algorithm uses consistency checks and Goldberg's smart-queue SSSP algorithm as described in this paper, and is implemented in Java. Assuming BarceLogicTools hasn't changed since [17], it uses no consistency checks, eager propagation, a depth first search SSSP based $\mathcal{O}(mn)$ propagation algorithm, and is implemented in C. The Jat solver is generally faster.

seems that our approach outperforms the others on these types of problems[3], this is no longer the case when Boolean constraints play a stronger role. In fact, BCLT is several times faster than Jat on many of the other types of difference logic problems in SMT-LIB. Upon profiling Jat, we found that on all the non-scheduling problems in SMT-LIB, Jat was spending the vast majority of its time doing unit propagation, whereas in the scheduling problems Jat was spending the vast majority of its time doing difference constraint propagation. Although it is at best difficult to account for the difference in implementations and programming language, this suggests that the techniques for difference constraint propagation presented in this paper are efficient, in particular for problems in which numerical constraints play a strong role.

6   Conclusion

We presented several improvements for difference constraint propagation in SMT solvers. We show that lazy constraint propagation is faster than eager constraint propagation, and that relevancy based early termination is helpful. We presented adaptations of state-of-the-art shortest paths algorithms to the difference constraint propagation context in the DPLL(T) framework. We showed

---

[3] Although we have few direct comparisons, this is suggested by the fact that BCLT did outperform the others in a recent contest on scheduling problems, and that our experiments indicate that our approach outperforms BCLT on the same problems.

experimentally that these improvements taken together make for a fast difference logic solver which is highly competitive on problems dominated by numerical constraints.

References

1. Boris V. Cherkassky and Andrew V. Goldberg. Negative-cycle detection algorithms. In *ESA '96: Proceedings of the Fourth Annual European Symposium on Algorithms*, pages 349–363, London, UK, 1996. Springer-Verlag.
2. T. H. Cormen, C. E. Leiserson, R. L. Rivest, and C. Stein. *Introduction to algorithms*. MIT Press, 1990.
3. S. Cotton. Satisfiability checking with difference constraints. Master's thesis, Max Planck Institute, 2005.
4. S. Cotton and O. Maler. Satisfiability modulo theory chains with DPLL(T). In *Verimag Technical Report* http://www-verimag.imag.fr/TR/TR-2006-4.pdf, 2006.
5. M. Davis, G. Logemann, and D. Loveland. A machine program for theorem proving. In *Communications of the ACM*, volume 5(7), pages 394–397, 1962.
6. M. Davis and H. Putnam. A computing procedure for quantification theory. *Journal of the ACM*, 7(1):201–215, 1960.
7. E. W. Dijkstra. A note on two problems in connexion with graphs. In *Numer. Math.*, volume 1, pages 269–271, 1959.
8. N. Eèn and N. S orensson. Minisat – a sat solver with conflict-clause minimization. In *SAT 2005*, 2005.
9. D. Frigioni, A. Marchetti-Spaccamela, and U. Nanni. Fully dynamic shortest paths and negative cycles detection on digraphs with arbitrary arc weights. In *European Symposium on Algorithms*, pages 320–331, 1998.
10. H. Ganzinger, G. Hagen, R. Nieuwenhuis, A. Oliveras, and C. Tinelli. DPLL(T): Fast decision procedures. In *CAV'04*, pages 175–188, 2004.
11. A. V. Goldberg. Shortests path algorithms: Engineering aspects. In *Proceedings of the Internation Symposium of Algorithms and Computation*, 2001.
12. E. Goldberg and Y. Novikov. Berkmin: A fast and robust SAT solver, 2002.
13. D.B. Johnson. Efficient algorithms for shortest paths in sparse networks. *J. Assoc. Comput. Mach.*, 24:1, 1977.
14. J. P. Marquez-Silva and K. A. Sakallah. Grasp – a new search algorithm for satisfiability. In *CAV'96*, pages 220–227, November 1996.
15. M. W. Moskewicz, C. F. Madigan, Y. Zhao, L. Zhang, and S. Malik. Chaff: Engineering an Efficient SAT Solver. In *DAC'01*, 2001.
16. P. Niebert, M. Mahfoudh, E. Asarin, M. Bozga, O. Maler, and N. Jain. Verification of timed automata via satisfiability checking. In *Lecture Notes in Computer Science*, volume Volume 2469, pages 225 – 243, Jan 2002.
17. R. Nieuwenhuis and A. Oliveras. DPLL(T) with Exhaustive Theory Propagation and its Application to Difference Logic. In *CAV'05*, volume 3576 of *LNCS*, pages 321–334, 2005.
18. R. Nieuwenhuis, A. Oliveras, and C. Tinelli. Abstract DPLL and abstract DPLL modulo theories. In *LPAR'04*, volume 3452 of *LNCS*, pages 36–50. Springer, 2005.
19. S. Ranise and C. Tinelli. The SMT-LIB format: An initial proposal. In *PDPAR*, July 2003.
20. R. E. Tarjan. Shortest paths. In *AT&T Technical Reports*. AT&T Bell Laboratories, 1981.

A Progressive Simplifier for Satisfiability Modulo Theories*

Hossein M. Sheini and Karem A. Sakallah

University of Michigan, Ann Arbor MI 48109, USA
{hsheini, karem}@umich.edu

Abstract. In this paper we present a new progressive cooperating simplifier for deciding the satisfiability of a quantifier-free formula in the first-order theory of integers involving combinations of sublogics, referred to as Satisfiability Modulo Theories (SMT). Our approach, given an SMT problem, replaces each non-propositional theory atom with a Boolean indicator variable yielding a purely propositional formula to be decided by a SAT solver. Starting with the most abstract representation (the Boolean formula), the solver gradually integrates more complex theory solvers into the working decision procedure. Additionally, we propose a method to simplify "expensive" atoms into suitable conjunctions of "cheaper" theory atoms when conflicts occur. This process considerably increases the efficiency of the overall procedure by reducing the number of calls to the slower theory solvers. This is made possible by adopting our novel inter-logic implication framework, as proposed in this paper. We have implemented these methods in our Ario SMT solver by combining three different theory solvers within a DPLL-style SAT solver: a Unit-Two-Variable-Per-Inequality (UTVPI) solver, an integer linear programming (ILP) solver, and a solver for systems of equalities with uninterpreted functions. The efficiencies of our proposed algorithms are demonstrated and exhaustively investigated on a wide range of benchmarks in hardware and software verification domain. Empirical results are also presented showing the advantages/limitations of our methods over other modern techniques for solving these SMT problems.

1   Introduction

Procedures to decide quantifier-free formulas in (a combination of) first-order theories, recently dubbed as Satisfiability Modulo Theories (SMT), have been used to solve problems in a wide range of applications mainly in hardware and software verification. Examples of such applications are RTL datapath verification [1], symbolic timing verification [2], and buffer over-run vulnerability detection [3]. In the past several years, there has been considerable progress in solving these types of problems by leveraging the recent advances in Boolean SAT and combining them with different solvers for conjunctions of theory atoms

---

* This work was funded in part by the National Science Foundation under ITR grant No. 0205288.

A. Biere and C.P. Gomes (Eds.): SAT 2006, LNCS 4121, pp. 184–197, 2006.

[4,5,6,7,8]. These combination strategies include the *layered approach* of [6] based on solving each particular set of theory atoms separately and in layers, ordered in their increasing expressiveness, and the *online approach* including the Mixed Logical Integer Linear Programming (MLILP) [9] and DPLL(T) [8] methods, where the SAT reasoning and learning techniques are applied to both propositional and theory literals simultaneously. Applied to simpler logics, such as to the Unit-Two-Variable-Per-Inequality (UTVPI) Logic in [9] or Difference Logic in [8], the online method utilizes incremental theory solvers while the propositional abstraction of the problem is being solved in order to reason about any single assignment to the theory atoms on-demand.

Recognizing the expressiveness relations among the theory atoms within the SMT problem, on one hand, and the unequal efficiencies of different theory solvers specialized in certain logics, on the other hand, in this paper, we introduce a new combined *inter-logic deduction* procedure that enables the SMT solver reason across different solvers. We subsequently propose the following two simplification schemes (collectively called *progressive*) in order to gradually increase the role of simpler theory solvers in the overall procedure:

1. **Conflict-induced inter-Logic Constraint Simplification,** where at each conflict, implications involving expensive theory atoms are simplified into a combination of cheaper atoms. Note that in order to be able to detect such implications, the "inter-logic deduction" and reasoning framework is utilized. Unlike many recent SAT-based SMT solvers [4,6] where at each conflict, the formula is augmented with constraints over its existing atoms, our simplification approach can introduce new theory atoms to yield the strongest representation of the conflicts/implications in terms of atoms in different logics. This consequently results in better pruning of the search space, lower propagation costs and ultimately considerable speed-up.

2. **Step-by-Step Formula Concretization,** where a more concrete representation of the problem is solved only if a solution to an abstract version is found. Unlike the layered approach where the satisfying solutions are passed down the layers, this method gradually incorporates theory solvers into the overall procedure upon finding the *first* SAT solution to each abstract formula. The abstract formulas are solved following the online approach.

Note that in this paper we are considering different logics within the theory of integer numbers. In case of combining different theories, such as the theory of arrays or the theory of list structures, a combination strategy such as Nelson-Oppen method [10] or Delayed Theory Combination approach [11] is adopted to communicate equalities between shared variables. Our progressive simplification and our inter-logic deduction methods are only applicable to those theory atoms that are convertible to each other.

The remainder of this paper is organized as follows. In Section 2 we formally define the SMT problem and in Section 3 we describe the cooperating theory solvers and our inter-logic deduction scheme. We introduce the two algorithms of our progressive simplifier in Section 4. Experimental results are reported in Section 5 and we conclude in Sections 6.

2   Satisfiability Modulo Theories

An SMT problem is the problem of deciding the satisfiability of a quantifier-free formula in (combinations of) first-order theories.

Following the SMT-LIB standards [12], a *language* of a theory within the first-order logic with equality consists of three disjoint sets of symbols: *individual variables* V, *n-ary function symbols* Fun, and *n-ary predicate symbols* Pr. A 0-ary function symbol is called a *constant*, denoted by c, and a 0-ary predicate symbol is called a *propositional variable*, denoted by P. A *term*, denoted by t, is defined inductively as a variable symbol, $v \in V$, or $f(t_1, \cdots, t_n)$ where $f \in Fun$. Furthermore, an *Atom* is $p(t_1, \cdots, t_n)$ where $p \in Pr$. A function (predicate) symbol is termed "uninterpreted" and denoted by $f^U \in Fun$ ($p^U \in Pr$), when we only know that it is *consistent*, i.e. two of its applications to the same arguments produce the same value. Hence, a *Functional Consistency constraint*, FC, of the following form is implicit due to the uninterpreted functions (predicates):

$$\begin{aligned} FC ::= \ & \forall f^U \in Fun, \ \forall p^U \in Pr, \ \forall t_{11}, \ldots, t_{1k}, t_{21}, \ldots, t_{2k} : \\ & (t_{11} = t_{21} \wedge \ldots \wedge t_{1k} = t_{2k}) \rightarrow f^U(t_{11}, \cdots, t_{1k}) = f^U(t_{21}, \cdots, t_{2k}) \\ & (t_{11} = t_{21} \wedge \ldots \wedge t_{1k} = t_{2k}) \rightarrow [\ p^U(t_{11}, \cdots, t_{1k}) \leftrightarrow p^U(t_{21}, \cdots, t_{2k}) \ ] \end{aligned} \quad (1)$$

We distinguish an *input language* to be a certain syntactic class of atoms that a procedure accepts as input. A pair of a theory and an input language is referred to as a *sublogic* or simply a *logic*. Table 1 contains the terms and atoms defined under the logics of interest in our SMT framework (all defined within the theory of integer numbers). Note that we interchangeably refer to atoms that are in the form of integer constraints, i.e. equality, UTVPI and non-UTVPI unrestricted constraints, as *integer constraints*.

Thus, an SMT formula, φ, is constructed by applying logical connectives $(\wedge, \vee, \neg)$ to the atoms in any of these logics. Replacing integer constraints, C, in the SMT formula with fresh propositional *indicator variables*, denoted by P_C,

Table 1. Logics of Interest within the theory of integer numbers

Logic	Input Language
Sentential (Propositional) Logic ($\mathcal{P}$)	Term ::= $-$ Atom ::= P
Equality Logic with Successors ($\mathcal{E}$)	Term ::= $c \mid v \mid t \mid t + c$ Atom ::= $t_i = t_j \mid t_i \neq t_j$
Equality Logic with Successors and Uninterpreted Functions ($\mathcal{E}\mathcal{U}\mathcal{F}$)	Term ::= $c \mid v \mid t \mid t + c \mid f^U(t_1, \cdots, t_n)$ Atom ::= $t_i = t_j \mid t_i \neq t_j \mid p^U(t_1, \cdots, t_n)$
Integer UTVPI Logic ($\mathcal{T}\mathcal{V}\mathcal{L}$)	Term ::= $c \mid v \mid t \mid t + c$ Atom ::= $a_i t_i + a_j t_j \leq b$ where $a_i, a_j \in \{0, \pm 1\}$; $b \in \mathbb{Z}$
Linear Integer Arithmetic ($\mathcal{L}\mathcal{I}\mathcal{A}$)	Term ::= $c \mid v \mid t \mid t + t \mid t - t$ Atom ::= $\sum_{i=1}^{n} a_i t_i \leq b$ where $a_i, b \in \mathbb{Z}$

results in the propositional *abstraction* of the problem, φ^{bool}, and the decomposition of the SMT formula, as follows:

$$\varphi = \varphi^{bool} \wedge \bigwedge_{C \in \mathscr{E}} (P_C \leftrightarrow C) \wedge FC \wedge \bigwedge_{C \in \mathscr{D}} (P_C \leftrightarrow C) \wedge \bigwedge_{C \in \mathscr{L}} (P_C \leftrightarrow C) \qquad (2)$$

where $\mathscr{E}$, $\mathscr{D}$ and $\mathscr{L}$ denote sets of equality, UTVPI and unrestricted integer constraints (atoms) within the SMT formula respectively. Additionally, the SMT formula φ can be represented at five different *levels of abstraction* as follows:

level	theory	formula
1	$\mathcal{P} \cup \mathcal{EUF} \cup \mathcal{TVL} \cup \mathcal{LIA}$	φ
2	$\mathcal{P} \cup \mathcal{EUF} \cup \mathcal{TVL}$	$\varphi^{bool} \wedge \bigwedge_{C \in \mathscr{E}} (P_C \leftrightarrow C) \wedge FC \wedge \bigwedge_{C \in \mathscr{D}} (P_C \leftrightarrow C)$
3	$\mathcal{P} \cup \mathcal{EUF}$	$\varphi^{bool} \wedge \bigwedge_{C \in \mathscr{E}} (P_C \leftrightarrow C) \wedge FC$
4	$\mathcal{P} \cup \mathcal{E}$	$\varphi^{bool} \wedge \bigwedge_{C \in \mathscr{E}} (P_C \leftrightarrow C)$
5	$\mathcal{P}$	φ^{bool}

$$(3)$$

Moreover, we would call a logic to be the *sub-logic* of another logic if all its atoms can be expressed in the more expressive logic. For instance, $\mathcal{TVL}$ is a sub-logic of $\mathcal{LIA}$ and $\mathcal{E}$ is a sub-logic of $\mathcal{EUF}$ and $\mathcal{TVL}$. Finally, an SMT formula, φ, is *satisfiable* if it evaluates to true under (the combination of) its theory structures.

3   Cooperating Theory Solvers

The most straightforward method to decide the satisfiability of an SMT formula, φ as in (2), is to enumerate all the satisfiable solutions to the Boolean formula, φ^{bool}, using a generic SAT solver [13] and to check the consistency of the corresponding set of theory literals using a specific theory solver. This *theory solver* is capable to solve a conjunction of all theory literals within the solution. This process terminates if a theory-consistent solution is found, implying that φ is satisfiable, or all solutions are rejected by the theory solver, implying otherwise.

Following a similar scheme, it is possible to *layer* a series of specialized theory solvers in an increasing solving capability, each utilized if a satisfying solution at the previous layer is found. If a theory solver finds a conflict, then this conflict is used to prune the search at the propositional level and the system backtracks to the highest (Boolean) layer; if it does not, the consistency of a more expressive set of theory atoms is determined at the next layer. This approach is demonstrated in Fig. 1. In order to find conflicts earlier in the search, an *incremental layered approach* is suggested in [6] where instead of complete Boolean assignments, "partial" ones are passed through layers to be checked for consistency. Note that in this framework, each theory solver is responsible for deciding the satisfiability of conjunctions of atoms in only one logic and does not interact with the other

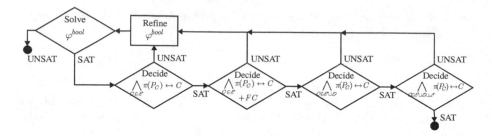

Fig. 1. The layered approach - $\pi(P_C)$ represents the assignment to P_C satisfying φ^{bool} (C represents an integer constraint in the form of equality, UTVPI or non-UTVPI)

theory solvers. Therefore, theory atoms are converted to ones in more expressive logics as the procedure advances to higher layers.

3.1 Online Theory Solver Integration

In this approach, as introduced in [14] and further improved in [8,9], a Boolean SAT solver is coupled with other theory solvers, each capable to decide the satisfiability of a conjunction of theory atoms in a particular logic independently and incrementally. The Boolean abstraction of the SMT formula, φ^{bool}, is decided by the SAT solver and the consistency of theory atoms is checked upon assignments to their corresponding indicator variables. This process is referred to as *online approach* because the satisfiability of the constraints are determined as theory atoms are "activated", i.e. added to the model. This is achieved by maintaining a satisfiable set of activated theory atoms, referred to as *active set,* by each theory solver. Upon activation of a new theory atom, its corresponding theory solver *incrementally* determines the satisfiability of its entire activated atoms that could result in the rejection or addition of the new atom to the active set. A SAT solver is responsible to maintain the satisfiability of the formula at the Boolean level and "orchestrate" the theory atom activations accordingly. Similar to the layered approach, in the online framework, the interactions among theory solvers are limited to the communications through the SAT solver.

A *hybrid* layered+online approach is suggested in [7] where the simpler and more efficient theory solvers, i.e. the UTVPI solver, are integrated within the SAT solver following the online approach while harder ones (i.e. the ILP solver) are utilized in separate layers.

3.2 Inter-logic Deduction Scheme

Noting the hierarchies among theory atoms in different logics and in order to detect the inter-logic implications and conflicts *online*, we propose an inter-logic interaction scheme that maintains a graph representing "all" the implications

Table 2. Summary of covered inter-logic implications

initiating logic	activated atom $\wedge$ [existing atom(s)] $\rightarrow$ implied atom		involved logic
$\mathcal{E}$	$t_i = t_j \ \wedge \ [\ ]$	$\rightarrow P_{t_i=t_j}$	$\mathcal{P}$
	$t_{1i} = t_{2i} \ \wedge \ [t_{1j} = t_{2j}, \cdots]$	$\rightarrow f^U(t_{1i}, \cdots) = f^U(t_{2i}, \cdots)$	$\mathcal{EUF}$
	$t_{1i} = t_{2i} \ \wedge \ [a_i t_{1i} + a_j t_j \leq b]$	$\rightarrow a_i t_{2i} + a_j t_j \leq b$	$\mathcal{TVL}$
$\mathcal{TVL}$	$a_i t_i + a_j t_j \leq b \ \wedge \ [\ ]$	$\rightarrow P_{a_i t_i + a_j t_j \leq b}$	$\mathcal{P}$
	$t_i - t_j \leq b \ \wedge \ [t_j - t_i \leq -b]$	$\rightarrow t_i = t_j + b$	$\mathcal{E}$
	$a_i t_i + a_j t_j \leq b \ \wedge$ $[-a_i . k . t_i + \sum_{l=1, l \neq i}^{n} a_l t_l \leq b']$ $k > 0$	$\rightarrow a_j . k . t_j + \sum_{l=1, l \neq i}^{n} a_l t_l \leq b''$ $b'' = b' + k.b$	$\mathcal{LIA}$
$\mathcal{LIA}$	$a_i . k . t_i + \sum_{l=1, l \neq i}^{n} a_l t_l \leq b \ \wedge$ $[-a_i t_i + a_j t_j \leq b']$ $a_i, a_j \in \{0, \pm 1\}, \ k > 0$	$\rightarrow a_j . k . t_j + \sum_{l=1, l \neq i}^{n} a_l t_l \leq b''$ $b'' = b + k.b'$	$\mathcal{TVL}$

across all involved theory atoms. The vertices of this graph are atoms in different logics and a new implication edge is incrementally added to the graph whenever:

- the current assignments enforce an assignment to an atom in order to maintain the satisfiability of the first-order formula, or
- the combination of two atoms yields another, possibly fresh, atom.

Note that even though all the edges corresponding to the first case are detected and added, not all combinations subject to the other case are checked. This does not affect the completeness of the overall process since these new implications are augmentations to an already complete method. The allowed combinations resulting in inter-logic implications due to the activation of an atom are listed in Table 2 and can be categorized as follows:

1. implying a theory atom if it is subsumed by the activated atom. This would also result in the implication of the corresponding Boolean indicator variable.
2. implying an equality atom due to the functional consistency constraint (1).
3. implying an integer constraint resulting from the non-negative linear combination of two active integer constraints (Only those combinations resulting in the elimination of at least one variable are allowed and two unrestricted integer constraints are not combined due to the complexity of the process).

Note that our proposed inter-logic deduction scheme can be applied in both layered and online frameworks. In the layered approach, the activated atoms are only added to the combined implication graph and are solved later at their particular layers. For instance, it is possible *not to solve* the unrestricted integer constraints as they are activated and only use them for inter-logic deductions. More specifically, the layered+online approach of [7] can be augmented with inter-logic deductions regardless of its theory integration scheme. Additionally, note that our proposed combined implication graph is a generalization of the Bool+Theory graphs of [9,15] extended to combinations of non-Boolean atoms.

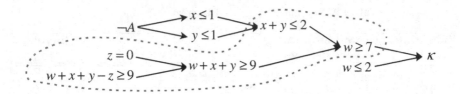

Fig. 2. Combined inter-logic implication graph for Example 1

4   SMT Progressive Simplification Algorithms

In this section, we describe our progressive simplifier for solving an SMT problem. The main objective in this simplifier is to take advantage of the efficiencies of the simpler theory solvers by "progressively" simplifying the SMT formula in terms of theory atoms in less expressive logics.

4.1   Conflict-Induced Inter-logic Constraint Simplification

The most straightforward approach for learning from conflicts in a SAT-based cooperating SMT solver is to prune that conflict from the propositional representation of the problem. Detected by a theory solver for a particular logic, conflicts in many SMT solvers [4,5,6,7,8] are only represented in terms of the atoms solely in that logic and learned as a constraint over Boolean indicator variables corresponding to those atoms. Specifically, if any of the theory solvers detects a conflict in its "active set", the reason for that inconsistency is refined from the Boolean abstraction of the problem by adding a conflict-induced learned constraint. However, by utilizing our inter-logic implication graph, it is possible to detect conflicts resulting from atoms in multiple logics and consequently learn a constraint comprising of atoms across different logics.

Example 1. Consider the following SMT problem comprising of $\mathcal{P}$, $\mathcal{TVL}$ and $\mathcal{LIA}$ atoms (refer to Fig. 2 for combined implication graph):

$$\varphi = (w+x+y-z \geq 9) \wedge (z = 0) \wedge [A \vee (x \leq 1 \wedge y \leq 1)] \wedge (w \leq 2 \vee w \leq 4 \vee w \leq 6)$$

where $(w + x + y - z \geq 9)$ and $(z = 0)$ should be satisfied in all cases. Suppose we first assign A to `false` so that both $(x \leq 1)$ and $(y \leq 1)$ should be satisfied. Selecting $(w \leq 2)$ to satisfy the third constraint would violate $(w+x+y-z \geq 9)$ to be detected by checking an $\mathcal{LIA}$ atom (refer to Fig. 2 for details). This results in learning $\neg[(w + x + y - z \geq 9) \wedge (z = 0) \wedge (x \leq 1) \wedge (y \leq 1) \wedge (w \leq 2)]$ across $\mathcal{TVL}$ and $\mathcal{LIA}$ logics. The process continues by selecting $(w \leq 4)$ and then $(w \leq 6)$ both resulting in conflicts, detected again by checking an $\mathcal{LIA}$ atom, and learning $\neg[(w + x + y \geq 9) \wedge (z = 0) \wedge (x \leq 1) \wedge (y \leq 1) \wedge (w \leq 4)]$ and $\neg[(w + x + y \geq 9) \wedge (z = 0) \wedge (x \leq 1) \wedge (y \leq 1) \wedge (w \leq 6)]$ respectively.

Detecting the involvement of expensive atoms, i.e. $\mathcal{LIA}$ or $\mathcal{EUF}$ atoms, in conflicts offers the opportunity to augment the SMT formula with information about those atoms in the language of their sub-logics. Our proposed *inter-logic simplification scheme* acknowledges the advantages of representing the constraints in less complex logics (i.e. $\mathcal{TVL}$ instead of $\mathcal{LIA}$) in order to use them for possible earlier implications in those theory solvers. Hence, all implications participating in the conflict, referred to as "conflicting implications", that involve atoms in two or more *related* logics, i.e. $\mathcal{TVL}$ and $\mathcal{LIA}$ or $\mathcal{E}$ and $\mathcal{EUF}$, are simplified into new constraints comprising of cheaper atoms in their sub-logics and representing the same conflicting implication.

Following this approach would ultimately result in a "progressive" simplification of the theory atoms in terms of atoms in their sub-logics. Note that this process is performed in addition to the general learning scheme and results in a constraint across logics. We apply this algorithm to $\mathcal{LIA}$ and $\mathcal{EUF}$ atoms, involved in a conflict, and simplify them to logical combinations of respectively $\mathcal{TVL}$ and $\mathcal{E}$ atoms to be added to the formula after each conflict.

Simplifying $\mathcal{LIA}$ atoms. If the analysis of the combined inter-logic implication graph determines that an unrestricted integer constraint is involved in a conflict, a new constraint is generated and conjoined to the formula, comprising of only $\mathcal{TVL}$ atoms and indicator variables. This constraint represents the same conflicting implication and is generated as follows:

1. **Traverse** the combined implication graph backward from the conflict.
2. **Locate** any unseen implications involving an unrestricted integer constraint, C_1, as an implicant. If none is found, **Terminate**. Note that we only consider implications of the form of $C \wedge \bigwedge_j U_j \to U$ where U and C represent UTVPI and unrestricted integer constraints, respectively.
3. If C_i is implied by other integer constraints, **Replace** it with its implicants.
4. If any *implied* unrestricted integer constraint is involved, **Goto** step 3.
5. **Replace** any existing unrestricted integer constraint, C_i, with its corresponding indicator variable, P_{C_i}.
6. **Conjoin** the final constraint of the form of $(\bigwedge_j U_j \wedge \bigwedge_i P_{C_i}) \to U$ to the SMT formula. **Goto** step 1.

Note that the final constraint could include some U_j's that were not originally in the formula (implied by other UTVPI constraints). These atoms are referred to as "unrepresented" atoms and in order to control the number of new atoms permanently added to the problem, only one unrepresented atom is allowed in the final generated constraint and the rest are replaced by their UTVPI implicants.

Example 2. Consider the conflict of Example 1, where the unrestricted integer constraint $(w + x + y \geq 9)$ is first detected to be involved in a conflicting implication: $(w + x + y \geq 9) \wedge (x + y \leq 2) \to (w \geq 7)$. Since $(w + x + y \geq 9)$ is not original, it is replaced by its implicants following step 3 of our algorithm resulting in $(w + x + y - z \geq 9) \wedge (z = 0) \wedge (x + y \leq 2) \to (w \geq 7)$. At step 5, the unrestricted integer constraint is replaced by its indicator variable, in this case **true** and therefore the constraint would be $(z = 0) \wedge (x + y \leq 2) \to (w \geq 7)$. The atoms involved in this process are highlighted in the combined implication graph of Fig. 2. Since in this constraint two unrepresented atoms, $(x + y \leq 2)$

and $(w \geq 7)$ exist, we replace $(x + y \leq 2)$ with its implicants $(x \leq 1)$ and $(y \leq 1)$ and the final constraint to be conjoined to the SMT formula would be:

$$(z = 0) \wedge (x \leq 1) \wedge (y \leq 1) \rightarrow (w \geq 7)$$

This constraint further prunes the infeasible space of the $\mathcal{TVL}$ theory solver and results in implying $(w \leq 4)$ and $(w \leq 6)$ to false as long as $(x \leq 1)$ and $(y \leq 1)$ are true. This ultimately implies A to true removing $(x \leq 1)$ and $(y \leq 1)$ from the solution with no further involvement of the $\mathcal{LIA}$ atoms and the ILP solver.

Note that this process is an augmentation to the offline refinement approach of [7] in the sense that it is applied to each conflict encountered in the online search and requires the analysis of the combined implication graph. Utilizing this method reduces the applications of the offline ILP solver by detecting the majority of the conflicts online.

Simplifying $\mathcal{EUF}$ atoms. A similar procedure to the one that simplifies unrestricted integer constraints is also applied to the functional consistency applications involved in the conflicts. If an equality atom is implied by the functional consistency constraint of (1) and is involved in a conflict, that implication is reduced into a constraint in $\mathcal{E}$ logic, which is then conjoined to the formula.

Example 3. Consider the following formula where f is an uninterpreted function:

$$\varphi = [f(x) \neq f(y) \wedge f(z) = f(x)] \wedge (1 \leq x \leq n) \wedge \bigwedge_{i=1}^{n} (x = i \rightarrow y = i)$$

In order to determine the (un)satisfiability of this formula, it is required to examine n pairs of equalities between x and y separately by calling the $\mathcal{EUF}$ theory solver, each being rejected in order to maintain the functional consistency of f. However, our method, after detecting the first conflict, say due to activating $x = 1$ and $y = 1$, conjoins the formula with the following consistency constraint:

$$(x = y) \rightarrow [f(x) = f(y)]$$

which is essentially the reduction of FC only for this particular conflicting instance. Since $f(x) \neq f(y)$, all remaining activations of $(x = i)/(y = i)$'s are rejected in the equality solver with no further calls to the costly $\mathcal{EUF}$ solver.

Note that our method is essentially an "on-demand" utilization of Ackermann reduction [16] on those functional consistency applications involved in a conflict.

4.2   Step-by-Step Formula Concretization

As discussed earlier, the SMT formula can be represented at different levels of abstraction. By solving each of these formulas individually and "sequentially", it is possible to progressively simplify the SMT problem through learning from conflicts in abstracted formulas earlier and cheaper. Note that the satisfiabilities of the abstracted formulas do not imply the satisfiability of the SMT problem. Thus, our proposed *progressive* procedure initially solves the formula at the highest level of abstraction. If satisfiable, it continues with solving

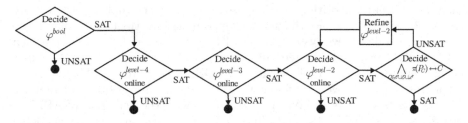

Fig. 3. progressive approach on online solvers. The problem is solved in two layers

the formula at the next level of abstraction and so forth. Our algorithm solves each abstracted formula of (3) in descending abstraction order, following the online solving approach. If any of the abstracted formulas is proved unsatisfiable, the SMT formula is UNSAT, otherwise the formula at the next lower abstraction level is solved. This procedure is demonstrated in the first four steps of Fig. 3. Note that all the learned constraints added to the abstract formulas remain valid as the representation of the formula becomes more concrete. This allows the overall procedure to progressively simplify the formula by pruning large parts of the search space detected as conflicts in abstract formulas earlier and less costly.

Note that our proposed progressive method is fundamentally different from the layered approach. Unlike the layered approach where the theory-consistency of activated atoms is explored separately and in a "depth-first" manner, i.e. each set of atoms in a particular logic is solved in isolation at its own layer (Fig. 1), in our progressive approach, each abstracted formula is solved once, following the online approach, after the first satisfiable solution to the formula at the previous abstraction level is found.

It is possible to combine the layered and progressive approaches based on the computational overheads of the involved theory solvers. The solvers with high overhead, i.e. ILP solver, are integrated in layered manner while the ones with lower overhead and with the possibility of simpler integration inside SAT, are combined in progressive manner. An example of such hybrid configuration is demonstrated in Fig. 3. In this example, the propositional abstraction of the formula is initially solved and if found satisfiable, the formula containing the consistency constraints associated with equality atoms (abstraction level 4) is solved, keeping the status of the SAT solver in terms of learned constraints, variable ordering, etc. unchanged. The process continues until it finds a SAT solution to the formula at the abstraction level 2, after which it moves to the next layer to establish the consistency of the $\mathcal{LIA}$ atoms. At this point the solver follows the layered+online approach of [7] augmented with inter-theory deduction and conflict-induced simplification techniques as described earlier.

5   Implementation and Experimental Evaluations

We implemented our inter-logic deduction scheme and progressive simplifier in our Ario SMT solver [7]. Ario utilizes MiniSAT [13] SAT solver and adopts the congruence closure algorithm of [17] to solve systems of equalities, the currifying method [17] to represent the uninterpreted functions, the transitivity closure algorithm of [18] to solve systems of UTVPI constraints and a Simplex/Branch-and-Bound solver to solve systems of unrestricted integer constraints. All experiments were conducted on a Pentium-IV 2.8GHz with 1 GB of RAM.

For this evaluation we used four sets of SMT benchmark suites as follows: 1) Wisconsin Safety Analyzer (WiSA) [19] dealing with the detection of API-level software exploits, 2) CIRC suite dealing with the verification of circuits modeling the equivalence of sum/multiplication of two integers and their bit-wise operation [6], 3) Pipelined Machine Verification (PMV) suite dealing with the verification of safety and liveness for XScale-like processor models [20], and 4) UCLID suite dealing with the verification of out-of-order microprocessors [21]. The WiSA and CIRC suites are formulated in $\mathcal{TVL}$ and $\mathcal{LIA}$ logics and PMV and UCLID suites are formulated in $\mathcal{P}$ and $\mathcal{EUF}$ logics. We compared Ario against the top three performers in the SMT Competition [22], namely MathSAT v3.3.1 [6,11] based on the incremental layered approach, YICES [23] the most recent version of ICS [5] and BarceLogicTools (BCLT) [24] following the online approach. Note that BCLT does not support $\mathcal{LIA}$ logic.

As demonstrated in Fig. 4, Ario performed relatively better than YICES and MathSAT in most of the benchmarks. This is mostly due to its hybrid theory combination strategy (layered+progressive) and its inter-logic deduction scheme which enables it to detect conflicts earlier with no expensive communications through layers.

The performance of the progressive simplifier of Ario, utilizing both algorithms of Section 4, is best demonstrated in the comparison between Ario and its non-progressive version[1], referred to as Ario-hnp (left column in Fig. 4). In WiSA suite, the step-by-step formula concretization was the main reason behind better performance of Ario against Ario-hnp. This is mainly because of the fact that, in this benchmark suite, the non-UTVPI integer constraints include variables not shared with any UTVPI integer constraint. This automatically disables the conflict-induced inter-logic constraint simplification technique because no combination of $\mathcal{TVL}$ and $\mathcal{LIA}$ atoms yields an $\mathcal{TVL}$ atom that can be used in the online solvers. On the other hand, on CIRC suite, the unrestricted integer constraints ($\mathcal{LIA}$ atoms) are actively involved in almost all the conflicts which makes their simplification into a constraint over $\mathcal{TVL}$ atoms considerably effective in improving the overall performance of the solver.

On the benchmarks in the $\mathcal{EUF}$ logic, i.e. PMV and UCLID suites, Ario and Ario-hnp performed relatively similar, incorrectly inferring that the progressive simplifier is not as effective on these benchmarks. The same can be observed by the comparison of Ario and BCLT, noting that both Ario and BCLT adopt a similar online theory combination approach. However, by comparing Ario-ho1 (Ario-hnp with only conflict-induced inter-theory constraint simplification) and Ario-ho2 (Ario-hnp with only step-by-step formula concretization) against Ario-hnp, as partially demonstrated in Fig. 5, it is evident that:

---

[1] As described in [7] augmented with an $\mathcal{EUF}$ solver and inter-logic deductions.

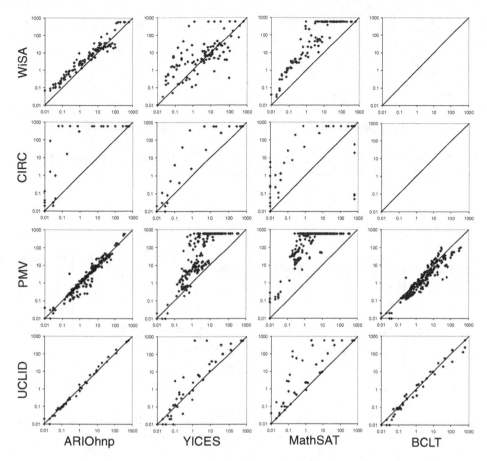

Fig. 4. Comparison of Ario hybrid progressive simplifier (X-axis) against Ario-hnp (no progressive simplifier), Yices, MathSAT and BarcelogicTools (Y-Axis). A dot above the diagonal line represents an instance that Ario performed better. (time-out: 600 s).

- on PMV benchmarks, the conflict-induced constraint simplification method is relatively effective but its resulted performance gains are canceled out by the adoption of the step-by-step formula concretization algorithm. This is due to the high number of unnecessary conflicts detected in abstract formulas that would have been covered by conflicts in more concrete formulas.
- on UCLID benchmarks, on the contrary, step-by-step formula concretization is more effective and constraint simplification imposes some overhead.

This is further shown in Table 3 which includes some statistics on a representative set of instances from these benchmark suites. This table shows that solving increasingly concretizations of the formula is mainly useful when a benchmark over a complex logic (say $\mathcal{EUF}$) can be instead solved in an abstract level. Examples of this case are the benchmarks "q2.14" which is $\mathcal{P}$ solvable and "cache.inv18" which is $\mathcal{P} + \mathcal{E}$ solvable although they contain uninterpreted functions. In such

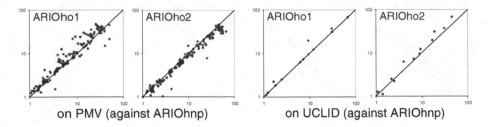

on PMV (against ARIOhnp)   on UCLID (against ARIOhnp)

Fig. 5. Comparison of Ario-ho1 (only algorithm of Section 4.1) and Ario-ho2 (only algorithm of Section 4.2) (X-axis) against Ario-hnp (Y-axis)

Table 3. Ario solving statistics on instances from PMV and UCLID suites

benchmark	% of FC simplified	# of conflict detected in: $\mathcal{P}$	$\mathcal{P} \cup \mathcal{E}$	$\mathcal{P} \cup \mathcal{E}\mathcal{U}\mathcal{F}$	run time (sec.) Ario	Ario-ho1	Ario-ho2	Ario-hnp
c10ni.i	7.28%	0	1127	18395	36.44	27.13	44.18	37.13
c9nidw.i	3.30%	0	794	18823	31.56	23.03	31.62	28.16
9stage.flush	5.29%	1	195	8312	4.37	4.27	6.47	7.46
fxs.bp.safety	5.57%	0	43	1258	0.61	0.48	0.68	0.62
cache.inv18	0%	4839	60728	NA	121.59	174.32	123.24	175.89
q2.14	0%	18543	NA	NA	8.9	9.44	8.81	9.62
ooo.rf10	19.02%	1	70	19641	11.83	9.25	16.9	19.86

benchmarks the Boolean abstraction of the formula contributes to a substantial pruning which can be achieved before processing expensive theory atoms.

6   Conclusions and Future Work

In this paper, we presented our novel inter-logic deduction scheme as well as our progressive simplifier comprising of two algorithms: conflict-induced inter-logic constraint simplification and step-by-step formula concretization. We thoroughly analyzed the performance of these algorithms on a wide range of benchmarks in various logics. While in many instances all these methods significantly added to the efficiency of the SMT solver, in some cases, some limitations were observed.

Consequently, we are planning to further investigate the interactions between theory solvers and work on methods to combine these algorithms, preserving all their advantages. Finally, in the SMT formulas with only a few atoms in complex logics (common in hardware and software verification applications), it is possible to cheaply simplify those atoms on-demand. Therefore, we are working on comprehensive and complete simplification methods that overcome methods based on the adoption of full-scale solvers for those logics. This could considerably reduce the complexity and add to the efficiency of the SMT solvers.

References

1. Brinkmann, R., Drechsler, R.: RTL-datapath verification using integer linear programming. In: ASP-DAC '02. (2002) 741–746
2. Amon, T., Borriello, G., Hu, T., Liu, J.: Symbolic timing verification of timing diagrams using presburger formulas. In: DAC '97. (1997) 226–231
3. Wagner, D., Foster, J.S., Brewer, E.A., Aiken, A.: A first step towards automated detection of buffer overrun vulnerabilities. In: Network and Distributed System Security Symposium. (2000) 3–17
4. Barrett, C.W., Berezin, S.: CVCLite: A new implementation of the cooperating validity checker category b. In: CAV. (2004) 515–518
5. Filliâtre, J.C., Owre, S., Rueß, H., Shankar, N.: ICS: Integrated canonizer and solver. In: CAV. (2001) 246–249
6. Bozzano, M., Bruttomesso, R., Cimatti, A., Junttila, T.A., van Rossum, P., Schulz, S., Sebastiani, R.: An incremental and layered procedure for the satisfiability of linear arithmetic logic. In: TACAS. (2005) 317–333
7. Sheini, H.M., Sakallah, K.A.: A scalable method for solving satisfiability of integer linear arithmetic logic. In: SAT. (2005) 241–256
8. Nieuwenhuis, R., Oliveras, A.: DPLL(T) with exhaustive theory propagation and its application to difference logic. In: CAV. (2005) 321–334
9. Sheini, H.M., Sakallah, K.A.: A SAT-based decision procedure for mixed logical/integer linear problems. In: CPAIOR. (2005) 320–335
10. Nelson, G., Oppen, D.C.: Simplification by cooperating decision procedures. ACM Trans. Program. Lang. Syst. 1(2) (1979) 245–257
11. Bozzano, M., Bruttomesso, R., Cimatti, A., Junttila, T.A., Ranise, S., van Rossum, P., Sebastiani, R.: Efficient satisfiability modulo theories via delayed theory combination. In: CAV. (2005) 335–349
12. Ranise, S., Tinelli, C.: The SMT-LIB format: An initial proposal. In: Workshop on Pragmatics of Decision Procedures in Automated Reasoning. (2003)
13. Eén, N., Sörensson, N.: An extensible SAT-solver. In: SAT. (2003) 502–518
14. Berezin, S., Ganesh, V., Dill, D.L.: An online proof-producing decision procedure for mixed-integer linear arithmetic. In: TACAS. (2003) 521–536
15. Iyer, M.K., Parthasarathy, G., Cheng, K.T.: Efficient conflict-based learning in an RTL circuit constraint solver. In: DATE '05. (2005) 666–671
16. Ackermann, W.: Solvable cases of the decision problem. In: Studies in Logic and the Foundations of Mathematics. North-Holland (1954)
17. Nieuwenhuis, R., Oliveras, A.: Proof-Producing Congruence Closure. In: Proceedings of the 16th Int'l Conf. on Term Rewriting and Applications. (2005) 453–468
18. Jaffar, J., Maher, M.J., Stuckey, P.J., Yap, R.H.C.: Beyond finite domains. In: Workshop on Principles and Practice of Constraint Programming. (1994) 86–94
19. Ganapathy, V., Seshia, S.A., Jha, S., Reps, T.W., Bryant, R.E.: Automatic discovery of API-level exploits. In: ICSE '05. (2005) 312–321
20. Manolios, P., Srinivasan, S.K.: Automatic verification of safety and liveness for XScale-like processor models using WEB refinements. In: DATE. (2004) 168–175
21. Lahiri, S.K., Seshia, S.A., Bryant, R.E.: Modeling and verification of out-of-order microprocessors in UCLID. In: FMCAD. (2002) 142–159
22. Barrett, C., de Moura, L., Stump, A.: SMT-COMP: Satisfiability Modulo Theories Competition. In: CAV. (2005) 20–23
23. de Moura, L.: YICES. http://fm.csl.sri.com/yices/ (2005)
24. Nieuwenhuis, R., Oliveras, A.: Decision procedures for SAT, SAT modulo theories and beyond. the BarcelogicTools. In: LPAR. (2005) 23–46

Dependency Quantified Horn Formulas: Models and Complexity

Uwe Bubeck[1] and Hans Kleine Büning[2]

[1] International Graduate School
Dynamic Intelligent Systems,
Universität Paderborn,
33098 Paderborn Germany
bubeck@upb.de
[2] Department of Computer Science,
Universität Paderborn,
33098 Paderborn Germany
kbcsl@upb.de

Abstract. Dependency quantified Boolean formulas ($DQBF$) extend quantified Boolean formulas with Henkin-style partially ordered quantifiers. It has been shown that this is likely to yield more succinct representations at the price of a computational blow-up from $PSPACE$ to $NEXPTIME$. In this paper, we consider dependency quantified Horn formulas ($DQHORN$), a subclass of $DQBF$, and show that the computational simplicity of quantified Horn formulas is preserved when adding partially ordered quantifiers.

We investigate the structure of satisfiability models for $DQHORN$ formulas and prove that for both $DQHORN$ and ordinary $QHORN$ formulas, the behavior of the existential quantifiers depends only on the cases where at most one of the universally quantified variables is zero. This allows us to transform $DQHORN$ formulas with free variables into equivalent $QHORN$ formulas with only a quadratic increase in length.

An application of these findings is to determine the satisfiability of a dependency quantified Horn formula Φ with $|\forall|$ universal quantifiers in time $O(|\forall| \cdot |\Phi|)$, which is just as hard as $QHORN$-SAT.

1 Introduction

The language of Quantified Boolean Formulas (QBF) offers a concise way to represent formulas which arise in areas such as planning, scheduling or verification [15, 17]. QBF formulas are usually assumed by definition to be in prenex form such that all quantifiers appear at the beginning, and that is also the input format generally required by QBF solvers. This does, however, impose a total ordering on the quantifiers where each existentially quantified variable depends on all preceding universal variables. Consider the following example:

$$\forall x_1 \left[(\forall x_2 \exists y_1 \phi(x_1, x_2, y_1)) \wedge (\forall x_3 \exists y_2 \psi(x_1, x_3, y_2)) \right]$$

A. Biere and C.P. Gomes (Eds.): SAT 2006, LNCS 4121, pp. 198–211, 2006.

In this non-prenex formula, the choice for y_1 depends on the values of x_1 and x_2, and y_2 depends on x_1 and x_3. Using quantifier rewriting rules, we can obtain the equivalent prenex formula

$$\forall x_1 \forall x_2 \exists y_1 \forall x_3 \exists y_2 \ \phi(x_1, x_2, y_1) \wedge \psi(x_1, x_3, y_2)$$

As above, y_1 depends on x_1 and x_2, but y_2 now depends on x_1, x_2 and x_3, so we lose some of the structural information inherent in the original formula.

Recent experimental studies [7, 8] have shown that this problem may have a considerable impact on the performance of QBF solvers. Accordingly, different solutions have lately been suggested to overcome the problem, e.g. by recovering lost information from the formula structure (in particular from the local connectivity of variables in common clauses) [2, 3], or by extending QBF solvers to directly handle non-prenex formulas [8].

Another solution has been proposed by Henkin [9] for first-order predicate logic. He has introduced partially ordered quantifiers, called *branching quantifiers* or simply *Henkin quantifiers*, as in the expression

$$\begin{pmatrix} \forall x_1 \forall x_2 \exists y_1 \\ \forall x_1 \forall x_3 \exists y_2 \end{pmatrix} \ \phi(x_1, x_2, y_1) \wedge \psi(x_1, x_3, y_2)$$

which correctly preserves the dependencies from our introductory example. Since the only relevant information is which universal quantifiers precede which existential quantifier, we can use a (typographically) simpler function-like notation as follows:

$$\forall x_1 \forall x_2 \exists y_1(x_1, x_2) \forall x_3 \exists y_2(x_1, x_3) \ \phi(x_1, x_2, y_1) \wedge \psi(x_1, x_3, y_2)$$

For each existential quantifier, we indicate the universal variables on which it depends. Without loss of information, we can assume that the prefix is in the form $\forall^* \exists^*$:

$$\forall x_1 \forall x_2 \forall x_3 \exists y_1(x_1, x_2) \exists y_2(x_1, x_3) \ \phi(x_1, x_2, y_1) \wedge \psi(x_1, x_3, y_2)$$

This notation has been introduced for quantified Boolean formulas by Peterson, Azhar and Reif [16] under the name *Dependency Quantified Boolean Formulas (DQBF)*.

Notice that partially ordered quantifiers do not only eliminate the aforementioned loss of information due to prenexing, caused by flattening a tree-like hierarchy of quantifiers and corresponding scopes into a linear ordering. The Henkin approach is significantly more general than the suggestions above, because it allows to express subtle dependencies where the hierarchy of quantifier scopes is no longer tree-like. For example, we could add an existential variable y_3 to our sample formula, such that y_3 depends on x_2 and x_3 as indicated in the following prefix:

$$\forall x_1 \forall x_2 \forall x_3 \exists y_1(x_1, x_2) \exists y_2(x_1, x_3) \exists y_3(x_2, x_3)$$

It is not clear how this prefix could be represented in a succinct QBF, even if we allow non-prenex formulas.

Partially-ordered quantification has been around for quite some time, but has not been widely used in combination with quantified Boolean formulas. This is probably due to the fact that *DQBF* is NEXPTIME-complete, as has been shown by Peterson, Azhar and Reif [16]. Assuming that NEXPTIME≠PSPACE, this means that there are *DQBF* formulas for which no equivalent *QBF* of polynomial length can be computed in polynomial space. It also means a jump in complexity compared to *QBF* which is PSPACE-complete. The latter is already considered quite hard, but continued research and the lifting of propositional SAT techniques to *QBF*s have recently produced interesting improvements (see, e.g., [3, 14, 18]) and have led to the emergence of more powerful *QBF*-SAT solvers [13]. In addition, tractable subclasses of *QBF* have been identified and investigated, e.g. *QHORN*, which contains all *QBF* formulas in conjunctive normal form (*CNF*) whose clauses have at most one positive literal. This subclass is important, because it is sufficient for expressing simple "if-then" statements, and because *QHORN* formulas may occur as subproblems when solving arbitrary *QBF* formulas [5].

In this paper, we consider dependency quantified Horn formulas (*DQHORN*), the dependency quantified equivalent to *QHORN*. Our main contribution is to prove that *DQHORN* is a tractable subclass of *DQBF* and is in fact just as difficult as *QHORN*. To be more precise, we present an algorithm which can determine the satisfiability of a *DQHORN* formula Φ with free variables, $|\forall|$ universal quantifiers and an arbitrary number of existential quantifiers in time $O(|\forall| \cdot |\Phi|)$.

We achieve this by investigating the interplay of existential and universal quantifiers with the help of satisfiability models. This concept has been introduced in [12], and Section 3 shows how it can be extended for *DQBF* formulas. We prove that for both *DQHORN* and ordinary *QHORN*, the behavior of the existential quantifiers depends only on the cases where at most one of the universally quantified variables is zero. In Section 4, we demonstrate how *DQBF* formulas with free variables can be transformed into equivalent *QBF* formulas by expanding the universal quantifiers. This expansion may cause an exponential blowup for arbitrary formulas. But the results from Section 3 allow us to avoid this for *DQHORN* formulas with free variables by applying a generalization of the special expansion method that we have presented in [4] for *QHORN*. Finally, an algorithm for solving *DQHORN*-SAT is developed in Section 5.

2   Preliminaries

In this section, we recall the basic terminology and notation for *QBF* and introduce *DQBF*.

A quantified Boolean formula $\Phi \in QBF$ in prenex form is a formula

$$\Phi = Q_1 v_1 ... Q_k v_k\, \phi(v_1, ..., v_k)$$

with quantifiers $Q_i \in \{\forall, \exists\}$ and a propositional formula $\phi(v_1, ..., v_k)$ over variables $v_1, ..., v_k$. We call $Q := Q_1 v_1 ... Q_k v_k$ the *prefix* and ϕ the *matrix* of Φ.

Variables which are bound by universal quantifiers are called *universal variables* and are usually given the names $x_1, ..., x_n$. Analogously, variables in the scope of an existential quantifier are *existential variables* and have names $y_1, ..., y_m$. We write $\Phi = Q \; \phi(\mathbf{x}, \mathbf{y})$ or simply $\Phi = Q \; \phi$.

Variables which are not bound by quantifiers are *free variables*. Formulas without free variables are said to be *closed*. If free variables are allowed, we indicate this with an additional star * after the name of the formula class. Accordingly, QBF is the class of closed quantified Boolean formulas, and QBF^* denotes the quantified Boolean formulas with free variables. We write $\Phi(\mathbf{z}) = Q \; \phi(\mathbf{x}, \mathbf{y}, \mathbf{z})$ or $\Phi(\mathbf{z}) = Q \; \phi(\mathbf{z})$ for a QBF^* formula with free variables $\mathbf{z} = (z_1, ..., z_r)$. A closed QBF formula is either true or false, whereas the truth value of a QBF^* formula depends on the value of the free variables. Two QBF^* formulas $\Psi_1(z_1, ..., z_r)$ and $\Psi_2(z_1, ..., z_r)$ are said to be equivalent ($\Psi_1 \approx \Psi_2$) if and only if $\Psi_1 \models \Psi_2$ and $\Psi_2 \models \Psi_1$, where semantic entailment $\models$ is defined as follows: $\Psi_1 \models \Psi_2$ if and only if for all truth assignments $t(\mathbf{z}) := (t(z_1), ..., t(z_r)) \in \{0, 1\}^r$ to the free variables $\mathbf{z} = (z_1, ..., z_r)$, we have $\Psi_1(t(\mathbf{z})) = 1 \; \Rightarrow \; \Psi_2(t(\mathbf{z})) = 1$.

For $DQBF$ formulas, we introduce a notation which allows us to quickly enumerate the dependencies for a given existential variable y_i ($1 \leq i \leq m$). We are using indices $d_{i,1}, ..., d_{i,n_i}$ which point to the n_i universals on which y_i depends. For example, given the existential quantifier $\exists y_4(x_3, x_5)$, we say that y_4 depends on $x_{d_{4,1}}$ and $x_{d_{4,2}}$ with $d_{4,1} = 3$ and $d_{4,2} = 5$.

With this notation, a dependency quantified Boolean formula $\Phi \in DQBF$ with universal variables $\mathbf{x} = (x_1, ..., x_n)$ and existential variables $\mathbf{y} = (y_1, ..., y_m)$ is a formula of the form

$$\Phi = \forall x_1 ... \forall x_n \exists y_1(x_{d_{1,1}}, ..., x_{d_{1,n_1}}) ... \exists y_m(x_{d_{m,1}}, ..., x_{d_{m,n_m}}) \; \phi(\mathbf{x}, \mathbf{y})$$

In Sections 4 and 5, we will also allow free variables, using the same notation and definition of equivalence as for QBF^*.

The class $DQHORN$ contains all $DQBF$ formulas in conjunctive normal form (CNF) whose clauses have at most one positive literal.

As stated in the following Definitions 1 and 2, the semantics of $DQBF$ is defined over *model functions*. A $DQBF$ formula is said to be true if for each existential variable y_i, there exists a propositional formula f_{y_i} over the universals $x_{d_{i,1}}, ..., x_{d_{i,n_i}}$ on which y_i depends, such that substituting the model functions for the existential variables (and dropping the existential quantifiers) leads to a universally quantified QBF formula which is true. The tuple $M = (f_{y_1}, ..., f_{y_m})$ of such functions is called a *satisfiability model*.

Definition 1. *For a dependency quantified Boolean formula $\Phi \in DQBF$ with existential variables $\mathbf{y} = (y_1, ..., y_m)$, let $M = (f_{y_1}, ..., f_{y_m})$ be a mapping which maps each existential variable y_i to a propositional formula f_{y_i} over the universal variables $x_{d_{i,1}}, ..., x_{d_{i,n_i}}$ on which y_i depends. Then M is a **satisfiability model** for Φ if the resulting QBF formula $\Phi[\mathbf{y}/M] := \Phi[y_1/f_{y_1}, ..., y_m/f_{y_m}]$, where simultaneously each existential variable y_i is replaced by its corresponding formula f_{y_i} and the existential quantifiers are dropped from the prefix, is true.*

Definition 2. *A dependency quantified Boolean formula is true if and only if it has a satisfiability model.*

The notion of satisfiability models has been originally introduced in [12] for *QBF* formulas. For *QBF*s, the last definition is actually a theorem, because their semantics is usually defined inductively without referring to model functions, which is not possible for *DQBF*s. In fact, the NEXPTIME-completeness of *DQBF* suggests that solving a *DQBF* formula involves finding and storing those functions. Fortunately, we will soon see that this is not a problem in the *DQHORN* case.

3   Satisfiability Models for *DQHORN* Formulas

We are not only interested in the mere existence of satisfiability models, but we also want to characterize their structure for certain classes of formulas. In this section, we will see that *DQHORN* formulas have satisfiability models of a very simple structure.

We begin with an observation: it is a well known fact about *propositional* Horn formulas, proved by Alfred Horn himself [10], that the intersection of two satisfying truth assignments is a satisfying truth assignment, too. If we represent truth assignments by sets which collect the variables that are assigned the value 1, the intersection of these assignments is given by the intersection of the corresponding sets of variables.

Now assume that a quantified Horn formula with two universal variables x_i and x_j is known to be satisfiable when $x_i = 0$ and $x_j = 1$ or when $x_i = 1$ and $x_j = 0$. That means there exist two truth assignments t_1 and t_2 to the existential variables such that the formula is satisfied in both cases. If we could lift the closure under intersection to the quantified case, it would mean that the intersection of t_1 and t_2 would satisfy the formula when both x_i and x_j are zero. This would imply that the satisfiability of a quantified Horn formula is determined only by those cases where at most one of the universal variables is zero.

Unfortunately, we have to obey the quantifier dependencies when choosing truth values for the existential variables, so we cannot simply intersect t_1 and t_2. Thus, lifting this result is obviously not so straightforward for the *QHORN* case, and even less straightforward for *DQHORN* with its sophisticated dependencies. What we need here is a way to characterize the behavior of the existentially quantified variables. As it turns out, satisfiability models are a suitable formalism for this and allow us to present a model-based proof which even works for *DQHORN*.

Since the number of zeros being assigned to the universal variables is an important criterion for our investigations, we first introduce some useful notation.

Definition 3. *By B_n^i, we denote the bit vector of length n where only the i-th element is zero, i.e. $B_n^i := (b_1, ..., b_n)$ with $b_i = 0$ and $b_j = 1$ for $j \neq i$. Moreover, we define the following relations on n-tuples of truth values:*

1. $Z_{\leq 1}(n) = \bigcup_i \{B_n^i\} \cup \{(1, ..., 1)\}$ (at most one zero)
2. $Z_{=1}(n) = \bigcup_i \{B_n^i\}$ (exactly one zero)
3. $Z_{\geq 1}(n) = \{(a_1, ..., a_n) \mid \exists i : a_i = 0\}$ (at least one zero)

For example, if $n = 3$, we have the following relations:

$$Z_{\leq 1}(3) = \{(0, 1, 1), (1, 0, 1), (1, 1, 0), (1, 1, 1)\}$$
$$Z_{=1}(3) = \{(0, 1, 1), (1, 0, 1), (1, 1, 0)\}$$
$$Z_{\geq 1}(3) = \{(0, 0, 0), (0, 0, 1), (0, 1, 0), (0, 1, 1), (1, 0, 0), (1, 0, 1), (1, 1, 0)\}$$

We omit the parameter n and simply write $Z_{\leq 1}$ (or $Z_{=1}$ resp. $Z_{\geq 1}$) when it is clear from the context. Usually, n equals the total number of the universal quantifiers in a given formula.

Let $\Phi = Q\phi(\mathbf{x}, \mathbf{y}) \in DQBF$. The definition of a satisfiability model in Section 2 requires that substituting the existentials $\mathbf{y}$ in Φ produces a formula $\Phi[\mathbf{y}/M]$ which is true. That means the matrix $\phi[\mathbf{y}/M]$ must be true for all possible assignments to the universals $\mathbf{x}$. But as motivated before, we want to focus only on the cases where at most one of the universals is assigned zero. Accordingly, we now introduce a special kind of satisfiability model which weakens the condition that all possible assignments are considered: a so-called $R_\forall$-partial satisfiability model is only required to satisfy $\phi[\mathbf{y}/M]$ for certain truth assignments to the universal variables which are given by a relation $R_\forall$.

Definition 4. *For a formula $\Phi = Q\phi(\mathbf{x}, \mathbf{y}) \in DQBF$ with universal variables $\mathbf{x} = (x_1, ..., x_n)$ and existential variables $\mathbf{y} = (y_1, ..., y_m)$, let $M = (f_{y_1}, ..., f_{y_m})$ be a mapping which maps each existential variable y_i to a propositional formula f_{y_i} over the universal variables $x_{d_{i,1}}, ..., x_{d_{i,n_i}}$ on which y_i depends. Furthermore, let $R_\forall(n)$ be a relation on the set of possible truth assignments to n universals. Then M is a $R_\forall$-**partial satisfiability model** for Φ if the formula $\phi[\mathbf{y}/M]$ is true for all $\mathbf{x} \in R_\forall(n)$.*

Consider the following example:

$$\Phi = \forall x_1 \forall x_2 \forall x_3 \exists y_1(x_1, x_2) \exists y_2(x_2, x_3) \, (x_1 \vee y_1) \wedge (x_2 \vee \neg y_1) \wedge (\neg x_2 \vee x_3 \vee \neg y_2)$$

Then Φ does not have a satisfiability model, but $M = (f_{y_1}, f_{y_2})$ with $f_{y_1}(x_1, x_2) = \neg x_1 \vee x_2$ and $f_{y_2}(x_2, x_3) = 0$ is a $Z_{\leq 1}$-partial satisfiability model for Φ, because $\phi[\mathbf{y}/M] = (x_1 \vee \neg x_1 \vee x_2) \wedge (x_2 \vee (x_1 \wedge \neg x_2)) \wedge (\neg x_2 \vee x_3 \vee 1) \approx x_2 \vee x_1$, which is true for all $\mathbf{x} = (x_1, x_2, x_3)$ with $\mathbf{x} \in Z_{\leq 1}$.

It is not surprising that the mere existence of a $Z_{\leq 1}$-partial satisfiability model does not imply the existence of a (total) satisfiability model - at least not in the general case. But as discussed before, we are going to prove that for $DQHORN$ formulas, the behavior of the formula for $\mathbf{x} \in Z_{\leq 1}$ does indeed completely determine its satisfiability. Accordingly, we now show: if we can find a $Z_{\leq 1}$-partial satisfiability model M to satisfy a $DQHORN$ formula whenever at most one of the universals is false, then we can also satisfy the formula for arbitrary truth assignments to the universals. We achieve this by using M to construct a (total) satisfiability model M^t.

Definition 5. *Let* $\Phi = Q\phi(\mathbf{x}, \mathbf{y}) \in DQHORN$ *with universal variables* $\mathbf{x} = (x_1, ..., x_n)$ *and existential variables* $\mathbf{y} = (y_1, ..., y_m)$, *and let* $M = (f_{y_1}, .., f_{y_m})$ *be a* $Z_{\leq 1}$-*partial satisfiability model for* Φ. *For each* $f_{y_i}(x_{d_{i,1}}, ..., x_{d_{i,n_i}})$ *in* M, *we define* $f_{y_i}^t$ *as follows:*

$$
\begin{aligned}
f_{y_i}^t(x_{d_{i,1}}, ..., x_{d_{i,n_i}}) := \quad &(\neg x_{d_{i,1}} \rightarrow f_{y_i}(0, 1, 1, ..., 1)) \\
\wedge \, &(\neg x_{d_{i,2}} \rightarrow f_{y_i}(1, 0, 1, ..., 1)) \\
\wedge \, &... \\
\wedge \, &(\neg x_{d_{i,n_i}} \rightarrow f_{y_i}(1, 1, ..., 1, 0)) \\
\wedge \, &f_{y_i}(1, ..., 1)
\end{aligned}
$$

Then we call $M^t = (f_{y_1}^t, ..., f_{y_m}^t)$ *the* **total completion** *of* M.

The intuition behind this definition is the following: for *each* argument which is zero, we consider the value of the original function when *only* this argument is zero. Then we return the conjunction (the intersection) of those original function values. Additionally, we have to intersect with $f_{y_i}(1, ..., 1)$. For example, $f_{y_i}^t(1, 0, 0, 1) = f_{y_i}(1, 0, 1, 1) \wedge f_{y_i}(1, 1, 0, 1) \wedge f_{y_i}(1, 1, 1, 1)$. In case all the arguments are 1, we simply return the value of the original function, i.e. $f_{y_i}^t(1, .., 1) = f_{y_i}(1, .., 1)$.

At the beginning of this section, we have mentioned that for propositional Horn formulas, the intersection of satisfying truth assignments is again a satisfying truth assignment. If you compare this to the previous definition (together with the following theorem), you will notice that we have just presented the generalized *DQHORN* version of it. The most important difference is that we now always intersect with $f_{y_i}(1, ..., 1)$. This takes care of the cases where certain universal variables are zero, but y_i does not depend on them due to the imposed quantifier dependencies.

Theorem 1. *Let* $\Phi = Q\phi(\mathbf{x}, \mathbf{y}) \in DQHORN$ *be a dependency quantified Horn formula with a* $Z_{\leq 1}$-*partial satisfiability model* $M = (f_{y_1}, .., f_{y_m})$. *Then the total completion of* M, *i.e.* $M^t = (f_{y_1}^t, ..., f_{y_m}^t)$ *as defined above, is a satisfiability model for* Φ.

Proof. We must show that $\phi[\mathbf{y}/M^t]$ is true for all truth assignments $t(\mathbf{x}) := (t(x_1), ..., t(x_n)) \in \{0, 1\}^n$ to the universal variables. Since $f_{y_j}^t(1, ..., 1) = f_{y_j}(1, ..., 1)$, we only need to consider $t(\mathbf{x}) \in Z_{\geq 1}$.

The proof is by induction on the number of zeros in $t(\mathbf{x})$. The induction base is the case $t(\mathbf{x}) \in Z_{=1}$. Then the definition of M^t implies that

$$
f_{y_j}^t(t(x_{d_{j,1}}), ..., t(x_{d_{j,n_j}})) = f_{y_j}(t(x_{d_{j,1}}), ..., t(x_{d_{j,n_j}})) \wedge f_{y_j}(1, ..., 1) = 1
$$

for all y_j. Now let $t(\mathbf{x}) = B_n^i$ be an assignment to the universals where $t(x_i) = 0$. In order to prove that every clause in $\phi[\mathbf{y}/M^t]$ is true for $t(\mathbf{x})$, we make a case distinction on the structure of Horn clauses. Any clause C belongs to one of the following cases:

1. C contains a positive existential variable y_j:

   Consider a clause of the form $C = y_j \vee \bigvee_{l \in L_y} \neg y_l \vee \bigvee_{l \in L_x} \neg x_l$. We assume that $i \notin L_x$, because $C[\mathbf{y}/M^t]$ is trivially true for $t(\mathbf{x})$ if $i \in L_x$.

   If $f_{y_j}(t(x_{d_{j,1}}), ..., t(x_{d_{j,n_j}})) = f_{y_j}(1, ..., 1) = 1$ then $f_{y_j}^t(t(x_{d_{j,1}}), ..., t(x_{d_{j,n_j}})) = 1$. Otherwise, without loss of generality, let $f_{y_j}(t(x_{d_{j,1}}), ..., t(x_{d_{j,n_j}})) = 0$. Then $f_{y_r}(t(x_{d_{r,1}}), ..., t(x_{d_{r,n_r}})) = 0$ for some $r \in L_y$, as M is a $Z_{\leq 1}$-partial satisfiability model. This implies $f_{y_r}^t(t(x_{d_{r,1}}), ..., t(x_{d_{r,n_r}})) = 0$, which makes $C[\mathbf{y}/M^t]$ true.

2. C contains a positive universal variable x_j:

   Consider a clause of the form $C = x_j \vee \bigvee_{l \in L_x} \neg x_l \vee \bigvee_{l \in L_y} \neg y_l$. The only interesting case to discuss is $i = j$. As above, M being a $Z_{\leq 1}$-partial satisfiability model implies that $f_{y_r}(t(x_{d_{r,1}}), ..., t(x_{d_{r,n_r}})) = 0$ for some $r \in L_y$. And this implies $f_{y_r}^t(t(x_{d_{r,1}}), ..., t(x_{d_{r,n_r}})) = 0$.

3. no positive literal in C:

   Consider a clause of the form $C = \bigvee_{l \in L_x} \neg x_l \vee \bigvee_{l \in L_y} \neg y_l$. We only need to discuss the case that $i \notin L_x$. Again, M being a $Z_{\leq 1}$-partial satisfiability model implies that we have $f_{y_r}(t(x_{d_{r,1}}), ..., t(x_{d_{r,n_r}})) = 0$ for some $r \in L_y$. This means $f_{y_r}^t(t(x_{d_{r,1}}), ..., t(x_{d_{r,n_r}})) = 0$.

For the induction step, we consider an assignment where $k > 1$ universals are false. Let $t(x_{i_1}) = 0, ..., t(x_{i_k}) = 0$ and $t(x_s) = 1$ for $s \neq i_1, ..., i_k$. In order to show that $\phi[\mathbf{y}/M^t]$ is true for $t(\mathbf{x})$, we can use the induction hypothesis and assume $\phi[\mathbf{y}/M^t]$ is true for $t_1(\mathbf{x}) = B_n^{i_k}$ as well as for $t_{k-1}(\mathbf{x})$ with $t_{k-1}(x_1) = 0, ..., t_{k-1}(x_{i_{k-1}}) = 0$ and $t_{k-1}(x_s) = 1$ for $s \neq i_1, ..., i_{k-1}$. That means, the case with k zeros $x_{i_1}, ..., x_{i_k}$ is reduced to the case where only x_{i_k} is zero and the case where $x_{i_1}, ..., x_{i_{k-1}}$ are zero. Then the definition of $f_{y_j}^t$ implies

$$f_{y_j}^t(t(x_{d_{j,1}}), ..., t(x_{d_{j,n_j}})) = f_{y_j}^t(t_1(x_{d_{j,1}}), ..., t_1(x_{d_{j,n_j}})) \wedge f_{y_j}^t(t_{k-1}(x_{d_{j,1}}), ..., t_{k-1}(x_{d_{j,n_j}}))$$

Again, we make a case distinction. It is actually very similar to the one from the induction base:

1. C contains a positive existential variable y_j:

   Consider a clause of the form $C = y_j \vee \bigvee_{l \in L_y} \neg y_l \vee \bigvee_{l \in L_x} \neg x_l$. We assume that $i_1, ..., i_k \notin L_x$, because otherwise, $C[\mathbf{y}/M^t]$ is trivially true for $t(\mathbf{x})$.

   If $f_{y_j}^t(t_1(x_{d_{j,1}}), ..., t_1(x_{d_{j,n_j}})) = 1$ and $f_{y_j}^t(t_{k-1}(x_{d_{j,1}}), ..., t_{k-1}(x_{d_{j,n_j}})) = 1$, we have $f_{y_j}^t(t(x_{d_{j,1}}), ..., t(x_{d_{j,n_j}})) = 1$.

   Otherwise, without loss of generality, let $f_{y_j}^t(t_1(x_{d_{j,1}}), ..., t_1(x_{d_{j,n_j}})) = 0$. Then the induction hypothesis implies that $f_{y_r}^t(t_1(x_{d_{r,1}}), ..., t_1(x_{d_{r,n_r}})) = 0$ for some $r \in L_y$, and we get $f_{y_r}^t(t(x_{d_{r,1}}), ..., t(x_{d_{r,n_r}})) = 0$.

2. C contains a positive universal variable x_j:

   Consider a clause of the form $C = x_j \vee \bigvee_{l \in L_x} \neg x_l \vee \bigvee_{l \in L_y} \neg y_l$. The only interesting case to discuss is $j \in \{i_1, ..., i_k\}$. Without loss of generality, we assume $j = i_k$.

   It follows from the induction hypothesis that $f_{y_r}^t(t_1(x_{d_{r,1}}), ..., t_1(x_{d_{r,n_r}})) = 0$ for some $r \in L_y$. Then $f_{y_r}^t(t(x_{d_{r,1}}), ..., t(x_{d_{r,n_r}})) = 0$.

3. no positive literal in C:

   Consider a clause of the form $C = \bigvee_{l \in L_x} \neg x_l \vee \bigvee_{l \in L_y} \neg y_l$. We only need to discuss the case that $i_1, ..., i_k \notin L_x$. Again, the induction hypothesis implies that we have $f_{y_r}^t(t_1(x_{d_{r,1}}), ..., t_1(x_{d_{r,n_r}})) = 0$ for some $r \in L_y$. Then $f_{y_r}^t(t(x_{d_{r,1}}), ..., t(x_{d_{r,n_r}})) = 0$.

   $\square$

4   From $DQBF^*$ to QBF^*: Eliminating Universals

4.1   The General Case

Quantifier expansion is a well-known technique for solving QBFs [1, 3]. As demonstrated in this section, it can be generalized to dependency quantified formulas and may be used to compute for any $DQBF^*$ formula an equivalent prenex QBF^* formula.

A universal quantifier $\forall x \ \phi(x)$ is just an abbreviation for $\phi(0) \wedge \phi(1)$, so we can expand it and make two copies of the original matrix, one for the universally quantified variable being false, and one for that variable being true. As explained in [3], existential variables which depend on that universal variable need to be duplicated as well. For example, in

$$\forall x_1 \forall x_2 \forall x_3 \exists y_1(x_1, x_2) \exists y_2(x_2, x_3) \ \phi(x_1, x_2, x_3, y_1, y_2)$$

the choice for y_1 depends on the value of x_1. We must therefore introduce two separate instances $y_{1,(0)}$ and $y_{1,(1)}$ of the original variable y_1, where $y_{1,(0)}$ is used in the copy of the matrix for $x_1 = 0$, and analogously $y_{1,(1)}$ for $x_1 = 1$. We obtain the expanded formula

$$\forall x_2 \forall x_3 \exists y_{1,(0)}(x_2) \exists y_{1,(1)}(x_2) \exists y_2(x_2, x_3) \ \phi(0, x_2, x_3, y_{1,(0)}, y_2) \wedge \phi(1, x_2, x_3, y_{1,(1)}, y_2)$$

We can do this successively to expand multiple universal quantifiers. Unlike the QBF^* case described in [3] and [4], we do not need to start with the innermost quantifier, because $DQBF^*$ formulas can always be written with a $\forall^* \exists^*$ prefix where the order of the universals is irrelevant. After expanding all universal quantifiers, we are left with a QBF^* formula - actually a very special one with a $\exists^*$ prefix. Obviously, this expansion leads to an exponential blowup of the original formula. In practice, we do not need to expand all universals. For our sample formula, the expansion of x_1 is sufficient, because the resulting formula can be written in QBF^* as

$$\forall x_2 \exists y_{1,(0)} \exists y_{1,(1)} \forall x_3 \exists y_2 \ \phi(0, x_2, x_3, y_{1,(0)}, y_2) \wedge \phi(1, x_2, x_3, y_{1,(1)}, y_2)$$

One could think of sophisticated strategies for selecting which universals must be expanded for a given $DQBF^*$ formula. In the general case, however, this cannot avoid exponential growth, therefore the following discussion will assume that all universal quantifiers are eliminated. Using the results from the previous section, we will show that this is not a problem for $DQHORN^*$ formulas, because the expanded formula is always small, even if all universals are expanded.

In the general case, for a $DQBF^*$ formula

$$\Phi(\mathbf{z}) = \forall x_1...\forall x_n \exists y_1(x_{d_{1,1}}, ..., x_{d_{1,n_1}})...\exists y_m(x_{d_{m,1}}, ..., x_{d_{m,n_m}}) \, \phi(\mathbf{x}, \mathbf{y}, \mathbf{z})$$

with universal variables $\mathbf{x} = x_1, ..., x_n$, existential variables $\mathbf{y} = y_1, ..., y_m$ and free variables $\mathbf{z}$, we obtain the expanded QBF^* formula

$$\Phi_{\exists QBF}(\mathbf{z}) \quad := \quad \exists y_{1,(0,...,0)} \exists y_{1,(0,...,0,1)}...\exists y_{1,(1,...,1,0)} \exists y_{1,(1,...,1)}$$

$$\vdots$$

$$\exists y_{m,(0,...,0)} \exists y_{m,(0,...,0,1)}...\exists y_{m,(1,...,1,0)} \exists y_{m,(1,...,1)}$$

$$\bigwedge_{t(\mathbf{x}) \in \{0,1\}^n} \phi(t(\mathbf{x}), y_{1,(t(x_{d_{1,1}}),...,t(x_{d_{1,n_1}}))}, ..., y_{m,(t(x_{d_{m,1}}),...,t(x_{d_{m,n_m}}))}, \mathbf{z})$$

We omit the formal proof that $\Phi(\mathbf{z}) \approx \Phi_{\exists QBF}(\mathbf{z})$, as it is quite obvious that $\Phi_{\exists QBF}$ is simply the formalization of the elimination algorithm we have just described.

Here is an example: the formula

$$\Phi(\mathbf{z}) = \forall x_1 \forall x_2 \forall x_3 \exists y_1(x_1, x_2) \exists y_2(x_2, x_3) \, \phi(x_1, x_2, x_3, y_1, y_2, \mathbf{z})$$

from above is expanded to

$$\Phi_{\exists QBF}(\mathbf{z}) = \exists y_{1,(0,0)} \exists y_{1,(0,1)} \exists y_{1,(1,0)} \exists y_{1,(1,1)} \exists y_{2,(0,0)} \exists y_{2,(0,1)} \exists y_{2,(1,0)} \exists y_{2,(1,1)}$$
$$\phi(0,0,0, y_{1,(0,0)}, y_{2,(0,0)}, \mathbf{z}) \wedge \phi(0,0,1, y_{1,(0,0)}, y_{2,(0,1)}, \mathbf{z})$$
$$\wedge \quad \phi(0,1,0, y_{1,(0,1)}, y_{2,(1,0)}, \mathbf{z}) \wedge \phi(0,1,1, y_{1,(0,1)}, y_{2,(1,1)}, \mathbf{z})$$
$$\wedge \quad \phi(1,0,0, y_{1,(1,0)}, y_{2,(0,0)}, \mathbf{z}) \wedge \phi(1,0,1, y_{1,(1,0)}, y_{2,(0,1)}, \mathbf{z})$$
$$\wedge \quad \phi(1,1,0, y_{1,(1,1)}, y_{2,(1,0)}, \mathbf{z}) \wedge \phi(1,1,1, y_{1,(1,1)}, y_{2,(1,1)}, \mathbf{z})$$

4.2 Special Case: $DQHORN^*$

We will now show that the expansion of universal quantifiers is feasible for $DQHORN^*$ formulas.

Definition 6. *Let* $\Phi \in DQHORN^*$ *with*

$$\Phi(\mathbf{z}) = \forall x_1...\forall x_n \exists y_1(x_{d_{1,1}}, ..., x_{d_{1,n_1}})...\exists y_m(x_{d_{m,1}}, ..., x_{d_{m,n_m}}) \, \phi(\mathbf{x}, \mathbf{y}, \mathbf{z})$$

be a dependency quantified Horn formula with universal variables $\mathbf{x} = x_1, ..., x_n$, *existential variables* $\mathbf{y} = y_1, ..., y_m$ *and free variables* $\mathbf{z}$.
Then we define the formula $\Phi_{\exists HORN}(\mathbf{z})$ *as*

$$\Phi_{\exists HORN}(\mathbf{z}) \quad := \quad \exists y_{1,(0,1,...,1)} \exists y_{1,(1,0,1,...,1)}...\exists y_{1,(1,...,1,0)} \exists y_{1,(1,...,1)}$$

$$\vdots$$

$$\exists y_{m,(0,1,...,1)} \exists y_{m,(1,0,1,...,1)}...\exists y_{m,(1,...,1,0)} \exists y_{m,(1,...,1)}$$

$$\bigwedge_{t(\mathbf{x}) \in Z_{\leq 1}(n)} \phi(t(\mathbf{x}), y_{1,(t(x_{d_{1,1}}),...,t(x_{d_{1,n_1}}))}, ..., y_{m,(t(x_{d_{m,1}}),...,t(x_{d_{m,n_m}}))}, \mathbf{z})$$

The only difference between the formula $\Phi_{\exists HORN}$ and the expansion $\Phi_{\exists QBF}$ for general $DQBF^*$ formulas is that for Horn formulas, not all possible truth assignments to the universally quantified variables have to be considered. Based on the results from Section 3, assignments where more than one universal variable is false are irrelevant for $DQHORN^*$ formulas.

For the formula

$$\Phi(\mathbf{z}) = \forall x_1 \forall x_2 \forall x_3 \exists y_1(x_1, x_2) \exists y_2(x_2, x_3)\, \phi(x_1, x_2, x_3, y_1, y_2, \mathbf{z})$$

from the example in Section 4.1, we have

$$\begin{aligned}
\Phi_{\exists HORN}(\mathbf{z}) = \;& \exists y_{1,(0,1)} \exists y_{1,(1,0)} \exists y_{1,(1,1)} \exists y_{2,(0,1)} \exists y_{2,(1,0)} \exists y_{2,(1,1)} \\
& \phi(0, 1, 1, y_{1,(0,1)}, y_{2,(1,1)}, \mathbf{z}) \wedge \phi(1, 0, 1, y_{1,(1,0)}, y_{2,(0,1)}, \mathbf{z}) \\
\wedge\; & \phi(1, 1, 0, y_{1,(1,1)}, y_{2,(1,0)}, \mathbf{z}) \wedge \phi(1, 1, 1, y_{1,(1,1)}, y_{2,(1,1)}, \mathbf{z})
\end{aligned}$$

Before we can prove that $\Phi_{\exists HORN}$ is indeed equivalent to Φ, we make a fundamental observation: for the special case that Φ is closed, i.e. there are no free variables, the satisfiability of $\Phi_{\exists HORN}$ implies the existence of a $Z_{\leq 1}$-partial satisfiability model for Φ.

Lemma 1. *Let $\Phi \in DQHORN$ be a dependency quantified Horn formula without free variables, and let $\Phi_{\exists HORN}$ be defined as above. If $\Phi_{\exists HORN}$ is satisfiable then Φ has a $Z_{\leq 1}$-partial satisfiability model.*

Proof. Let t be a satisfying truth assignment to the existentials in $\Phi_{\exists HORN}$. This assignment t provides us with all the information needed to construct a $Z_{\leq 1}$-partial satisfiability model for Φ. The idea is to assemble the truth assignments to the individual copies $y_{i,(x_{d_{i,1}},...,x_{d_{i,n_i}})}$ of an existential variable y_i into a common model function. We achieve this with the following definition:

$$\begin{aligned}
f_{y_i}(x_{d_{i,1}}, ..., x_{d_{i,n_i}}) = \;& (\bar{x}_{d_{i,1}} \wedge x_{d_{i,2}} \wedge ... \wedge x_{d_{i,n_i}} \rightarrow t(y_{i,(0,1,...,1)})) \\
\wedge\; & (x_{d_{i,1}} \wedge \bar{x}_{d_{i,2}} \wedge x_{d_{i,3}} \wedge ... \wedge x_{d_{i,n_i}} \rightarrow t(y_{i,(1,0,1,...,1)})) \\
\wedge\; & ... \\
\wedge\; & (x_{d_{i,1}} \wedge ... \wedge x_{d_{i,n_i-1}} \wedge \bar{x}_{d_{i,n_i}} \rightarrow t(y_{i,(1,...,1,0)})) \\
\wedge\; & (x_{d_{i,1}} \wedge ... \wedge x_{d_{i,n_i}} \rightarrow t(y_{i,(1,...,1)}))
\end{aligned}$$

Now, the f_{y_i} form a $Z_{\leq 1}$-partial satisfiability model for Φ, because for all $\mathbf{x} = (x_1, ..., x_n)$ with $\mathbf{x} \in Z_{\leq 1}$, we have $f_{y_i}(x_{d_{i,1}}, ..., x_{d_{i,n_i}}) = t(y_{i,(x_{d_{i,1}},...,x_{d_{i,n_i}})})$, and $\phi(x_1, ..., x_n, t(y_{1,(x_{d_{1,1}},...,x_{d_{1,n_1}})}), ..., t(y_{m,(x_{d_{m,1}},...,x_{d_{m,n_m}})})) = 1$ due to the satisfiability of $\Phi_{\exists HORN}$.   □

Using Lemma 1 in combination with Theorem 1, it is now easy to show that $\Phi_{\exists HORN}$ is equivalent to Φ.

Theorem 2. *For $\Phi \in DQHORN^*$ and $\Phi_{\exists HORN}$ as defined above, it holds that $\Phi \approx \Phi_{\exists HORN}$.*

Proof. The implication $\Phi(\mathbf{z}) \models \Phi_{\exists HORN}(\mathbf{z})$ is obvious, as the clauses in $\Phi_{\exists HORN}$ are just a subset of the clauses in $\Phi_{\exists QBF}$, which in turn is equivalent to Φ.

The implication $\Phi_{\exists HORN}(\mathbf{z}) \models \Phi(\mathbf{z})$ is more interesting. Assume $\Phi_{\exists HORN}(\mathbf{z}^*)$ is satisfiable for some fixed $\mathbf{z}^*$. With the free variables fixed, we can treat both $\Phi_{\exists HORN}(\mathbf{z}^*)$ and $\Phi(\mathbf{z}^*)$ as closed formulas and apply Lemma 1 and the results from Section 3 as follows:

According to Lemma 1, the satisfiability of $\Phi_{\exists HORN}(\mathbf{z}^*)$ implies that $\Phi(\mathbf{z}^*)$ has a $Z_{\leq 1}$-partial satisfiability model. On this partial model, we can apply the total expansion from Definition 5 and Theorem 1 to obtain a (total) satisfiability model. The fact that $\Phi(\mathbf{z}^*)$ has a satisfiability model implies that $\Phi(\mathbf{z}^*)$ is satisfiable. $\qquad\square$

We immediately obtain the following corollary:

Corollary 1. *For any dependency quantified Horn formula $\Phi \in DQHORN^*$ with free variables, there exists an equivalent formula $\Phi_{\exists HORN} \in QHORN^*$ without universal quantifiers. The length of $\Phi_{\exists HORN}$ is bounded by $|\forall| \cdot |\Phi|$, where $|\forall|$ is the number of universal quantifiers in Φ, and $|\Phi|$ is the length of Φ.*

5   Solving *DQHORN*-*SAT

We can take advantage of the fact that the transformation we have just presented produces $QHORN^*$ formulas without universal variables. The absence of universals allows us to easily determine their satisfiability, because a formula of the form $\Psi(\mathbf{z}) = \exists y_1 ... \exists y_m \, \psi(y_1, ..., y_m, \mathbf{z})$ is satisfiable if and only if its matrix $\psi(y_1, ..., y_m, \mathbf{z})$ is satisfiable. The latter is a purely propositional formula, so we can apply existing SAT solvers for propositional Horn.

We then obtain the following algorithm for determining the satisfiability of a formula $\Phi \in DQHORN^*$:

1. Transform Φ into $\Phi_{\exists HORN}$ according to Definition 6. This requires time $O(|\forall| \cdot |\Phi|)$ and produces a formula of length $|\Phi_{\exists HORN}| = O(|\forall| \cdot |\Phi|)$.
2. Determine the satisfiability of $\phi_{\exists HORN}$, which is the purely propositional matrix of $\Phi_{\exists HORN}$. It is well known [6] that SAT for propositional Horn formulas can be solved in linear time, in this case $O(|\phi_{\exists HORN}|) = O(|\forall| \cdot |\Phi|)$.

In total, this requires time $O(|\forall| \cdot |\Phi|)$, which is just as hard as $QHORN^*$-SAT [11].

6   Conclusion

We have introduced the class of dependency quantified Horn formulas $DQHORN^*$ and have shown that it is a tractable subclass of $DQBF^*$. We have demonstrated that the tractability of $DQHORN^*$ is due to an interesting effect that the Horn property has on the behavior of the quantifiers, a phenomenon which is preserved when adding partially ordered quantifiers. Based on this result, we have been able to prove that

- any dependency quantified Horn formula $\Phi \in DQHORN^*$ of length $|\Phi|$ with free variables, $|\forall|$ universal quantifiers and an arbitrary number of existential quantifiers can be transformed into an equivalent quantified Horn formula of length $O(|\forall| \cdot |\Phi|)$ which contains only existential quantifiers.
- $DQHORN^*$-SAT can be solved in time $O(|\forall| \cdot |\Phi|)$.

This shows that the class $DQHORN^*$ is no more difficult than $QHORN^*$, but apparently does not provide significant increases in expressive power either. $DQHORN^*$ should, however, not be considered as an isolated subclass of $DQBF^*$. Just like ordinary $QHORN^*$ formulas are important as subproblems when solving arbitrary QBF^* formulas [5], our findings on $DQHORN^*$ should prove useful for handling more general classes of $DQBF^*$ formulas. And since the latest trend of enabling QBF solvers to directly handle non-prenex formulas [8] constitutes a special case of partially-ordered quantification with tree-like dependencies, our results might also be applied in non-prenex QBF solvers for cutting Horn branches.

In addition, the tractability of $DQHORN^*$ shows that adding partially ordered quantifiers does not necessarily lead to a computational blow-up as in the general case with $DQBF^*$. Further research should therefore explore the complexity and expressiveness of other subclasses and special cases, in particular tree-like dependencies as mentioned above.

References

[1] A. Ayari and D. Basin. *QUBOS: Deciding Quantified Boolean Logic using Propositional Satisfiability Solvers.* Proc. 4th Intl. Conf. on Formal Methods in Computer-Aided Design (FMCAD'02). Springer LNCS 2517, 2002.

[2] M. Benedetti. *Quantifier Trees for QBFs.* Proc. 8th Intl. Conf. on Theory and Applications of Satisfiability Testing (SAT'05). Springer LNCS 3569, 2005.

[3] A. Biere. *Resolve and Expand.* Proc. 7th Intl. Conf. on Theory and Applications of Satisfiability Testing (SAT'04). Springer LNCS 3542, 2005.

[4] U. Bubeck, H. Kleine Büning, and X. Zhao. *Quantifier Rewriting and Equivalence Models for Quantified Horn Formulas.* Proc. 8th Intl. Conf. on Theory and Applications of Satisfiability Testing (SAT'05). Springer LNCS 3569, 2005.

[5] S. Coste-Marquis, D. Le Berre, and F. Letombe. *A Branching Heuristics for Quantified Renamable Horn Formulas.* Proc. 8th Intl. Conf. on Theory and Applications of Satisfiability Testing (SAT'05). Springer LNCS 3569, 2005.

[6] W. Dowling and J. Gallier. *Linear-Time Algorithms for Testing the Satisfiability of Propositional Horn Formulae.* J. of Logic Programming, 1(3):267–284, 1984.

[7] U. Egly, M. Seidl, H. Tompits, S. Woltran, and M. Zolda. *Comparing Different Prenexing Strategies for Quantified Boolean Formulas.* Proc. 6th Intl. Conf. on Theory and Applications of Satisfiability Testing (SAT'03). Springer LNCS 2919, 2004.

[8] E. Giunchiglia, M. Narizzano, and A. Tacchella. *Quantifier structure in search based procedures for QBFs.* Proc. Design, Automation and Test in Europe (DATE '06), 2006.

[9] L. Henkin. *Some remarks on infinitely long formulas.* In: Infinitistic Methods (Warsaw, 1961) 167–183, 1961.

[10] A. Horn. *On sentences which are true of direct unions of algebras.* Journal of Symbolic Logic, 16(1):14–21, 1951.

[11] H. Kleine Büning and T. Lettmann. *Propositional Logic: Deduction and Algorithms.* Cambridge University Press, Cambridge, UK, 1999.

[12] H. Kleine Büning, K. Subramani, and X. Zhao. *On Boolean Models for Quantified Boolean Horn Formulas.* Proc. 6th Intl. Conf. on Theory and Applications of Satisfiability Testing (SAT'03). Springer LNCS 2919, 2004.

[13] D. Le Berre, M. Narizzano, L. Simon, and A. Tacchella. *The second QBF solvers comparative evaluation.* Proc. 7th Intl. Conf. on Theory and Applications of Satisfiability Testing (SAT'04). Springer LNCS 3542, 2004.

[14] R. Letz. *Advances in Decision Procedures for Quantified Boolean Formulas.* Proc. IJCAR Workshop on Theory and Applications of Quantified Boolean Formulas, 2001.

[15] M. Mneimneh and K. Sakallah. *Computing Vertex Eccentricity in Exponentially Large Graphs: QBF Formulation and Solution.* Proc. 6th Intl. Conf. on Theory and Applications of Satisfiability Testing (SAT'03). Springer LNCS 2919, 2004.

[16] G. Peterson, S. Azhar, and J. Reif. *Lower Bounds for Multiplayer Non-Cooperative Games of Incomplete Information.* Computers and Mathematics with Applications, 41(7-8):957–992, 2001.

[17] J. Rintanen. *Constructing Conditional Plans by a Theorem-Prover.* Journal of Artificial Intelligence Research, 10:323–352, 1999.

[18] L. Zhang and S. Malik. *Towards Symmetric Treatment of Conflicts and Satisfaction in Quantified Boolean Satisfiability Solver.* Proc. 8th Intl. Conf. on Principles and Practice of Constraint Programming (CP'02), 2002.

On Linear CNF Formulas

Stefan Porschen, Ewald Speckenmeyer, and Bert Randerath

Institut für Informatik, Universität zu Köln, D-50969 Köln, Germany
{porschen, esp, randerath}@informatik.uni-koeln.de

Abstract. In the present paper we introduce the class of *linear* CNF formulas generalizing the notion of linear hypergraphs. Clauses of a linear formula intersect in at most one variable. We show that SAT for the general class of linear formulas remains NP-complete. Moreover we show that the subclass of exactly linear formulas is always satisfiable. We further consider the class of uniform linear formulas and investigate conditions for the formula graph to be complete. We define a formula hierarchy such that one can construct a 3-uniform linear formula belonging to the ith level such that the clause-variable density is of $\Omega(2.5^{i-1}) \cap O(3.2^{i-1})$. Finally, we introduce the subclasses $\text{LCNF}_{\geq k}$ of linear formulas having only clauses of length at least k, and show that SAT remains NP-complete for $\text{LCNF}_{\geq 3}$.

Keywords: linear CNF formula, satisfiability, edge colouring, NP-completeness, linear hypergraph, latin square.

1 Introduction

A prominent concept in hypergraph research are *linear* hypergraphs [1] having the special property that its hyperedges have pairwise at most one vertex in common. A hypergraph is called *loopless* if no hyperedge has length one. A long-standing open problem for linear hypergraphs is the Erdös-Farber-Lovàsz conjecture [4] stating that for each loopless linear hypergraph over n vertices there exists an edge n-coloring such that hyperedges of non-empty intersection are colored differently. In this paper we introduce the class of *linear* CNF formulas generalizing the concept of linear hypergraphs. In a linear formula each pair of distinct clauses has at most one variable in common.

The motivation for our work basically is the abstract interest in the structure and the complexity of linear formulas w.r.t. SAT. We thus take a theoretical point of view in this paper. However, the class of linear formulas may be useful for applications with objects exhibiting only weak interdependencies in the sense that the corresponding CNF encoding yields only sparse overlapping clauses.

By reduction from the well known SAT problem it can be shown that SAT restricted to linear CNF formulas remains NP-complete. The reduction relies on introducing new variables for variables occuring in clauses having at least two variables in common with a different clause. The truth values of the original variable and the corresponding new one must be forced to be identical. This can easily be achieved by convenient binary clauses. In case of linear CNF formulas C

A. Biere and C.P. Gomes (Eds.): SAT 2006, LNCS 4121, pp. 212–225, 2006.

without clauses of length at most two it causes serious problems forcing variables to have the same value in every satisfying truth assignment (model) of C.

For a while we had the impression that linear CNF formulas without clauses of length two or less are all satisfiable. Early experiments supported this hypothesis but all attempts to prove it failed. In order to disprove the hypothesis we tried to construct linear CNF formulas with a highest possible dependency between the clauses. The first class of candidates, the exactly linear CNF formulas, where every two clauses have exactly one common variable supported the hypothesis, for the famous König-Hall theorem yields satisfiability in this case, see Section 3, Theorem 2. So, we had to exploit further structural properties of the linear hypergraphs underlying the linear formulas. We restricted to k-CNF formulas. For ease of explaining let $k = 3$. We first constructed an exactly linear 3-CNF formula such that for the corresponding hypergraph its intersection graph which forms a clique is isomorphic to the vertex graph in which each two vertices are joined by an edge if they occur together in a hyperedge. Such formulas have as many clauses as variables. In case of $k = 3$ there are seven clauses and variables, see Section 4, p. 219. Lemma 4 formulates conditions to be satisfied by such formulas, called k-blocks, for arbitrary values of $k \geq 3$. Taking the signature of the 3-block with the corresponding 7 variables, a 7-block (fragment) is constructed yielding linear 3-CNF formula with 43 variables and 133 clauses. Here we additionally use a result about the existence of certain latin squares as stated in Section 4, Prop. 3. On the basis of this (monotone) formula skeleton we have generated more than $3.5 \cdot 10^8$ formulas by randomly assigning different polarities to the variables in the skeleton of this linear 3-CNF formula. 488 of these formulas were unsatisfiable using the solver [2], i.e., only one of 70,000 of these formulas was unsatisfiable. From one of these unsatisfiable formulas we extracted a minimal unsatisfiable subformula with 43 variables and 81 clauses. Eliminating an arbitrary clause $(x \vee y \vee z)$ from this formula yields three backbone variables of the remaining (satisfiable) formula, which can be used to show the NP-completeness of SAT for linear 3-CNF formulas by padding up clauses of length 2 by such complemented backbone variables. For this we extracted an even smaller formula see Section 4, p. 223. Recall that a backbone variable x in a satisfiable formula C, by definition, has the same truth value in each model of C (cf. e.g. [10]). Note that random 3-CNF formulas with constant clause-variable density w.h.p. are linear up to a logarithmic number of clauses. Unfortunately, this observation does not help showing that linear formulas may have backbone variables which forms the core of showing the major NP-completeness result of Theorem 4.

Finally, in Section 5 we formulate some open problems.

2   Preliminary Facts

To fix notation, let CNF denote the set of formulas (free of duplicate clauses) in conjunctive normal form over propositional variables $x_i \in \{0, 1\}$. A variable x induces a positive literal (variable x) or a negative literal (negated variable:

$\overline{x}$). The *complement* of a literal l is $\overline{l}$. Each formula $C \in$ CNF is considered as a set of its clauses $C = \{c_1, \ldots, c_{|C|}\}$ having in mind that it is a conjunction of these clauses. Each clause $c \in C$ is a disjunction of different literals, and is also represented as a set $c = \{l_1, \ldots, l_{|c|}\}$. A clause $c \in C$ is called *unit* iff $|c| = 1$. A literal in C is called *pure* iff its complement does not occur in C. For a given formula C, clause c, by $V(C), V(c)$ we denote the set of variables occuring (negated or unnegated) in C resp. c. The *satisfiability problem (SAT)* takes as input a formula $C \in$ CNF and asks whether there is a truth (value) assignment $t : V(C) \rightarrow \{0, 1\}$ such that at least one literal in each clause of c is set to 1, in which case C is said to be *satisfiable*, and t is a *model* of C. For convenience we allow the empty set to be a formula: $\emptyset \in$ CNF which is satisfiable. From now on the notion *formula* means a (duplicate free) element of CNF.

A *hypergraph* is a pair $H = (V, E)$ where $V = V(H)$ is a finite set, the *vertex set* and $E = E(H)$ is a family of subsets of V the *(hyper)edge set* such that for each $x \in V$ there is an edge containing it. If $|e| \geq 2$ holds for all edges of a hypergraph it is called *loopless*. A hypergraph H is called *k-uniform* if for each edge holds $|e| = k$ and k is a fixed positive integer. For a vertex x of a hypergraph $H = (V, E)$, let $E_x = \{e \in E : x \in e\}$ be the set of all edges containing x. Then $\omega_H(x) := |E_x|$ denotes the *degree* of vertex x in H, we simply write $\omega(x)$ when there is no danger of confusion. H is called *j-regular* if there is a positive integer j and each vertex has degree j in H. We call $\|E\| := \sum_{e \in E} |e|$ the *length* of the hypergraph which is a useful constant. The next equation, throughout refered to as the *length condition of H*, is obvious, but useful: $\|E\| = \sum_{e \in E} |e| = \sum_{x \in V} \omega(x)$. A hypergraph is called *linear* if $(*) : |e \cap e'| \leq 1, e \neq e'$, and is called *exactly* linear if in $(*)$ equality holds for each pair of distinct hyperedges. Let LIN (resp. XLIN) denote the class of all linear (resp. exactly linear) (finite) hypergraphs. There are some useful graphs that can be assigned to a hypergraph $H = (V, E)$. First, the *intersection graph* G_E of H. It has a vertex for each hyperedge and two vertices are joined by an edge in G_E if the corresponding hyperedges have a non-empty intersection; let each edge of G_E be labeled by the vertices in the corresponding intersection of hyperedges. Further, the *vertex graph* G_V with vertex set V. x and x' are joined by an edge in G_V iff there is a hyperedge in E containing x and x', let each edge of G_V be labeled by the corresponding hyperedges. Clearly, for each $e \in E$ the induced subgraph $G_V|e$ of G_V is isomorphic to the complete graph $K_{|e|}$. The *incidence graph* of a hypergraph $H = (V, E)$ is the bipartite graph whose vertex set is $V \cup E$. Each vertex is joined to all hyperedges containing it.

Definition 1. $C \in$ CNF *is called* linear *if*

(1) C contains no pair of complementary unit clauses and
(2) $(*)$: for all $c_1, c_2 \in C : c_1 \neq c_2$ holds $|V(c_1) \cap V(c_2)| \leq 1$.
$C \in$ CNF *is called* exactly linear *if it is linear and equality holds in $(*)$.*
Let (X)LCNF denote the class of all (exactly) linear formulas.

Clearly linear formulas that do not have property (1) are unsatisfiable. Due to condition (1) a linear formula C directly corresponds to a linear hypergraph H_C by disregarding all negations of variables which correspond to the vertices and the clauses to the hyperedges; we call H_C the *underlying* hypergraph of C. A monotone formula by definition having no negated variables thus is identical to its underlying hypergraph. So, we are justified to call a linear formula C *j-regular, resp. k-uniform* if H_C is j-regular resp. k-uniform. Similarly, the *incidence graph* I_C resp. *the intersection graph* G_C of C are identified by the corresponding graphs of H_C, the *variable graph* $G_{V(C)}$ of C is defined to be the vertex graph of H_C. Reversely, to a given linear hypergraph H there corresponds a family $\mathcal{C}(H)$ of linear formulas such that H is the underlying hypergraph of each $C \in \mathcal{C}$. It is easy to see that $\mathcal{C}(H)$ (up to permutations of vertices in the hyperedges) has size $2^{\|E(H)\|}$ if $E(H)$ is the edge set of H.

Lemma 1. *For $C \in$ LCNF, with $n := |V(C)|$ holds $|C| \le n + \binom{n}{2}$.*

PROOF. Let $V(C) = \{x_1, \ldots, x_n\}$. C can have at most n unit clauses which are independent of the remaining formula, because otherwise by the pigeonhole principle there exists a pair of complementary unit clauses. Since C is linear each pair of variables (x_i, x_j), with $j > i$, can occur in exactly one clause of C, yielding $\binom{n}{2}$ possible clauses of length at least 2 by the pigeonhole principle completing the proof.   □

Theorem 1. SAT *remains* NP-*complete when restricted to the class* LCNF.

PROOF. We provide a polynomial time reduction from CNF-SAT to LCNF-SAT. Let $C \in$ CNF be arbitrary. We recursively transform C step by step due to the following procedure:

begin
while there are two clauses $c, c' \in C$ such that $|V(c) \cap V(c')| \ge 2$ **do**:
for each variable $x \in V(c) \cap V(c')$ introduce new variables x_1, x_1'
remove c and c' from C
add the new clauses c_1, c_1' obtained from c, c' by replacing each $x \in V(c) \cap V(c')$
by x_1 in c and by x_1' in c' such that the polarities remain the same as in c, c'
for each $x \in V(c) \cap V(c')$ add the three clauses $\{\overline{x}, x_1\}, \{\overline{x_1}, x_1'\}, \{\overline{x_1'}, x\}$
end

Clearly, the transformation of C by the procedure above takes polynomial time. Moreover it is obvious that the resulting formula C' is linear because all variables occuring in the intersection of any two clauses is recursively replaced by a new variable. It remains to verify that C is satisfiable iff C' is satisfiable. This can be seen immediately by observing that the clauses added in the last step ensure equivalence of the replaced variables with the original ones correspondingly, as they are equivalent to the implicational chain: $x \Rightarrow x_1 \Rightarrow x_1' \Rightarrow x$ implying $x \Leftrightarrow x_1 \Leftrightarrow x_1'$ independently for each triple x, x_1, x_1'. Note that these equivalences are independent of the polarities of the corresponding literals as long as the new variables are assigned the same polarities as that of the substituted ones in the corresponding clause. It is not hard to see that via these equivalences one can construct a model of C' from a model of C and vice versa finishing the proof.   □

The reduction given above adds 2-clauses to a non-linear input formula forcing the newly introduced variables all to be assigned the same truth value in every model of C'. Therefore, if we consider the subclass $\mathrm{LCNF}_{\geq 3}$ of LCNF where each formula contains only clauses of length at least 3, then the reduction above does not work. So, the question arises whether SAT restricted to $\mathrm{LCNF}_{\geq 3}$ is NP-complete, too. Below we will reconsider that point, but let us mention here already that in case of formulas in $\mathrm{LCNF}_{\geq 3}$ forcing introduced variables, as done by the previously given reduction, to have the same value becomes a serious problem, and its solution requires a lot of structural insight into this class of formulas. So the NP-completeness proof of SAT for $\mathrm{LCNF}_{\geq 3}$ in the last section of the paper relying on properties of certain exactly linear formulas and uniform linear formulas described in the next two sections.

Some of the combinatorial structure of a linear formula is reflected by its underlying hypergraph. So, before treating the class of exactly linear hypergraphs and formulas in the next section, let us collect some elementary relations holding for arbitrary linear hypergraphs (the proof can be found in [12]):

Lemma 2. For $H = (V, E) \in \mathrm{LIN}$ with $n := |V|, m := |E| \geq 1$ holds:

(i) $\forall e \in E$ holds $m \geq 1 - |e| + \sum_{x \in e} \omega(x)$,
(ii) $m(m - 1) \geq \sum_{x \in V} \omega(x)(\omega(x) - 1)$,
(iii) $\forall x \in V$ holds $n \geq 1 - \omega(x) + \sum_{e \in E_x} |e|$,
(iv) $n(n - 1) \geq \sum_{e \in E} |e|(|e| - 1)$.

3   Exactly Linear Hypergraphs and Formulas

Let $H = (V, E)$ be an exactly linear hypergraph with $n := |V|$ and $m := |E|$, hence $G_E = K_m$. A basic result is the following:

Proposition 1. For every $H \in \mathrm{XLIN}$ holds $m \leq n$.   □

The result is a special case of the Fisher-inequality [13]. A short indirect proof of which can be found in [11]. Obviously, due to Proposition 1, the Erdös-Farber-Lovàsz conjecture holds for the class of exactly linear hypergraphs.

The above assertion has direct impact on SAT for exactly linear formulas. Restated for formulas, the result above tells us that these formulas and all its subformulas have deficiency $m - n$ at most 0 corresponding to *matching formulas* as introduced in [6]. A matching argument showing satisfiability of deficiency 0 formulas has already been used in [15]. With similar arguments we obtain:

Theorem 2. Every $C \in \mathrm{XLCNF}$ is satisfiable, and a model for C can be determined in $O(\sqrt{n}\|C\|)$ time.

PROOF. Recall that $C \in \mathrm{XLCNF}$ by definition has no pair of complementary unit clauses therefore $H_C \in \mathrm{XLIN}$, similarly every subformula $C' \subseteq C$ is exactly linear, and contains no pair of complementary unit clauses, hence for each $C' \subseteq C$ holds $H_{C'} \in \mathrm{XLIN}$. Now consider I_C the bipartite incidence graph of C with vertex set partition $V(C) \cup C$. It is easy to see, that every subset $C' \subseteq C$ has the neighbourhood $N_I(C') = V(C') \subseteq V(C)$ in I_C. Because of $|C'| \leq |V(C')| = |N_I(C')|$ for every subset $C' \subseteq C$, we can apply the classic Theorem of König-Hall [7,8] for bipartite graphs stating that there exists a matching in I_C covering the component C of the vertex set. In terms of the formula, this means that there is a set of variables, corresponding to the vertices of the matching edges such that each of it is joined uniquely to a clause of C such that no clause is left out. Since these variables are all distinct the corresponding literals can independently be set to true yielding a model of C.

To verify the time bound first observe that for given $C \in$ XLCNF, I_C can be constructed in $O(\|C\|)$ time using appropriate data strutures. Next regard the bipartite matching problem on I_C as formulated previously as a network flow problem: Assign to each of I_C an orientation from the variable partition to the clause partition. Introduce a source vertex joined to each variable vertex by exactly one directed edge, similarly, introduce a sink vertex t such thta each clause vertex gets exactly one directed arc terminating in t, no further edges are added. Even in [5] provided an algorithm for finding a maximum flow which can easily be seen to be equivalent to a König-Hall matching in I_C covering the clause partition. That algorithms runs in $O(\sqrt{p}q)$ time if the network has p vertices and q edges. Because I_C has $\|C\|$ edges and at most $n + m \le 2n$ vertices, the network has at most $\|C\| + 2n$ edges thus in summary, we obtain $O(\sqrt{n}\|C\|)$ as running time for finding a model of $C \in$ XLCNF with n variables.   □

4   Uniform Linear Formulas: Key for NP-Completeness Proof

Observe that a linear formula has the property that each pair of variables occurs at most once. Let $P(C) := \{p_1, \ldots, p_s\}$ denote all pairs of literals that occur in a linear formula $C = \{c_1, \ldots, c_m\}$. Consider the bipartite graph $G_{P(C)}$ associated with C having vertex set bipartition $P(C) \cup C$ and each literal pair p is joined to the unique clause of C it belongs to, hence the degree of each p is 1. In case of a k-uniform formula C, $k \ge 2$, each clause-vertex in $G_{P(C)}$ has degree $k(k-1)/2$. Hence, if C has n variables, we have $s \le \binom{n}{2}$ and on the other hand $s = m \cdot k(k-1)/2$ implying $m \le \frac{n(n-1)}{k(k-1)}$. We only have $s = \binom{n}{2}$ if each pair occurs exactly once, i.e., if the variable graph $G_{V(C)}$ is a clique . So, we have proven:

Lemma 3. *For $C \in$ LCNF k-uniform with n variables always holds $|C| \le \frac{n(n-1)}{k(k-1)}$. And equality holds iff $G_{V(C)}$ is complete.*   □

Satisfiability of linear formulas can be characterized in terms of matchings in $G_{P(C)}$: Clearly, $P(C)$ itself is a 2-uniform linear formula, and in case $P(C)$ is satisfiable then also C is satisfiable. More generally, by the pigeonhole principle holds $s \ge m$. And the fact that each subformula $C' \subseteq C$ again satisfies $|P(C')| \ge |C'|$ enables us once more to apply the König-Hall Theorem providing existence of a matching M of cardinality m covering the clause-vertices in $G_{P(C)}$. Now it is not hard to see that C is satisfiable iff there exists a matching M as above with the additional property that the 2-CNF subformula of $P(C)$ consisting of those literal pairs p that are incident to edges of M is satisfiable. So, if C

has m clauses there are exactly $[k(k-1)/2]^m$ König-Hall matchings in $G_{P(C)}$, and an unsatisfiable formula C forces all $[k(k-1)/2]^m$ subformulas of $P(C)$ of cardinality m selected by the corresponding matchings to be unsatisfiable. Observe that the case $k = 2$ is specific in the sense that it exhibits exactly one König-Hall matching. Therefore it is easy to construct an unsatisfiable linear 2-uniform formula. A shortest one consists of 6 clauses $C = \{c_1, \ldots, c_6\}$ where c_1, c_2, c_3 are determined by $x \Rightarrow y \Rightarrow z \Rightarrow \overline{x}$ and c_4, c_5, c_6 are determined by $\overline{x} \Rightarrow u \Rightarrow v \Rightarrow x$. This is not surprising as in a certain sense the SAT-complexity of a 2-uniform formula is exhibited by its linear part: Suppose C is 2-uniform but not linear and let c, c' be two clauses such that $V(c) = V(c') = \{x, y\}$. Then we claim that c, c' can be removed from C without affecting satisfiability status of C. Indeed, we have three cases: (1) $c = \{x, y\}, c' = \{\overline{x}, y\}$, then c, c' are always satisfied by x. (2) $c = \{\overline{x}, y\}, c' = \{x, \overline{y}\}$, meaning $x \Leftrightarrow y$ a condition according to which the resulting formula can be evaluated. And (3) $c = \{x, y\}, c' = \{\overline{x}, \overline{y}\}$, similarly meaning $x \Leftrightarrow \overline{y}$ which can be handled as before yielding the claim.

To construct an unsatisfiable 3-uniform linear formula "at hand" seems not to be an easy task. Below we will provide a scheme for finding such formulas also revealing that unsatisfiable formulas are very sparsely distributed. On the other hand, it will turn out that finding one such formula answers the earlier stated question whether SAT remains NP-complete also for the class $\mathrm{LCNF}_{\geq 3}$. For obtaining that answer it is useful to consider the combinatorially somehow extreme class of linear formulas C containing each pair of variables exactly once, in other words the variable graph is a clique K_n, for n variables in C. For k-uniform linear hypergraphs, i.e., the monotone case, this situation is also known as a Steiner triple system $S(2, k, n)$ [1]. So we derive some necessary algebraic existence conditions for a k-uniform linear hypergraph $H = (V, E)$ with complete vertex graph G_V. The degree of each vertex x in G_V then is given by

$$\deg_V(x) = \sum_{e \in E_x} (|e| - 1) = (k-1)\omega(x) = n - 1$$

therefore $\omega(x) = \frac{n-1}{k-1}$ for each vertex, hence H is regular. By the length condition for H we immediately derive $k|E| = n\frac{n-1}{k-1}$ hence recovering the assertion of Lemma 3. More generally we have (the proof can be found in [12]):

Proposition 2. *If a k-uniform linear hypergraph, $k \geq 3$, with n vertices admits a complete vertex graph then necessarily $n \in M_1 \cup M_2$ where*

$$M_1 = \{k + jk(k-1)|j \in \mathbb{N}\}, \quad M_2 = \{1 + jk(k-1)|j \in \mathbb{N}\}, \quad M_1 \cap M_2 = \varnothing$$

For $k = 3$ the conditions in Prop. 2 are equivalent to $6|n-3$ or $6|n-1$ which have shown (non-constructively) also to be sufficient by Kirkman, resp. Hanani according to [1]. For k a prime power and n sufficiently large the above conditions also are sufficient [16]. Some specific k-uniform linear hypergraphs admitting complete vertex graphs are listed as the corresponding Steiner triple systems in [1], also confer [9] for a more complete presentation. Although the Hanani result ensures indirectly that there exist very dense linear formulas we have no systematic way to explicitly construct and to investigate them. To circumvent that problem we next provide a scheme for explicitly constructing k-uniform formulas of highest possible clause-variable density. To that end it is instructive first to consider k-uniform exactly linear formulas having a complete variable graph. Clearly, a formula containing only one k-clause is exactly linear and satisfies $G_{V(C)} = K_k$ thus we require formulas of at least two k-clauses.

Definition 2. *A k-uniform formula $B \in$ XLCNF with $|B| > 1$ is called a k-block if $G_{V(C)} = K_{|V(C)|}$. Let $\mathcal{B}_k$ denote the set of all k-blocks, and $n(k) :=$ $1 + k(k-1)$. Any subset of a k-block is called a k-block fragment.*

Lemma 4. *For k-uniform $C \in$ XLCNF, with $|C| > 1$, $k \geq 3$, the following assertions are equivalent:*

(i) C is a k-block,
(ii) $|V(C)| = |C|$,
(iii) $\omega(x) = k$, for each $x \in V(C)$. Each k-block has $n(k)$ variables.

PROOF. Obviously it suffices to consider the monotone case as we only touch the combinatorial hypergraph structure disregarding any logical aspect. We first show (i) implies (ii): If $C \in \mathcal{B}_k$ then $G_{V(C)}$ is a clique, and for each variable x we have $\deg(x) = (k-1)\omega(x) = n - 1$ where $n := |V(C)|$. Therefore $\omega(x) = \frac{n-1}{k-1}$. Moreover, as C is exactly linear also the intersection graph is complete, hence $\deg(c) = k(\frac{n-1}{k-1} - 1) = |C| - 1$. Since $(*)$ $n\frac{n-1}{k-1} = \|C\| = |C|k$ we derive $k(\frac{n-1}{k-1} - 1) = \frac{n(n-1)}{k(k-1)} - 1$ which is equivalent to $(n-k)(n - [1 + k(k-1)]) = 0$ having the roots $n = k$ corresponding to $|C| = 1$ and $n = 1 + k(k-1)$. For the latter case we have $\frac{n-1}{k-1} = k = \omega(x)$, for each $x \in V(B)$. Therefore from $(*)$ we immediately obtain $|C| = n$.

(ii) $\Rightarrow$ (iii): If the formula is regular meaning $\forall x : \omega(x) = j$ then $nj = kn$ by the length condition thus $j = k$, and we are done. If the formula is not regular, we see by the length condition $\sum_{x \in V(C)} \omega(x) = kn$ that if $\omega(x) \leq k$ for each x then already $\omega(x) = k$ for each $x \in V(C)$. So assume there is a variable x with $r := \omega(x) > k \geq 3$. Then C contains at least all r clauses in $C(x)$ each having length k. Suppose there was no further clause then $n = 1 + r(k-1) = |C| = r$ which has a solution in r only for $k = 1$. So there is at least one further k-clause c contained in C. Because c must contain exactly one variable of each clause in $C(x)$ we get that $r \leq k$ so we obtain $\omega(x) = k$ for each $x \in V(C)$.

(iii) $\Rightarrow$ (i): From the degree relation in the variable graph $G_{V(C)}$ we see $\forall x \in V(C) : \deg_{G_{V(C)}}(x) = k(k-1)$ if $\omega(x) = k$ for each $x \in V(C)$ thus the variable graph is $k(k-1)$ regular. Similarly, from the degree relation in the clause graph we obtain $\forall c \in C : \deg_{G_C}(c) = k(k-1) = |C|-1$ hence $|C| = 1+k(k-1)$. Finally, by the length condition we see $nk = \|C\| = k|C|$ thus $n = |C| = 1 + k(k-1)$. Therefore $G_{V(C)}$ with n vertices is $n-1$-regular, so is complete and by definition C is a k-block.   □

Thus in a k-block each clause has length k and each variable occurs exactly k times, moreover the number of variables equals the number of its clauses. As an example consider a monotone 3-block :

$$c_1 := \{x, y_1, y_2\}$$
$$c_2 := \{x, a_{11}, a_{12}\}$$
$$c_3 := \{x, a_{21}, a_{22}\}$$
$$c_4 := \{y_1, a_{11}, a_{21}\}$$
$$c_5 := \{y_1, a_{12}, a_{22}\}$$
$$c_6 := \{y_2, a_{11}, a_{22}\}$$
$$c_7 := \{y_2, a_{12}, a_{21}\}$$

Although we can construct k-blocks also for $k = \{4, 5, 6\}$, we are not aware whether a k-block really exists for arbitrary values of $k \geq 7$. The next result relates that question to the number of latin squares for a given positive integer that mutually satisfy a certain condition. Recall that a *latin square of order* $s \in \mathbb{N}$ is a $s \times s$-matrix where each row and each column contains each element of $S = \{1, \ldots, s\}$ exactly once (cf. e.g. [14,3]), as an examples for $s = 5$ consider the first two of the following matrices:

$$L_5 = \begin{pmatrix} 1\,2\,3\,4\,5 \\ 5\,1\,2\,3\,4 \\ 2\,3\,4\,5\,1 \\ 3\,4\,5\,1\,2 \\ 4\,5\,1\,2\,3 \end{pmatrix}, \quad L_5' = \begin{pmatrix} 1\,2\,3\,4\,5 \\ 4\,5\,1\,2\,3 \\ 3\,4\,5\,1\,2 \\ 5\,1\,2\,3\,4 \\ 2\,3\,4\,5\,1 \end{pmatrix}, \quad \begin{pmatrix} (1,1)\,(2,2)\,(3,3)\,(4,4)\,(5,5) \\ (5,4)\,(1,5)\,(2,1)\,(3,2)\,(4,3) \\ (2,3)\,(3,4)\,(4,5)\,(5,1)\,(1,2) \\ (3,5)\,(4,1)\,(5,2)\,(1,3)\,(2,4) \\ (4,2)\,(5,3)\,(1,4)\,(2,5)\,(3,1) \end{pmatrix}$$

Two latin squares $L = (l_{ij})_{1 \leq i,j \leq s}, L' = (l_{ij}')_{1 \leq i,j \leq s}$ of order s are said to be *orthogonal* iff the pairs (l_{ij}, l_{ij}') are distinct for all $1 \leq i, j \leq s$. L_5, L_5' above are orthogonal as the third matrix above providing all corresponding pairs indicates. A set of latin squares is called *mutually orthogonal*, if each different pair of its elements are orthogonal.

Proposition 3. *A k-block exists if there is a set $\mathcal{L}$ of $k - 2$ latin squares each of order $k - 1$ such that each $K, L \in \mathcal{L}$: $K \neq L$ satisfy the following condition: For each row r of K and each row r' of L holds:*

$$(*) \; \forall 1 \leq q < p \leq k - 1 : r_p = r_p' \Rightarrow r_q \neq r_q'$$

PROOF. Considering k-blocks it suffices to consider its underlying hypergraph corresponding to the monotone case. So, a monotone k-block B has $n(k)$ clauses each of length k, let $c_0 := \{x, y_1, \ldots, y_{k-1}\}$ be its first clause, called the *leading clause*. As each variable occurs in k different clauses of B, there are $k - 1$ further clauses containing x, namely determined by the $(k-1) \times (k-1)$-variable matrix:

$$A_k = \begin{pmatrix} a_{11} & a_{12} & \cdots & a_{1k-1} \\ a_{21} & a_{22} & \cdots & a_{2k-1} \\ \vdots & \vdots & \vdots & \vdots \\ a_{k-11} & a_{k-12} & \cdots & a_{k-1k-1} \end{pmatrix}$$

such that the ith clause contains x and all variables in the ith row of A_k. Observe that the subformula X consisting of all clauses containing x already has

$n(k) = 1 + k(k-1)$ variables that means all remaining $(k-1)(k-1)$ clauses of B can only contain these variables. We collect these clauses in $(k-1)$ subblocks $Y_i, 1 \leq i \leq k-1$, each consisting of $(k-1)$ clauses such that each clause of subblock Y_i contains variable y_i. W.l.o.g. Y_1 can be constructed by filling the remaining positions in the ith clause of Y_1 with the variables in the ith row of A_k^T, the transpose of A. Subblock Y_1 is shown in the left matrix below:

$$Y_1 = \begin{pmatrix} y_1 & a_{11} & a_{21} & \cdots & a_{k-11} \\ y_1 & a_{12} & a_{22} & \cdots & a_{k-12} \\ \vdots & \vdots & \vdots & \vdots & \vdots \\ y_1 & a_{1k-1} & a_{2k-1} & \cdots & a_{k-1k-1} \end{pmatrix}, \quad Y_i = \begin{pmatrix} y_i & a_{1i_{11}} & a_{2i_{12}} & \cdots & a_{k-1i_{1k-1}} \\ y_i & a_{1i_{21}} & a_{2i_{22}} & \cdots & a_{k-1i_{2k-1}} \\ \vdots & \vdots & \vdots & \vdots & \vdots \\ y_i & a_{1i_{k-11}} & a_{2i_{k-12}} & \cdots & a_{k-1i_{k-1k-1}} \end{pmatrix}$$

Observe that the formula $X \cup Y_1$ is exactly linear. Each of the remaining subblocks $Y_i, 2 \leq i \leq k-1$, w.l.o.g. looks as shown in the right above where $I = (i_{pq})_{1 \leq p,q \leq k-1}$ is a latin square of order $k-1$. Obviously, for each i, $X \cup Y_1 \cup Y_i$ is exactly linear. However, to ensure that $Y_i, Y_j, 2 \leq i < j \leq k-1$ satisfy mutually exact linearity the corresponding matrices I, J must satisfy the following condition (*) above clearly guaranteeing that no pair of variables in A_k occurs twice in any clause of $X \cup Y_1 \cup \cdots \cup Y_{k-1}$. Moreover, as then we have $n(k)$ k-clauses we have place capacity for exactly $n(k)k(k-1)/2$ variable pairs which is identical to the number of variab;le pairs we can build over $n(k)$ variables. Therefore by the pigeonhole principle each pair of variables indeed occurs exactly once in case (*) holds. □

Condition (*) and orthogonality for latin squares are incomparable in the sense that in general neither orthogonality implies (*) nor (*) implies orthogonality. Determining $s-1$ latin squares of order s that are mutually orthogonal resp. satisfy (*) is an extremely hard combinatorial task about which there is not much known. E.g. regarding orthogonality, it is known that if the order s is a prime power one can find $s-1$ orthogonal latin squares. Otherwise one can find at least $p^t - 1$ orthogonal latin squares, where p^t is the smallest prime-power in the prime factorization of s [14]. Consequently, if 2 is the smallest prime power one can find at most one orthogonal latin square. However we can also ensure existence of at least one latin square for each k providing the next density result:

Theorem 3. *For each $k \geq 3$ such that $\mathcal{B}_k \neq \emptyset$ one can explicitly construct for each $i \in \mathbb{N}$ a k-uniform linear formula $C^i(k)$, $k \geq 3$, such that $\frac{|C^i(k)|}{|V(C^i(k))|} \in \Omega(2.5^{i-1}) \cap O(3.2^{i-1})$.*

PROOF. To prove the assertion we describe a construction scheme based on Prop. 3: Let k be such that $\mathcal{B}_k$ is not empty and let B^1 be a corresponding monotone k-block as constructed in the proof of Prop. 3. Clearly B^1 has $n(k)$ variables and clauses. Next we build a monotone clause c_{B^1} of length $n(k)$ containing all variables of $V(B^1)$ with x as the first variable and use it as *signature* for our block B^1 canonically as described above. Now we interpret c_{B^1} as the leading clause of a $n(k)$-block fragment denoted as B^2. We only can ensure a block fragment because we do not know whether $\mathcal{B}_{n(k)}$ is non-empty. In any case we obviously can add $n(k)$-clauses to c_{B^1} such that the subblock X for B^2 is complete, each of its clauses again is regarded as the signature of another k-block, which pairwise have only variable x in common. Building X we obtain the $(n(k)-1) \times (n(k)-1)$ variable matrix $A_{n(k)}$ for our $n(k)$-block the transpose of which delivers the next subblock Y_1 of B^2 as shown in the proof of Prop. 3. Now we claim that we can always find at least one latin square of order $n(k) - 1$ satisfying condition $(*)$ above yielding another subblock Y_2 of B^2: Simply perform a cyclic shift of order i to the ith column of $A_{n(k)}^T$ for $0 \le i \le n(k) - 2$ guaranteeing linearity of B^2 as is easy to verify. By construction follows that each clause of B^2 delivers $n(k)$ blocks B^1 which pairwise have at most one variable in common thus expanding $C^2(k)$ that means resolving the signatures into k-blocks from $\mathcal{B}_k$ yields a k-uniform linear formula of $n^{(2)}(k) = n(n(k))$ variables and $[1 + 3(n(k) - 1)]n(k)$ clauses. The procedure described can be continued inductively by constructing a $n^{(i)}(k)$-block fragment $C^i(k)$ consisting of $1 + 3(n^{(i-1)}(k) - 1)$ clauses each of length $n^{(i-1)}(k)$ and regarded as the signature of an $n^{(i-1)}(k)$-block fragment $C^{i-1}(k)$ such that again all these signature-clauses have exactly one variable in common yielding a hierarchie $B^i, i \ge 1$, where $B^1 := B \in \mathcal{B}_k$. Expanding B^i thus provides a k-uniform linear formula $C^i(k)$ of $n^{(i)}(k)$ variables and $[1 + 3(n^{(i-1)}(k) - 1)]|C^{i-1}(k)|$ k-clauses. Again yielding a hierarchy of k-uniform linear formulas $C^i(k), i \ge 1$ where $C^1(k) := B$.

It remains to settle the claim on the clause-variable density $d^i(k) := \frac{|C^i(k)|}{|V(C^i(k))|}$, which is shown by induction on $i \ge 1$. For $i = 1$ we have $C^1(k) := B \in \mathcal{B}(k)$ thus $d^1(k) = n(k)/n(k) = 1$. Now assume the claim holds for all positive integers $\le i$ for fixed $i \ge 2$. Then

$$
\begin{aligned}
d^{i+1}(k) &= \frac{1 + 3[n^{(i)}(k) - 1]}{1 + n^{(i)}(k)[n^{(i)}(k) - 1]} \, |C^i(k)| \\
&\le \frac{1 + 3[n^{(i)}(k) - 1]}{n^{(i)}(k) - 1} \cdot \frac{|C^i(k)|}{|V(C^i(k))|} \\
&= (3 + [n^{(i)}(k)]^{-1}) d^i(k) \quad < \quad 3.2 \, d^i(k)
\end{aligned}
$$

because $n^{(i)}(k) \ge n(k) \ge 7$, and by the induction hypothesis we obtain $d^{i+1}(k) \in O(3.2^i)$. Similarly, for the remaining bound we derive:

$$
\begin{aligned}
d^{i+1}(k) &= \frac{1 + 3[n^{(i)}(k) - 1]}{1 + n^{(i)}(k)[n^{(i)}(k) - 1]} \, |C^i(k)| \\
&> \frac{3[n^{(i)}(k) - 1]}{[n^{(i)}(k)]^{-1} + n^{(i)}(k) - 1} \cdot \frac{|C^i(k)|}{|V(C^i(k))|} \\
&= \frac{1}{1/3 + 1/[3n^{(i)}(k)(n^{(i)}(k) - 1)]} d^i(k) \quad > \quad 2.5 \, d^i(k)
\end{aligned}
$$

where again for the last inequality we used $n^{(i)}(k) \ge n(k) \ge 7$ from which the claim follows by the induction hypothesis. $\square$

We have run some numerical experiments by randomly assigning polarities to the literals in the 3-uniform linear formula $C^2(3)$ containing 133 clauses and 43 variables constructed as shown in the proof above. The experiments supplement the intuition that unsatisfiable formulas are distributed very sparsely: Among 354442000 formulas over the monotone $C^2(3)$ we only found 488 unsatisfiable ones. From one such unsatisfiable formula we extracted a minimal unsatisfiable formula C consisting of 81 clauses, 43 variables. From C, in turn, we extracted a smaller satisfiable formula Γ shown below of 69 clauses and 43 variables containing only 0 as a backbone variable.

```
( -42,   21,   14) (   -0,   40, -39) (   -0,   22, -21) (   -0,    1,    2) (   -0,   26, -25)
(  -0,   24,   23) (   42,    0, -41) (   40,   -1, -34) (   41,   40,   38) (   40, -33,   -2)
(   0,   37, -38) (   42, -40,   37) (   -2, -36,   37) ( -40,   19,   18) (    2,   19, -26)
( -33,   11,   19) ( -37,   19, -13) ( -40,   22, -16) (    1,   22,   28) ( -37,   22,   15)
( -23,   22, -20) ( -19, -22, -24) ( -24,   21,   20) (   39, -24,   17) ( -19,    1,   25)
(  25,   28, -30) (   25, -32,   17) (    1,   13,   -7) ( -37, -25,   -7) ( -22,   -2, -29)
(  -2, -28,   21) (   -2,   -9,   16) (   24, -25,   -2) (   -2,   20, -27) ( -22, -34,   10)
(  -1,   16,   10) (    7,   10,   12) ( -40, -10, -28) (   34, -28, -16) (    0,   28,   27)
(  30,   26,   27) (   39,   27,   -9) (    1, -18, -12) (   41, -27, -12) (    0, -10,   -9)
(  12,    9,   -8) ( -15,   21, -39) ( -19,   21,   23) ( -27, -21,   -1) (   24,   -1,   30)
(  39,   -1,   33) (   -1, -17, -11) (    9, -21,   33) (   33,   32,   36) (    0, -33,   34)
( -27, -34,   13) ( -17,   13,   15) ( -34,   35, -32) (    8, -20, -32) (    1, -20,   26)
(  32,   26,   14) (   28, -26,   29) ( -17,   29, -35) (    0,   16,   15) ( -13,   16, -18)
( -16,   17, -14) (    0, -17,   18) (    2, -23, -30) ( -18,   15, -14)
```

Theorem 4. SAT *remains NP-complete when restricted to* LCNF$_{\geq 3}$.

PROOF. Let CNF$_{\geq 3}$ be the set of all CNF formulas containing only clauses of length at least 3, then clearly SAT is NP-complete for CNF$_{\geq 3}$. We now provide a polynomial time reduction from CNF$_{\geq 3}$-SAT to LCNF$_{\geq 3}$-SAT. The new reduction is a modification of the procedure used in the proof of Theorem 1. So let $C \in$ CNF$_{\geq 3}$ be arbitrary and perform the latter procedure on C. Recall that to transform C into a linear formula C' the procedure replaces overlapping variables by new variables and forces the new variables to be equivalent with the original ones via implicational chains that are added as 2-clauses. These 2-clauses are the only one in the resulting formula C' in case $C \in$ CNF$_{\geq 3}$.

For obtaining LCNF$_{\geq 3}$ to be NP-complete it remains to get rid of the 2-clauses adequately which is done as follows: For each such 2-clause c_i add a 3-uniform linear pattern Γ_i to C' such that $\forall i : V(\Gamma_i) \cap V(C') = \varnothing$ and $V(\Gamma_i) \cap V(\Gamma_j) = \varnothing, i \neq j$. Finally, let x_i be a backbone variable in Γ_i which is forced to be set in Γ_i according to $l(x_i)$, then replace c_i by $c_i \cup \overline{l(x_i)}$. That last step ensures that there are no more 2-clauses in the resulting formula and moreover the added literals must be set to false ensuring that the constructed 3-clauses can take their tasks as providing equivalences with originally overlapping variables in C. Since the Γ_i's are independently satisfiable we are done.   □

5   Concluding Remarks and Open Problems

The class of linear formulas introduced here has been shown to be NP-complete w.r.t. SAT for the general case. The reduction used essentially relies on 2-clauses, and it therefore did not cover the subset of linear formulas that contain only clauses of length at least 3. To prove that SAT remains complete also for this latter class, we provided an enlargment of the earlier reduction by certain patterns of 3-clauses that could eliminate the 2-clauses. However, the question whether SAT is NP-complete for the classes of linear formulas C containing only clauses of length at least k, for each fixed $k \geq 4$ is still open.

Intuitively, one might believe that for large k each k-uniform linear formula should be satisfiable. In turn that meant there was a smallest k_0 for which satisfiability holds implying that every formula containing only clauses of length at least k_0 was satisfiable as longer clauses could be shortened yielding a k_0-uniform linear formula which is satisfiable. However, there are two arguments against that intuition: First observe that the bipartite graph $G_{P(C)}$ admits $[k(k-1)/2]^m$ many König-Hall matchings if the input formula C has m clauses and is k-uniform, which is rapidly growing in k. The 2-CNF formula $P(C)$ extracted from C has $m \cdot k(k-1)/2$ clauses and thus admits $O([m \cdot k(k-1)/2]^m)$ subformulas of cardinality m. Recall that, for C to be unsatisfiable each subformula of $P(C)$ selected by a König-Hall matching must be unsatisfiable. Hence the density of these, let's say, König-Hall subformulas in all subformulas of cardinality m is of order $O(m^m)$ which is independent of k. Second, for larger k formulas of clause-variable density exponentially in i can exist. Clearly, both arguments do not replace a proof but make clear that it is quite likely that it might be hard to find a k for which all k-uniform linear formulas are satisfiable.

Otherwise, detecting a first unsatisfiable one means NP-completeness w.r.t. SAT for the corresponding class by the same padding argument as in the proof of Theorem 4. In conclusion we obtain:

Theorem 5. *For $k \geq 4$,* SAT *remains* NP-*complete restricted to the class* LCNF$_{\geq k}$ *iff there exists an unsatisfiable k-uniform linear formula.*   □

Besides linear CNF formulas having the defining property that each pair of distinct clauses has at most one common variable one can consider *r-intersecting formulas*. Restricting to the class k-CNF of formulas containing only clauses of length at most k one could define for each $r \leq k$ the r-intersecting subclasses k-CNF$_r$ defined as the collection of formulas C in k-CNF such that

$$\forall c, c' \in C, \ c \neq c' : \ |V(c) \cap V(c')| \leq r$$

E.g. for $k = 3$, the class of formulas is left where each two clauses $c \neq c'$ have variable-intersection either 0 or 2. The case 0 or 3 is trivial as one only has to detect whether there exist three variables over which the formula contains all 8 polarity patterns, which is the only case that such a formula can be unsatisfiable. By the Fisher-inequality also $m \leq n$ follows for *exactly r-intersecting* hypergraphs of m edges and n vertices, i.e., above $\leq$ is replaced by $=$. Thus, arguing again with König-Hall as in the exactly linear (= exactly 1-intersecting) case ensures that also exactly r-intersecting formulas are always satisfiable (as long as they contain no pair of complementary unit clauses). However, studying SAT-complexity for the 0 or 2 case is left as an open problem.

Finally, we have some implications towards polynomial time solvability regarding SAT of certain classes of linear formulas C that are characterized by the graph $G_{P(C)}$. Observe that the extracted 2-CNF $P(C)$ is linear and if it is satisfiable then also C is satisfiable. Otherwise $P(C)$ contains an unsatisfiable linear subformula which is determined by implicational double-chains of the form

$$x \Rightarrow l_1 \Rightarrow l_2 \Rightarrow \cdots \Rightarrow l_{p_1} \Rightarrow \overline{x}, \quad \overline{x} \Rightarrow l'_1 \Rightarrow l'_2 \Rightarrow \cdots \Rightarrow l'_{p_2} \Rightarrow x$$

where l_i $1 \leq i \leq p_1$, resp. l'_i, $1 \leq i \leq p_2$ are literals over distinct variables, the length of the double-chain is $p := p_1 + p_2 + 2$ as it is equivalent to p linear 2-clauses. Defining the class $\mathrm{LCNF}(p)$ consisting of all linear formulas such that $P(C)$ has a longest double-implicational chain of length p, we can decide satisfiability for members of $\mathrm{LCNF}(p)$ in $O(poly(p)n^{2p})$ time. A simple corresponding algorithm proceeds as follows: Observe that an input formula $C \in \mathrm{LCNF}(p)$ is unsatisfiable if there is a subformula C' of C of cardinality p for which each König-Hall matching in $G_{P(C')}$ selects an unsatisfiable subformula of $P(C')$; by usual matching algorithms that can be checked in polynomial time. Thus checking all $O(m^p)$ p-subformulas C' yields $O(poly(p)m^p)$ time. Since $m \leq n^2$ the claim follows.

Acknowledgement. We would like to thank M. Gärtner for implementations. Further, we are grateful to the anonymous referees for their helpful comments.

References

1. C. Berge, Hypergraphs, North-Holland, Amsterdam, 1989.
2. M. Böhm and E. Speckenmeyer, A Fast Parallel SAT-Solver – Efficient Workload Balancing, Annals of Mathematics and Artificial Intelligence 17 (1996) 381-400.
3. B. Bollobas, Combinatorics, Set Systems, Hypergraphs, Families of Vectors and Combinatorial Probability, Cambridge University Press, 1986.
4. P. Erdös, Problems and results in Graph Theory, Congressus Numerantium 15 (1976) 169-192.
5. S. Even and O. Kariv, An $O(n^{2.5})$ Algorithm for Maximum Matching in General Graphs, Proc. of 16-th Annual Symposium on Foundations of Computer Science, IEEE, pp. 100-112, 1975.
6. J. Franco, A. v. Gelder, A perspective on certain polynomial-time solvable classes of satisfiability, Discrete Appl. Math. 125 (2003) 177-214.
7. P. Hall, On representatives of subsets, J. London Math. Soc. 10 (1935) 26-30.
8. D. König, Graphen und Matrizen, Math. Fiz. Lapok 38 (1931) 116-119.
9. C. C. Lindner, and A. Rosa (Eds.), Topics on Steiner Systems, Annals of Discrete Math. 7, North-Holland 1980.
10. R. Monasson, R. Zecchina, S. Kirkpatrick, B. Selman, and L. Troyanski, Determining Computational Complexity from Characteristic 'Phase Transitions', Nature 400 (1999) 133-137.
11. R. Palisse, A short proof of Fisher's inequality, Discrete Math. 111 (1993) 421-422.
12. S. Porschen, E. Speckenmeyer, and B. Randerath, On linear CNF formulas, Techn. Report zaik2006-520, Univ. Köln, 2006.
13. H. J. Ryser, An extension of a theorem of de Bruijn and Erdös on combinatorial designs, J. Algebra 10 (1968) 246-261.
14. H. J. Ryser, Combinatorial Mathematics, Carus Mathematical Monographs 14, Mathematical Association of America, 1963.
15. C. A. Tovey, A Simplified NP-Complete Satisfiability Problem, Discrete Appl. Math. 8 (1984) 85-89.
16. R. M. Wilson, An existence theory for pairwise balanced designs, II, J. Combin. Theory A 13 (1972) 246-273.

A Dichotomy Theorem for Typed Constraint Satisfaction Problems

Su Chen[1], Tomasz Imielinski[1], Karin Johnsgard[2],
Donald Smith[1], and Mario Szegedy[1]

[1] Rutgers University, Piscataway NJ 08854, USA
[2] Monmouth University, West Long Branch NJ 07764, USA

Abstract. This paper is a contribution to the general investigation into how the complexity of constraint satisfaction problems (CSPs) is determined by the form of the constraints. Schaefer proved that the Boolean generalized CSP has the dichotomy property (i.e., all instances are either in P or are NP-complete), and gave a complete and simple classification of those instances which are in P (assuming P $\neq$ NP) [20]. In this paper we consider a special subcase of the generalized CSP. For this CSP subcase, we require that the variables be drawn from disjoint Boolean domains. Our relation set contains only two elements: a monotone multiple-arity Boolean relation R and its complement $\overline{R}$. We prove a dichotomy theorem for these monotone function CSPs, and characterize those monotone functions such that the corresponding problem resides in P.

1   Context and Related Work

Since Ladner's demonstration that if P $\neq$ NP, then there exist infinitely many problems in NP that are neither in P nor are NP-complete [19], there have been attempts to classify broad subclasses of NP by complexity. Therefore it is of striking interest that several (natural) subclasses of NP have been shown to exhibit dichotomy: that is, all members of the subclass either are in P, or are NP-complete [20], [16], [8], [3], [9], [18], [6], [4], [2]. (Of course if P = NP there is no "dichotomy.") In a number of such cases, it has been shown that there is a surprisingly compact set of conditions which determine whether a given member of the subclass is in P (is not NP-complete). There are also attempts to identify the largest subclass of NP for which such a dichotomy holds (see notably [13], [10], [2]).

A natural subclass of NP in which to carry out such investigations is that of Constraint Satisfaction Problems (CSPs). The generalized CSP case is important from a classification standpoint because it contains both 2SAT (in P) and 3SAT (NP-complete); it is important pragmatically because it contains special cases arising from applications throughout mathematics and computer science. In its most general sense, a CSP is a problem of the form: For each of the given variables, find an assignment of a value (from the appropriate specified domain) in such a way that all members of the given set of specified constraints are met. If all the variables have the same two-valued (Boolean) domain and the

A. Biere and C.P. Gomes (Eds.): SAT 2006, LNCS 4121, pp. 226–239, 2006.

constraints require satisfying logical propositions, this is the Boolean generalized satisfiability problem (see for example [14]).

Formally, the generalized CSP may be described as follows.

An **instance** consists of a specification of the following data:

- A set of domains $\{D_j | j \in J\}$, and for each domain D_j an associated set of variables V_j.
- A set of relations $\{R_i | i \in I\}$ over tuples of the domains. More precisely, for each $i \in I$ there is associated:
  - a positive integer Q_i [the *arity* (or *rank*) of R_i];
  - a Q_i-tuple of domain indices $\boldsymbol{j_i} = \big(j(i,1), j(i,2), \ldots, j(i, Q_i)\big) \in J^{Q_i}$ [this tuple is called the *signature* of R_i];
  - a relation $R_i \subseteq D_{j(i,1)} \times D_{j(i,2)} \times \ldots \times D_{j(i,Q_i)}$ over the corresponding domains.
- A set of constraint relations $\{C_i | i \in I\}$, where C_i is a relation over the variable sets having the same arity and signature as R_i. That is, $C_i \subseteq V_{j(i,1)} \times V_{j(i,2)} \times \ldots \times V_{j(i,Q_i)}$. [A *constraint* is a particular tuple from some C_i.]

The corresponding **question** is: Is there an assignment of variables to domain elements $\tau : \bigcup_{j \in J} V_j \to \bigcup_{j \in J} D_j$ with $\tau|_{V_j} \mapsto D_j$ so that for all $i \in I$ and for all $\boldsymbol{v} = (v_1, v_2, \ldots, v_{Q_i}) \in C_i$ we have $\big(\tau(v_1), \tau(v_2), \ldots, \tau(v_{Q_i})\big) \in R_i$?

(For this problem to be in NP all sets must be finite, assumed throughout.)

For the Boolean generalized CSP, there is a single two-element domain $D = \{0, 1\}$. Schaefer proved that the Boolean generalized CSP has the dichotomy property (i.e., all instances either in P or are NP-complete), and moreover gave a complete and simple classification of those instances which are in P (assuming $P \neq NP$) [20]. A powerful algebraic approach by polymorphisms, confirming and extending those results, was taken in [18] and [4]. That method does not appear to be applicable to the case in this paper.

A dichotomy result for satisfiability in which there is a single *three*-element domain was recently proved in [3], using methods from universal algebra. For single domains of larger cardinality, the problem is open, despite investigation and conjecture (see e.g. [13], [6], [2]). An overview of CSP complexity via limiting the types of constraints is in [7].

Interesting explorations into classification have been possible when limiting the number of occurences of each variable [17], [11], [10], [12]. Using the equivalence of CSP and homorphisms noted in [13], complexity classifications results have been obtained by [16] and [15].

There has recently been some initial exploration into the generalized CSP. Here the various variables are allowed to be drawn from distinct domains (as is often the case in applications). In [5] the authors demonstrate that treating distinct domains as one can disguise the complexity of the problem.

2   Typed Constraint Satisfaction Problem

In this paper we consider a special subcase of the generalized CSP. As in [5] we allow (in our case, *require*) distinct domains, but our domains are all two-valued (Boolean). Our relation set contains only two elements: a Q-arity Boolean relation R, and its complement $\overline{R} = \{0,1\}^Q \backslash R$.

It is convenient in this special case to think of R as being the set of Q-tuples mapped by a Boolean function $g : \{0,1\}^Q \to \{0,1\}$ to 1. So $\overline{R}$ is the analogous set of tuples for $\overline{g}$. We limit our attention to functions g that are monotone increasing. (Monotone functions and their properties are described in section 3.)

For a function $g : \{0,1\}^Q \to \{0,1\}$, an **instance of the problem SAT**$[g]$ consists of *disjoint* Boolean variable sets $V_1, V_2, \ldots, V_Q$ and two Boolean vector sets $\mathcal{P} = \{\boldsymbol{p_j} | j \in 1 \ldots P\}, \mathcal{N} = \{\boldsymbol{n_k} | k \in 1 \ldots N\}$, where $\mathcal{P}, \mathcal{N} \subseteq V_1 \times V_2 \times \ldots \times V_Q$. We wish to know whether $\Phi_g(\mathcal{P}, \mathcal{N}) = \bigwedge_{j=1}^{P} g(\boldsymbol{p_j}) \wedge \bigwedge_{k=1}^{N} \overline{g}(\boldsymbol{n_k})$ is satisfiable. (Clearly the answer is *yes* if $\mathcal{P} = \emptyset$ or if $\mathcal{N} = \emptyset$, so such instances are regarded as **trivial**. Likewise, the answer is *no* if $\mathcal{P} \cap \mathcal{N} \neq \emptyset$, in which case it is **inconsistent**.) We refer to an instance as a **typed constraint satisfaction problem (CSP) for the function** g.

This problem is clearly in NP; we wish to characterize those *monotone* functions for which SAT$[g]$ is in P (and those for which it is NP-complete).

For a non-constant monotone increasing Boolean function g, the minimal DNF [and CNF] of g contains positive literals only. Likewise, the minimal CNF of $\overline{g}$ contains only negated literals. We therefore regard the set $\{g(\boldsymbol{p_j}) = 1\}_{j=1}^{P}$ as the set of **positive constraints**, and the set $\{\overline{g}(\boldsymbol{n_k}) = 1\}_{k=1}^{N}$ as the set of **negative constraints**. Likewise, $\mathcal{P}$ is the set of **positive constraint vectors**, $\mathcal{N}$ the **negative constraint vectors**.

Lemma 1. *For a monotone function* $g: \{0,1\}^Q \to \{0,1\}$, *let* $g^\wedge: \{0,1\}^{Q+1} \to \{0,1\}$ *be defined by* $g^\wedge(x_1, \ldots, x_{Q+1}) = g(x_1, \ldots, x_Q) \wedge x_{Q+1}$. *Define* $g^\vee$ *similarly, and let* g^* *be defined by* $g^*(x_1, \ldots, x_{Q+1}) = g(x_1, \ldots, x_Q)$.

If SAT$[g]$ is in P, then g^*, $g^\wedge$, *and* $g^\vee$ *also have this property. If SAT$[g]$ is NP-complete, so are the corresponding satisfiability problems for* g^*, $g^\wedge$, *and* $g^\vee$.

Proof of Lemma 1. The statements are obvious for g^*.

Let $g : \{0,1\}^Q \to \{0,1\}$, and consider any instance of SAT$[g^\wedge]$: We have disjoint sets of Boolean variables $V_1, V_2, \ldots, V_{Q+1}$, and two sets of Boolean vectors $\mathcal{P} = \{\boldsymbol{p_j} | j \in 1 \ldots P\}, \mathcal{N} = \{\boldsymbol{n_k} | k \in 1 \ldots N\} \subseteq V_1 \times V_2 \times \ldots \times V_{Q+1}$. Consider the satisfiability of $\Phi_{g^\wedge} = \bigwedge_{j=1}^{P} g^\wedge(\boldsymbol{p_j}) \wedge \bigwedge_{k=1}^{N} \overline{g^\wedge}(\boldsymbol{n_k})$.

For this proof only, we adopt the following notation for a "truncated" vector. For any $\boldsymbol{v} = (v_1, v_2, \ldots, v_{Q+1}) \in V_1 \times V_2 \times \ldots V_{Q+1}$, write $\boldsymbol{v}^{\text{trunc}}$ for $(v_1, v_2, \ldots, v_Q)$. So we may write $g^\wedge(\boldsymbol{v}) = g(\boldsymbol{v}^{\text{trunc}}) \wedge v_{Q+1}$, and $\overline{g^\wedge}(\boldsymbol{v}) = \overline{g}(\boldsymbol{v}^{\text{trunc}}) \vee \overline{v_{Q+1}}$.

For any $s \in V_{Q+1}$, let $\{\boldsymbol{p_{j_l}} | l \in 1 \ldots P_s\}$ be the subset of positive constraint vectors containing s. Then the conjunction of corresponding positive constraints is: $\bigwedge_{l=1}^{P_s} g^\wedge(\boldsymbol{p_{j_l}}) = \bigwedge_{l=1}^{P_s} \left(g(\boldsymbol{p_{j_l}}^{\text{trunc}}) \wedge s \right) = \left[\bigwedge_{l=1}^{P_s} g(\boldsymbol{p_{j_l}}^{\text{trunc}}) \right] \wedge s$.

Similarly, for $s \in V_{Q+1}$, let $\{\boldsymbol{n_{k_m}} | m \in 1 \ldots N_s\}$ be the subset of negative constraint vectors containing s. Then the conjunction of corresponding constraints is: $\bigwedge_{m=1}^{N_s} \overline{g^\wedge}(\boldsymbol{n_{k_m}}) = \bigwedge_{m=1}^{N_s} \left(\overline{g}(\boldsymbol{n_{k_m}}^{\text{trunc}}) \vee \overline{s} \right) = \left[\bigwedge_{m=1}^{N_s} \overline{g}(\boldsymbol{n_{k_m}}^{\text{trunc}}) \right] \vee \overline{s}$.

Consider any truth assignment that maps each element of V_{Q+1} to 1. By the above, this reduces our satisfiability problem to $\Phi_g = \bigwedge_{j=1}^{P} g^{\wedge}(p_j{}^{\text{trunc}}) \wedge \bigwedge_{k=1}^{N} \overline{g^{\wedge}}(n_k{}^{\text{trunc}})$, an instance of SAT$[g]$.

In fact, the instance $\Phi_{g^{\wedge}}$ may be satisfiable in cases when Φ_g is not: For any element s of V_{Q+1} which occurs only in negative constraints, map s to 0. Since $\left[\bigvee_{m=1}^{N_s} \overline{g} \left(n_{k_m}{}^{\text{trunc}} \right) \right] \vee \overline{s}$ is trivially satisfied, we can eliminate some of the negative constraints from Φ_g to obtain Ψ_g (if distinct, Ψ_g has fewer constraints).

It follows that if SAT$[g]$ may be determined in polynomial time, then so may SAT$[g^{\wedge}]$. Using the same arguments, if SAT$[g]$ is NP-complete, then for a given (non-trivial) instance $\Theta_g(\mathcal{P}, \mathcal{N})$ we can create a corresponding instance $\Theta_{g^{\wedge}}(\mathcal{P} \times \{s\}, \mathcal{N} \times \{s\})$ (here $V_{Q+1} = \{s\}$). In any satisfying truth assignment for $\Theta_{g^{\wedge}}$, s must be mapped to 1 (because there is at least one positive constraint). So $\Theta_{g^{\wedge}}$ is satisfiable iff Θ_g is. (The argument for $g^{\vee}$ is symmetric.) $\qquad\square$

We say a function g of Q variables **depends on its ith variable** if there are vectors $a_1 \in \{0,1\}^{i-1}$ and $a_2 \in \{0,1\}^{Q-i}$ so that $g(a_1, 0, a_2) \neq g(a_1, 1, a_2)$.

Corollary 1. *Let g be a monotone function of arity at least two. If the minimal CNF (or the DNF) of g contains a single-literal clause, or if g does not depend on some variable, then we may derive from g a function h of fewer variables, such that SAT$[g]$ and SAT$[h]$ are either both in P, or are both NP-complete. After a finite number of reductions, we derive some function which cannot be further reduced.*

Let us define the set $\mathcal{M}$ to be all monotone increasing functions which cannot be reduced under this process. We wish to show:

Theorem 1. *For any monotone increasing Boolean function g, $\mathbf{SAT}[g]$ is either polynomial-time decidable, or is NP-complete.*

Consider a monotone increasing Boolean function g. By the above, it is no loss of generality to assume that g is in $\mathcal{M}$.

Then SAT$[g]$ is in P if (and only if, assuming P $\neq$ NP) g is constant, or is the identity function, or is one of the following functions or the dual thereof, up to permutation on the variables:

$$Triangle(x_1, x_2, x_3) = (x_1 \vee x_2) \wedge (x_1 \vee x_3) \wedge (x_2 \vee x_3)$$
$$Box(x_1, x_2, x_3, x_4) = (x_1 \vee x_2) \wedge (x_2 \vee x_3) \wedge (x_3 \vee x_4) \wedge (x_4 \vee x_1)$$
$$Path(x_1, x_2, x_3, x_4) = (x_1 \vee x_2) \wedge (x_2 \vee x_3) \wedge (x_3 \vee x_4) \ .$$

(The names *Triangle*, *Box*, and *Path* are based on the hypergraph representations of these functions. *Triangle* corresponds to the majority operation.)

3   Essential Terminology: Monotone Boolean Functions

If $a = (a_1, a_2, \ldots, a_Q)$ and $b = (b_1, b_2, \ldots, b_Q)$ are vectors in $\{0,1\}^Q$, then we say that $a \leq b$ if for all $i \in \{1, 2, \ldots, Q\}$ we have $a_i \leq b_i$.

A Boolean function $g\colon \{0,1\}^Q \to \{0,1\}$ is said to be **monotone (increasing)** if for any $a, b \in \{0,1\}^Q$ such that $a \le b$, we have $g(a) \le g(b)$.

For $g, h\colon \{0,1\}^Q \to \{0,1\}$, we write that $\mathbf{g} \le \mathbf{h}$ if for all $a \in \{0,1\}^Q$ we have $g(a) \le h(a)$. Note that in this case, $g^{-1}(1) \subseteq h^{-1}(1)$. For this reason, it is said that $\mathbf{g}$ **implies h**. We write $\mathbf{g} < \mathbf{h}$ if $g \le h$ and $g \ne h$.

It is sometimes convenient to write $\mathbf{x^0}$ for x, and $\mathbf{x^1}$ for $\bar{x}$.

If $X = \{x_1, x_2, \ldots, x_Q\}$ is a set of Boolean variables, an **implicant** on X is a non-empty conjunction of literals $m(x_1, x_2, \ldots, x_Q) = x_{i_1}^{a_1} \wedge x_{i_2}^{a_2} \wedge \ldots \wedge x_{i_P}^{a_P}$, where $P \ge 1$, $a = (a_1, a_2, \ldots, a_P) \in \{0,1\}^P$, and $1 \le i_1 < i_2 < \ldots < i_P \le Q$. For a Boolean function $g\colon \{0,1\}^Q \to \{0,1\}$, an implicant $m(x_1, x_2, \ldots, x_Q)$ is said to be **an implicant of the function** g if both m implies g [i.e., $m^{-1}(1) \subseteq g^{-1}(1)$], and $m^{-1}(1) \ne \emptyset$. An implicant m of g is said to be a **prime implicant of** g if there is no implicant m' of g such that $m' < m$. A prime implicant of g corresponds to a subset S of X, minimal under the property that some assignment for S forces the function value to be 1, regardless of the values for the remaining variables.

Symmetrically: An **implicate** on X is a non-empty ordered disjunction of literals from X. We say that M is an *implicate of a Boolean function* g if both g implies M and $M^{-1}(0) \ne \emptyset$, and a *prime implicate of* g if there is no implicate M' of g so that $M' < M$.

When a non-constant g is monotone (increasing), each implicant of the minimal DNF of g is a prime implicant consisting of positive literals only, and likewise for the CNF [21].

4   A Class of Monotone Functions g with SAT$[g]$ in P

Consider the set $\mathcal{M}_2 \subseteq \mathcal{M}$ consisting of the following three classes:

1. (Constant) $g, g^{\mathrm{dual}}\colon \{0,1\} \to \{0,1\}$ where $g \equiv 0$ and $g^{\mathrm{dual}} \equiv 1$;
2. (Identity) $g\colon \{0,1\} \to \{0,1\}$ by $g(x) = x$ (this is self-dual);
3. (Positive pair decomposition) $g, g^{\mathrm{dual}}\colon \{0,1\}^Q \to \{0,1\}$ have non-empty CNF and DNF decompositions so that:

$$g\,(x_1, x_2, \ldots, x_Q) = \bigwedge_{d=1}^{D} (y_{(d,1)} \vee y_{(d,2)}) = \bigvee_{c=1}^{C} (z_{(c,1)} \wedge z_{(c,2)}) \ , \ \text{and}$$

$$g^{\mathrm{dual}}(x_1, x_2, \ldots, x_Q) = \bigvee_{d=1}^{D} (y_{(d,1)} \wedge y_{(d,2)}) = \bigwedge_{c=1}^{C} (z_{(c,1)} \vee z_{(c,2)}) \ , \ \text{where}$$

- In both the CNF and DNF expressions, the pairs cover the variable set.
- Variables within any pair (in DNF or CNF) are distinct.
- Any two pairs within a decomposition may be assumed to be distinct.

Three examples of functions with positive pair decomposition are listed in Theorem 1. In fact these three (and their duals) are, up to permutation of the variables, the *entire* set of such functions (Lemma 5).

Lemma 2. *For each g in $\mathcal{M}_2$, the satisfiability problem $SAT[g]$ is in P.*

Proof of Lemma 2. For $g \colon \{0,1\}^Q \to \{0,1\}$ in $\mathcal{M}_2$, consider an instance of $SAT[g]$: $\Phi(\mathcal{P}, \mathcal{N}) = \bigwedge_{j=1}^{P} g(\boldsymbol{p_j}) \wedge \bigwedge_{k=1}^{N} \overline{g}(\boldsymbol{n_k})$.

If $g \equiv 0$, the instance is satisfiable iff $\mathcal{P} = \emptyset$; $g \equiv 1$ is satisfiable iff $\mathcal{N} = \emptyset$.

If $g \colon \{0,1\} \to \{0,1\}$ is the identity, the instance is satisfiable iff $\mathcal{P} \cap \mathcal{N} = \emptyset$.

For a function g with positive pair decomposition, say $g(x_1, x_2, \ldots, x_Q) = \bigwedge_{d=1}^{D} (y_{(d,1)} \vee y_{(d,2)}) = \bigvee_{c=1}^{C} (z_{(c,1)} \wedge z_{(c,2)})$, we observe that $\overline{g}(x_1, x_2, \ldots, x_Q) = \bigvee_{d=1}^{D} (\overline{y_{(d,1)}} \wedge \overline{y_{(d,2)}}) = \bigwedge_{c=1}^{C} (\overline{z_{(c,1)}} \vee \overline{z_{(c,2)}})$. So $\Phi(\mathcal{P}, \mathcal{N})$ is an instance of 2-SAT, and therefore its satisfiability can be determined in polynomial time. $\square$

A set $\mathcal{S}$ of Boolean functions is said to be **closed with respect to disjoint conjunction** if, for any function $g \colon \{0,1\}^Q \to \{0,1\}$ in $\mathcal{S}$, the function $g^{\wedge}(x_1, \ldots, x_{Q+1}) = g(x_1, \ldots, x_Q) \wedge x_{Q+1}$, is also in $\mathcal{S}$. We define **closure under disjoint disjunction** similarly. We say $\mathcal{S}$ is **closed with respect to immaterial variables** if for each $g \in \mathcal{S}$ of arity Q, the function $g^{*}(x_1, \ldots, x_{Q+1}) = g(x_1, \ldots, x_Q)$, is also in $\mathcal{S}$.

Let $\mathrm{Cl}(\mathcal{S})$ denote the closure of the set $\mathcal{S}$ under disjoint conjunction, disjoint disjunction, and immaterial variables.

It follows immediately from lemmas 1 and 2 that:

Theorem 2. *$\mathrm{Cl}(\mathcal{M})$ contains all monotone increasing Boolean functions. For each function g in $\mathrm{Cl}(\mathcal{M}_2)$, the satisfiability problem $SAT[g]$ is in P.*

5   Monotone Functions so $SAT[g]$ Is NP-Complete

For a non-constant monotone function g, a conjunction $x_{i_1} \wedge x_{i_2} \wedge \ldots \wedge x_{i_P}$ defines a prime implicant of g iff the set of variables $\{x_{i_j}\}_{j=1}^{P}$ is minimal under the property that assigning each of these variables the value 1, ensures that the value of g is 1. Such a variable set will be called a **1-set of** g. Likewise, a disjunction $x_{i_1} \vee x_{i_2} \vee \ldots \vee x_{i_P}$ defines a prime implicate of g iff the set $\{x_{i_j}\}_{j=1}^{P}$ is minimal under the property that assigning each of these variables the value 0, ensures that the value of g is 0. Such a variable set will be called a **0-set of** g.

Note the functions in $\mathcal{M}_2 \subseteq \mathcal{M}$ are either constant, or so that the maximum cardinality over all 1-sets and 0-sets is no more than two.

We now consider $\mathcal{M}_3 = \mathcal{M} - \mathcal{M}_2$. Necessarily, $\mathcal{M}_3$ is the set of monotone functions g so that 1) g depends on each of its variables; 2) g has no 1-set and no 0-set of cardinality exactly one; and 3) g has either some 1-set or some 0-set of cardinality at least three. (So for any function f in $\mathcal{M}_3$, the *minimum* cardinality for 1-sets and 0-sets is two, and the *maximum* is at least three.)

Corollary 2. *For any monotone increasing Boolean function g, either g is in $\mathrm{Cl}(\mathcal{M}_2)$ (in which case $SAT[g]$ is in P) or g is in $\mathrm{Cl}(\mathcal{M}_3)$.*

Thus, to establish a dichotomy, it would be enough to show that:

Theorem 3. *For all g in $\mathcal{M}_3$, the satisfiability problem $SAT[g]$ is NP-complete.*

We first require an intermediate result. By the **restricted 3SAT problem** (or **R-3SAT**) we shall mean instances of 3SAT $\bigwedge_{i=1}^{Q}(\alpha_i \vee \beta_i \vee \gamma_i)$, where $\alpha_i \in A \cup A^{\neg}$, $\beta_i \in B \cup B^{\neg}$, $\gamma_i \in C \cup C^{\neg}$ for disjoint sets A, B, and C.

Lemma 3. *The restricted 3SAT problem, R-3SAT, is NP-complete.*

Proof of Lemma 3. Let $\Phi = \bigwedge_{i=1}^{Q}(\alpha_i \vee \beta_i \vee \gamma_i)$ be an instance of the unrestricted 3SAT problem, with $\alpha_i, \beta_i, \gamma_i \in X \cup X^{\neg}$; let $X = \{x_1, x_2, \ldots, x_n\}$ be the variable set of Φ. We will define an equivalent instance for R-3SAT. First we create disjoint variable sets $A = \{a_1, a_2, \ldots, a_n\}$, $B = \{b_1, b_2, \ldots, b_n\}$, $C = \{c_1, c_2, \ldots, c_n\}$.

We need to express the constraint that $a_j = b_j = c_j$. Let $N = \{0,1\}^3 - \{(0,0,0),(1,1,1)\}$, and recall the notation $x^0 \equiv x$, $x^1 \equiv \overline{x}$. For each $j \in \{1 \ldots n\}$, define $G_j = \bigwedge_{(v_1,v_2,v_3) \in N}(a_j^{1-v_1} \vee b_j^{1-v_2} \vee c_j^{1-v_3})$.

Also, from Φ we create a CNF Φ' by replacing each disjunct $\alpha_i \vee \beta_i \vee \gamma_i$ by $\alpha_i' \vee \beta_i' \vee \gamma_i'$ so that each literal in $X \cup X^{\neg}$ is replaced by the appropriate copy in $A \cup A^{\neg}$, $B \cup B^{\neg}$, or $C \cup C^{\neg}$. (For example, replace $\overline{x_3} \vee x_1 \vee \overline{x_2}$ by $\overline{a_3} \vee b_1 \vee \overline{c_2}$.)

We see that Φ is satisfiable iff the instance $G_1 \wedge G_2 \wedge \ldots \wedge G_n \wedge \Phi'$ of R-3SAT is satisfiable. It follows that R-3SAT is NP-complete.   □

It remains for any function g in $\mathcal{M}_3$ to define a map carrying each instance Φ of R-3SAT to an instance $\Psi_{\Phi,g}$ of SAT$[g]$ so that Φ satisfiable $\Leftrightarrow \Psi_{\Phi,g}$ satisfiable.

The result we need is based solely on the properties of g, and does not depend on any instance of SAT$[g]$. For this reason, we may temporarily set aside the language of satisfiability, and concentrate solely on the 1-sets and the 0-sets of g. A methodology for studying collections of subsets is provided by hypergraphs.

6   Hypergraphs and Monotone Functions

A [**nonempty**] **hypergraph** is a ordered pair (X, E), where X is a finite [nonempty] set and E is a collection of non-empty subsets of X whose union is X. Elements of X are referred to as **nodes** of the hypergraph, and elements of E as **edges**. A hypergraph is said to be **simple** if no edge contains any distinct edge. (A general source for hypergraph terminology is [1].)

If (X, E) is a hypergraph, a **blocker** of the hypergraph is some subset of the nodes which intersects every edge $e \in E$. A blocker is considered **minimal** if no proper subset of the blocker is itself a blocker.

If $H = (X, E)$ is a hypergraph, then the **transversal** (or **blocker**) **hypergraph** of the hypergraph is a simple hypergraph $\text{Tr}(H) = (X, B)$, where B consists of all minimal blockers of (X, E). In the case that the original hypergraph H is simple, $\text{Tr}(\text{Tr}(H)) = H$.

Lemma 4. *Consider a simple hypergraph with non-empty edge set, with the additional property that neither the hypergraph nor its transversal hypergraph contains any single element edge. Then:*

- *For any edge e, the complement of e, together with any node of e, is a blocker.*
- *For any node, there is an edge which does not contain the node.*

- *For any node n and edge e not containing n, there is some minimal blocker containing n and some node of e.*

Proof of Lemma 4. Let $H = (X, E)$ be such a hypergraph.
Let e be an edge; by definition, e is a non-empty subset of X, so let $n \in E$. Consider the node set $Y_{(e,n)} = (X - e) \cup \{n\}$. We claim that $Y_{(e,n)}$ is a blocker: Any edge of H that does not intersect e is entirely contained in $Y_{(e,n)}$. Any edge $e' \neq e$ intersecting e can't be contained in e (since H is simple), and so must contain some node outside of e, hence in $Y_{(e,n)}$. Finally, $n \in e \cap Y_{(e,n)}$.

Because the transversal hypergraph contains no single element sets, for any node there is some edge not containing the node.

Consider a fixed node n and an edge $e \in E$ so that $n \notin e$. By the above, $Y_{(e,n)}$ is a blocker. It therefore contains some minimal blocker b. Since b must intersect e, and since n is the only node of e in $Y_{(e,n)} \supseteq b$, necessarily $n \in b$.   □

A node n in a hypergraph is a **basepoint for an odd alternating circuit** if there is an node sequence $n = n_0, n_1, \ldots, n_{2k+1} = n$ (where $k \geq 1$) such that:

- Any two consecutive nodes are distinct;
- For $i \in \{1, \ldots, k+1\}$, some minimal blocker b_i contains n_{2i-2} and n_{2i-1};
- For $i \in \{1, \ldots, k\}$, some edge e_i contains n_{2i-1} and n_{2i}.

Analogously, an **even alternating chain from n to m** is a node sequence $n = n_0, n_1, \ldots, n_{2k} = m$ (where $k \geq 1$) such that any two consecutive nodes are distinct; for $i \in \{1, \ldots, k\}$, some minimal blocker b_i contains both n_{2i-2} and n_{2i-1} and some edge e_i contains both n_{2i-1} and n_{2i}.

Theorem 4. *Consider a simple hypergraph with non-empty edge set, with the additional property that neither the hypergraph nor its transversal hypergraph contains any single element edge. Then each node of the hypergraph is the basepoint for an odd alternating circuit.*

Proof of Theorem 4. Fix a node n_*. Define a subset S_* of the nodes consisting of n_* together with all nodes $n \in S_*$ so that there is an even alternating chain from n_* to n. We claim first that S_* intersects all edges of E, i.e., S_* is a blocker:

Clearly any edge containing n_* intersects S_*. So now consider an edge e such that $n_* \notin e$. By Lemma 4, there is some minimum blocker $b \in B$ containing n_* and some node n' in e. Since e necessarily contains another node n'', the sequence n_*, n', n'' is an even alternating chain. It follows that $n'' \in S_*$, so e intersects S_*.

Since S_* is a blocker, it contains a minimal blocker $b \in B$; we know $|b| \geq 2$. In the special case that $b = \{n_*, n\}$, we see that n_* is a basepoint for the odd alternating circuit obtained by appending $n_{2k+1} = n_*$ to the even alternating chain from n_* to n. Otherwise, b contains two nodes n and n', neither of which is n_*. By the definition of S_*, there is an even alternating chain β from n_* to n, and one γ from n_* to n'. Since $b \in B$ contains both n and n', we obtain the desired odd circuit by concatenating the sequences β and γ^{-1}.   □

For a non-constant monotone function, we consider the collection of 1-sets.

Observation 1. If $g : \{0,1\}^Q \to \{0,1\}$ is monotone and $\boldsymbol{a} = (a_1, \ldots, a_Q)$ is such that $g(\boldsymbol{a}) = 1$, then $X_{\boldsymbol{a}} = \{x_i | a_i = 1\}$ contains a 1-set of g.

Proof of Observation 1. Using the notation $x^0 \equiv x$, $x^1 \equiv \bar{x}$: Define a conjunction $m_{\boldsymbol{a}} = x_1^{1-a_1} \wedge x_2^{1-a_2} \wedge \ldots \wedge x_Q^{1-a_Q}$. Since $m_{\boldsymbol{a}}^{-1}(1) = \{\boldsymbol{a}\}$ and $g(\boldsymbol{a}) = 1$, $m_{\boldsymbol{a}}$ is an implicant of g. If it is a prime implicant, $X_{\boldsymbol{a}}$ is the desired 1-set. Otherwise, there is $\boldsymbol{b} < \boldsymbol{a}$ such that $m_{\boldsymbol{b}}$ is a prime implicant of g, and $X_{\boldsymbol{b}} \subset X_{\boldsymbol{a}}$ is a 1-set. □

Consequence 1. If $g : \{0,1\}^Q \to \{0,1\}$ is monotone and depends on its Pth variable, then there is some 1-set of g containing the Pth variable. (So the 1-sets cover the variable set of g.)

Proof of Consequence 1. Choose vectors $\boldsymbol{a} = (a_1, \ldots, a_{P-1}, 1, a_{P+1}, \ldots, a_Q)$ and $\boldsymbol{a}' = (a_1, \ldots, a_{P-1}, 0, a_{P+1}, \ldots, a_Q)$ so that $g(\boldsymbol{a}) = 1$ and $g(\boldsymbol{a}') = 0$. By Observation 1, $X_{\boldsymbol{a}}$ contains a 1-set $X_{\boldsymbol{b}}$ of g. We observe that $b_P = 1$; else, $\boldsymbol{b} \leq \boldsymbol{a}'$, and since $g(\boldsymbol{a}') = 0$ and g is monotone, it follows that $g(\boldsymbol{b}) = 0$, contradicting $X_{\boldsymbol{b}}$ being a 1-set. □

Consequence 2. Let g be a non-constant monotone increasing function that depends on each of its variables X, and let E be collection of the 1-sets of g. Then (X, E) is a simple hypergraph.

A subset S of X is a [minimal] blocker for the hypergraph if and only if S [is minimal such that it] has the property that for any $\boldsymbol{a}$ so that $\{a_i = 0 | x_i \in S\}$, we have $g(\boldsymbol{a}) = 0$.

Thus, if B is the collection of 0-sets of g: $Tr(X, E) = (X, B)$ is a simple hypergraph. The hypergraph of 1-sets of g, and the hypergraph of 0-sets of g, are transverse hypergraphs (blockers) to one another.

Proof of Consequence 2. By Consequence 1, E covers X, so (X, E) is a hypergraph. It is simple because no 1-set can contain another.

Consider a blocker S, and any $\boldsymbol{a}$ such that $\{a_i = 0 | x_i \in S\}$. If $g(\boldsymbol{a}) = 1$, we know (by Observation 1) that $X - S$ contains a 1-set of g. This is impossible, since by definition the blocker S must intersect any 1-set of g.

Now consider a subset $T \subseteq X$ such that for any $\boldsymbol{a}$ with $\{a_i = 0 | x_i \in T\}$, we have $g(\boldsymbol{a}) = 0$. Suppose that some 1-set m of g does not intersect T. Define $\boldsymbol{b}$ so that $b_i = 1$ iff $x_i \notin T$. By the property of T, $g(\boldsymbol{b}) = 0$, but since $b_i = 1$ for each $x_i \in m$ (a 1-set), $g(\boldsymbol{b}) = 1$. This is a contradiction, so T is a blocker. □

Therefore, by duality, we may reverse the roles of 0-set and 1-set, 1 and 0, in the observations above. It follows from Theorem 4 and from Consequence 2:

Theorem 5. *Let g be an element of $\mathcal{M}$ of arity at least two: a monotone increasing Boolean function which depends on each of its variables, having neither a 1-set (prime implicant) nor a 0-set (prime implicate) of cardinality one.*

Then for any variable v of g, there is a sequence $v_0 = v$, v_1, $\ldots$, $v_{2p+1} = v$ (where $p \geq 1$) on the variables of g so that:

- *Consecutive elements of the sequence differ;*
- *Each consecutive pair $\{v_{2i}, v_{2i+1}\}$ is contained in some 0-set;*
- *Each consecutive pair $\{v_{2i+1}, v_{2i+2}\}$ is contained in some 1-set.*

Recall the set of functions with positive pair decomposition; for such a function, the hypergraph of 1-sets and its transverse hypergraph of 0-sets are both graphs. It is now easy to show that:

Lemma 5. *The set of functions with positive pair decomposition consists of the functions* Triangle, Box, *and* Path *and their duals (up to variable permutation).*

Proof of Lemma 5. Let $g(x_1, x_2, \ldots, x_Q) = \bigwedge_{d=1}^{D}(y_{(d,1)} \vee y_{(d,2)}) = \bigvee_{c=1}^{C}(z_{(c,1)} \wedge z_{(c,2)})$. Consider the hypergraph of 0-sets and its traverse hypergraph of 1-sets; both are graphs, neither has isolated vertices. Note that no single variable (node) can occur in every pair (edge) in the CNF (or the DNF), else this single variable is a blocker. So $C, D \geq 2$, $Q \geq 3$.

Note that for any c, the pair $\{z_{(c,1)}, z_{(c,2)}\}$ is a 1-set, and therefore intersects each 0-set $\{y_{(d,1)}, y_{(d,2)}\}$. In other words, for any edge in the hypergraph, its endpoints are a vertex cover for the transverse hypergraph. So suppose we have $d_i \neq d_j$ so that $y_1 = y_{(d_i,1)}$, $y_2 = y_{(d_i,2)}$, $y_3 = y_{(d_j,1)}$, $y_4 = y_{(d_j,2)}$ are all distinct (the hypergraph of 0-sets contains two edges with no common vertices). (Note that if there are no such disjoint pairs, $Q = 3$ and the function is Triangle.) Since each pair of z's must intersect both $\{y_1, y_2\}$ and $\{y_3, y_4\}$, there are at most four ways to do this, so $Q \leq 4$.

Note that if no variable occurs more than once in the CNF, the function is $(y_1 \vee y_2) \wedge (y_3 \vee y_4) = \text{Box}^{\text{dual}}$. Otherwise, we observe that no variable x_i occurs in more than two pairs $\{y_{(d,1)}, y_{(d,2)}\}$, or in more than two pairs $\{z_{(c,1)}, z_{(c,2)}\}$.

Consider any node n in the hypergraph of 0-sets (respectively, 1-sets). There is some minimal blocker not containing n. Any blocker which does not contain n must necessarily contain all neighbors of n. Since the transverse hypergraph is a graph, n may have no more than two neighbors.

So we may assume some variable x occurs in exactly two pairs $\{x, y\}$ and $\{x, z\}$ in the DNF (two edges in the hypergraph). Let us consider the special case that $\{x, y\}$ also occurs as a pair in the CNF (a blocker); then every edge must be incident on either x or y. Note x and y each must have exactly two neighbors. If the remaining neighbor of y is z this is Triangle, else it is Path.

Now consider the remaining case, that $\{x, y\}$ and $\{x, z\}$ are pairs in the DNF (edges) and that $\{x, y\}$ is not a pair in the CNF (blocker). Note that $\{y, z\}$ is a blocker (since x cannot be), and there must be at least one more edge, incident on either y or z. Some edge is not blocked by $\{x, y\}$, so there must be an edge $\{z, w\}$ (and blocker $\{x, y\}$). So this function must be either Box or Path.   □

7   Defining a Map from R-3SAT to SAT[g]

Our goal is to show that for g in $\mathcal{M}_3$, the problem SAT[g] is NP-complete. We now fix a Boolean function g on Q variables in $\mathcal{M}_3$, and some 0-set [resp. 1-set] M^* of g of size at least three.

Given an instance $\Phi = \bigwedge_{l=1}^{L}(\alpha_l \vee \beta_l \vee \gamma_l)$ of the R-3SAT problem, we will construct disjoint Boolean variable sets $V_1, V_2, \ldots, V_Q$ and two sets of Boolean vectors $\mathcal{P} = \{p_j\}_{j=1}^{P}$, $\mathcal{N} = \{n_k\}_{k=1}^{N} \subseteq V_1 \times V_2 \times \ldots \times V_Q$, so that $\Psi_\Phi(\mathcal{P}, \mathcal{N})$ is satisfiable if and only if Φ is satisfiable.

We shall denote the input variables of g by $X = \{x_i\}_{i=1}^{Q}$ and its index set by I. Without loss of generality, we may assume that the variables of g occuring in the specified 0-set M^* are $\{x_i\}_{i=1}^{H}$, where $H = |M^*| \geq 3$.

We represent the three disjoint variable sets of the R-3SAT instance by W_1, W_2, and W_3. For notational simplicity, we shall define $W_i = \emptyset$ for $i \in I$, $i > 3$.

Definition 1. *For $i \in I$, define the variable set V_i as follows:*

- $W_i \cup W_i^\neg \subseteq V_i$;
- *Add two variables, t_i and f_i, to V_i;*
- *For each $j \in I$ so that $j \neq i$, add a new variable d_i^j to V_i;*
- *For each variable $w \in (W_1 \cup W_2 \cup W_3) - W_i$, add variables e_i^w and $e_i^{\overline{w}}$ to V_i.*

To explain the roles of these new variables, it is helpful to state first the goals of this construction.

Goals. In any truth assignment τ for the variables in $V_1 \cup \ldots \cup V_Q$ which satisfies the entire set of constraints we are about to define, we would like the following three conditions to hold:

1. For all $i \in I$, $\tau(t_i) = 1$ and $\tau(f_i) = 0$;
2. For all $l \in \{1, 2, \ldots, L\}$, $\tau(\alpha_l) \vee \tau(\beta_l) \vee \tau(\gamma_l) = 1$;
3. For each variable $w \in W_1 \cup W_2 \cup W_3$, it is not the case that $\tau(w) = \tau(\overline{w}) = 1$.

With these goals in mind, we remark that the only function of the dummy variables $\{d_i^j \mid j \neq i\}$ introduced in Definition 1 is to create constraints that will enforce condition 1. In turn, the variables $\{t_i, f_i\}_{i=1}^{Q}$, used together with the variables inherited from Φ, allow us to create constraints to enforce condition 2. The role of the variables e_i^w and $e_i^{\overline{w}}$ is to create constraints to enforce condition 3 (these constraints will also use $\{t_i, f_i\}_{i=1}^{Q}$ and the variables inherited from Φ).

Definition 2. *For $i \in I$, define $p_i = (d_1^i, \ldots, d_{i-1}^i, t_i, d_{i+1}^i, \ldots, d_Q^i)$ and $n_i = (d_1^i, \ldots, d_{i-1}^i, f_i, d_{i+1}^i, \ldots, d_Q^i)$. Define constraints $g(p_i) = 1$ and $\overline{g}(n_i) = 1$.*

Now we can use these artificial variables to "force" any desired component of a constraint vector to have a particular value, to be specifically 0 or 1.

Observation 2. Let $J \subseteq I$ be the set of indices corresponding to some 0-set of g, and let K be a non-empty subset of J. Consider $a = (a_1, a_2, \ldots, a_Q) \in \{0,1\}^Q$ such that: 1) $g(a) = 1$; 2) for all $i \in I - J$, $a_i = 1$; and 3) for all $j \in J - K$, $a_j = 0$. In this case, there is some $k \in K$ so that $a_k = 1$.

The analogous statement for 1-sets must therefore hold also.

Proof of Observation 2. If $a_j = 0$ for all $j \in J$, then $g(a) = 0$, which is not true. So $a_m = 1$ for some m in J; necessarily $m \in K$. $\qquad\square$

Observation 3. Let $J \subseteq I$ be the set of indices corresponding to some 0-set of g. Then for any $\boldsymbol{a}$ so that 1) $a_i = 1$ for each $i \notin J$, and 2) there is some $j \in J$ so that $a_j = 1$, we have that $g(\boldsymbol{a}) = 1$. (Analogously for 1-sets.)

Proof of Observation 3. This is immediate from the minimality of 0-sets.   □

Definition 3. *Recall that H is the integer so that the variables of g occuring in the 0-set M^* are $\{x_i\}_{i=1}^H$, where $H = |M^*| \geq 3$. Also recall that $\Phi = \bigwedge_{l=1}^L (\alpha_l \vee \beta_l \vee \gamma_l)$ is our instance of the R-3SAT problem.*
  For $l \in \{1, 2, \ldots, L\}$, define $\boldsymbol{p}_{Q+l} = (\alpha_l, \beta_l, \gamma_l, f_4, \ldots, f_H, t_{H+1}, \ldots, t_Q)$. Add the constraint $g(\boldsymbol{p}_{Q+l}) = 1$.

These constraints allow us to meet our first two goals.

For our final and most difficult goal, we will need Observations 2 and 3, as well as theorem 5. Our method is to create a set of constraints such that, for each $w \in W_1 \cup W_2 \cup W_3$, we ensure the existence of a sequence $(i_1, i_2, \ldots, i_{2P_w})$ on I so that for the corresponding sequence of variables $(w, e_{i_1}^{\overline{w}}, e_{i_2}^w, \ldots, e_{i_{2P_w}-1}^{\overline{w}}, e_{i_{2P_w}}^w, \overline{w})$ we have that, for any truth assignment τ satisfying the new system:

- At least one of $\{\tau(w), \tau(e_{i_1}^{\overline{w}})\}$ is 0;
- For $j \in \{1, \ldots, P_w\}$, at least one of each pair $\{\tau(e_{i_{2j-1}}^{\overline{w}}), \tau(e_{i_{2j}}^w)\}$ is 1;
- For $j \in \{1, \ldots, P_w - 1\}$, at least one of each pair $\{\tau(e_{i_{2j}})^w, \tau(e_{i_{2j+1}}^{\overline{w}})\}$ is 0;
- At least one of $\{\tau(e_{i_{2P_w}}^w), \tau(\overline{w})\}$ is 0.

This sequence ensures $\tau(w) = 1 \Rightarrow \tau(\overline{w}) = 0$; also, if $\tau(\overline{w}) = 1$, then $\tau(w) = 0$.

Definition 4. *For each pair (k, l) (where $k, l \in I$, $k < l$) such that x_k, x_l belong to some 1-set of g, let $J_{k,l}$ denote the index set of some 1-set containing $\{x_k, x_l\}$. For each combination of such pair $\{k, l\}$ and some $w \in (W_1 \cup W_2 \cup W_3) - (W_k \cup W_l)$, define the vector $\boldsymbol{n}_{(k,l,w,+)} = (n_1, \ldots, n_Q)$ as follows. Set $n_k = e_k^w \in V_k$; set $n_l = e_l^{\overline{w}} \in V_l$; for each index j so that $j \in J_{k,l} - \{k, l\}$, set $n_j = t_j$; and for each remaining index $i \in I - J_{k,l}$, set $n_i = f_i$.*
  Also define a vector $\boldsymbol{n}_{(k,l,w,-)}$, which differs from $\boldsymbol{n}_{(k,l,w,+)}$ only in that $n_k = e_k^{\overline{w}}$ and $n_l = e_l^w$. Define negative constraints $\overline{g}(\boldsymbol{n}_{(k,l,w,+)}) = 1$ and $\overline{g}(\boldsymbol{n}_{(k,l,w,-)}) = 1$.
  In the case that $k \in \{1, 2, 3\}$ (and similarly for $l \in \{1, 2, 3\}$), then in addition define the vector $\boldsymbol{n}_{(k,l,w,\top)}$, which differs from $\boldsymbol{n}_{(k,l,w,+)}$ only in that $n_k = w$; and $\boldsymbol{n}_{(k,l,w,\perp)}$, for which $n_k = \overline{w}$. Add negative constraints $\overline{g}(\boldsymbol{n}_{(k,l,w,\top)}) = 1$ and $\overline{g}(\boldsymbol{n}_{(k,l,w,\perp)}) = 1$.
  Positive constraints are defined analogously. For example, for each pair (k, l) of variables from some 0-set of g, define $\boldsymbol{p}_{(k,l,w,+)}$ so that $p_k = e_k$, $p_l = e_l^{\overline{w}}$, $p_j = f_j$ for all $j \in J_{k,l} - \{k, l\}$, and $p_i = t_i$ for all $i \in I - J_{k,l}$. The constraint is $g(\boldsymbol{p}_{(k,l,w,+)}) = 1$.

These constraints ensure the existence of the desired sequence above.

8   Satisfiability Problem SAT[g] Is NP-Complete

Let Φ be an instance of R-3SAT, and Ψ_Φ the instance of SAT[g] derived from Φ as above. We claim that Φ is satisfiable if and only if Ψ_Φ is satisfiable.

First suppose that there is a truth τ assignment for $W_1 \cup W_2 \cup W_3$ which satisfies Φ. Then we can extend this truth assignment to τ' for $V_1 \times V_2 \times \ldots \times V_k$ in such a way that the constraints of Ψ_Φ are satisfied as follows:

- For each $i \in I$, set $\tau'(f_i) = 0$ and $\tau'(t_i) = 1$.
- Because g depends on each of its variables, for each $i \in I$, we may choose constants $\{a^i_j \in \{0,1\} | j \in I - \{i\}\}$ so that $g(a^i_1, \ldots, a^i_{i-1}, 0, a^i_{i+1}, \ldots, a^i_Q) = 0$ and $g(a^i_1, \ldots, a^i_{i-1}, 1, a^i_{i+1}, \ldots, a^i_Q) = 1$. For $i \in I$, $j \neq i$, let $\tau'(d^j_i) = a^j_i$.
- For each $i \in I$ and for each variable $w \in (W_1 \cup W_2 \cup W_3) - W_i$, set $\tau'(e^w_i) = \tau(w)$ and $\tau'(e^{\overline{w}}_i) = 1 - \tau(w)$.

Note that the constraints of type 1 are satisfied, and that, by Observation 3, the constraints of type 2 are satisfied. We have extended the truth assignment to the artificial variables e^w_i and $e^{\overline{w}}_i$ in such a way that for any cardinality two subset $\{x_k, x_l\}$ of a 1-set (respectively, 0-set), the only two components of the corresponding type 3 constraint vectors which are not in $\{t_i\} \cup \{f_i | i \in I\}$ have opposite truth values—so necessarily, there is at least one 1 and at least one 0. By Observation 3, this is sufficient to ensure that the constraint of type 3 is fulfilled.

Now suppose that there is a truth assignment t for the $V = V_1 \cup V_2 \cup \ldots \cup V_Q$ satisfying Ψ_Φ. Since all of the type 2 constraints of Ψ_Φ are satisfied, when we take the restriction of V to $(W_1 \cup W_1{}^-) \cup (W_2 \cup W_2{}^-) \cup (W_3 \cup W_3{}^-)$, all of the terms $\alpha_l \vee \beta_l \vee \gamma_l$ hold. However, it is possible that for some $w \in W_1 \cup W_2 \cup W_3$ we have both w and $\overline{w}$ with a truth assignment of 0. In such a case, we can simply randomly choose one of $\{w, \overline{w}\}$ to have truth assignment 1. Each $\alpha_l \vee \beta_l \vee \gamma_l$ must still hold.

This shows that SAT$[g]$ is NP-complete when $g \in \mathcal{M}_3$.

References

[1] C. Berge. *Hypergraphs: Combinatorics of Finite Sets*. North-Holland, 1989.
[2] Andrei A. Bulatov. Tractable conservative constraint satisfaction problems. Submitted to *ACM Transactions on Computational Logic*.
[3] Andrei A. Bulatov. A dichotomy theorem for constraints on a three-element set. In *Proceedings of 43rd Annual IEEE Symposium of Foundation of Computer Science (FOCS'02)*, pages 649–658, 2002.
[4] Andrei A. Bulatov. Tractable conservative constraint satisfaction problems. In *Proceedings of 18th IEEE Symposium on Logic in Computer Science (LICS'03)*, pages 321–330, 2003.
[5] Andrei A. Bulatov and Peter Jeavons. An algebraic approach to multi-sorted constraints. In *Proceedings of 9th International Conference on Principles and Practice of Constraint Programming (CP'03)*, pages 183–198, 2003.
[6] Andrei A. Bulatov, Andrei A. Krokhin, and Peter Jeavons. Constraint satisfaction problems and finite algebras. In *Proceedings of 27th International Colloquium on Automata, Languages, and Programming (ICALP'00)*, volume 1853 of *Lecture Notes in Comp. Sci.*, pages 272–282, 2000.

[7] David Cohen and Peter Jeavons. *Handbook of Constraint Programming*, chapter 6: The complexity of constraint languages. Elsevier, 2006. (To appear).

[8] Nadia Creignou. A dichotomy theorem for maximum generalized satisfiability problems. *J. Comput. Syst. Sci.*, 51(3):511–522, 1995.

[9] Nadia Creignou, Sanjeev Khanna, and Madhu Sudan. *Complexity Classifications of Boolean Constraint Satisfaction Problems*, volume 7 of *SIAM Monographs on Discrete Mathematics and Applications*. SIAM, 2001.

[10] Víctor Dalmau and Daniel K. Ford. Generalized satisfiability with limited occurrences per variable: A study through delta-matroid parity. In *Mathematical Foundations of Computer Science 2003 (MFCS'03)*, pages 358–367, 2003.

[11] Tomás Feder. Fanout limitations on constraint systems. *Theor. Comput. Sci.*, 255(1-2):281–293, 2001.

[12] Tomás Feder and Daniel Ford. Classification of bipartite boolean constraint satisfaction through delta-matroid intersection. *Electronic Colloquium on Computational Complexity (ECCC)*, TR05(016), 2005. To appear in *SIAM J. Discrete Math.*

[13] Tomás Feder and Moshe Y. Vardi. The computational structure of monotone monadic SNP and constraint satisfaction: A study through Datalog and group theory. *SIAM Journal on Computing*, 28:57–104, 1998.

[14] Michael R. Garey and David S. Johnson. *Computers and Intractability: A Guide to the Theory of NP-Completeness*. W.H. Freeman and Co., 1979.

[15] Martin Grohe. The complexity of homomorphism and onstraint satisfaction problems seen from the other side. In *44th Symposium on Foundations of Computer Science (FOCS'03)*, pages 552–561, 2003.

[16] Pavol Hell and Jaroslav Nešetřil. On the complexity of H-coloring. *J. Comb. Theory, Series B*, 48:92–110, 1990.

[17] Gabriel Istrate. Looking for a version of Schaefer's dichotomy theorem when each variable occurs at most twice. Technical Report TR 652, U. Rochester, CS Dept., 1997.

[18] Peter G. Jeavons. On the algebraic structure of combinatorial problems. *Theor. Comput. Sci.*, 200(1-2):185–204, 1998.

[19] Richard E. Ladner. On the structure of polynomial time reducibility. *J. ACM*, 22(1):155–171, 1975.

[20] Thomas J. Schaefer. The complexity of satisfiability problems. In *Proceedings of 10th Annual ACM Symposium on Theory of Computing (STOC'78)*, pages 216–226, 1978.

[21] Ingo Wegener. *Complexity of Boolean Functions*. John Wiley and Sons, 1987.

A Complete Calculus for Max-SAT*

María Luisa Bonet[1], Jordi Levy[2], and Felip Manyà[2]

[1] Dept. Llenguatges i Sistemes Informàtics (LSI),
Universitat Politècnica de Catalunya (UPC),
Jordi Girona, 1-3, 08034 Barcelona, Spain
[2] Artificial Intelligence Research Institute (IIIA),
Spanish Scientific Research Council (CSIC),
Campus UAB, 08193 Bellaterra, Spain

Abstract. Max-SAT is the problem of finding an assignment minimizing the number of unsatisfied clauses of a given CNF formula. We propose a resolution-like calculus for Max-SAT and prove its soundness and completeness. We also prove the completeness of some refinements of this calculus. From the completeness proof we derive an exact algorithm for Max-SAT and a time upper bound.

1 Introduction

The Max-SAT problem for a CNF formula ϕ is the problem of finding an assignment of values to variables that minimizes the number of unsatisfied clauses in ϕ. Max-SAT is an optimization counterpart of SAT and is NP-hard.

The most competitive exact Max-SAT solvers [1,2,3,7,9,11,12,13] implement variants of the following branch and bound (BnB) schema: Given a CNF formula ϕ, BnB explores the search tree that represents the space of all possible assignments for ϕ in a depth-first manner. At every node, BnB compares the upper bound (UB), which is the best solution found so far for a complete assignment, with the lower bound (LB), which is the sum of the number of clauses unsatisfied by the current partial assignment plus an underestimation of the number of clauses that will become unsatisfied if the current partial assignment is completed. If $LB \geq UB$ the algorithm prunes the subtree below the current node and backtracks to a higher level in the search tree. If $LB < UB$, the algorithm tries to find a better solution by extending the current partial assignment by instantiating one more variable. The solution to Max-SAT is the value that UB takes after exploring the entire search tree.

The amount of inference performed by BnB at each node of the proof tree is limited compared with DPLL-style SAT solvers. Since unit propagation is unsound for Max-SAT,[1] when branching is applied on a literal l, BnB just

---

* This research has been partially founded by the CICYT research projects iDEAS (TIN2004-04343), Mulog (TIN2004-07933-C03-01/03) and SofSAT (TIC2003-00950).
[1] The multiset of clauses $\{a, \overline{a} \vee b, \overline{a} \vee \overline{b}, \overline{a} \vee c, \overline{a} \vee \overline{c}\}$ has a minimum of one unsatisfied clause. However, setting a to true (by unit propagation) leads to a non-optimal assignment falsifying two clauses.

A. Biere and C.P. Gomes (Eds.): SAT 2006, LNCS 4121, pp. 240–251, 2006.

removes the clauses containing l and deletes the occurrences of $\bar{l}$. The new unit clauses derived as a consequence of deleting the occurrences of $\bar{l}$ are not propagated as in DPLL. To mitigate that problem some simple inference rules have been incorporated into state-of-the-art Max-SAT solvers: (i) the pure literal rule [1,6,11,13,14]; (ii) the dominating unit clause rule first proposed in [8], and applied in [2,6,8,11]; (iii) the almost common clause rule, first proposed in [4] and extended to weighted Max-SAT in [2]; that rule was called neighborhood resolution in [5] and used as a preprocessing technique in [2,6,10]; and (iv) the complementary unit clause rule [8]. All these rules, which are sound but not complete, have proved to be useful in practice.

The main objective of this paper is to make a step forward in the study of resolution-like inference rules for Max-SAT by defining a sound and complete resolution rule. That rule should subsume the previous rules, and provide a general framework that should allow us to define complete refinements of resolution and devise faster Max-SAT solvers.

In the context of SAT, a *sound* rule has to preserve satisfiability, like resolution does. However, in Max-SAT this is not enough; rules have to preserve the number of unsatisfied clauses for every possible assignment. Therefore, the way we apply the rule is different. To obtain a sound calculus, instead of *adding* the conclusion, which would make the number of unsatisfied clauses increase, we *replace* the premises of the rule by its conclusion. Then, the resolution rule $x \vee A, \bar{x} \vee B \vdash A \vee B$ is not sound for Max-SAT, because an assignment satisfying x and A, and falsifying B, would falsify one of the premises, but would satisfy the conclusion. So the number of unsatisfied clauses would not be preserved for every truth assignment.

The most natural variant of a sound resolution rule for Max-SAT was defined in [5]:

$$\frac{\begin{array}{c} x \vee A \\ \bar{x} \vee B \end{array}}{\begin{array}{c} A \vee B \\ x \vee A \vee \overline{B} \\ \bar{x} \vee \overline{A} \vee B \end{array}}$$

However, two of the conclusions of this rule are not in clausal form, and the application of distributivity:

$$\frac{\begin{array}{c} x \vee a_1 \vee \ldots \vee a_s \\ \bar{x} \vee b_1 \vee \ldots \vee b_t \end{array}}{\begin{array}{c} a_1 \vee \ldots \vee a_s \vee b_1 \vee \ldots \vee b_t \\ x \vee a_1 \vee \ldots \vee a_s \vee \overline{b_1} \\ \ldots \\ x \vee a_1 \vee \ldots \vee a_s \vee \overline{b_t} \\ \bar{x} \vee b_1 \vee \ldots \vee b_t \vee \overline{a_1} \\ \ldots \\ \bar{x} \vee b_1 \vee \ldots \vee b_t \vee \overline{a_s} \end{array}}$$

results into an unsound rule. As we show in the next section, obtaining a sound rule requires a more sophisticated adaptation of the resolution rule.

This paper proceeds as follows. First, in Section 2 we define Max-SAT resolution and prove its soundness. Despite of the similitude of the inference rule with the classical resolution rule, it is not clear how to simulate classical inferences with the new rule. To obtain a complete strategy, we need to apply the new rule widely to get a saturated set of clauses, as described in Section 3. In Section 4 we prove the completeness of the new rule, and in Section 5 we prove that this result extends to ordered resolution. Finally, in Section 6 we deduce an exact algorithm and give a worst-case time upper bound in Section 7.

2   The Max-SAT Resolution Rule and Its Soundness

In Max-SAT we use multisets of clauses instead of just sets. For instance, the multiset $\{a, \overline{a}, \overline{a}, a \vee b, \overline{b}\}$, where a clause is repeated, has a minimum of two unsatisfied clauses.

Max-SAT resolution, like classical resolution, is based on a unique inference rule. In contrast to the resolution rule, the premises of the Max-SAT resolution rule are *removed* from the multiset after applying the rule. Moreover, apart from the classical conclusion where a variable has been cut, we also conclude some additional clauses that contain one of the premises as sub-clause.

Definition 1. *The* Max-SAT resolution *rule is defined as follows:*

$$
\begin{array}{l}
x \vee a_1 \vee \ldots \vee a_s \\
\overline{x} \vee b_1 \vee \ldots \vee b_t \\
\hline
a_1 \vee \ldots \vee a_s \vee b_1 \vee \ldots \vee b_t \\
x \vee a_1 \vee \ldots \vee a_s \vee \overline{b_1} \\
x \vee a_1 \vee \ldots \vee a_s \vee b_1 \vee \overline{b_2} \\
\ldots \\
x \vee a_1 \vee \ldots \vee a_s \vee b_1 \vee \ldots \vee b_{t-1} \vee \overline{b_t} \\
\overline{x} \vee b_1 \vee \ldots \vee b_t \vee \overline{a_1} \\
\overline{x} \vee b_1 \vee \ldots \vee b_t \vee a_1 \vee \overline{a_2} \\
\ldots \\
\overline{x} \vee b_1 \vee \ldots \vee b_t \vee a_1 \vee \ldots \vee a_{s-1} \vee \overline{a_s}
\end{array}
$$

This inference rule is applied to multisets of clauses, and replaces the premises of the rule by its conclusions.

We say that the rule cuts *the variable* x.

The tautologies concluded by the rule are removed from the resulting multiset. Similarly, repeated literals in a clause are also removed.

Definition 2. *We write* $\mathcal{C} \vdash \mathcal{D}$ *when the multiset of clauses* $\mathcal{D}$ *can be obtained from the multiset* $\mathcal{C}$ *applying the rule finitely many times. We write* $\mathcal{C} \vdash_x \mathcal{D}$ *when this sequence of applications only cuts the variable* x.

The Max-Sat resolution rule concludes many more clauses than the classical version. However, when the two premises share literals, some of the conclusions are tautologies, hence removed. In particular we have $x \vee A, \overline{x} \vee A \vdash A$. Moreover, as we will see when we study the completeness of the rule, there is no need to cut the conclusions of a rule among themselves. Finally, we will also see that the size of the worst-case proof of a set of clauses is similar to the size for classical resolution.

Notice that the instance of the rule not only depends on the two clauses of the premise and the cut variable (like in resolution), but also on the order of the literals. Notice also that, like in classical resolution, this rule concludes a new clause not containing the variable x, except when this clause is a tautology.

Example 1. The Max-SAT resolution rule removes clauses after using them in an inference step. Therefore, it could seem that it can not simulate classical resolution when a clause needs to be used more than once, like in:

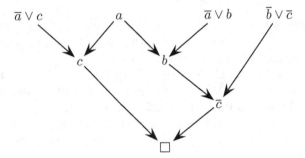

However, this is not the case. We can derive the empty clause as follows (where already used clauses are put into boxes):

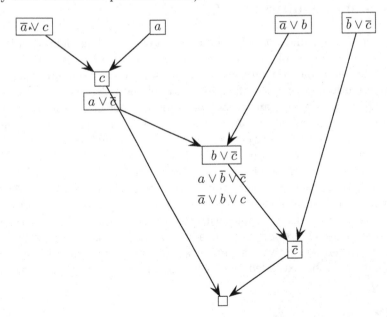

More precisely, we have derived a, $\bar{a} \vee b$, $\bar{a} \vee c$, $\bar{b} \vee \bar{c} \vdash \square$, $a \vee \bar{b} \vee \bar{c}$, $\bar{a} \vee b \vee c$, where any truth assignment satisfying $\{a \vee \bar{b} \vee \bar{c}, \bar{a} \vee b \vee c\}$ minimizes the number of falsified clauses in the original formula.

Notice that the structure of the classical proof and the Max-SAT resolution proof is quite different. It seems difficult to adapt a classical resolution proof to get a Max-SAT resolution proof, and it is an open question if this is possible without increasing substantially the size of the proof.

Theorem 1 (Soundness). *The Max-SAT resolution rule is sound. i.e. the rule preserves the number of unsatisfied clauses for every truth assignment.*

PROOF: For every assignment I, we will prove that the number of clauses that I falsifies in the premises of the inference rule is equal to the number of clauses that it falsifies in the conclusions.

Let I be any assignment. I can not falsify both upper clauses, since it satisfies either x or $\bar{x}$.

Suppose I satisfies $x \vee a_1 \vee \ldots \vee a_s$ but not $\bar{x} \vee b_1 \vee \ldots \vee b_t$. Then I falsifies all b_j's and sets x to true. Now, suppose that I satisfies some a_i. Say a_i is the first of such elements. Then I falsifies $\bar{x} \vee b_1 \vee \ldots \vee b_t \vee a_1 \vee \ldots \vee a_{i-1} \vee \overline{a_i}$ and it satisfies all the others in the set below. Suppose now that I falsifies all a_i's. Then, it falsifies $a_1 \vee \ldots a_s \vee b_1 \vee \ldots \vee b_t$ but satisfies all the others.

If I satisfies the second but not the first, then it is the same argument.

Finally, suppose that I satisfies both upper clauses. Suppose that I sets x to true. Then, for some j, b_j is true and I satisfies all the lower clauses since all of them have either b_j or x. ∎

3   Saturated Multisets of Clauses

In this Section we define *saturated* multisets of clauses. This definition is based on the classical notion of sets of clauses closed by (some restricted kind of) inference, in particular, on sets of clauses closed by cuts of some variable. In classical resolution, given a set of clauses and a variable, we can saturate the set by cutting the variable exhaustively, obtaining a superset of the given clauses. If we repeat this process for all the variables, we get a complete resolution algorithm, i.e. we obtain the empty clause whenever the original set was unsatisfiable. Our completeness proof is based on this idea. However, notice that the classical saturation of a set w.r.t. a variable is unique, whereas in Max-SAT, it is not (see Remark 1). In fact, it is not even a superset of the original set. Moreover, in general, if we saturate a set w.r.t. a variable, and then w.r.t. another variable, we obtain a set that is not saturated w.r.t. both variables. Fortunately, we still keep a good property: given a multiset of clauses saturated w.r.t. a variable x, if there exists an assignment satisfying all the clauses not containing x, then it can be extended (by assigning x) to satisfy all the clauses (see Lemma 4).

Definition 3. *A multiset of clauses C is said to be* saturated *w.r.t. x if for every pair of clauses $C_1 = x \vee A$ and $C_2 = \overline{x} \vee B$, there is a literal l such that l is in A and $\overline{l}$ is in B.*

A multiset of clauses C' is a saturation *of C w.r.t. x if C' is saturated w.r.t. x and $C \vdash_x C'$, i.e. C' can be obtained from C applying the inference rule cutting x finitely many times.*

The following is a trivial equivalent version of the definition.

Lemma 1. *A multiset of clauses C is saturated w.r.t. x if, and only if, every possible application of the inference rule cutting x only introduces clauses containing x (since tautologies get eliminated).*

We assign a function $P : \{0,1\}^n \rightarrow \{0,1\}$ to every clause, and a function $P : \{0,1\}^n \rightarrow \mathbb{N}$ to every multiset of clauses as follows.

Definition 4. *For every clause $C = x_1 \vee \ldots \vee x_s \vee \overline{x}_{s+1} \vee \ldots \vee \overline{x}_{s+t}$ we define its* characteristic function *as $P_C(\boldsymbol{x}) = (1 - x_1) \ldots (1 - x_s)\, x_{s+1} \ldots x_{s+t}$.*

For every multiset of clauses $C = \{C_1, \ldots, C_m\}$, we define its characteristic *function as $P_C = \Sigma_{i=1}^m P_{C_i}(\boldsymbol{x})$.*

Notice that the set of functions $\{0,1\}^n \rightarrow \mathbb{N}$, with the order relation: $f \leq g$ if for all x, $f(x) \leq g(x)$, defines a partial order between functions. The strict part of this relation, i.e. $f < g$ if for all x, $f(x) \leq g(x)$ and for some x, $f(x) < g(x)$, defines a strictly decreasing partial order.

Lemma 2. *Let P_C be the characteristic function of a multiset of clauses C. For every assignment I, $P_C(I)$ is the number of clauses of C falsified by I.*

The inference rule replaces a multiset of clauses by another with the same characteristic function.

Lemma 3. *For every multiset of clauses C and variable x, there exists a multiset C' such that C' is a saturation of C w.r.t. x.*

Moreover, this multiset C' can be computed applying the inference rule to any pair of clauses $x \vee A$ and $\overline{x} \vee B$ satisfying that $A \vee B$ is not a tautology, using any ordering of the literals, until we can not apply the inference rule any longer.

PROOF: We proceed by applying nondeterministically the inference rule cutting x, until we obtain a saturated multiset. We only need to prove that this process terminates in finitely many inference steps, i.e that there does not exist infinite sequences $C = C_0 \vdash C_1 \vdash \ldots$, where at every inference we cut the variable x and none of the sets C_i are saturated.

At every step, we can divide C_i into two multisets: D_i with all the clauses that do not contain x, and $\mathcal{E}_i$ with the clauses that contain the variable x (in positive or negative form). When we apply the inference rule we replace two clauses of $\mathcal{E}_i$ by a multiset of clauses, where one of them, say A, does not contain x. Therefore, we obtain a distinct multiset $C_{i+1} = D_{i+1} \cup \mathcal{E}_{i+1}$, where $D_{i+1} = D_i \cup \{A\}$. Since A is not a tautology the characteristic function P_A is not zero for some value.

Then, since $P_{C_{i+1}} = P_{C_i}$ and $P_{D_{i+1}} = P_{D_i} + P_A$, we obtain $P_{\mathcal{E}_{i+1}} = P_{\mathcal{E}_i} - P_A$. Therefore, the characteristic function of the multiset of clauses containing x strictly decreases after every inference step. Since the order relation between characteristic functions is strictly decreasing, this proves that we can not perform infinitely many inference steps.   ∎

Remark 1. Although every multiset of clauses is saturable, its saturation is not unique. For instance, the multiset $\{a,\ \overline{a} \vee b,\ \overline{a} \vee c\}$ has two possibles saturations w.r.t. a: the multiset $\{b,\ \overline{b} \vee c,\ a \vee \overline{b} \vee \overline{c},\ \overline{a} \vee b \vee c\}$ and the multiset $\{c,\ b \vee \overline{c},\ a \vee \overline{b} \vee \overline{c},\ \overline{a} \vee b \vee c\}$.

Another difference with respect to classical resolution is that we can not saturate a set of clauses simultaneously w.r.t. two variables by saturating w.r.t. one, and then w.r.t. the other. For instance, if we saturate $\{\overline{a} \vee c,\ a \vee b \vee c\}$ w.r.t. a, we obtain $\{b \vee c,\ \overline{a} \vee \overline{b} \vee c\}$. This is the only possible saturation of the original set. If now we saturate this multiset w.r.t. b, we obtain again the original set $\{\overline{a} \vee c,\ a \vee b \vee c\}$. Therefore, it is not possible to saturate this multiset of clauses w.r.t. a and b simultaneously.

Lemma 4. *Let C be a saturated multiset of clauses w.r.t. x. Let D be the subset of clauses of C not containing x. Then, any assignment I satisfying D (and not assigning x) can be extended to an assignment satisfying C.*

PROOF: We have to extend I to satisfy the whole C. In fact we only need to set the value of x. If x has a unique polarity in $C \setminus D$, then the extension is trivial ($x = true$ if x always occurs positively, and $x = false$ otherwise). If, for any clause of the form $x \vee A$ or $\overline{x} \vee A$, the assignment I already satisfies A, then any choice of the value of x will work. Otherwise, assume that there is a clause $x \vee A$ (similarly for $\overline{x} \vee A$) such that I sets A to false. We set x to true. All the clauses of the form $x \vee B$ will be satisfied. For the clauses of the form $\overline{x} \vee B$, since C is saturated, there exists a literal l such that $l \in A$ and $\overline{l} \in B$. This ensures that, since I falsifies A, $I(l) = false$ and I satisfies B.   ∎

4   Completeness of Max-SAT Resolution

Now, we prove the main result of this paper, the completeness of Max-SAT resolution. The main idea is to prove that we can get a complete algorithm by successively saturating w.r.t. all the variables. However, notice that after saturating w.r.t. x_1 and then w.r.t. x_2, we get a multiset of clauses that is not saturated w.r.t. x_1. Therefore, we will use a variant of this basic algorithm: we saturate w.r.t x_1, then we remove all the clauses containing x_1, and saturate w.r.t x_2, we remove all the clauses containing x_2 and saturate w.r.t x_3, etc. Using Lemma 4, we prove that, if the original multiset of clauses was unsatisfiable, then with this process we get the empty clause. Even better, we get as many empty clauses as the minimum number of unsatisfied clauses in the original formula.

Theorem 2 (Completeness). *For any multiset of clauses $\mathcal{C}$, we have*

$$\mathcal{C} \vdash \underbrace{\Box, \ldots, \Box}_{m}, \mathcal{D}$$

where $\mathcal{D}$ is a satisfiable multiset of clauses, and m is the minimum number of unsatisfied clauses of $\mathcal{C}$.

PROOF: Let $x_1, \ldots, x_n$ be any list of the variables of $\mathcal{C}$. We construct two sequences of multisets $\mathcal{C}_0, \ldots, \mathcal{C}_n$ and $\mathcal{D}_1, \ldots, \mathcal{D}_n$ such that

1. $\mathcal{C} = \mathcal{C}_0$,
2. for $i = 1, \ldots, n$, $\mathcal{C}_i \cup \mathcal{D}_i$ is a saturation of $\mathcal{C}_{i-1}$ w.r.t. x_i, and
3. for $i = 1, \ldots, n$, $\mathcal{C}_i$ is a multiset of clauses not containing $x_1, \ldots, x_i$, and $\mathcal{D}_i$ is a multiset of clauses containing the variable x_i.

By lemma 3, this sequences can effectively be computed: for $i = 1, \ldots, n$, we saturate $\mathcal{C}_{i-1}$ w.r.t. x_i, and then we partition the resulting multiset into a subset $\mathcal{D}_i$ containing x_i, and another $\mathcal{C}_i$ not containing this variable.

Notice that, since $\mathcal{C}_n$ does not contain any variable, it is either the empty multiset $\emptyset$, or it only contains (some) empty clauses $\{\Box, \ldots, \Box\}$.

Now we are going to prove that the multiset $\mathcal{D} = \bigcup_{i=1}^{n} \mathcal{D}_i$ is satisfiable by constructing an assignment satisfying it. For $i = 1, \ldots, n$, let $\mathcal{E}_i = \mathcal{D}_i \cup \ldots \cup \mathcal{D}_n$, and let $\mathcal{E}_{n+1} = \emptyset$. Notice that, for $i = 1, \ldots, n$,

1. the multiset $\mathcal{E}_i$ only contains the variables $\{x_i, \ldots, x_n\}$,
2. $\mathcal{E}_i$ is saturated w.r.t. x_i, and
3. $\mathcal{E}_i$ decomposes as $\mathcal{E}_i = \mathcal{D}_i \cup \mathcal{E}_{i+1}$, where all the clauses of $\mathcal{D}_i$ contain x_i and none of $\mathcal{E}_{i+1}$ contains x_i.

Now, we construct a sequence of assignments $I_1, \ldots, I_{n+1}$, where I_{n+1} is the empty assignment, hence satisfies $\mathcal{E}_{n+1} = \emptyset$. Now, I_i is constructed from I_{i+1} as follows. Assume by induction hypothesis that I_{i+1} satisfies $\mathcal{E}_{i+1}$. Since $\mathcal{E}_i$ is saturated w.r.t. x_i, and decomposes into $\mathcal{D}_i$ and $\mathcal{E}_{i+1}$, by lemma 4, we can extend I_{i+1} with an assignment for x_i to obtain I_i satisfy $\mathcal{E}_i$. Iterating, we get that I_1 satisfies $\mathcal{E}_1 = \mathcal{D} = \bigcup_{i=1}^{n} \mathcal{D}_i$.

Concluding, since by the soundness Theorem 1 the inference preserves the number of falsified clauses for every assignment, $m = |\mathcal{C}_n|$ is the minimum number of unsatisfied clauses of $\mathcal{C}$.   ∎

5   Complete Refinements

In classical resolution we can assume a given total order on the variables $x_1 > x_2 > \ldots > x_n$ and restrict inferences $x \vee A, \overline{x} \vee B \vdash A \vee B$ to satisfy x is maximal in $x \vee A$ and in $\overline{x} \vee B$. This refinement of resolution is complete, and has some advantages: the set of possible proofs is smaller, thus its search is more efficient.

The same result holds for Max-SAT Resolution:

Theorem 3 (Completeness of Ordered Max-SAT Resolution). *Max-SAT resolution with the restriction that the cut variable is maximal on the premises is complete.*

PROOF: The proof is similar to Theorem 2. First, given the ordering $x_1 > x_2 > \ldots > x_n$, we start by computing the saturation w.r.t. x_1 and finish with x_n. Now, notice that, when we saturate C_0 w.r.t. x_1 to obtain $C_1 \cup D_1$, we only cut x_1, and this is the biggest variable. Then, when we saturate C_1 w.r.t. x_2 to obtain $C_2 \cup D_2$, we have to notice that the clauses of C_1, and the clauses that we could obtain from them, do not contain x_1, and we only cut x_2 which is the biggest variable in all the premises. In general, we can see that at every inference step performed during the computation of the saturations (no matter how they are computed) we always cut a maximal variable. We only have to choose the order in which we saturate the variables coherently with the given ordering of the variables. ∎

Corollary 1. *For any multiset of clauses C, and for every ordering $x_1 > \ldots > x_n$ of the variables, we have*

$$C \vdash_{x_1} C_1 \vdash_{x_2} \cdots \vdash_{x_n} \underbrace{\square, \ldots, \square}_{m}, D$$

where D is a satisfiable multiset of clauses, m is the minimum number of unsatisfied clauses of C, and in every inference step the cut variable is maximal.

6   An Algorithm for Max-SAT

From the proof of Theorem 2, we can extract the following algorithm:

```
input:  C
C_0 := C
for i := 1 to n
        C := saturation(C_{i-1}, x_i)
        ⟨C_i, D_i⟩ := partition(C, x_i)
endfor
m := |C_n|
I := ∅
for i := n downto 1
        I := I ∪ [x_i ↦ extension(x_i, I, D_i)]
output: m, I
```

Given an initial multiset of clauses C, this algorithm obtains the minimum number m of unsatisfied clauses and an optimal assignment I for C.

The function *partition*(C, x) computes a partition of C into the subset of clauses containing x and the subset of clauses not containing x.

The function *saturation*(C, x) computes a saturation of C w.r.t. x. As we have already said, the saturation of a multiset is not unique, but the proof of Theorem 2 does not depends on which particular saturation we take. Therefore, this computation can be done with "don't care" nondeterminism.

The function *extension*(x, I, D) computes a truth assignment for x such that, if I assigns the value true to all the clauses of D containing x, then the function returns false, if I assigns true to all the clauses of D containing $\overline{x}$, then returns true. According to Lemma 4 and the way the D_i's are computed, I evaluates to true all the clauses containing x or all the clauses containing $\overline{x}$.

The order on the saturation of the variables can be also freely chosen, i.e. the sequence $x_1, \ldots x_n$ can be any enumeration of the variables.

7   Efficiency

In classical resolution, we know that there are formulas that require exponentially long refutations on the number of variables, and even on the size of the formula, but no formula requires more than 2^n inference steps to be refuted, being n the number of variables. We don't have a better situation in Max-SAT resolution. Moreover, since we can have repeated clauses, and need to generate more than one empty clause, the number of inference steps is not bounded by the number of variables. It also depends on the number of original clauses. The following theorem states an upper bound on the number of inference steps, using the strategy of saturating variable by variable:

Theorem 4. *For any multiset C of m clauses on n variables, we can deduce $C \vdash \Box, \ldots, \Box, D$, where D is satisfiable, in less than $n \cdot m \cdot 2^n$ inference steps.*

Moreover, the search of this proof can be also done in time $\mathcal{O}(m\, 2^n)$.

PROOF: Let n be the number of variables, and m the number of original clauses. Instead of the characteristic function of a clause, we will assign to every clause C a weight $w(C)$ equal to the number of assignments to the n variables that falsify the clause. The weight of a multiset of clauses is then the sum of the weights of its clauses. Obviously the weight of a clause is bounded by the number of possible assignments $w(C) \leq 2^n$, being $w(C) = 0$ true only for tautologies. Therefore, the weight of the original multiset is bounded by $m\, 2^n$.

Like for the characteristic function, when $C \vdash D$, we have $w(C) = w(D)$.

A similar argument to Lemma 3 can be used to prove that we can obtain a saturation D of any multiset C w.r.t. any variable x in less than $w(C)$ many inference steps. If we compute the weight of the clauses containing x and of those not containing x separately, we see that in each inference step, the first weight strictly decreases while the second one increases. Therefore, the saturation w.r.t. the first variable needs no more than $m\, 2^n$ inference steps.

When we partition C into a subset containing x and another not containing x, both subsets will have weight smaller than $w(C)$, so the weight of C when we start the second round of saturations will also be bounded by the original weight. We can repeat the same argument for the saturation w.r.t. the n variables, and conclude that the total number of inference steps is bounded by $n\,m\,2^n$.

The proof of completeness for ordered Max-SAT resolution, does not depends on which saturation we compute. Each inference step can be computed in time $\mathcal{O}(n)$. This gives the worst-case time upper bound.   ∎

8   Conclusions

We have defined a complete resolution rule for Max-SAT which subsumes the resolution-like rules defined so far. To the best of our knowledge, this is the first complete logical calculus defined for Max-SAT. We have also proved the completeness of the ordered resolution refinement, described an exact algorithm and computed a time upper bound.

In a longer version of this paper, we have extended the contributions to weighted Max-SAT and we have found formulas that require exponential refutations. There remain many interesting directions to follow both from a theoretical and practical perspective. For example, define further complete refinements, and use the rule to derive equivalent encodings of a given instance and study the impact on the performance of exact and non-exact Max-SAT solvers.

References

1. T. Alsinet, F. Manyà, and J. Planes. Improved branch and bound algorithms for Max-SAT. In *Proc. of the 6th Int. Conf. on the Theory and Applications of Satisfiability Testing, SAT'03*, 2003.
2. T. Alsinet, F. Manyà, and J. Planes. A Max-SAT solver with lazy data structures. In *Proc. of the 9th Ibero-American Conference on Artificial Intelligence, IBERAMIA'04*, number 3315 in LNCS, pages 334–342, Puebla, México, 2004. Springer.
3. T. Alsinet, F. Manyà, and J. Planes. Improved exact solver for weighted Max-SAT. In *Proc. of the 8th Int. Conf. on Theory and Applications of Satisfiability Testing, SAT'05*, number 3569 in LNCS, pages 371–377, St. Andrews, Scotland, 2005. Springer.
4. N. Bansal and V. Raman. Upper bounds for MaxSat: Further improved. In *Proc. of the 10th Int. Symposium on Algorithms and Computation, ISAAC'99*, number 1741 in LNCS, pages 247–260, Chennai, India, 1999. Springer.
5. J. Larrosa and F. Heras. Resolution in Max-SAT and its relation to local consistency in weighted CSPs. In *Proc. of the 19th Int. Joint Conference on Artificial Intelligence, IJCAI'05*, pages 193–198, Edinburgh, Scotland, 2005. Morgan Kaufmann.
6. C. M. Li, F. Manyà, and J. Planes. Exploiting unit propagation to compute lower bounds in branch and bound Max-SAT solvers. In *Proc. of the 11th Int. Conf. on Principles and Practice of Constraint Programming, CP'05*, number 3709 in LNCS, pages 403–414, Sitges, Spain, 2005. Springer.

7. C. M. Li, F. Manyà, and J. Planes. Detecting disjoint inconsistent subformulas for computing lower bounds for Max-SAT. In *Proc. of the 21st National Conference on Artificial Intelligence, AAAI'06*, Boston, USA, 2006.
8. R. Niedermeier and P. Rossmanith. New upper bounds for maximum satisfiability. *Journal of Algorithms*, 36(1):63–88, 2000.
9. H. Shen and H. Zhang. Study of lower bound functions for MAX-2-SAT. In *Proc. of the 19th National Conference on Artificial Intelligence, AAAI'04*, pages 185–190, San Jose, California, USA, 2004.
10. H. Shen and H. Zhang. Improving exact algorithms for MAX-2-SAT. *Annals of Mathematics and Artificial Intelligence*, 44(4):419–436, 2005.
11. Z. Xing and W. Zhang. Efficient strategies for (weighted) maximum satisfiability. In *Proc. of the 10th Int. Conf. on Principles and Practice of Constraint Programming, CP'04*, number 3258 in LNCS, pages 690–705, Toronto, Canada, 2004. Springer.
12. Z. Xing and W. Zhang. An efficient exact algorithm for (weighted) maximum satisfiability. *Artificial Intelligence*, 164(2):47–80, 2005.
13. H. Zhang, H. Shen, and F. Manya. Exact algorithms for Max-SAT. In *4th Int. Workshop on First-Order Theorem Proving, FTP'03*, Valencia, Spain, 2003.
14. H. Zhang, H. Shen, and F. Manya. Exact algorithms for Max-SAT. *Electronic Notes in Theoretical Computer Science*, 86(1), 2003.

On Solving the Partial MAX-SAT Problem[*]

Zhaohui Fu[**] and Sharad Malik

Department of Electrical Engineering
Princeton University
Princeton, NJ 08544, USA
{zfu, sharad}@Princeton.EDU

Abstract. Boolean Satisfiability (SAT) has seen many successful applications in various fields such as Electronic Design Automation and Artificial Intelligence. However, in some cases, it may be required/preferable to use variations of the general SAT problem. In this paper, we consider one important variation, the Partial MAX-SAT problem. Unlike SAT, Partial MAX-SAT has certain constraints (clauses) that are marked as relaxable and the rest are hard, i.e. non-relaxable. The objective is to find a variable assignment that satisfies all non-relaxable clauses together with the maximum number of relaxable ones. We have implemented two solvers for the Partial MAX-SAT problem using a contemporary SAT solver, zChaff. The first approach is a novel diagnosis based algorithm; it iteratively analyzes the UNSAT core of the current SAT instance and eliminates the core through a modification of the problem instance by adding relaxation variables. The second approach is encoding based; it constructs an efficient auxiliary counter that constrains the number of relaxed clauses and supports binary search or linear scan for the optimal solution. Both solvers are complete as they guarantee the optimality of the solution. We discuss the relative strengths and thus applicability of the two solvers for different solution scenarios. Further, we show how both techniques benefit from the persistent learning techniques of incremental SAT. Experiments using practical instances of this problem show significant improvements over the best known solvers.

1 Introduction

In the last decade Boolean Satisfiability (SAT) has seen many great advances, including non-chronological backtracking, conflict driven clause learning, efficient Boolean Constraint Propagation (BCP) and UNSAT core generation. As a consequence, many applications have been able to successfully use SAT as a decision procedure to determine if a specific instance is SAT or UNSAT. However, there are many other variations of the SAT problem that go beyond this decision procedure use of SAT solvers. For example, the MAX-SAT Problem [8] seeks the maximum number of clauses that can be satisfied. This paper examines a generalization of this problem referred to as Partial MAX-SAT [15,2].

---

[*] This research is supported by a grant from the Air Force Research Laboratory.

[**] Z. Fu is on leave from the Department of Computer Science, National University of Singapore, Singapore 117543.

A. Biere and C.P. Gomes (Eds.): SAT 2006, LNCS 4121, pp. 252–265, 2006.

Partial MAX-SAT [15,2] (PM-SAT) sits between the classic SAT problem and MAX-SAT. While the classic SAT problem requires *all* clauses to be satisfied, PM-SAT relaxes this requirement by having certain clauses marked as *relaxable* or *soft* and others to be *non-relaxable* or *hard*. Given n relaxable clauses, the objective is to find an assignment that satisfies all non-relaxable clauses together with the maximum number of relaxable clauses (i.e. a minimum number k of these clauses get *relaxed*). PM-SAT can thus be used in various optimization tasks, e.g. multiple property checking, FPGA routing, university course scheduling, etc. In these scenarios, simply determining that an instance is UNSAT is not enough. We are interested in obtaining a best way to make the instance satisfiable allowing for *some* clauses to be unsatisfied.

The difference between PM-SAT and MAX-SAT [8] is that every clause in MAX-SAT can be relaxed, which clearly makes MAX-SAT a special case of PM-SAT. Though decision versions of both problems are NP-Complete [5], PM-SAT is clearly more versatile.

1.1 Previous Work

PM-SAT was first defined by Miyazaki *et al.* [15] during their work on optimization of database queries in 1996. In the same year, Kautz *et al.* [10] proposed the first heuristic algorithm based on local search for solving this problem. Later in 1997 Cha *et al.* [2] proposed another local search technique to solve the PM-SAT problem in the the context of university course scheduling.

In 2005, Li used two MinCostSat solvers, *eclipse-stoc* [12] and *wpack* [12], for the transformed PM-SAT problem in FPGA routing. MinCostSat is a SAT problem which minimizes the cost of the satisfying assignment. For example, assigning a variable to be true usually incurs a positive cost while assigning it to be false incurs no cost. The objective is to find a satisfying assignment with minimum total cost. By inserting a slack variable [12] to each of the relaxable clauses, Li transforms the PM-SAT problem into a MinCostSat problem with each slack variable having a unit cost. *eclipse-stoc* is a general purpose MinCostSat solver and *wpack* is specialized for FPGA routing benchmarks. Li demonstrated some impressive results using *wpack* in his thesis. However, both *eclipse-stoc* and *wpack* are based on local search techniques and hence are not complete solvers, i.e. the solver provides no guarantee on the optimality of the solution.

Argelich and Manyà uses a branch and bound approach for the over constrained MAX-SAT problems [1]. However, as we will show in Section 4, branch and bound based algorithms, including *bsolo*, do not work well on the PM-SAT problem.

1.2 Our Contribution

In this paper, we propose two practically efficient approaches to solve the PM-SAT problem optimally. Both approaches use the state-of-the-art SAT solver zChaff [16] with certain extensions.

1. Diagnosis Based. The first approach is based on the ability of SAT solvers to provide an UNSAT core [21] for unsatisfiable instances. This core is a subset of original clauses that are unsatisfiable by themselves and in some sense can be considered to be the "cause" of the unsatisfiability. This core is generated as a byproduct of the proof of the unsatisfiability. The UNSAT core is analyzed and each relaxable clause appearing in the core is augmented with a distinct relaxation variable. Additional

clauses are added to the original SAT instance to ensure the *one-hot* property of these relaxation variables. This augmentation essentially *eliminates* this core from the SAT instance. The procedure continues until the SAT instance is satisfiable. We give a proof of the optimality of the final solution using these relaxation variables and the one-hot property.

2. Encoding Based. The second approach constructs an efficient auxiliary logic counter, i.e. an adder and comparator, to constrain the number of clauses that can be relaxed simultaneously. It then uses either *binary search* or *linear scan* techniques to find the minimum number of k (out of n) clauses that need to be relaxed. The logic counter is carefully designed such that maximum amount of learned information can be re-used across different invocations of the decision procedure.

2   Diagnosis Based Approach: Iterative UNSAT Core Elimination

Being the best solver in the Certified UNSAT Track of SAT 2005 Competition, zChaff is very efficient in generating UNSAT cores. Our diagnosis based approach takes full advantage of this feature. It iteratively identifies the reason of the unsatisfiability of the instance, i.e. the UNSAT core [21], and uses relaxation variables to eliminate these UNSAT cores one by one until the instance becomes satisfiable.

Definition 1. *An unsatisfiable core is a subset of the* original *CNF clauses that are unsatisfiable by themselves.*

Modern SAT solvers provide the UNSAT core as a byproduct of the proof of unsatisfiability [21].

2.1   The Optimal Algorithm with Proof

The diagnosis based approach is illustrated in Algorithm 1. We use *CNF* to represent the original SAT instance and $V(CNF)$ is the set of all Boolean variables and $C(CNF)$ is the set of all clauses. An UNSAT core UC is a set of clauses, i.e. $UC \subseteq C(CNF)$. A clause $c \in C(CNF)$ consists of a set of literals. A literal l is just a Boolean variable, $v \in V(CNF)$, with positive or negative phase, i.e. $l = v$ or $l = v'$.

Given an UNSAT core UC, for each relaxable clause $c \in UC$, a distinct relaxation variable is added to this clause, i.e. c is replaced by $c \cup \{v\}$. Setting this variable to true makes the associated clause satisfied (and hence relaxed). An UNSAT core is said to be *eliminated* when at least one of its clauses is satisfied (relaxed) by a relaxation variable setting to be true.

Let S be the set of relaxation variables from UNSAT core UC, the one-hot constraint over a set S of Boolean variables requires that one and only one of the variables in S is assigned to be true and the other $|S| - 1$ variables must be false. The number of clauses added due to the one-hot constraint is $\frac{|S| \times (|S|-1)}{2} + 1$. For example, with $S = \{a, b, c\}$ and the one-hot constraint clauses are $(a' + b')(b' + c')(a' + c')(a + b + c)$.

One relaxable clause is relaxed during each UNSAT iteration of the `while` loop in Algorithm 1. The algorithm stops after exactly k iterations, where k is the minimum number of clauses to be relaxed. R is the subset of S consisting of all relaxation variables set to 1, i.e. the corresponding clauses are relaxed.

Algorithm 1. Iterative UNSAT Core Elimination

1: $S := \emptyset$
2: **while** SAT solver returns UNSATISFIABLE **do**
3:      Let UC be the UNSAT core provided by the SAT solver
4:      $S := \emptyset$
5:      **for all** Clause $c \in UC$ **do**
6:          **if** c is relaxable **then**
7:              Allocate a new relaxation variable v
8:              $c := c \cup \{v\}$
9:              $S := S \cup \{v\}$
10:         **end if**
11:     **end for**
12:     **if** $S = \emptyset$ **then**
13:         Return CNF UNSATISFIABLE
14:     **else**
15:         Add clauses enforcing the *One-Hot* constraint for S to the SAT solver
16:         $\mathcal{S} := \mathcal{S} \cup S$
17:     **end if**
18: **end while**
19: $R := \{v \mid v \in \mathcal{S}, v = 1\}; \ k := |R|$
20: Return Satisfying Assignment, k, R.

Theorem 1. *Algorithm 1 finds the minimum number of clauses to be relaxed.*

Proof. Consider the interesting case where the original problem is always satisfiable with relaxation of certain relaxable clauses (Otherwise Algorithm 1 returns unsatisfiable in Line 12). Suppose Algorithm 1 stops after exactly k iterations, i.e. relaxing k clauses. Clearly, Algorithm 1 has encountered a total number of k UNSAT cores (one in each iteration), which we denote them by a set $U, |U| = k$. Note that original problem instance contains at least k UNSAT cores, even though each iteration starts with a new modified problem instance. Now suppose that there exists some optimal solution that relaxes a set $M, |M| < k$, clauses to make the original problem satisfiable. Obviously, relaxing all clauses in M eliminates all the UNSAT cores in U. However, since $|M| < k = |U|$ and by the Pigeon Hole Principle there must be at least one clause $c \in M$ those relaxation eliminates two or more UNSAT cores $(u_1, u_2, \ldots) \in U$ and $c \in u_1, c \in u_2$. Without the loss of generality, let us assume that Algorithm 1 encounters u_1 first in the while loop. So every clause including c in u_1 is added with a relaxation variable. Let v be the relaxation variable added to c. Now there exists an assignment that can eliminate both u_1 and u_2 by setting v to be true. Due to the completeness of our SAT solver, the UNSAT core u_2 should never be encountered, which leads to a contradiction. This contradiction is caused by the assumption that $|M| < k = |U|$. Hence we have to conclude that $|M| = k = |U|$, i.e. each relaxed clause in M can eliminate at most one UNSAT core in U. □

It is worth mentioning that the UNSAT core extraction is not compulsory. One could add a relaxation variable to each relaxable clause and require this batch of relaxation variables to be one-hot for every iteration in which the problem remains unsatisfiable.

This naive approach is still capable of finding the minimum number of clauses to be relaxed. However, recall that the one-hot constraint requires $O(|S|^2)$ additional clauses where S is the set of relaxable clauses in the UNSAT core. Therefore it is impractical to enforce the one-hot constraint on the relaxation variables for all relaxable clauses. For example, the PM-SAT instance might have 100 relaxation clauses while only 3 appear in the UNSAT core. The naive approach adds $\frac{100 \times 99}{2} + 1 = 4951$ clauses while the diagnosis based approach adds only $\frac{3 \times 2)}{2} + 1 = 4$ clauses. The diagnosis based approach exploits the availability of the UNSAT core to keep the number of relaxation variables and one-hot constraint clauses small.

2.2   An Illustrative Example

It is worth mentioning that Algorithm 1 does not require the UNSAT core UC to be minimal. Furthermore, the UNSAT cores encountered by Algorithm 1 need not be disjoint. The following example shows a simple CNF formula that contains two overlapping cores. Suppose we have four Boolean variables x_1, x_2, x_3 and x_4. Relaxable clauses are shown with square brackets and $\odot$ denotes the *resolution* operator.

$$(x_1' + x_2')(x_1' + x_3)(x_1' + x_3')(x_2' + x_4)(x_2' + x_4')[x_1][x_2]$$

This CNF formula is unsatisfiable since $(x_1' + x_2')[x_1][x_2]$ form an UNSAT core because

$$(x_1' + x_2') \odot [x_1] \odot [x_2] = (x_2') \odot [x_2] = ()$$

Note that whether a clause is relaxable or not does not affect the resolution. Recall that a UNSAT core is a set of original clauses that are unsatisfiable and they resolve to an empty clause (), which can never be satisfied. The only relaxable clauses in this core are $[x_1][x_2]$. So in the first iteration we add two distinct relaxation variables r_1 and r_2 to each of them respectively and enforce r_1 and r_2 to be one-hot. The resulting CNF formula is

$$(x_1' + x_2')(x_1' + x_3)(x_1' + x_3')(x_2' + x_4)(x_2' + x_4')[x_1 + r_1][x_2 + r_2](r_1' + r_2')(r_1 + r_2)$$

Note that clauses due to the one-hot constraint are not relaxable. However, the relaxable clauses are still marked as relaxable even after inserting relaxation variables. This is because, as we will show, one relaxation variable may not be enough to make the instance satisfiable. The current CNF formula is still unsatisfiable as

$$(x_1' + x_3) \odot (x_1' + x_3') \odot [x_1 + r_1] \odot (r_1' + r_2') \odot [x_2 + r_2] = (x_2)$$
$$(x_2) \odot (x_2' + x_4') = (x_4')$$
$$(x_2) \odot (x_2' + x_4) = (x_4)$$
$$(x_4') \odot (x_4) = ()$$

So in the second iteration, we add another two relaxation variables r_3 and r_4 to the relaxable clauses $[x_1 + r_1][x_2 + r_2]$ in the core. Together with clauses due to the one-hot constraint of r_3 and r_4, the CNF formula becomes

$$(x_1' + x_2')(x_1' + x_3)(x_1' + x_3')(x_2' + x_4)(x_2' + x_4')[x_1 + r_1 + r_3][x_2 + r_2 + r_4]$$
$$(r_1' + r_2')(r_1 + r_2)(r_3' + r_4')(r_3 + r_4)$$

This formula is satisfiable with the following assignment

$$x_1 = 0, \quad x_2 = 0, \quad x_3 = 1, \quad x_4 = 1,$$
$$r_1 = 1, \quad r_2 = 0, \quad r_3 = 0, \quad r_4 = 1.$$

Based on the above satisfying assignment, both $[x_1][x_2]$ should be relaxed to make the problem satisfiable, i.e. $k = n = 2$. Note that there is no constraint among the relaxation variables added in different iterations and the one-hot constraint only applies to all relaxation variables added due to the same UNSAT core *in one iteration*. The number of the relaxation variables needed only depends on the number of relaxation clauses in the current UNSAT core[1], and not the total number of relaxation clauses in the entire SAT instance. In Section 4 we will see some cases where the total number of relaxable clauses is large and our diagnosis based approach still performs well on these cases.

The iterative core elimination requires the SAT solver to be able to provide the UN-SAT core (or proof) as part of answering UNSAT. This feature does incur some overhead. For example, the SAT solver needs to record the resolution trace for each learned clause[2]. Even when a learned clause is deleted, which happens very frequently in most state-of-the-art SAT solvers, the resolution trace for that particular learned clause cannot be deleted because it might be used to resolve other learned clauses that are not yet deleted. In case of an unsatisfiable instance, we need all the resolution information so that we could trace back from the conflict to the original clauses, which then form the UNSAT core. Recording the resolution trace not only slows down the search speed, but also uses a large amount of memory, which could otherwise be used for learned clauses.

3   Encoding Based Approach: Constructing an Auxiliary Counter

With highly optimized state-of-the-art SAT solvers like zChaff [16], Berkmin [6], Siege [17] and MiniSat [4] , the most straightforward way is to translate the PM-SAT problem directly into a SAT instance. Such an implementation is likely to be efficient since the translated SAT instance takes advantages of all the sophisticated techniques used in a contemporary SAT solver. Furthermore, this approach requires very little or no modification to the SAT solver itself and hence could continuously benefit from the advances in SAT.

However, conventional SAT solvers do not support integer arithmetic, which is necessary in PM-SAT for expressing the constraint of $\leq k$ *clauses left unsatisfied*. We use an auxiliary logic counter [11] to represent this $\leq k$ condition, whose output is a Boolean variable and the entire counter could then be translated into CNF in a straightforward way. There are various ways of constructing such an an auxiliary logic counter [11]. Xu considers four types of these logic counters, namely chain counter, hierarchical tree counter, routing counter and sorting counter in her work on *subSAT* [20]. *subSAT* is a

---

[1] Recall that we assume there must exist at least one relaxation clause in every core since otherwise the problem is unsatisfiable even if we relax all relaxable clauses.

[2] Each learned clause is the result of a series of resolutions of other clauses, both learned or original. But ultimately, each learned conflict clause is the result of a series of resolutions using only the original clauses.

MAX-SAT problem but with the assumption that $k \ll n$, where n here is the total number of clauses. In other words, the problem becomes satisfiable with very small number (usually $k < 5$) of clauses removed (relaxed). In the *subSAT* implementation, one *mask* variable (which is equivalent to our relaxation variable) is added to each of the n clauses and they constrain that only $\leq k$ mask variables can be true by using one of the four logic counters as mentioned above. The chain counter method creates a $\lceil \lg k + 1 \rceil$ bit adder for each clause and concatenates them together. The final output from the last adder is constrained to be $\leq k$. The hierarchical tree counter creates a tree using $\lceil \lg k + 1 \rceil$ bit adders as internal nodes that sum up all n mask variables and gives a $\lceil \lg k + 1 \rceil$ bit output at the root of the tree. The routing counter implements k k-to-n decoders with k inputs all set to be 1. The sorting counter uses a sorting circuit with k max operators (range from n bit to $n - k + 1$ bit) to move the 1s to one side of the output and then checks the kth bit of the output. Xu states that the first two counters (chain and tree counters) are more efficient than the others in terms of the amount of additional logic.

3.1   An Efficient Hierarchical Tree Adder

The most significant differences between our proposed encoding based approach and the *subSAT* approach are that our hierarchical tree adder is independent of k and we do not assume $k \ll n$. In addition to the linear scan for minimum k, we also use a binary search on $[0, n]$ for the minimum k (*subSat* only uses linear scan due to their assumption of $k \ll n$). We design our tree adder to be independent of the value k for two obvious reasons. First, we only need to construct the adder once at the beginning and re-use it during each iteration of the binary search, as compared to constructing the adder $\lg n$ times for binary search and k times for linear scan. Second, using the idea of incremental SAT [18], all clauses associated with the adder can be kept intact since they are always consistent with the problem. Maintaining the learned information is very important to the performance of most contemporary SAT solvers. Unfortunately, all the above 4 types of auxiliary counters proposed by Xu are dependent on k, particularly for the routing and sorting counters.

We propose a hierarchical tree adder that is independent of k using elementary adders, e.g. half adder and full adders. Figure 1 gives an example of such an adder with $n = 9$. It can be shown that the total number of additional 2-input logic gates is $\leq 5n$ as follows. Consider starting with $n \geq 3$ bit input, we use a full adder to sum up 3 bits while returning a sum bit and a carry bit [9]. The sum bit needs to be added with the other $n - 3$ bits left and the carry bit will only be used in next level. So each full adder reduces the number of inputs left by 2 and $\lfloor \frac{n}{2} \rfloor$ full adders are sufficient for the first level. Note the last sum bit becomes the least significant bit of the final sum. In the second level, we consider all the carry bits from the previous level and there are at most $\lfloor \frac{n}{2} \rfloor$ of them. Similar results extends to the third level and so on. So the total number of full adders is:

$$\lfloor \frac{n}{2} \rfloor + \lfloor \frac{n}{4} \rfloor + \lfloor \frac{n}{8} \rfloor + \ldots + 1 \leq n$$

Each full adder requires 5 2-input logic gates (2 AND, 2XOR and 1OR gate), which gives the total number of additional logic gates $\leq 5n$. Note that we can sometimes replace a

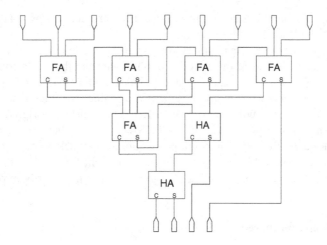

Fig. 1. An efficient hierarchical tree adder that sums the number of 1s from the $n = 9$ bit input (top) and gives a 4 bit binary value (bottom). The first level uses 4 full adders (FA); the second level needs 1 full adder and 1 half adder (HA); the third level just needs 1 half adder. S and C are the sum and carry bits of the adder respectively.

full adder by a half adder due to simplification by constant value (0), as shown in the second and third levels in Figure 1.

There is an important distinction between our hierarchical tree adder in Figure 1 and the one used by Xu [20]. Instead of using full/half adders as internal nodes of the tree, Xu uses a $\lceil \lg k + 1 \rceil$ bit adder for each of the internal nodes, which introduces a large amount of redundancy with relatively large k. For example, the first level inputs to the adder are at most 1 and in a binary representation of $\lceil \lg k + 1 \rceil$ bits, at least $\lceil \lg k + 1 \rceil - 1$ bits are just 0s. Our hierarchical tree adder is free of such redundancy due to the judicious use of full/half adders.

The hierarchical tree adder outputs a $\lceil \lg n + 1 \rceil$ bit binary value, which is then compared against a given value k using a logic comparator that outputs true if and only if the sum is less than or equal to k. Note that this logic comparator is dependent on k for efficiency reasons[3]. This hierarchical tree adder with comparator provides us an efficient platform for searching the minimum k. Generally binary search has advantages over linear scan on the benchmarks with $k > \lg n$.

It is worth mentioning that when this logic counter (adder with comparator) are translated into CNF clauses, the hierarchical tree adder generates many more clauses than the comparator does. In general, the number of CNF clauses from the adder is $O(n)$ while from the comparator is $O(\lg n)$. For each iteration during the binary search or linear scan, only $O(\lg n)$ original CNF clauses with related learned clauses need to be changed. The remaining clauses include both original and learned clauses corresponding to the original problem instance and the adder. The learned clauses capture the logic

---

[3] The resulting circuit is equivalent to performing the constant propagation on a general logic comparator with any given k.

relationship among the Boolean variables used in the problem instance and the adder and they cannot be learned without the adder.

One disadvantage of using an auxiliary counter is the introduction of a large number of XOR gates. Each full adder consists of two XOR gates and the entire counter results in $2n$ XOR gates. Though the number of additional logic gates is only linear in n, the situation could get worse when many XOR gates are chained together. For example, the least significant bit of the sum comes from an XOR chain of length $\lceil \lg n \rceil$. XOR chains are well known to cause poor performance of SAT solvers. One main reason is that unlike AND/OR gates, Boolean constraint propagation over XOR gates is very limited. This complication makes this approach no longer efficient for solving problems with very large n. It is worth mentioning that a large n does not necessarily imply a large k though obviously $k \le n$.

3.2   A Discussion on Incremental SAT

Incremental SAT was first formalized by Strichman [18]. It is the process of solving a series of SAT instances $\varphi_1, \varphi_2, \ldots, \varphi_n$. The consecutive SAT instances, φ_i and φ_{i+1}, are *similar*, i.e. only a small number of clauses (and variables) are different. Given the solution of φ_i, we could solve φ_{i+1} *incrementally* by only updating the different clauses while keeping most learned clauses in φ_i, which are still consistent with φ_{i+1}, intact. Maintaining the maximum amount of the learned clauses, i.e. recording the most visited search space, is a great advantage than starting from scratch each time.

A key issue in the implementation of incremental SAT is the efficient updating from instance φ_i to φ_{i+1}, which usually includes both deletion and addition of original clauses. Addition of new original clauses is trivial. However, deletion of original clauses implies the additional deletion of all learned clauses related to the deleted original clauses in order to maintain the integrity of the clauses database. This deletion can be performed efficiently with the use of group IDs. A group ID indicates a particular group, to which the clause belongs. The group IDs of a learned clause is the *union* of all the group IDs from the clauses used to generate this learned clauses (through resolution). Deletion according to a particular group ID removes all clauses (both original and learned) having this ID.

For the encoding based approach, we utilize the incremental SAT feature of zChaff and group all CNF clauses associated with the comparator using the same group ID. This implementation enable us to only change a very small fraction of all clauses (both original and learned) that are related to the comparator for each different value of k during binary search or linear scan. All clauses associated with the adder are independent of k and hence remain unchanged throughout the entire incremental SAT. Recall that the adder corresponds to many more clauses than the comparator does.

However, unlike the encoding based approach, the diagnosis based approach requires us to update the original clauses by inserting some relaxation variables. This makes it harder to use the incremental SAT algorithm. However, we can still group all the hard constraint clauses in a group and reuse all learned clauses that are generated within this group. In other words, we only delete the learned clauses associated with the relaxable constraint clauses.

4 Experimental Results

We implemented the *subSat* approach for PM-SAT using the chain counter and hierarchical tree counter proposed by Xu [20] for comparison. In addition, we translate our PM-SAT benchmarks into MinCostSat instances so that we can have an extensive comparison using other general purpose solvers. Recall that a MinCostSat problem is a SAT problem with a cost function for each satisfying assignment. We add a unique relaxation variable to each relaxable clause in PM-SAT and the cost of this relaxation variable is 1. All other variables have a cost of 0. Non-relaxable clauses remain unchanged in the above translation. The resulting problem is now a MinCostSat instance where the minimum cost corresponds to the minimum number of relaxation variables setting to be 1, which in turn implies that minimum number of clauses are relaxed. We then use *Scherzo* [3], *bsolo* [13] and *cplex* [7] to solve the translated MinCostSat problem. *Scherzo* is a well known branch-and-bound solver for Binate/Uniate Covering Problem (BCP/UCP) that incorporates many state-of-the-art techniques, including Maximum Independent Set [5] based lower bounding, branch variable selection and various search pruning rules. The BCP problem is essentially a MinCostSat problem [12] with a specific cost function. UCP has the additional restriction that all variables appear in only one phase. But unfortunately, *Scherzo* is not able to solve any of the benchmarks in the following tables. *bsolo* is another state-of-the-art branch-and-bound BCP/UCP solver based on the SAT solver GRASP [14]. *cplex* is the cutting edge commercial Linear Programming (LP) solver that is also capable of finding integer solutions efficiently.

All the experiments are conducted on a Dell PowerEdge 700 running Linux Fedora core 1.0 (g++ GCC 3.3.2) with single Pentium 4 2.8GHz, 1MB L2 cache CPU on 800MHz main bus.

4.1 FPGA Routing Benchmarks

We conduct our experiments mainly on industrial benchmarks. Table 1 shows the results of industrial examples resulting from a SAT based FPGA router. Each relaxable clause corresponds to a net-arc (single source, single destination) in the routing problem. Relaxation of clauses in the unsatisfiable SAT instance to make it satisfiable represent finding the fewest number of net-arcs which, if re-routed elsewhere, e.g. route-around, would allow the remaining set of net-arcs to be routed simultaneously.

Table 1. Performance comparison on FPGA routing benchmarks. Timeout for all solvers: 1 hour. * indicates server times out, the best solution found is reported.

Bench-mark	Num. Vars.	Num. Cls.	Rlx. Cls.	Min. k	Diagnosis Core Rmv	Encoding Binary	Linear	subSat Chain	Tree	Gen. Solver bsolo	cplex
FPGA_27	3953	13537	27	3	1.85	2.13	1.65	2.64	2.85	21.29	3*
FPGA_31	17869	65869	31	1	380.83	88.68	309.75	393.26	860.18	4*	12*
FPGA_32	2926	9202	32	3	0.89	1.10	0.95	1.12	1.18	6.56	3*
FPGA_33	9077	32168	33	3	18.65	19.25	26.44	27.02	27.93	61.5	4*
FPGA_39	6352	22865	39	4	31.22	7.76	7.15	8.83	8.48	59.07	6*
FPGA_44	6566	22302	44	3	10.12	9.00	8.36	11.75	12.80	6*	5*

The benchmark name in Table 1 shows the number of net-arcs in the actual FPGA routing problem. For example, the first row shows the result of benchmark FPGA_27, which has 27 net-arcs to be routed. The PM-SAT instance consists of 3953 Boolean variables with 13537 clauses. Among these clauses, 27 clauses are marked as relaxable, which corresponds to the 27 net-arcs to be routed. The optimal solution is a relaxation of $k = 3$ clauses (out of 27) that makes the entire problem satisfiable. The diagnosis based approach takes 1.85 seconds to find the optimal solution. The binary search and linear scan of the encoding based approach take 2.13 and 1.65 seconds respectively. The *subSat* approach using linear scan with the chain counter and hierarchical tree counter need 2.64 and 2.85 seconds respectively. For the translated MinCostSat problem, *bsolo* takes 21.29 seconds. *cplex* only reports a solution of $k = 3$ but it cannot prove its optimality. *Scherzo* could not report any solution found within the 1 hour time limit for all our benchmarks and hence is omitted from all of our tables.

4.2 Multiple Property Checking Benchmarks

Table 2 shows the results of multiple property checking using circuits from ISCAS85 and ITC99 benchmarks. Relaxable clauses are the properties (assertions) that assume each output signal of the entire circuit to be 1 or 0. The non-relaxable clauses are translated from the circuit structure. The corresponding PM-SAT instance is to find the maximum number of outputs that can be 1 or 0 (satisfying the property). Benchmarks that are satisfiable without any relaxation are excluded from the tables. All benchmarks start with a c are from the iscas85 family and the rests are from the itc99 family.

Table 2. Performance comparison on multiple property checking benchmarks. Benchmarks end with _1 (_0) are asserted to be 1 (0). Timeout for all solvers: 1 hour. * indicates server times out, the best solution found is reported.

Bench-mark	Num. Vars.	Num. Cls.	Rlx. Cls.	Min. k	Diagnosis Core Rmv	Encoding Binary	Encoding Linear	subSat Chain	subSat Tree	Gen. Solver bsolo	Gen. Solver cplex
c2670_1	1426	3409	140	7	0.05	0.07	0.06	0.25	0.34	1.05	106.40
c5315_1	2485	6816	123	10	0.09	0.18	0.13	0.83	1.13	15.35	208.80
c6288_1	4690	11700	32	2	343.71	81.98	192.27	185.67	169.94	2*	3*
c7552_1	4246	10814	108	5	2.64	1.41	1.17	1.62	1.77	5*	1909.63
b14_1	10044	28929	245	1	0.19	0.53	0.48	0.79	0.93	4.45	1182.85
b15_1	8852	26060	449	2	0.26	1.08	1.07	1.23	1.85	7.55	1308.47
b17_1	32229	94007	1445	6	1.65	14.93	14.92	67.73	90.85	65.88	65*
b20_1	20204	58407	512	2	0.50	2.62	2.90	1.97	5.59	10.88	5*
b21_1	20549	59532	512	2	0.49	2.59	2.89	2.36	5.47	12.63	1522.35
b22_1	29929	86680	757	4	0.96	6.19	5.43	11.78	18.52	25.53	31*
c7552_0	4246	10814	108	6	1.57	2.54	1.99	2.07	1.95	2369.5	6*
b15_0	8852	26060	449	3	0.22	0.14	0.79	1.83	2.68	19.37	260.75
b17_0	32229	94007	1445	13	4.54	13.74	6.85	90.64	173.17	17*	13*

4.3   Randomized UNSAT Benchmarks

Table 3 shows the results of classic UNSAT benchmarks with randomly chosen relaxable clauses. These benchmarks are from the `fvp-unsat-2.0` (verification of superscalar microprocessors) family by Velev [19]. Note that all the benchmarks in Table 3 have $k = 1$, which makes it inefficient to use binary search.

Table 3. Performance comparison on randomized UNSAT benchmarks. The encoding based approach using binary search is inefficient since $k = 1$ for all benchmarks and hence omitted. Timeout for all solvers: 1 hour. * indicates server times out, the best solution found is reported.

| Bench- | Num. | Num. | Rlx. | Min. | Diagnosis | Encoding | subSat | | Gen. Solver | |
mark	Vars.	Cls.	Cls.	k	Core Rmv	Linear	Chain	Tree	bsolo	cplex
2pipe	892	6695	6695	1	5.12	22.34	34.82	65.54	868.34	1*
3pipe	2468	27533	5470	1	4.96	18.18	19.97	31.43	1*	1*
4pipe	5237	80213	802	1	8.45	8.65	8.81	11.32	1*	1*
5pipe	9471	195452	19474	1	18.91	305.69	273.79	367.93	1*	1*
6pipe	15800	394739	15828	1	107.33	383.36	463.81	424.35	1*	1*

Table 1, Table 2 and Table 3 clearly show that both approaches constantly outperform the best known solvers. For benchmarks with a large number of relaxable clauses, e.g. `b17`, `b20` and `b22` in Table 2 and all benchmarks in Table 3, the diagnosis based approach has obvious advantage over the search approach (either binary or linear), which suffers from the large auxiliary adder. With most other benchmarks like `c6288` in Table 2 and `FPGA_39` in Table 1, whose number of relaxable clauses is small, the encoding based search approach is faster. As we can see from the tables that there is no significant difference between binary search and linear scan used in the encoding based approach. However, we still believe that for instances with large k, binary search is a better option. In addition, for the benchmarks with small k value, the performance of the *subSat* approach is comparable with our encoding based approach. This is because the $\lceil \lg k + 1 \rceil$ bit adder used in *subSat* is not much larger than a full adder or half adder for very small k, e.g. $k = 2$. However, the performance *subSat* of degrades dramatically with relative large k, e.g. benchmark `b17_0` in Table 2.

It is interesting to see that all SAT based approaches (all of our approaches, *subSat* and *bsolo*) generally outperform the non-SAT based branch-and-bound methods like *scherzo* and *cplex*. One possible reason is that these industrial benchmarks have more implications and conflicts, than the typical UCP/BCP or ILP instances.

Both our diagnosis based and encoding based approaches benefit from the incremental SAT and so does our implementation of the *subSat*. Among these three, the encoding based approach gains the most improvements due to the incremental SAT as it solves very *similar* SAT instances. However, we could not provide additional results for this due to the page limit.

Note that though both our diagnosis based and encoding based methods use zChaff as an underlying SAT solver, each of them has a customized zChaff solver based on the features needed. The zChaff solver used in the encoding based approach is relatively faster than the one used in the diagnosis based approach as the latter one has a significant overhead of bookkeeping the information for constructing an UNSAT core. Further, different approaches solve different numbers of underlying SAT instances. The binary search always makes $\lceil \lg n + 1 \rceil$ SAT calls while the core elimination approach and the linear search make $k + 1$ such SAT calls. The overall time used as presented in all tables includes the time used by these intermediate SAT instances. Usually the SAT instance becomes more and more difficult as we approach k, which is because the corresponding instance becomes more and more constrained.

Usually the diagnosis based approach has an advantage for instances with large n *and small* k *due to the absence of the overhead caused by the hierarchical adder. For instances with relatively small* n, *the encoding based approach is faster and particularly binary/linear search should be used when* k *is large/small.*

5 Conclusions and Future Directions

We have presented two complete and efficient approaches specialized for solving the PM-SAT problem, which arises from various situations including multiple property checking, FPGA routing, etc. These two specialized solvers significantly outperform the best known solvers.

Some key features about these two approaches can be summarized as follows:

1. The diagnosis based approach uses an iterative core elimination technique, which does not require any auxiliary structure and hence is independent of the total number of relaxable clauses. This approaches iteratively identifies the UNSAT core of the problem and relaxes it by inserting relaxation variables to the relaxable clauses in the core. We provide a proof of optimality for this approach.
2. The encoding based approach uses an auxiliary counter implemented as an efficient hierarchical tree adder with a logic comparator to constrain the number of true relaxation variables, and hence the number of relaxed clauses, during the search. The hierarchical tree adder only needs to be constructed once at the beginning and the SAT solver can re-use most of the learned clauses for the instances generated during each search iteration using incremental SAT techniques. Both binary and linear search are supported in this approach.

We believe that there is still room for improvement. One such area is the further tuning of zChaff solver according to different characteristics of the SAT instances generated by each of the two approaches. The other area is to optimize the UNSAT core generation process, e.g. reducing overhead, minimizing UNSAT core, etc.

Acknowledgement

We would like to thank Richard Rudell, Olivier Coudert and Vasco Manquinho for providing us various solvers and benchmarks.

References

1. J. Argelich and F. Manyà. Solving over-constrained problems with SAT technology. *Lecture Notes in Computer Science (LNCS): Theory and Applications of Satisfiability Testing: 8th International Conference*, 3569:1–15, 2005.
2. B. Cha, K. Iwama, Y. Kambayashi, and S. Miyazaki. Local search algorithms for partial MAXSAT. In *Proceedings of the Fourteenth National Conference on Artificial Intelligence*, pages 263–268, 1997.
3. O. Coudert. On solving covering problems. In *Proceedings of the 33rd Design Automation Conference*, pages 197–202, 1996.
4. N. Eén and N. Sörensson. MiniSat – A SAT solver with conflict-clause minimization. In *Proceedings of the International Symposium on the Theory and Applications of Satisfiability Testing*, 2005.
5. M. R. Garey and D. S. Johnson. *Computers and Intractability: A Guide to the Theory of NP-completeness*. Freemand & Co., New York, 1979.
6. E. Goldberg and Y. Novikov. Berkmin: A fast and robust SAT solver. In *Proceedings of the Design Automation and Test in Europe*, pages 142–149, 2002.
7. ILOG. Cplex homepage, http://www.ilog.com/products/cplex/, 2006.
8. D. S. Johnson. Approximation algorithms for combinatorial problems. *Journal of Computer and System Sciences*, 9:256–278, 1974.
9. R. H. Katz. *Contemporary Logic Design*. Benjamin Cummings/Addison Wesley Publishing Company, 1993.
10. H. Kautz, B. Selman, and Y. Jiang. A general stochastic approach to solving problems with hard and soft constraints,, In D. Du, J. Gu, and P. M. Pardalos, editors, The Satisfiability Problem: Theory and Applications, 1996.
11. I. Koren. *Computer Arithmetic Algorithms*. Prentice Hall, 1993.
12. X. Y. Li. *Optimization Algorithms for the Minimum-Cost Satisfiability Problem*. PhD thesis, Department of Computer Science, North Carolina State University, Raleigh, North Carolina, 2004. 162 pages.
13. V. Manquinho and J. Marques-Silva. Search pruning techniques in SAT-based branch-and-bound algorithms for the binate covering problem. *IEEE Transactions on Computer-Aided Design of Integrated Circuits and Systems*, 21:505–516, 2002.
14. J. Marques-Silva and K. A. Sakallah. GRASP: A search algorithm for propositional satisfiability. *IEEE Transactions on Computers*, 48:506–521, 1999.
15. S. Miyazaki, K. Iwama, and Y. Kambayashi. Database queries as combinatorial optimization problems. In *Proceedings of the International Symposium on Cooperative Database Systems for Advanced Applications*, pages 448–454, 1996.
16. M. W. Moskewicz, C. F. Madigan, Y. Zhao, L. Zhang, and S. Malik. Chaff: Engineering an efficient SAT solver. In *Proceedings of the 38th Design Automation Conference*, 2001.
17. L. Ryan. Efficient algorithms for clause-learning SAT solvers. Master's thesis, Simon Fraser University, 2004.
18. O. Strichman. Prunning techniques for the SAT-based bounded model checking problem. In *Proceedings of the 11th Conference on Correct Hardware Design and Verification Methods*, 2001.
19. M. Velev. Sat benchmarks, http://www.ece.cmu.edu/~mvelev/, 2006.
20. H. Xu. *subSAT: A Formulation for Relaxed Satisfiability and its Applications*. PhD thesis, Department of Electrical and Computer Engineering, Carnegie Mellon University, Pittsburgh, Pennsylvania, 2004. 160 pages.
21. L. Zhang and S. Malik. Validating SAT solvers using an independent resolution-based checker: Practical implementations and other applications. In *Proceedings of the Design Automation and Test in Europe*, 2003.

MAX-SAT for Formulas with Constant Clause Density Can Be Solved Faster Than in $\mathcal{O}(2^n)$ Time

Evgeny Dantsin and Alexander Wolpert

Roosevelt University, 430 S. Michigan Av., Chicago, IL 60605, USA
{edantsin, awolpert}@roosevelt.edu

Abstract. We give an exact deterministic algorithm for MAX-SAT. On input CNF formulas with constant clause density (the ratio of the number of clauses to the number of variables is a constant), this algorithm runs in $\mathcal{O}(c^n)$ time where $c < 2$ and n is the number of variables. Worst-case upper bounds for MAX-SAT less than $\mathcal{O}(2^n)$ were previously known only for k-CNF formulas and for CNF formulas with small clause density.

1 Introduction

When solving MAX-SAT, we can search for an approximate solution or we can search for an exact solution. Approximate algorithms are typically faster than exact ones. In particular, there are polynomial-time approximate algorithms for MAX-SAT, but they are limited by thresholds on approximation ratios (some of these algorithms achieve the thresholds, e.g. [9]). There are also exponential-time algorithms that give arbitrarily good approximation, but when a high precision is required they are not considerably faster than exact algorithms, see e.g. [6,8].

In this paper we deal with exact algorithms for MAX-SAT and worst-case upper bounds on their runtime. Beginning in the early 1980s [11], many such upper bounds were obtained, and now we have a spectrum of MAX-SAT upper bounds for various classes of input formulas such as 2-CNF, 3-CNF, k-CNF, formulas with constant clause density, etc. The upper bounds usually depend on the number of variables, the number of clauses, the maximum number of satisfiable clauses, etc. Majority of exact MAX-SAT algorithms use the DPLL approach, but other approaches are used too, for example local search.

We give exact deterministic algorithms that solve MAX-SAT for formulas with no restriction on clause length (the general case of MAX-SAT) and prove worst-case upper bounds on their runtime. Previously known upper bounds for this general case were obtained in [4,10,2,12,14]. The most recent "record" bounds are given in [5]:

$$\mathcal{O}(1.3247^m \cdot |F|)$$
$$\mathcal{O}(1.3695^K + |F|)$$

where $|F|$ is the size of input formula F, m is the number of clauses in F, and K is the maximum number of satisfiable clauses (or the input parameter of

A. Biere and C.P. Gomes (Eds.): SAT 2006, LNCS 4121, pp. 266–276, 2006.

the decision version of MAX-SAT). Note that these bounds are better than the trivial upper bound $\mathcal{O}(2^n \cdot |F|)$, where n is the number of variables, only for small values of the clause density ($m/n < 3$). The algorithms in [5] use DPLL techniques; the proof of the bounds is based on case analysis and recurrence relations.

The contribution of this paper is twofold:

- Our bounds for MAX-SAT are better than $\mathcal{O}(2^n)$ for formulas with constant clause density, i.e., with $m/n < c$ where c is an arbitrary constant. Moreover, they are better than $\mathcal{O}(2^n)$ even if m/n is a slowly growing function.
- Our algorithms are based on a new method of solving MAX-SAT: we search for a partial assignment that satisfies a sufficient number of clauses and then we try to extend this assignment to satisfy more clauses.

Note that our method uses memoization which often occurs in recent algorithms for SAT and MAX-SAT, e.g. [3] and [13]. In particular, the randomized algorithm in [13] solves MAX-kSAT using DPLL combined with memoization. Its upper bound on the expected runtime is less than $\mathcal{O}(2^n)$ for k-CNF formulas with constant clause density.

Structure of the paper. Basic definitions and notation are given in Sect. 2. In Sect. 3 we discuss the idea of our algorithms and prove key lemmas. We describe the main algorithm in Sect. 4 and we prove an upper bound on its runtime in Sect. 5. In Sect. 6 we describe a modification of the main algorithm for the case when K is not too large compared to m. Sect. 7 summarizes our results.

2 Definitions and Notation

We deal with Boolean formulas in conjunctive normal form (CNF formulas). By a *variable* we mean a Boolean variable that takes truth values true or false. A *literal* is a variable x or its negation $\neg x$. A *clause* C is a set of literals such that C contains no complementary literals. When dealing with the satisfiability problem, a formula is typically viewed as a set of clauses. In the contexts of (weighted) MAX-SAT, it is more natural to view formulas as multisets. Therefore, we define a *formula* to be a multiset of clauses.

For a formula F, we write $|F|$ to denote the number of clauses in this multiset (we use letter m for this number). We use letters V and n to denote, respectively, the set of variables in F and its cardinality. The *clause density* of F is the ratio m/n. For any positive number Δ, we write $\mathcal{F}(\Delta)$ to denote the set of formulas such that their clause density is at most Δ.

We assign truth values to all or some variables in F. Let U be a subset of the variables of F. An *assignment* to the variables in U is a mapping from U to {true, false}. This mapping is extended to literals: each literal $\neg x$ is mapped to the complement of the truth value assigned to x. We say that a clause C is *satisfied* by an assignment a if a assigns true to at least one literal in C. The formula F is *satisfied* by A if every clause in F is satisfied by A.

For assignment a, we write $F(a)$ to denote the formula obtained from F as follows: any clause that contains a true literal is removed from F, all false literals are deleted from the other clauses. The empty clause is false; the empty formula is true.

The MAX-SAT problem is stated as follows: Given a formula, find an assignment that maximizes the number of satisfied clauses. We also consider the decision version of MAX-SAT: given a formula F and an integer K, is there an assignment that satisfies at least K clauses?

We use $\mathcal{O}^*$ notation that extends big-Oh notation and suppresses all polynomial factors. For functions $f(n_1, \ldots, n_k)$ and $g(n_1, \ldots, n_k)$ we write

$$f(n_1, \ldots, n_k) = \mathcal{O}^*(g(n_1, \ldots, n_k))$$

if for some multivariate polynomial p, we have

$$f(n_1, \ldots, n_k) = \mathcal{O}(p(n_1, \ldots, n_k) \cdot g(n_1, \ldots, n_k)).$$

We write $\log x$ to denote $\log_2 x$. The entropy function is denoted by H:

$$H(x) = -x \log x - (1 - x) \log(1 - x).$$

3   Idea of Algorithms and Key Lemmas

In the next sections we describe and analyze two algorithms for MAX-SAT. Both are based on an idea outlined in this section. This idea is suggested by Lemma 2 and Lemma 3. The former is a key observation behind our approach. Its weaker version was used by Arvind and Schuler in [1] for their quantum satisfiability-testing algorithm. The second lemma is a well known fact that any formula has an assignment satisfying at least a half of its clauses.

We consider the decision version of MAX-SAT. To solve this problem, we enumerate partial assignments and try to extend them to satisfy at least K clauses.

1. **Enumeration of partial assignments.** We select a set of partial assignments such that for some δ
   - each of them assigns truth values to δn variables;
   - each of them satisfies at least δK clauses.

   It follows from Lemma 2 that this set contains an assignment a that can be extended to satisfy at least K clauses (if K clauses in F can be satisfied at all). Lemma 2 also shows that this set (containing a) can be constructed by processing $\mathcal{O}(2^{\delta n})$ partial assignments.

2. **Extension of partial assignments.** When processing a partial assignment, we try to extend it to satisfy more clauses in F. There are at most $m - \delta K$ unsatisfied clauses left and we need to satisfy at least $(1 - \delta)K$ of them. We consider two possibilities:
   - *There are "many" unsatisfied clauses.* Then $(1-\delta)K$ clauses are satisfied due to the fact that any formula has an assignment satisfying at least a half of its clauses.

– *There are "few" unsatisfied clauses.* This case reduces to solving MAX-SAT for "short" subformulas of F: we could test the extension if for each "short" subformula S, we knew the maximum number M of clauses that can be satisfied in S. We prepare all such pairs (S, M) in advance and store them in an array. This array is sorted according to some order on formulas. Using binary search, we look up any pair (S, M) that we need.

3. **Tradeoff for δ.** The runtime and space of our algorithms depend on the choice of δ. When δ decreases, the number of assignments we need to enumerate decreases too. On the other hand, the larger δ is, the less time is needed for preparing the array. Our analysis in Sect. 5 gives a good tradeoff for δ.

Algorithm $\mathcal{A}$ defined in Sect. 4 implements the approach above. Algorithm $\mathcal{B}$ defined in Sect. 4 is another version of this approach. Loosely speaking, in this version we choose δ so that we do not have to build the table, i.e., for the chosen δ we have "many" unsatisfied clauses for every partial assignment. It is natural to apply Algorithm $\mathcal{B}$ when K is not too large compared to m.

Below we prove three lemmas needed for the algorithms and their analysis. Lemma 1 is nearly obvious (a reincarnation of the well known fact that the average of d numbers is greater than or equal to at least one of these numbers). This lemma is used in Lemma 2 which plays a key role in our approach. Lemma 3 is well known but we include it in the paper for convenience of our exposition.

Lemma 1. *Let r and $r_1, \ldots, r_d$ be real numbers ($d \geq 2$). If*

$$r_1 + \ldots + r_d \geq r \tag{1}$$

then there exists $i \in \{1, \ldots, d\}$ such that

$$r_1 + \ldots + r_{i-1} + r_{i+1} + \ldots + r_d \geq \left(1 - \frac{1}{d}\right) r \tag{2}$$

Proof. The average of $r_1, \ldots, r_d$ is greater than or equal to at least one of these numbers. That is, there exists a number r_i such that

$$\frac{s}{d} \geq r_i \quad \text{where} \quad s = r_1 + \ldots + r_d.$$

Using this fact and the fact that $s \geq r$, we have

$$s - r_i \geq s - \frac{s}{d} \geq \left(1 - \frac{1}{d}\right) s \geq \left(1 - \frac{1}{d}\right) r. \qquad \square$$

Lemma 2 (on partial assignments). *Let F be a formula and A be an assignment that satisfies at least K clauses in F. Let V denote the set of variables in F. Consider a partition of V into d subsets:*

$$V = V_1 \cup \ldots \cup V_d$$

where $d \geq 2$ and $V_i \cap V_j = \emptyset$ for all i and j. For each i from 1 to d, let A_i denote the restriction of A to V_i and let F_i be the multiset of clauses satisfied by A_i, i.e.,

$$F_i = \{C \in F \mid C \text{ is satisfied by } A_i\}$$

Then there exists $i \in \{1, \ldots, d\}$ such that

$$|F_1 \cup \ldots \cup F_{i-1} \cup F_{i+1} \cup \ldots \cup F_d| \geq (1 - 1/d) K$$

Proof. We prove this lemma using Lemma 1. Let S denote the multiset of all clauses in F satisfied by A. This multiset can be represented as the union of d multisets:

$$S = S_1 \cup \ldots \cup S_d$$

where $S_j = F_j - (F_1 \cup \ldots \cup F_{j-1})$. Since $|S| \geq K$, we have $|S_1| + \ldots + |S_d| \geq K$. Applying Lemma 1, we obtain that there exists i such that

$$|S_1| + \ldots + |S_{i-1}| + |S_{i+1}| + \ldots + |S_d| \geq (1 - 1/d) K. \tag{3}$$

The multisets $S_1, \ldots, S_d$ are pairwise disjoint, therefore

$$|S_i| + \ldots + |S_{i-1}| + |S_{i+1}| + \ldots + |S_d| = |S_i \cup \ldots \cup S_{i-1} \cup S_{i+1} \cup \ldots \cup S_d|. \tag{4}$$

By the definition of multisets S_j, we have $S_j \subseteq F_j$ for every j and consequently

$$S_1 \cup \ldots \cup S_{i-1} \cup S_{i+1} \cup \ldots \cup S_d \subseteq F_1 \cup \ldots \cup F_{i-1} \cup F_{i+1} \cup \ldots \cup F_d. \tag{5}$$

Taking (3), (4), and (5) together we get

$$|F_1 \cup \ldots \cup F_{i-1} \cup F_{i+1} \cup \ldots \cup F_d| \geq$$
$$|S_i \cup \ldots \cup S_{i-1} \cup S_{i+1} \cup \ldots \cup S_d| =$$
$$|S_i| + \ldots + |S_{i-1}| + |S_{i+1}| + \ldots + |S_d| \geq$$
$$(1 - 1/d) K. \qquad \square$$

Lemma 3 (well known fact). *Any formula F has an assignment that satisfies at least a half of its clauses.*

Proof. Consider an arbitrary assignment a. If a satisfies less than a half of clauses in F, we flip the values of the variables in a. It is easy to see that the new assignment satisfies at least a half of clauses in F. $\qquad \square$

4   Main Algorithm

In this section we describe the main algorithm based on the approach discussed informally in Sect. 3.

Algorithm $\mathcal{A}$

Input: Formula F with m clauses over n variables, integers K and d such that $2 \leq d \leq n$ and $m/2 \leq K \leq m$.

Output: "yes" if there is an assignment that satisfies at least K clauses, "no" otherwise.

1. **Partition.** We partition the set V of variables in F into d subsets $V_1, \ldots, V_d$ of (approximately) same size. More exactly, we partition V into d subsets such that each of them has either $\lfloor n/d \rfloor$ or $\lceil n/d \rceil$ elements.

2. **Database preparation.** We prepare a "database" needed for the next steps. Namely, we build tables (arrays) $T_1, \ldots, T_d$ defined as follows:

   (a) Let $F|_{V_j}$ denote the formula obtained from F by removing all clauses that contain no variables from V_j and deleting variables from $V - V_j$ in the remaining clauses. By a *short subformula* of $F|_{V_j}$ we mean any subformula of $F|_{V_j}$ that has length at most $2K/d$.

   (b) We define T_j to be a table consisting of all pairs (S, M) where S is a short subformula of $F|_{V_j}$ and M is the maximum number of satisfiable clauses in S. For every S, we compute M by brute-force, which takes time $\mathcal{O}^*(2^{n/d})$.

   (c) Fixing an order on short subformulas of $F|_{V_j}$, we sort each table T_j according to this order. Therefore, given a short subformula, we can find the corresponding record in logarithmic time.

3. **Partial assignments.** For each $j \in \{1, \ldots, d\}$ we consider all assignments to the variables in $V - V_j$. Given such an assignment, we check whether it satisfies at least $(1 - 1/d)K$ clauses. For each assignment that passes this test, we perform the next step.

4. **Extension.** Let a be a partial assignment selected at the previous step. Let K_a denote the number of clauses satisfied by a. If $K_a \geq K$, return "yes". Otherwise, we try to extend a to an assignment that satisfies at least K clauses. Consider the formula $F(a)$ obtained from F by substitution of the truth values corresponding to a. To extend a, we need to satisfy $K - K_a$ clauses in $F(a)$. Two cases are possible:

   (a) *Case 1.* If $|F(a)| \geq 2(K - K_a)$ then by Lemma 3 there exists an assignment that satisfies at least $K - K_a$ clauses in $F(a)$. Therefore, a can be extended to an assignment that satisfies at least K clauses. Return "yes".

   (b) *Case 2.* The formula $F(a)$ is shorter than $2(K - K_a)$. We need to check whether there is an assignment that satisfies at least $K - K_a$ clauses in $F(a)$. Note that $F(a)$ must occur in the table T_j. Therefore, we can find (in logarithmic time) the maximum number of satisfiable clauses in $F(a)$. If this number is at least $K - K_a$ then return "yes".

5. **"No" answer.** The output is "no" if we failed to find an appropriate assignment as described above.

Note that the algorithm can be also implemented using DPLL combined with memoization: partial assignments can be enumerated in the DPLL manner and the tables can be built on the run with memoization.

5   Runtime of Algorithm $\mathcal{A}$

In this section we give a worst-case upper bound on the runtime of Algorithm $\mathcal{A}$ (Theorem 1). Applying this algorithm to solve MAX-SAT, we get essentially the same upper bound (Corollary 1). Then we show that if we restrict input formulas to formulas with constant clause density, we have the $\mathcal{O}(c^n)$ upper bound where $c < 2$ (Corollary 2).

Theorem 1. *If Algorithm $\mathcal{A}$ is run with d such that*

$$d \geq 4 \quad and \quad d \geq 2\left(\frac{m}{n}\log\left(\frac{ed}{2}\right) + 1\right) \tag{6}$$

where m and n are, respectively, the number of clauses and the number of variables in the input formula, then the runtime of Algorithm $\mathcal{A}$ is

$$\mathcal{O}^*\left(2^{n(1-1/d)}\right).$$

Proof. First, we note that for any m and n, there exists d that meets the inequalities in (6). Consider the execution of Algorithm $\mathcal{A}$ on input formula F with any $K \geq m/2$ and with d chosen according to (6). After partitioning of the variables in F into d subsets, we build d database tables. Each table contains $\sum_{i=1}^{2K/d}\binom{m}{i}$ records and can be built in time

$$\mathcal{O}^*\left(\sum_{i=1}^{2K/d}\binom{m}{i} \cdot 2^{n/d}\right).$$

After building the tables, we take d identical steps. At each step, we enumerate all assignments to $n(1 - 1/d)$ variables. For each assignment a, we compute the formula $F(a)$ and (if needed) look it up using binary search in the corresponding sorted table. Since the table size is $2^{\mathcal{O}^*(m)}$, the lookup can be done in polynomial time. Therefore, the overall runtime of Algorithm $\mathcal{A}$ is bounded by

$$\mathcal{O}^*\left(\sum_{i=1}^{2K/d}\binom{m}{i} \cdot 2^{n/d}\right) + \mathcal{O}^*\left(2^{(1-1/d)n}\right). \tag{7}$$

We show that for any d that meets (6), the first term in this sum is asymptotically smaller than the second term. The sum of binomial coefficients in (7) can be approximated using the binary entropy function, see e.g. [7, exercise 9.42], so we can approximate the first term in (7) as

$$\mathcal{O}^*\left(2^{H(2K/md)\,m + n/d}\right).$$

Since the function H is increasing on the interval $(0, \frac{1}{2}]$, it follows from $d \geq 4$ and $K \leq m$ that

$$H\left(\frac{2K}{md}\right) \leq H\left(\frac{2}{d}\right).$$

Using the well known inequality $\ln(1 + x) \leq x$, we have

$$H\left(\frac{2}{d}\right) = -\frac{2}{d}\log\left(\frac{2}{d}\right) - \left(1 - \frac{2}{d}\right)\log\left(1 - \frac{2}{d}\right)$$

$$= \frac{2}{d}\log\left(\frac{d}{2}\right) + \left(1 - \frac{2}{d}\right)\log e \, \ln\left(1 + \frac{2}{d-2}\right)$$

$$\leq \frac{2}{d}\log\left(\frac{d}{2}\right) + \frac{2}{d}\log e$$

$$= \frac{2}{d}\log\left(\frac{ed}{2}\right).$$

Therefore, (7) is not greater than

$$\mathcal{O}^*\left(2^{(2m/d)\,\log(ed/2)\,+\,n/d}\right) + \mathcal{O}^*\left(2^{(1-1/d)\,n}\right)$$

It remains to observe that the second term in this sum dominates over the first term if

$$\frac{2m}{d}\log\left(\frac{ed}{2}\right) + \frac{n}{d} \leq n - \frac{n}{d},$$

which is equivalent to the condition on d in (6):

$$d \geq 2\left(\frac{m}{n}\log\left(\frac{ed}{2}\right) + 1\right). \qquad \square$$

Corollary 1. *There is an exact deterministic algorithm that solves MAX-SAT in $\mathcal{O}^*(2^{n(1-1/d)})$ time, where d meets condition (6).*

Proof. We repeatedly apply Algorithm $\mathcal{A}$ to find K such that the algorithm returns "yes" for K and returns "no" for $K + 1$. This can be done using either binary search or straightforward enumeration. $\qquad \square$

Remark 1. What value of d minimizes the upper bound in Corollary 1? We can approximate the optimum value of d as follows:

$$d = \mathcal{O}\left(\frac{m}{n}\log\left(\frac{m}{n} + 2\right)\right) \qquad (8)$$

It is straightforward to check that this approximation meets condition (6).

Corollary 2. *There is an exact deterministic algorithm that solves MAX-SAT for formulas with constant clause density in $\mathcal{O}(c^n)$ time where $c < 2$. More exactly, for any constant Δ, there is a constant $c < 2$ such that on formulas in $\mathcal{F}(\Delta)$, the algorithm runs in $\mathcal{O}(c^n)$ time.*

Proof. The $\mathcal{O}^*(2^{n(1-1/d)})$ upper bound can be written as

$$\mathcal{O}^*\left(2^{n(1-1/d)}\right) = \mathcal{O}\left(2^{n-n/d+\mathcal{O}(\log n)}\right) \qquad (9)$$

If $\Delta = m/n$ is a constant, it immediately follows from (6) that d can be taken as a constant too. Therefore n/d is asymptotically larger than $\mathcal{O}(\log n)$, which yields the $\mathcal{O}(c^n)$ upper bound. $\qquad \square$

Remark 2. Note that the proof above can be used to obtain the $\mathcal{O}(2^n)$ bound for some classes $\mathcal{F}(\Delta)$ where Δ is not a constant. For example, we can allow Δ to be $\mathcal{O}(\sqrt{n})$. Using the approximation (8), we have

$$d = \mathcal{O}(\sqrt{n} \log n).$$

Then n/d is asymptotically larger than $\mathcal{O}(\log n)$ and, therefore, (9) gives the $\mathcal{O}(2^n)$ bound.

6   Algorithm That Needs No Tables

An obvious disadvantage of Algorithm $\mathcal{A}$ is that the algorithm has to build an exponential-size database. We could avoid it if we chose d so that $F(a)$ is always "long" enough. Then a can be extended due to Lemma 3 and no lookup is needed. Using this observation, we define Algorithm $\mathcal{B}$ that does not require building tables. Theorem 2 gives a worst-case upper bound on its runtime:

$$\mathcal{O}^* \left(2^{n\left(1 - \frac{m-K}{K}\right)} \right).$$

Algorithm $\mathcal{B}$

Input: Formula F with m clauses over n variables, integer K such that

$$m/2 < K \le m\frac{n}{n+1}. \tag{10}$$

Output: "yes" if there is an assignment that satisfies at least K clauses, "no" otherwise.

1. **Partition.** We take an integer d defined by

$$d = \left\lceil \frac{K}{m-K} \right\rceil.$$

   Then, similarly to the partition step in Algorithm $\mathcal{A}$, we partition the set V of variables in F into d subsets $V_1, \ldots, V_d$. Note that it follows from (10) that $d \le n$.
2. **Partial assignments.** This step is similar to the partial assignment step in Algorithm $\mathcal{A}$: for each $j \in \{1, \ldots, d\}$ we enumerate all assignments to the variables in $V - V_j$. For each such assignment, we check whether it satisfies at least $(1 - 1/d)K$ clauses. In Theorem 2 below we show that any assignment a that satisfies at least $(1 - 1/d)K$ clauses can be extended to an assignment that satisfies at least K clauses. Therefore, as soon as we find a satisfying $(1 - 1/d)K$ clauses, we return "yes".
3. **"No" answer.** If we failed to find an appropriate assignment at the previous step, "no" is returned.

Theorem 2. *Algorithm $\mathcal{B}$ is correct for all K that meet (10). Its runtime is*

$$\mathcal{O}^*\left(2^{n\left(1-\frac{m-K}{K}\right)}\right)$$

where m and n are, respectively, the number of clauses and the number of variables in the input formula.

Proof. Let a be a partial assignment selected at Step 3 and K_a be the number of clauses satisfied by a. Consider $F(a)$ that has $m - K_a$ clauses. If $F(a)$ is "long" enough, namely

$$m - K_a \geq 2(K - K_a), \tag{11}$$

then it follows from Lemma 3 that a can be extended to an assignment satisfying at least K clauses (cf. Case 1 in Algorithm $\mathcal{A}$). It remains to show that the choice of d in the algorithm guarantees (11).

Since $d = \lceil K/(m - K) \rceil$, we have

$$d \geq K/(m - K).$$

This inequality can be rewritten as

$$m - \frac{K}{d} \geq K.$$

Adding K to both sides we get

$$m + K - \frac{K}{d} \geq 2K.$$

Since $K_a \geq (1 - 1/d)K$,

$$m + K_a \geq 2K.$$

Subtracting $2K_a$ from both sides, we obtain (11).

Clearly, the runtime of the algorithm is determined by the time needed to enumerate all assignments to the variables in each $V - V_j$. Since each such subset of variables contains at most $\lceil (1-1/d)n \rceil$ variables, we obtain the claimed runtime. $\qquad\qquad\square$

7   Summary of Results

1. We give an exact deterministic algorithm that solves MAX-SAT for CNF formulas with no limit on clause length. Its runtime is

$$\mathcal{O}^*\left(2^{n(1-1/d)}\right) \quad \text{where} \quad d = \mathcal{O}\left(\frac{m}{n}\log\left(\frac{m}{n}+2\right)\right).$$

2. This algorithm solves MAX-SAT for formulas with constant clause density in time

$$\mathcal{O}(c^n) \quad \text{where} \quad c < 2.$$

3. We give another exact deterministic algorithm that solves MAX-SAT in time

$$\mathcal{O}^*\left(2^{n\left(1-\frac{m-K}{K}\right)}\right)$$

where K is the maximum number of satisfiable clauses. This algorithm is faster than the first one when K is not too large compared to m.

Acknowledgement

We thank anonymous referees for their useful comments.

References

1. V. Arvind and R. Schuler. The quantum query complexity of 0-1 knapsack and associated claw problems. In *Proceedings of the 14th Annual International Symposium on Algorithms and Computation, ISAAC 2003*, volume 2906 of *Lecture Notes in Computer Science*, pages 168–177. Springer, December 2003.
2. N. Bansal and V. Raman. Upper bounds for MaxSat: Further improved. In *Proceedings of the 14th Annual International Symposium on Algorithms and Computation, ISAAC'99*, volume 1741 of *Lecture Notes in Computer Science*, pages 247–258. Springer, December 1999.
3. P. Beame, R. Impagliazzo, T. Pitassi, and N. Segerlind. Memoization and DPLL: Formula caching proof systems. In *Proceedings of the 18th Annual IEEE Conference on Computational Complexity, CCC 2003*, pages 225–236, 2003.
4. L. Cai and J. Chen. On fixed-parameter tractability and approximability of NP optimization problems. *Journal of Computer and System Sciences*, 54(3):465–474, 1997.
5. J. Chen and I. Kanj. Improved exact algorithm for Max-Sat. *Discrete Applied Mathematics*, 142(1-3):17–27, August 2004.
6. E. Dantsin, M. Gavrilovich, E. A. Hirsch, and B. Konev. MAX SAT approximation beyond the limits of polynomial-time approximation. *Annals of Pure and Applied Logic*, 113(1-3):81–94, December 2001.
7. R. Graham, D. Knuth, and O. Patashnik. *Concrete Mathematics: A Foundation for Computer Science*. Addison-Wesley, 2nd edition, 1994.
8. E. A. Hirsch. Worst-case study of local search for MAX-k-SAT. *Discrete Applied Mathematics*, 130(2):173–184, 2003.
9. H. Karloff and U. Zwick. A 7/8-approximation algorithm for MAX 3SAT? In *Proceedings of the 38th Annual IEEE Symposium on Foundations of Computer Science, FOCS'97*, pages 406–415, 1997.
10. M. Mahajan and V. Raman. Parameterizing above guaranteed values: Maxsat and maxcut: An algorithm for the satisfiability problem of formulas in conjunctive normal form. *Journal of Algorithms*, 31(2):335–354, 1999.
11. B. Monien and E. Speckenmeyer. Upper bounds for covering problems. Technical Report Bericht Nr. 7/1980, Reihe Theoretische Informatik, Universität-Gesamthochschule-Paderborn, 1980.
12. R. Niedermeier and P. Rossmanith. New upper bounds for maximum satisfiability. *Journal of Algorithms*, 36(1):63–88, July 2000.
13. R. Williams. On computing k-CNF formula properties. In *Proceedings of the 6th International Conference on Theory and Applications of Satisfiability Testing, SAT 2003*, volume 2919 of *Lecture Notes in Computer Science*, pages 330–340. Springer, May 2003.
14. H. Zhang, H. Shen, and F. Manyà. Exact algorithms for MAX-SAT. *Electronic Notes in Theoretical Computer Science*, 86(1), May 2003.

Average-Case Analysis for the MAX-2SAT Problem

Osamu Watanabe and Masaki Yamamoto

Dept. of Math. and Comput. Sci., Tokyo Inst. of Technology, Japan
watanabe@is.titech.ac.jp

Abstract. We propose a "planted solution model" for discussing the average-case complexity of the MAX-2SAT problem. We show that for a large range of parameters, the planted solution (more precisely, one of the planted solution pair) is the optimal solution for the generated instance with high probability. We then give a simple linear time algorithm based on a message passing method, and we prove that it solves the MAX-2SAT problem with high probability under our planted solution model.

1 Introduction

We discuss the average-case analysis of the difficulty of MAX-2SAT problems, the simplest variation of MAX-SAT problems, where input CNF formulas are restricted to those consisting of only clauses with *two* literals. It is known that even the MAX-2SAT problem is NP-hard (even hard to approximate) though the 2SAT problem is in P. On the other hand, it seems that there are some algorithms/heuristics that solve MAX-2SAT problems quite well *on average*. Unfortunately, not so much theoretical investigation has been made for the average-case complexity of the MAX-2SAT problem. One example is the theoretical analysis of random 2CSP instances (including 2SAT) of Scott and Sorkin [SS03], where they showed, among other results, some deterministic algorithm solving MAX-2CSP in polynomial-time on average for random sparse instances, i.e., formulas with linear number of clauses, and the authors ask for algorithms solving MAX-2SAT on dense instances. It should be remarked here that the MAX-2SAT problem is not the same as the ordinary 2SAT problem, which is different from kSAT problems for $k \geq 3$. For example, even though it has been proved that some algorithm performs well for certain random 2SAT instances, this does not mean at all that it works for similar random MAX-2SAT instances.

For discussing the average-case complexity of MAX-2SAT problem, we propose one simple probability model for generating MAX-2SAT instances, thereby giving one instance distribution for the MAX-2SAT problem. Our model is one of the planted solution models. We also demonstrate that a simple linear-time algorithm can solve the MAX-2SAT with high probability when input formulas are given under this distribution with probability parameters in a certain range. Our parameter range is for a dense regime; we could prove that our algorithm solves the MAX-2SAT problem with high probability for random formulas with $O(n^{1.5} \ln n)$ clauses; it is an interesting open problem to show some efficient

A. Biere and C.P. Gomes (Eds.): SAT 2006, LNCS 4121, pp. 277–282, 2006.

algorithm for sparse formulas. (Note that our distribution is different from the one considered in [SS03].)

In this short note, we explain our probability model in Section 2 and a simple algorithm for MAX-2SAT in Section 3. Due to space constraint, we omit most of the technical details and some references, which can be found in [WY06].

2   A Planted Solution Model for MAX-2SAT

In this section, we define a planted solution model for MAX-2SAT. We begin with introducing some notions and notations for discussing the MAX-SAT problem. Throughout this paper, we use n to denote the number of variables and use $x_1, \ldots, x_n$ for denoting Boolean variables. A *2CNF formula* we consider here is a formula defined as a conjunction of clauses of two literals, where each clause is "syntactically" of the form $(x_i \vee x_j)$, $(x_i \vee \overline{x}_j)$, $(\overline{x}_i \vee x_j)$, or $(\overline{x}_i \vee \overline{x}_j)$, for some $1 \leq i < j \leq n$, and $(x_i \vee x_i)$, $(x_i \vee \overline{x}_i)$, or $(\overline{x}_i \vee \overline{x}_i)$ for some i, $1 \leq i \leq n$. An *assignment* is a function t mapping $\{x_1, \ldots, x_n\}$ to $\{-1, +1\}$; $t(x_i) = +1$ (resp., $t(x_i) = -1$) means to assign true (resp., false) to a Boolean variable x_i. An assignment is also regarded as a sequence $\boldsymbol{a} = (a_1, a_2, \ldots, a_n)$ of ± 1's, where $a_i = t(x_i)$ for each i, $1 \leq i \leq n$. For a given CNF formula F, its *optimal assignment* is an assignment satisfying the largest number of clauses in F. Now our MAX-2SAT problem is, for a given 2CNF formula of the above form, to find its optimal assignment, more precisely, any one of the optimal assignments.

We explain our probability model for generating MAX-2SAT instances, more specifically, a way of generating a 2-CNF formula over n variables $x_1, \ldots, x_n$. Our model is a "planted solution model", a method for generating a problem instance so that a target solution, which is also generated in some way, is the answer to this instance w.h.p. Here we generate a sequence $\boldsymbol{a} = (a_1, \ldots, a_n)$ uniformly at random; let $\boldsymbol{a}'$ be its *complement assignment* $(-a_1, \ldots, -a_n)$. Then we use a pair of $\boldsymbol{a}$ and $\boldsymbol{a}'$ as a *planted solution pair*. For constructing a formula, we generate each clause satisfied by both assignments with probability p, and it is added to the formula. Since there are n^2 such clauses, the number of clauses of this type added to the formula is *on average* pn^2. In order to make the formula unsatisfiable, we also generate each clause that is unsatisfied by $\boldsymbol{a}$ (resp., $\boldsymbol{a}'$) with probability $r < p$. Again *on average* the formula has $rn(n+1)/2$ clauses that are not satisfied by $\boldsymbol{a}$ (resp., by $\boldsymbol{a}'$). Hence, the generated formula has *on average* $pn^2 + rn(n+1)$ clauses and each assignment of the planted solution pair fails to satisfy $rn(n+1)/2$ clauses *on average*. As our first theorem we show that one of the planted solution pair is indeed an optimal assignment w.h.p. for random formulas if $p > 4r$ for any sufficiently large p.

At this point, we explain our motivation for the above model and some related works. A reason for using a planted solution pair is for producing clauses so that each literal appears with the same probability. The same approach has been proposed [AJM05] for generating hard sat. instances for kSAT problems. The important difference here is the point that inconsistent clauses are also added with probability $r < p$. This is for generating unsatisfiable formulas; otherwise, i.e., if only satisfiable formulas were generated, the problem would be trivially easy because the 2SAT problem is in P, which is different from the other kSAT problem $k \geq 3$. (For example, even distribution on only satisfiable formulas can be useful for analyzing MAX-3SAT algorithms; but not at all for MAX-2SAT!) One open problem here is whether our approach can be also used for MAX-kSAT problems for $k \geq 3$ in general. Clearly, all pair of clauses would appear with the same probability if $p = r$; but then we may not be able to guarantee the optimality of planted solution pairs. The idea of adding inconsistent clauses with smaller probability $r < p$ has been proposed in [Yam06], where, based on this idea, some test instance generating algorithm for MAX-2SAT is proposed. Although the optimality of a planted solution pair is proved for $p > 4r$, it may be still possible to have the same result for much closer p and r.

Now we show our first result.

Theorem 1. *For any $\delta > 0$, if probability parameters p and r satisfies $p \geq 4r$ and $p \geq c_1 \frac{\ln(n/\delta)}{n}$, then for a randomly generated formula F under our planted solution model with parameters p and r, with probability $\geq 1 - \delta$, one of the two planted solution pair is the optimal assignment for F; furthermore, there is no optimal solution other than the planted solution pair.*

Remark. This bound for p, implying $\Omega(n \log n)$ expected number of clauses, is necessary for our model where each clause is selected independently. On the other hand, we can consider more balanced random generation, where each literal appears exactly pn times in clauses consistent to both planted solutions and rn times in clauses inconsistent to one of them. In this case, we may only need $p = c/n$ for some sufficiently large constant c, resulting linear number of clauses.

Proof. Consider p, r, and n satisfying the condition of the theorem. Here we explain by fixing one planted solution pair; for simplicity, consider a pair of all $+1$ assignment $\boldsymbol{a}^+$ and all -1 assignment $\boldsymbol{a}^-$. Let F be a randomly generated formula for this planted solution pair. Our goal is to show that w.h.p. either $\boldsymbol{a}^+$ or $\boldsymbol{a}^-$ satisfies the most number of clauses in F, which cannot be achieved by any other assignment.

For our discussion, we consider a directed graph $G = (V, E)$ naturally defined as follows: $V = V_+ \cup V_-$, where $V_+ = \{v_{+1}, \ldots, v_{+n}\}$ and $V_- = \{v_{-n}, \ldots, v_{-1}\}$. E consists of two directed edges *corresponding to* a clause $(\ell_i \vee \ell_j)$, where $i < j$, in F. For example, for a clause $(x_i \vee x_j)$, E has two directed edges (v_{-i}, v_{+j}) and (v_{-j}, v_{+i}), each of which corresponds to $(\overline{x}_i \rightarrow x_j)$ and $(\overline{x}_j \rightarrow x_i)$; clauses of the other type define two corresponding directed edges in E similarly. Due to our syntactic restriction (see also a remark below), the obtained graph G has no multiple edge nor self-loop. (*Remark.* In this proof, we ignore clauses like $(x_i \vee \overline{x}_i)$ because such clauses are always satisfied and has no meaning in this analysis.)

Consider any assignment t to $x_1, \ldots, x_n$. We regard this also as an assignment to V; i.e., $t(v_{+i}) = t(x_i)$ and $t(v_{-i}) = -t(v_{+i})$. In general, an assignment t to V satisfying $t(v_{-i}) = -t(v_{+i})$ for all i is called a *legal assignment*. It is easy to see that a clause $(\ell \vee \ell')$ is unsatisfied by t if and only if its two corresponding directed edges are, under the corresponding legal assignment, from a vetex assigned $+1$ to a vertex assigned -1, which we call *unsatisfied edges*. That is, the number of unsatisfied clauses is the same as the half of that of unsatisfied edges. Thus, for proving the theorem, we estimate the number of unsatisfied edges under an arbitrary legal assignment to V.

First, we estimate the number of unsatisfied edges within $G[V_+]$ and $G[V_-]$, which are subgraphs of G induced respectively by V_+ and V_-, by the well-known fact that a random graph is almost surely an expander. Note that $G[V_+]$ (resp., $G[V_-]$) can be regarded as a "random graph", i.e., a graph with vertex set V_+ (resp., V_-) with a directed edge randomly generated with probability p between every ordered pair of distinct vertices. Thus, by a standard argument, we may assume that, for some ϵ' that will be fixed at the end, $H[V_+]$ and $H[V_-]$ both have the following "expansion property": For each $U \in \{V_+, V_-\}$ and for every $S \subset U$, we have

$$\|E(S, U - S)\| \geq (1/2 - \epsilon')pn\|S\| \quad \text{and} \quad \|E(U - S, S)\| \geq (1/2 - \epsilon')pn\|S\|.$$

(*Remark.* Here we need $p \geq c\ln(n/\delta)/n$ for some constant c so that this assumption holds with prob. $> 1 - \delta$.)

Now consider any legal assignment t to V that is different from our two planted solutions. By h we denote the number of unsatisfied edges of G under t. On the other hand, let $h_0 = \min\{|E \cap (V_+ \times V_-)|, |E \cap (V_- \times V_+)|\}$; that is, h_0 is the number of unsatisfied edges by a better assignment among our two planted solutions. From now on, we estimate h and show that $h > h_0$ w.h.p.

Let A_+ and B_+ be subsets of V_+ assigned $+1$ and -1 respectively under t; on the other hand, let A_- and B_- be subsets of V_- assigned -1 and $+1$ respectively. Note that $|A_+| = |A_-|$ and $|B_+| = |B_-|$, and let a and b be the number of $|A_+|$ and $|B_+|$ respectively; we may assume that $a, b \geq 1$. Below we consider the case of $a \leq b$ and show that $h > \|E \cap (V_- \times V_+)\|$ holds w.h.p.

For edges in V_+ and V_-, we see from the above expansion property that the number of unsatisfied edges within each of V_+ and V_- is respectively at least $(1/2 - \epsilon')pn \cdot a$; that is, $\|E(B_+, V_+ - B_+)\|$, $\|E(A_-, V_- - A_-)\| \leq (1/2 - \epsilon')pna$. Consider then edges between V_+ and V_-; here we estimate only the number of unsatisfied edges from V_- to V_+, which are exactly those from B_- to B_+. Thus, we have $h \geq (1 - 2\epsilon')pna + |E \cap (B_- \times B_+)|$, where we decompose the last term as follows.

$$\|E \cap (B_- \times B_+)\| = \|E \cap (V_- \times V_+)\| - \|E \cap (A_- \times V_+)\| - \|E \cap (B_- \times A_+)\|$$
$$\geq \|E \cap (V_- \times V_+)\| - \|E \cap (A_- \times V_+)\| - \|E \cap (V_- \times A_+)\|.$$

Hence, for our goal, it suffices to show that

$$(1 - 2\epsilon')pna - \|E \cap (A_- \times V_+)\| - \|E \cap (V_- \times A_+)\| \quad - (*)$$

is positive. Again with an argument similar to the one showing the expansion property, we show that both $\|E \cap (A_- \times V_+)\|$ and $\|E \cap (V_- \times A_+)\|$ are close to their expectations w.h.p. and they are respectively less than $(1 + \epsilon'')rna$. Hence,

procedure MPalgo_for_MAX2-SAT (F);
// An input $F = C_1 \wedge \cdots C_m$ is a 2CNF formula over variables $x_1, \ldots, x_n$.
// Let $S = S_+ \cup S_-$, where $S_+ = \{+1, \ldots, +n\}$ and $S_- = \{-n, \ldots, -1\}$.
begin
    construct $G = (V, E)$;   // See the text for the explanation.
    set $b(v_s)$ to 0 for all $s \in S$;
    $b(v_{+1}) \leftarrow +1$; $b(v_{-1}) \leftarrow -1$;   // This is for the assumption that $x_1 = +1$.
    repeat MAXSTEP times **do** {
        for each $i \in \{2, \ldots, n\}$ **in parallel do** {

$$b(v_{+i}) \leftarrow \sum_{v_s \in N(v_{+i})} \min(0, b(v_s)); \quad b(v_{-i}) \leftarrow \sum_{v_s \in N(v_{-i})} \min(0, b(v_s));$$

$$b(v_{+i}) \leftarrow b(v_{+i}) - b(v_{-i}); \quad b(v_{-i}) \leftarrow -b(v_{+i}); \qquad — (1)$$

        }
        if $\text{sign}(b(v_i))$ is stabilized for all $i \in \{2, \ldots, n\}$ **then break**;
        $b(v_{+1}) \leftarrow 0$; $b(v_{-1}) \leftarrow 0$;   — (2)
    }
    output$(+1, \text{sign}(b(v_{+2})), \ldots, \text{sign}(b(v_{+n})))$;
end-procedure

Fig. 1. A message passing algorithm for the MAX-2SAT problem

we have $(*) > ((1 - 2\epsilon')p - 2(1 + \epsilon'')r)na$. Here by using $\epsilon' = \epsilon'' = 1/8$, we can show that the righthand side is positive if $p \geq 4r$.   □

3   A Simple Algorithm

For our probability model for the average-case analysis of MAX-2SAT, we show in this section that a simple algorithm can solve MAX-2SAT on average when parameters p and r are in a certain but nontrivial range. The algorithm is a message passing algorithm stated in Figure 1; this algorithm is motivated by a modification of Pearl's belief propagation algorithm for graph partitioning problems [OW05].

We explain the outline of the algorithm. Below we use i and j to denote unsigned (i.e., positive) indices in $\{1, \ldots, n\}$, whereas s and t are used for signed indices in S. The algorithm is executed on a directed graph $G = (V, E)$ that is constructed from a given formula F in essentially the same way as in the proof of Theorem 1. V is a set of $2n$ vertices v_s, $s \in S = \{-n, -(n - 1), \ldots, -1, +1, \ldots, +(n - 1), +n\}$, and E consists of two directed edges corresponding to each clause $(\ell_i \vee \ell_j)$ of F, where $i < j$; on the other hand, only one edge is added to E for each clause of type $(\overline{x}_i \vee x_i)$. Note that graph G has no multiple edge, while it may have some self-loops. Let $N(u)$ denote the set of vertices v having a directed edge to u.

The algorithm computes a "belief" $b(v_s)$ at each vertex v_s, an integral value indicating whether the Boolean variable $x_{|s|}$ should be assigned true (i.e., $+1$) or false (i.e., -1). More specifically, for an optimam assignment, the algorithm suggests, for each x_i, to assign $x_i = +1$ if the final value of $b(v_{+i})$ is positive and $x_i = -1$ if it is negative. Note that $b(v_{-i}) = -b(v_{+i})$; we may regard $b(v_{-i})$

as a belief for $\overline{x}_i$. These belief values are initially set to 0 except for one pair of vertices, e.g., v_{+1} and v_{-1} that are assigned $+1$ or -1 initially. In the algorithm of Figure 1, $b(v_{+1})$ (resp., $b(v_{-1})$) is set to $+1$ (resp., -1), which considers the case that x_1 is true in the optimal assignment. Clearly we need to consider the other case; that is, the algorithm is executed again with the initial assignment $b(v_{+1}) = -1$ and $b(v_{-1}) = +1$, and one of the obtained assignments satisfying more clauses is used as an answer. Now consider the execution of the algorithm. The algorithm updates beliefs based on messages from the other vertices. At each iteration, the belief of each vertex v_{+i} (resp., v_{-i}) is recomputed based on the last belief values of its neighbor vertices. More specifically, if there is an edge from v_{+i} to v_s, and $b(v_s)$ is negative, then this negative belief is sent to v_{+i} (from v_s) and used for computing the next belief of v_{+i}. The edge $v_{+1} \to v_s$ corresponds to a clause $(x_i \to \ell_{|s|})$ (where $\ell_{|s|}$ is the literal corresponding to v_s), and the condition that $b(v_s) < 0$ means that the literal $\ell_{|s|}$ is assigned false (under the current belief). Thus, in order to satisfy the clause $(x_i \to \ell_{|s|})$, we need to assign false to x_i. This is the reason for the message from v_s. Belief $b(v_{+i})$ at this iteration is defined as the sum of these messages. It should be remarked here that all belief values are updated *in parallel*; that is, updated beliefs are not used when updating the other beliefs in the same iteration, but those computed at the previous iteration are used. This update is repeated until no belief value is changed its sign after one updating iteration *or* the number of iterations reaches to a bound MAXSTEP. This is the outline of our algorithm. It is easy to see that each iteration can be executed in time $\mathcal{O}(n + m)$.

Now we state our theoretical analysis. For the analysis, we modify the algorithm a bit: (i) set MAXSTEP $= 2$; that is, beliefs are updated only twice, (ii) execute a statement (1) only after the second iteration, and (iii) insert a statement (2). For this algorithm, we can prove the following.

Theorem 2. *For any $\delta > 0$, if $n \geq c_2 \ln(n/\delta)/p^2$, or roughly, $p = \Omega(n^{-1/2} \ln n)$, then the algorithm, executed with two different initial values for $b(v_{+1})$ for instances generated under our planted solution model, yields one of the planted solution pair with probability $1 - \delta$.*

References

[AJM05]   D. Achlioptas, H. Jia, and C. Moore, Hiding satisfying assignments: two are better than one, *J. AI Research* 24, pp.623–639, 2005.

[OW05]   M. Onsjö and O. Watanabge, Simple algorithms for graph partition problems, Res. Report C-212, Dept. of Math. and Comput. Sci., Tokyo Tech., 2005. `http://www.is.titech.ac.jp/research/research-report/C/index.html`

[SS03]   A.D. Scott and G.B. Sorkin, Faster algorithms for MAX CUT and MAX CSP, with polynomial expected time for sparse instances, in *Proc. APPROX and RANDOM 2003*, LNCS 2764, pp.382–395, 2003.

[WY06]   O. Watanabe and M. Yamamoto, Res. Report, Dept. of Math. and Comput. Sci., Tokyo Tech., 2006, in preparation.

[Yam06]   M. Yamamoto, Generating instances for MAX2SAT with optimal solutions, *Theory of Computing Systems*, to appear.

Local Search for Unsatisfiability

Steven Prestwich[1] and Inês Lynce[2]

[1] Cork Constraint Computation Centre,
Department of Computer Science, University College, Cork, Ireland
s.prestwich@cs.ucc.ie
[2] IST/INESC-ID, Technical University of Lisbon, Portugal
ines@sat.inesc-id.pt

Abstract. Local search is widely applied to satisfiable SAT problems, and on some classes outperforms backtrack search. An intriguing challenge posed by Selman, Kautz and McAllester in 1997 is to use it instead to prove unsatisfiability. We investigate two distinct approaches. Firstly we apply standard local search to a reformulation of the problem, such that a solution to the reformulation corresponds to a refutation of the original problem. Secondly we design a greedy randomised resolution algorithm that will eventually discover proofs of any size while using bounded memory. We show experimentally that both approaches can refute some problems more quickly than backtrack search.

1 Introduction

Most SAT solvers can be classed either as *complete* or *incomplete*, and the complete algorithms may be based on resolution or backtracking. Resolution provides a complete proof system by refutation [24]. The first resolution algorithm was the Davis-Putnam (DP) procedure [6] which was then modified to the Davis-Putnam-Logemann-Loveland (DPLL) backtracking algorithm [7]. Because of its high space complexity, resolution is often seen as impractical for real-world problems, but there are problems on which general resolution proofs are exponentially smaller than DPLL proofs [4]. Incomplete SAT algorithms are usually based on *local search* following early work by [14,26], but metaheuristics such as genetic algorithms may also be applied. On some large satisfiable problems, local search finds a solution much more quickly than complete algorithms, though it currently compares rather badly with backtracking algorithms on industrial benchmarks.

An interesting question is: can local search be applied to *unsatisfiable* problems? Such a method might be able to refute (prove unsatisfiable) SAT problems that defy complete algorithms. This was number five of the ten SAT challenges posed by Selman, Kautz and McAllester in 1997: *design a practical stochastic local search procedure for proving unsatisfiability* [25]. While substantial progress has been made on several challenges, this one remains wide open [18], and we explore two distinct ways of attacking it. It was suggested in [18] that local search could be applied to a space of incomplete proof trees, and our first approach uses a related idea: we apply standard local search to a space of (possibly incorrect) proof graphs each represented by a clause list. In order to exploit current local

A. Biere and C.P. Gomes (Eds.): SAT 2006, LNCS 4121, pp. 283–296, 2006.

search technology, this is done via a new reformulation of the original SAT problem. Our second approach is a new local search algorithm that explores a space of resolvent multisets, and aims to derive the empty clause non-systematically.

The paper is organised as follows. In Section (2) we SAT-encode the meta-problem of finding a proof for a given SAT problem. Section (3) describes the new local search algorithm. Section (4) discusses related work. Finally, Section (5) concludes the paper.

2   Local Search on a Reformulated Problem

Our first approach is to apply existing local search algorithms to a reformulation of the original SAT problem: a solution to the reformulation corresponds to a refutation of the original problem. A potential drawback with this approach is the sheer size of the reformulation, and it is noted in [18] that *a key issue is the need to find smaller proof objects*. Much of our effort is therefore devoted to reducing the size of the reformulation.

2.1   Initial Model

Suppose we have an unsatisfiable SAT problem with n variables $v_1 \ldots v_n$ and m clauses, and we want to prove unsatisfiability using no more than r resolvents. We can represent the proof as an ordered list of $m + r$ clauses, with the first m clauses being those of the problem, and each of the other clauses being a resolvent of two earlier clauses (which we call the *parents* of the resolvent). The final resolvent must be empty. This meta-problem can be SAT-encoded as follows.

Define meta-variables x_{ikp} (where $i < k$ and $p \in \{0, 1\}$) to be true (T) iff clause k in the list is a resolvent of clause i and another unspecified clause in the list, using an unspecified variable occurring in i as either a positive ($p = 1$) or negative ($p = 0$) literal. Define $u_{kv} = T$ iff clause k was the result of a resolution using variable v. Define $o_{ivp} = T$ iff literal v/p occurs in clause i: $v/1$ denotes literal v, and $v/0$ literal $\bar{v}$, and we shall refer to p as the *sign* of the literal. Define $d_{kvq} = T$ iff variable v in resolvent k occurs in a literal v/q in the parent clause. For example if clause 10 ($\bar{v}_{36} \vee v_{37}$) is resolved with clause 12 ($v_{36} \vee v_{38}$) to give clause 17 ($v_{37} \vee v_{38}$) then the following meta-variables are all true: $o_{36,10,0}$, $o_{37,10,1}$, $o_{36,12,1}$, $o_{38,12,1}$, $o_{37,17,1}$, $o_{38,17,1}$, $x_{10,17,0}$, $x_{12,17,1}$, $u_{17,36}$, $d_{17,37,0}$, $d_{17,38,1}$. There are $O(r^2 + rm + rn + mn)$ meta-variables.

The meta-clauses are as follows, with their space complexities in terms of number of literals. We represent the SAT problem by the following unary meta-clauses for all literals v/p [not] occurring in clauses i:

$$o_{ivp} \ [\bar{o}_{ivp}] \qquad O(mn) \tag{1}$$

Each resolvent must be the resolvent of one earlier clause in the list using a variable positively and one negatively. We use two sets of meta-clauses to ensure that at least one, and no more than one, earlier clause is used:

$$\bigvee_i x_{ikp} \qquad O(r(m+r)) \tag{2}$$

$$\bar{x}_{ikp} \vee \bar{x}_{jkp} \qquad O(r(m+r)^2) \tag{3}$$

Exactly one variable is used to generate a resolvent:

$$\bigvee_v u_{kv} \qquad O(nr) \tag{4}$$

$$\bar{u}_{kv} \vee \bar{u}_{kw} \qquad O(n^2 r) \tag{5}$$

If k is a resolvent using v then v does not occur in k:

$$\bar{u}_{kv} \vee \bar{o}_{kvp} \qquad O(nr) \tag{6}$$

If k is the resolvent of i and another clause using a variable with sign p, and k is a resolvent using variable v, then v/p occurs in i:

$$\bar{x}_{ikp} \vee \bar{u}_{kv} \vee o_{ivp} \qquad O(nr(m+r)) \tag{7}$$

Every literal in k occurs in a literal of at least one of its parent clauses:

$$\bar{o}_{kvp} \vee d_{kv0} \vee d_{kv1} \qquad O(nr) \tag{8}$$

$$\bar{d}_{kvq} \vee \bar{o}_{kvp} \vee \bar{x}_{ikq} \vee o_{ivp} \qquad O(nr(m+r)) \tag{9}$$

(Variables d_{kvq} were introduced to avoid referring to both parent clauses of k in a single meta-clause, which would increase the space complexity.) If i is a parent clause of k using a variable occurring with sign p in i, v/p occurs in i, and v was not used in the resolution generating k, then v/p occurs in k:

$$\bar{x}_{ikp} \vee \bar{o}_{ivp} \vee u_{kv} \vee o_{kvp} \qquad O(nr(m+r)) \tag{10}$$

If i is a parent clause of k using a variable occurring with sign p in i, and $v/\bar{p}$ occurs in i, then $v/\bar{p}$ occurs in k:

$$\bar{x}_{ikp} \vee \bar{o}_{iv\bar{p}} \vee o_{kv\bar{p}} \qquad O(nr(m+r)) \tag{11}$$

The last resolvent is empty:

$$\bar{o}_{m+r\ vp} \qquad O(n) \tag{12}$$

Tautologous resolvents are excluded (we assume that the original problem contains no tautologies):

$$\bigvee_p \bar{o}_{kvp} \qquad O(nr) \tag{13}$$

Every resolvent is used in a later resolution, so for $m \le i < k \le m+r$:

$$\bigvee_k \bigvee_p x_{ikp} \qquad O(r^2) \tag{14}$$

This meta-encoding has $O(nr(m+r)^2 + n^2 r)$ literals which can be reduced in several ways as follows.

2.2   Model Reduction by Unit Resolution

The model can be reduced by observing that many meta-variables appear in unary meta-clauses, and can therefore be eliminated. For every variable v and every original clause i we have a meta-variable o_{ivp} that occurs in a unary meta-clause (either $\bar{o}_{ivp}$ or o_{ivp} depending on whether literal v/p occurs in i). All of these $O(mn)$ meta-variables can be eliminated by resolving on the unary meta-clauses (1). For any such unary clause l: any clause $A \vee l$ is a tautology and can be removed; any clause $A \vee \bar{l}$ can be replaced by A via unit resolution and subsumption; and clause l itself can then be deleted by the pure literal rule. This leaves $O(r^2 + mr + nr)$ meta-variables. If we are searching for a short proof then mn is the dominant term in the number of meta-variables, so we have eliminated most of them. The complexity of some sets of meta-clauses is reduced by unit resolution. Suppose that λ is the mean clause length divided by n, so that $mn\lambda$ is the size in literals of the original problem. Then after unit resolution (7,8,9) become $O(nr(m(1-\lambda)+r))$ and (10,11) become $O(nr(m\lambda+r))$. The total space complexity of (7,8,9,10,11) is still $O(nr(m+r))$ but we will eliminate some of these below.

2.3   Model Reduction by Weakening Rule

We can greatly reduce the space complexity by allowing the proof to use the *weakening rule*, in which any literals may be added to a clause as long as this does not create a tautology. This allows us to remove some of the largest sets of meta-clauses: (7,8,9), removing the $nr(m(1-\lambda)+r)$ terms from the space complexity. This is a significant reduction: in a SAT problem without tautologous clauses $\lambda \leq 0.5$ but a typical problem will have a much smaller value, for example in a random 3-SAT problem $\lambda = \frac{3}{n}$, and the meta-encoding with $r = 10$ of a 600-variable problem from the phase transition is reduced from approximately 92,000,000 to 300,000 clauses (in both cases applying all our other reduction techniques). Removing these meta-clauses allows new literals to be added to a resolvent. For example $A \vee x$ and $B \vee \bar{x}$ may be the parent clauses of $A \vee B \vee C$ for some disjunction C. The d_{kvq} variables no longer occur in any meta-clauses and can be removed. The total space complexity is now $O(r(m+r)^2 + n^2r + nr(m\lambda+r))$ literals.

2.4   Model Reduction by Allowing Multiple Premises

We can eliminate the $r(m+r)^2$ complexity term by dropping meta-clauses (3), allowing a resolvent to have more than one parent clause of a given sign. For example $A \vee x$, $A \vee x$ and $B \vee \bar{x}$ may be parent clauses of $A \vee B$. When combined with the use of the weakening rule we obtain the more general: $x \vee A_i$ and $\bar{x} \vee B_j$ may all be parent clauses of $\bigvee_i A_i \vee \bigvee_j B_j$ (assuming that this clause is non-tautologous) because each of the possible resolvents can be extended to this clause via weakening. The total space complexity is now $O(n^2r + nr(m\lambda + r))$ literals.

2.5   Model Reduction by Ladder Encoding

We can eliminate the n^2r term by replacing meta-clauses (5) with a *ladder encoding* of the at-most-one constraint, adapted from [11]. Because we are using local search we are not concerned with the propagation properties of the encoding, so we can omit some clauses from the original ladder encoding. Define $O(nr)$ new variables l_{kv} and add ladder validity clauses $l_{kv} \vee \bar{l}_{k\,v-1}$ and channelling clauses $\bar{u}_{kv} \vee \bar{l}_{k\,v-1}$ and $\bar{u}_{kv} \vee l_{kv}$. These clauses are sufficient to prevent any pair u_{kv}, u_{kw} from both being true. This set of meta-clauses has only $O(nr)$ literals instead of the $O(n^2r)$ of (5), so the total space complexity is now $O(nr(m\lambda + r))$. In summary, the reduced meta-encoding contains $O(r^2 + mr + nr)$ variables and $O(nr^2 + rs)$ literals where $s = mn\lambda$ is the size in literals of the original SAT problem. We conjecture that this cannot be reduced further.

2.6   Discussion of the Meta-encoding

A useful property of the model is that we can overestimate r, which we would not normally know precisely in advance. This is because for any proof of length r there exists another proof of length $r + 1$. Suppose a proof contains a clause $i : x \vee A$ and a later clause $j : \bar{x} \vee B$, which are resolved to give $k : A \vee B \vee C$ where A, B, C are (possibly empty) disjunctions of literals and C is introduced by weakening. Then between j and k we can insert a new clause $k' : x \vee A \vee B \vee C$ derived from i, j and derive k from i, k'. If a appears in C then first remove it; we can always remove literals introduced by weakening without affecting the correctness of the proof.

Whereas local search on a SAT problem can prove satisfiability but not unsatisfiability, local search on the meta-encoding can prove unsatisfiability but not satisfiability. We can apply any standard local search algorithm for SAT to a meta-encoded problem. Many such algorithms have a property called *probabilistic approximate completeness* (PAC) [15]: the probability of finding a solution tends to 1 as search time tends to infinity. PAC has been shown to be an important factor in the performance of practical local search algorithms [15], and we expect it to be important also in proving unsatisfiability. If we use a PAC local algorithm then it will eventually refute any unsatisfiable problem, given sufficient time and assuming that we set the proof length r high enough.

2.7   Experiments

The local search algorithm we use is RSAPS [16], a state-of-the-art dynamic local search algorithm that has been shown to be robust over a variety of problem types using default runtime parameters. Its performance can sometimes be improved by parameter tuning but in our experiments the difference was not great, nor did any other local search algorithm we tried perform much better. All experiments in this paper were performed on a 733 MHz Pentium II with Linux.

We will make use of two unsatisfiable problems. One is derived from the well-known pigeon hole problem: place $n + 1$ pigeons in n holes, such that no hole receives more than one pigeon. The SAT model has variables v_{ij} for pigeons i and holes j. Clauses $\bigvee_j v_{ij}$ place each pigeon in at least one hole, and clauses $\bar{v}_{ij} \vee \bar{v}_{i'j}$ prevent more than one pigeon being placed in any hole. The 2-hole problem, which we denote by HOLE2, has a refutation of size 10. The other problem is one we designed and call HIDER:

$$
\begin{array}{llll}
a_1 \vee b_1 \vee c_1 & \bar{a}_1 \vee d \vee e & \bar{b}_1 \vee d \vee e & \bar{c}_1 \vee d \vee e \\
a_2 \vee b_2 \vee c_2 & \bar{a}_2 \vee \bar{d} \vee e & \bar{b}_2 \vee \bar{d} \vee e & \bar{c}_2 \vee \bar{d} \vee e \\
a_3 \vee b_3 \vee c_3 & \bar{a}_3 \vee d \vee \bar{e} & \bar{b}_3 \vee d \vee \bar{e} & \bar{c}_3 \vee d \vee \bar{e} \\
a_4 \vee b_4 \vee c_4 & \bar{a}_4 \vee \bar{d} \vee \bar{e} & \bar{b}_4 \vee \bar{d} \vee \bar{e} & \bar{c}_4 \vee \bar{d} \vee \bar{e}
\end{array}
$$

From these clauses we can derive $d \vee e$, $\bar{d} \vee e$, $d \vee \bar{e}$ or $\bar{d} \vee \bar{e}$ in 3 resolution steps each. For example resolving $(a_1 \vee b_1 \vee c_1)$ with $(\bar{a}_1 \vee d \vee e)$ gives $(b_1 \vee c_1 \vee d \vee e)$; resolving this with $(\bar{b}_1 \vee d \vee e)$ gives $(c_1 \vee d \vee e)$; and resolving this with $(\bar{c}_1 \vee d \vee e)$ gives $(d \vee e)$. From these 4 resolvents we can obtain d and $\bar{d}$ (or e and $\bar{e}$) in 2 resolution steps. Finally, we can obtain the empty clause in 1 more resolution step, so this problem has a refutation of size 15. We designed HIDER to be hidden in random 3-SAT problems as an unsatisfiable sub-problem with a short refutation. All its clauses are ternary and no variable occurs more than 12 times. In a random 3-SAT problem from the phase transition each variable occurs an expected 12.78 times, so these clauses blend well with the problem, and a backtracker has no obvious reason to focus on the new variables. Moreover, a resolution refutation of HIDER requires the generation of quaternary clauses, which SATZ's *compactor* preprocessor [19] does not generate. We combine both HIDER and HOLE2 with a random 3-SAT problem by renumbering their variables so that they are distinct from the 3-SAT variables, then taking the union of the two clause sets. We denote such a combination of two problems A and B by $A + B$, where A's variables are renumbered. We performed experiments to obtain preliminary answers to several questions.

What is the effect of the weakening rule on local search performance? To allow weakening in the refutation we may remove some meta-clauses as described in Section 2.3. Dropping clauses from a SAT problem can increase the solution density, which sometimes helps local search to solve the problem, but here it has a bad effect. RSAPS was able to refute HIDER in a few seconds (it is trivial to refute by DPLL) using the meta-encoding without weakening, but with weakening it did not terminate in a reasonable time. This was surprising for such a tiny problem: perhaps the meta-encoding is missing some vital ingredient such as a good set of implied clauses.

What is the effect of allowing unused resolvents? We tested the effect of dropping meta-clauses (14), which are optional (and do not affect the space complexity). Removing them allows a resolvent to be added to the proof then not used further. In experiments omitting them made HIDER much harder to refute, presumably by not penalising the construction of irrelevant chains of resolvents.

What effect does the allowed proof length have on local search performance? On HIDER under the original meta-encoding (without weakening) RSAPS finds the refutation very hard with r set to its minimum value of 15. However, as r increases the runtime decreases to a few seconds. The number of flips decreases as r increases, but the increasing size of the model means that using larger r eventually ceases to pay off.

Can local search on a reformulation beat DPLL on an unsatisfiable problem? We combined HOLE2 with f600, a fairly large, satisfiable, random 3-SAT problem from the phase transition region.[1] HOLE2+f600 has a refutation of size 10 and we set $r = 20$. ZChaff [21] aborted the proof after 10 minutes, whereas RSAPS found a refutation in a median of 1,003,246 flips and 112 seconds, over 100 runs. However, SATZ refutes the problem almost instantly.

3   Local Search on Multisets of Resolvents

A drawback with the reformulation approach is that it is only practical for proofs of relatively small size. Our second approach is to design a new local search algorithm that explores multisets of resolvents, and can in principle find a proof of any size while using only bounded memory. To determine how much memory is required we begin by reviewing a theoretical result from [8].

Given an unsatisfiable SAT formula ϕ with n variables and m clauses, a general resolution refutation can be represented by a series of formulae $\phi_1, \ldots, \phi_s$ where ϕ_1 consists of some or all of the clauses in ϕ, and ϕ_s contains the empty clause. Each ϕ_i is obtained from ϕ_{i-1} by (optionally) deleting some clauses in ϕ_{i-1}, adding the resolvent of two remaining clauses in ϕ_{i-1}, and (optionally) adding clauses from ϕ. The *space* of a proof is defined as the minimum k such that each ϕ_i contains no more than k clauses.

Intuitively each ϕ_i represents the set of *active* clauses at step i of the proof. Inactive clauses are not required for future resolution, and after they have been used as needed they can be deleted. It is proved in [8] that the space k need be no larger than $n + 1$: possibly fewer clauses than in ϕ itself.

The *width* of a proof is the length (in literals) of the largest clause in the proof. Any non-tautologous clause must have length no greater than n, so this is a trivial upper bound for the width used for our algorithm. However, short proofs are also narrow [3] so in practice we may succeed even if we restrict resolvent length to some small value. This may be useful for saving memory on large problems.

Thus we can in principle find a large refutation using a modest amount of working memory. But finding such a proof may not be easy. We shall use the above notions as the basis for a novel local search algorithm that performs a randomised but biased search in the space of formulae ϕ_i. Each ϕ_i will be of the same constant size, and derived from ϕ_{i-1} by the application of resolution or the replacement of a clause by one taken from ϕ. We call our algorithm RANGER (RANdomised GEneral Resolution).

---

[1] Available at http://www.cs.ubc.ca/~hoos/SATLIB/

3.1 The Algorithm

The RANGER architecture is shown in Figure 1. It has six parameters: the size k of the ϕ_i, the width w, three probabilities p_i, p_t, p_g and the formula ϕ.

```
1   RANGER(φ, pᵢ, pₜ, p_g, w, k):
2      i ← 1 and φ₁ ← {any k clauses from φ}
3      while φᵢ does not contain the empty clause
4         with probability pᵢ
5            replace a random φᵢ clause by a random φ clause
6         otherwise
7            resolve random φᵢ clauses c, c' giving r
8            if r is non-tautologous and |r| ≤ w
9               with probability p_g
10                 if |r| ≤ max(|c|, |c'|) replace the longer of c, c' by r
11              otherwise
12                 replace a random φᵢ clause by r
13           with probability pₜ
14              apply any satisfiability-preserving transformation to φ, φᵢ
15           i ← i + 1 and φᵢ₊₁ ← {the new formula}
16     return UNSATISFIABLE
```

Fig. 1. The RANGER architecture

RANGER begins with any sub-multiset $\phi_1 \subseteq \phi$ (we shall interpret ϕ, ϕ_i as multisets of clauses). It then performs iterations i, each either replacing a ϕ_i clause by a ϕ clause (with probability p_i), or resolving two ϕ_i clauses and placing the result r into ϕ_{i+1}. In the latter case, if r is tautologous or contains more than w literals then it is discarded and $\phi_{i+1} = \phi_i$. Otherwise a ϕ_i clause must be removed to make room for r: either (with probability p_g) the removed clause is the longer of the two parents of r (breaking ties randomly), or it is randomly chosen. In the former case, if r is longer than the parent then r is discarded and $\phi_{i+1} = \phi_i$. At the end of the iteration, any satisfiability-preserving transformation may (with probability p_t) be applied to ϕ, ϕ_{i+1} or both. If the empty clause has been derived then the algorithm terminates with the message "unsatisfiable". Otherwise the algorithm might not terminate, but a time-out condition (omitted here for brevity) may be added.

Local search algorithms usually use *greedy* local moves that reduce the value of an objective function, or *plateau* moves that leave it unchanged. However, they must also allow non-greedy moves in order to escape from local minima. This is often controlled by a parameter known as *noise* (or *temperature* in simulated annealing). But what is our objective function? Our goal is to derive the empty clause, and a necessary condition for this to occur is that ϕ_i contains at least some small clauses. We will call a local move *greedy* if it does not increase the number of literals in ϕ_i. This is guaranteed on line 10, so increasing p_g increases the greediness of the search, reducing the proliferation of large resolvents. There may be better forms of greediness but this form is straightforward, and in experiments it significantly improved performance on some problems.

3.2 A Convergence Property

We show that RANGER has the PAC property, used here to mean that given sufficient time it will refute any unsatisfiable problem:

Theorem. *For any unsatisfiable SAT problem with n variables and m clauses,* RANGER *is PAC if $p_i > 0$, $p_i, p_t, p_g < 1$, $w = n$ and $k \geq n + 1$.*

Proof. Firstly, any proof of the form in [8] can be transformed to one in which (i) all the ϕ_i have exactly k clauses, possibly with some duplications, and (ii) ϕ_{i+1} is derived from ϕ_i by replacing a clause either by a ϕ clause or a resolvent of two ϕ_i clauses. Take any proof and expand each set ϕ_i to a multiset ϕ_i' by adding arbitrary ϕ clauses, allowing duplications. Suppose that ϕ_{i+1} was originally derived from ϕ_i by removing a (possibly non-empty) set S_1 of clauses, adding the resolvent r of two ϕ_i clauses, and adding a (possibly non-empty) set S_2 of clauses. Then ϕ_{i+1}' can be derived from ϕ_i' by removing a multiset S_1' of clauses, adding r, and adding another multiset S_2'. Because all the ϕ_i' are of the same size k it must be true that $|S_1'| + 1 = |S_2'|$. Then we can derive ϕ_{i+1}' from ϕ_i' by replacing one S_1' clause by r, then then replacing the rest by S_2' clauses.

Secondly, any transformed proof may be discovered by RANGER from an arbitrary state ϕ_i'. Suppose that the proof begins from a multiset ϕ^*. Then ϕ_i' may be transformed into ϕ^* in at most k moves (they may already have clauses in common), each move being the replacement of a ϕ_i' clause by a ϕ^* clause. From ϕ^* the transformed proof may then be recreated by RANGER, which at each move may perform any resolution or replacement. $\square$

3.3 Subsumption and Pure Literal Elimination

Lines 13–14 provide an opportunity to apply helpful satisfiability-preserving transformations to ϕ or ϕ_i or both. We apply the subsumption and pure literal rules in several ways:

- Randomly choose two ϕ_i clauses c, c' containing a literal in common. If c subsumes c' then replace c' by a random ϕ clause.
- Randomly choose two ϕ clauses c, c' containing a literal in common. If c subsumes c' then delete c'.
- Randomly choose a ϕ clause c and a ϕ_i clause c' containing a literal in common. If c strictly subsumes c' then replace c' by c.
- If a randomly-chosen ϕ_i clause c contains a literal that is pure in ϕ then replace c by a randomly-chosen ϕ clause.
- If a randomly-chosen ϕ clause c contains a literal that is pure in ϕ then delete c from ϕ.

Each of these rules is applied once per RANGER iteration with probability p_t. Using ϕ_i clauses to transform ϕ, a feature we shall call *feedback*, can preserve useful improvements for the rest of the search. (We believe that for these particular transformations we can set $p_t = 1$ without losing completeness, but we defer the proof until a later paper.) Note that if ϕ is reduced then this will soon be reflected in the ϕ_i via line 5 of the algorithm.

The space complexity of RANGER is $O(n + m + kw)$. To guarantee the PAC property we require $w = n$ and $k \geq n+1$ so the complexity becomes at least $O(m + n^2)$. In practice we may require k to be several times larger, but a smaller value of w is often sufficient. Recall that the meta-encoding has space complexity $O(nr^2 + rs)$ where s is the size of the original problem and r the proof length. Thus for short proofs the meta-encoding may be economical, but RANGER's space complexity has the important advantage of being independent of the length of the proof.

A note on implementation. We maintain a data structure that records the locations in ϕ and ϕ_i of two clauses containing each of the $2n$ possible literals. Two locations in this structure are randomly updated at each iteration and used during the application of resolution, subsumption and the pure literal rule. The implementation could no doubt be improved by applying the pure literal rule and unit resolution as soon as possible, but our prototype relies on these eventually being applied randomly.

3.4   Experiments

Again we performed experiments to answer some questions.

Does RANGER perform empirically like a local search algorithm? Though RANGER has been described as a local search algorithm and has the PAC property, it is very different from more standard local search algorithms. It is therefore interesting to see whether its runtime performance is similar to a standard algorithm. Figure 2 shows run-length distributions of the number of iterations required for RANGER to refute HIDER+f600, with $p_i = 0.1$, $p_t = 0.9$, $k = 10m$, and 250 runs per curve. With no greed ($p_g = 0.00$) there is heavy-tailed behaviour. With maximum greed ($p_g = 1.00$, not shown) a refutation cannot be found because HIDER's refutation contains quaternary resolvents, which require non-greedy moves to derive from the ternary clauses. With high greed ($p_g = 0.99$) the median result is worse but there is no heavy tail. The best results are with a moderate amount of greed (such as $p_g = 0.50$): there is no heavy-tailed behaviour and the median result is improved.

What is the effect of space on RANGER's performance? Though a low-space proof may exist, performance was often improved by allowing more space: $10m$ usually gave far better results than the theoretical minimum of $n + 1$.

What is the effect of feedback on RANGER's performance? Feedback was observed to accelerate refutation on some problems, especially as ϕ is sometimes reduced to a fraction of its original size.

Can RANGER beat a non-trivial complete SAT algorithm on an unsatisfiable problem? It refutes HOLE2+f600 in about 0.15 seconds: recall that RSAPS on the meta-encoding took over 100 seconds, which beat ZChaff but

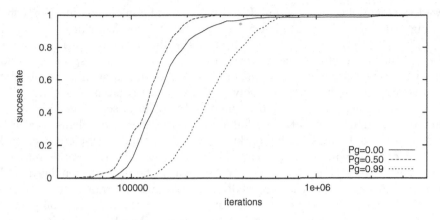

Fig. 2. Three run-length distributions for RANGER

not SATZ. RANGER also refutes HIDER+f600 in about 1 second, easily beating several backtrackers which take at least tens of minutes: ZChaff, SATZ, Siege, POSIT and Minisat (if we renumber the variables, because Minisat branches on the variables in reverse lexicographic order). On this problem RANGER performs roughly 130,000 iterations per second.

How does RANGER perform on benchmarks? It can refute the automotive product configuration problems of [27] in seconds or minutes, depending on the instance, but these are easier for current backtrackers. It is interesting to note that these problems are reduced to a fraction of their original size (approximately $\frac{1}{20}$) by RANGER's feedback mechanism. RANGER also refutes aim-100-2_0-no-1 in a few seconds, whereas Rish & Dechter's DR resolution algorithm [23] takes tens of minutes, as does the Tableau backtracker. But their resolution/backtrack hybrid algorithms take under 1 second, as does the *compactor* algorithm alone. On unsatisfiable random 3-SAT problems RANGER performs very poorly: an interesting asymmetry, given that local search performs well on *satisfiable* random problems. The DR algorithm refutes the dubois20/21 benchmarks quite quickly while RANGER finds them very hard. RANGER refutes ssa0432-003 in about 40 minutes, backtrackers take seconds, DR takes a long time, and a resolution/backtrack hybrid takes 40 seconds. RANGER, DR and the hybrids take a long time on bf0432-007, but current backtrackers find it easy. These results are mixed, but in future work we hope to find a useful class of SAT problems on which RANGER is the algorithm of choice. These problems should be unsatisfiable, fairly large, not susceptible to backtrack search, and require resolution proof of non-trivial width.

4   Related Work

As of 2003 no work had been done on using local search to prove unsatisfiability [18], and we are unaware of any such work since. However, some research may be viewed as moving in this direction. Local search can be made complete by using learning techniques, for example GSAT with dynamic backtracking [12], learn-SAT [22] and Complete Local Search [9]. But the aim of these algorithms seems to be to improve performance on satisfiable problems, not to speed up proof of unsatisfiability. Learning algorithms may also require exponential memory in the worst case, though in practice polynomial memory is often sufficient.

Backtracking algorithms have been modified with local search techniques, to improve performance on both satisfiable and unsatisfiable problems. Recently [20] proposed randomly selecting backtrack points within a complete backtrack search algorithm. Search restarts can also be seen as a form of randomization within backtrack search, and have been shown to be effective on hard SAT instances [13]. The search is repeatedly restarted whenever a cutoff value is reached. The algorithm proposed is not complete, since the restart cutoff point is kept constant. But in [2] search restarts were combined with learning for solving hard, real-world instances of SAT. This latter algorithm is complete, since the backtrack cutoff value increases after each restart. Local search has also been used to finding a good variable ordering, which is then used to speed up a DPLL proof of unsatisfiability [5].

Hybrid approaches have also been tried for the more general class of QBF formulas. WalkQSAT [10] has two main components. The first is the QBF engine, which performs a backjumping search based on conflict and solution directed backjumping. The second is the SAT engine, which is a slightly adapted version of the WalkSAT local search algorithm used as an auxiliary search procedure to find satisfying assignments quickly. The resulting solver is incomplete as it can terminate without a definite result. WalkMinQBF [17] has also two main components. The first is a local search algorithm that attempts to find an assignment to the universal variables that is a witness for unsatisfiability. The second is a complete SAT solver that tests the truth or falsity of SAT formulas that result from assigning the universal variables. WalkMinQBF is also incomplete: it outputs *unsatisfiable* if a certificate of unsatisfiability is found, otherwise it outputs *unknown*.

5   Conclusion

We proposed two distinct ways in which local search can be used to prove unsatisfiability, and demonstrated that there exist problems on which they outperform backtracking (and in some cases systematic resolution) algorithms. As far as we know, this is the first work reporting progress on the fifth SAT challenge of [25]. In experiments with both methods we noted an interesting trend: that short and low-space proofs are harder to find by local search. It is therefore advisable to allow greater length and space than theoretically necessary.

The more successful of our two approaches used randomised general resolution with greedy heuristics and other techniques. It is perhaps surprising that this relatively short-sighted algorithm beats a sophisticated dynamic local search algorithm, though they explore different search spaces and cannot be directly compared. In future work we hope to improve the first approach by modifying the reformulation, and the second by finding improved heuristics.

Powerful proof systems such as general resolution can in principle be used to solve harder problems than more simple systems. In practice such systems are rarely used, partly because of their excessive memory consumption, but also because no good strategy is known for applying the inference rules in order to find a small proof. In fact there may be no such strategy [1], and we suggest that a non-systematic approach is an interesting research direction for such proof systems.

Acknowledgements. Thanks to Eli Ben-Sasson for advice on resolution, and to the anonymous referees for many helpful comments. This material is based in part upon works supported by the Science Foundation Ireland under Grant No. 00/PI.1/C075, and by Fundação para a Ciência e Tecnologia under research project POSC/EIA/61852/2004.

References

1. M. Alekhnovich, A. Razborov. Resolution is Not Automizable Unless W[P] is Tractable. *Forty-Second IEEE Symposium on FOCS*, 2001, pp. 210–219.
2. L. Baptista, J. P. Marques-Silva. Using Randomization and Learning to Solve Hard Real-World Instances of Satisfiability. *Sixth International Conference on Principles and Practice of Constraint Programming, Lecture Notes in Computer Science* vol. 1894, Springer 2000, pp. 489–494.
3. E. Ben-Sasson, A. Wigderson. Short Proofs are Narrow — Resolution Made Simple. *Journal of the ACM* vol. 48 no. 2, 2001, pp. 149–169.
4. E. Ben-Sasson, R. Impagliazzo, A. Wigderson. Near-Optimal Separation of Treelike and General Resolution. *Combinatorica* vol. 24 no. 4, 2004, pp. 585–603.
5. F. Boussemart, F. Hemery, C. Lecoutre, L. Saïs. Boosting Systematic Search by Weighting Constraints. *Sixteenth European Conference on Artificial Intelligence*, IOS Press, 2004, pp. 146–150.
6. M. Davis, H. Putnam. A Computing Procedure for Quantification Theory. *Journal of the Association of Computing Machinery* vol. 7 no. 3, 1960.
7. M. Davis, G. Logemann, D. Loveland. A Machine Program for Theorem Proving. *Communications of the ACM* vol. 5, 1962, pp. 394–397.
8. J. L. Esteban, J. Torán. Space Bounds for Resolution. *Information and Computation* vol. 171 no. 1, 2001, pp. 84–97.
9. H. Fang, W. Ruml. Complete Local Search for Propositional Satisfiability. *Nineteenth National Conference on Artificial Intelligence*, AAAI Press, 2004, pp. 161–166.
10. I. P. Gent, H. H. Hoos, A. G. D. Rowley, K. Smyth. Using Stochastic Local Search to Solve Quantified Boolean Formulae. *Ninth International Conference on Principles and Practice of Constraint Programming, Lecture Notes in Computer Science* vol. 2833, Springer, 2003, pp. 348–362.
11. I. P. Gent, P. Prosser. SAT Encodings of the Stable Marriage Problem With Ties and Incomplete Lists. *Fifth International Symposium on Theory and Applications of Satisfiability Testing*, 2002.

12. M. L. Ginsberg, D. McAllester. GSAT and Dynamic Backtracking. *International Conference on Principles of Knowledge and Reasoning*, 1994, pp. 226–237.
13. C. P. Gomes, B. Selman, H. Kautz. Boosting Combinatorial Search Through Randomization. *Fifteenth National Conference on Artificial Intelligence*, AAAI Press, 1998, pp. 431–437.
14. J. Gu. Efficient Local Search for Very Large-Scale Satisfiability Problems. *Sigart Bulletin* vol. 3, no. 1, 1992, pp. 8–12.
15. H. H. Hoos. On the Run-Time Behaviour of Stochastic Local Search Algorithms. *Sixteenth National Conference on Artificial Intelligence*, AAAI Press, 1999, pp. 661–666.
16. F. Hutter, D. A. D. Tompkins, H. H. Hoos. Scaling and Probabilistic Smoothing: Efficient Dynamic Local Search for SAT. *Eighth International Conference on Principles and Practice of Constraint Programming, Lecture Notes in Computer Science* vol. 2470, Springer, 2002, pp. 233–248.
17. Y. Interian, G. Corvera, B. Selman, R. Williams. Finding Small Unsatisfiable Cores to Prove Unsatisfiability of QBFs. *Ninth International Symposium on AI and Mathematics*, 2006.
18. H. A. Kautz, B. Selman. Ten Challenges *Redux*: Recent Progress in Propositional Reasoning and Search. *Ninth International Conference on Principles and Practice of Constraint Programming, Lecture Notes in Computer Science* vol. 2833, Springer, 2003, pp. 1–18.
19. C. M. Li, Anbulagan. Look-Ahead Versus Look-Back for Satisfiability Problems. *Principles and Practice of Constraint Programming, Proceedings of the Third International Conference, Lecture Notes in Computer Science* vol. 1330, Springer-Verlag 1997, pp. 341–355.
20. I. Lynce, J. P. Marques-Silva. Random Backtracking in Backtrack Search Algorithms for Satisfiability. *Discrete Applied Mathematics*, 2006 (in press).
21. M. Moskewicz, C. Madigan, Y. Zhao, L. Zhang, S. Malik. Chaff: Engineering an Efficient SAT Solver. *Thirty Eighth Design Automation Conference*, Las Vegas, 2001, pp. 530–535.
22. E. T. Richards, B. Richards. Non-Systematic Search and No-Good Learning. *Journal of Automated Reasoning* vol. 24 no. 4, 2000, pp. 483–533.
23. I. Rish, R. Dechter. Resolution Versus Search: Two Strategies for SAT. *Journal of Automated Reasoning* vol. 24 nos. 1–2, 2000, pp. 225–275.
24. J. A. Robinson. A Machine-Oriented Logic Based on the Resolution Principle. *Journal of the Association for Computing Machinery* vol. 12 no. 1, 1965, pp. 23–41.
25. B. Selman, H. A. Kautz, D. A. McAllester. Ten Challenges in Propositional Reasoning and Search. *Fifteenth International Joint Conference on Artificial Intelligence*, Morgan Kaufmann, 1997, pp. 50–54.
26. B. Selman, H. Levesque, D. Mitchell. A New Method for Solving Hard Satisfiability Problems. *Tenth National Conference on Artificial Intelligence*, MIT Press, 1992, pp. 440–446.
27. C. Sinz, A. Kaiser, W. Küchlin. Formal Methods for the Validation of Automotive Product Configuration Data. *Artificial Intelligence for Engineering Design, Analysis and Manufacturing*, vol. 17 no. 2, special issue on configuration, 2003.

Efficiency of Local Search

Andrei A. Bulatov and Evgeny S. Skvortsov

School of Computing Science, Simon Fraser University, Burnaby, Canada
{abulatov, evgenys}@cs.sfu.ca

Abstract. The Local Search algorithm is one of the simplest heuristic algothms for solving the MAX-SAT problem. The goal of this paper is to estimate the relative error produced by this algorithm being applied to random 3-CNFs with fixed density ϱ. We prove that, for any ϱ, there is a constant c such that a weakened version of Local Search that we call One-Pass Local Search almost surely outputs an assignment containing $cn + o(n)$ unsatisfied clauses. Then using a certain assumtion we also show this for Local Search. Although the assumption remains unproved the results well matches experiments.

1 Introduction

In the Local Search (LS) algorithm we start with a random assignment to a CNF, and then on each step we choose at random a variable such that flipping this variable increases the number of satisfied clauses, or stop if such a variable does not exist. LS is one of the oldest algorithms for solving the SAT and MAX-SAT problems. Numerous variations of this method have been proposed starting from the late eighties, see for example [4,8].

It is well known that the worst-case performance of 'pure' LS is not very good: the only known lower bound for local optima of a k-CNF is $\frac{k}{k+1}m$, m the number of clauses [5]. Many other approximation algorithms, for example Goemans and Williamson's [3], Karloff and Zwick's [6] algorithms and their improvements (see for instance [2]), guarantee much better results than LS. However, it is also an empirical fact that the *expected* performance of LS is much better than $\frac{k}{k+1}m$ bound. This is also supported by the result by Koutsoupias and Papadimitriou [7] that for almost all satisfiable 3-CNFs LS almost surely finds a satisfying assignment.

The goal of this paper is to study the expected performance of LS on random 3-CNFs of the form $\Phi(n, \varrho n)$, ϱ fixed, where n is the number of variables and ϱn is the number of clauses. We consider also a weakened version of LS, the One-Pass LS (OLS) algorithm, in which every variable is visited only once in a certain order, and if flipping the variable increases the number of satisfied clauses then it flipped, otherwise it is never considered again. For OLS we build a model similar to the 'card-game' from [1] and then use Wormald's theorem [9] to show that for any ϱ there is a constant c such that almost surely OLS finds an assignment satisfying $cn + o(n)$ clauses. Then we extend this framework to LS and obtain the same result under a certain assumption that remains unproved.

A. Biere and C.P. Gomes (Eds.): SAT 2006, LNCS 4121, pp. 297–310, 2006.

INPUT: a random 3-CNF Φ

OUTPUT: the number of unsatisfied assignments in the resulting formula Φ'

Step 1   **compute** the set W of variables X such that $A(X) > B(X)$
Step 2   **while** $W \neq \emptyset$ **do**
Step 2.1   **choose** a variable X from W uniformly at random
Step 2.2   **for every** clause $C \in \Phi$ containing X or $\neg X$ **replace** X with $\neg X$ and
                 $\neg X$ with X
Step 2.3   **compute** the set W of variables X such that $A(X) > B(X)$
                 endwhile
Step 3   **output** the number of clauses in Φ with 3 negative literals

Fig. 1. The Local Search algorithm

2   Preliminaries

2.1   MAX-SAT and Local Search

First we give formal definitions of the main objects involved.

In the MAX-SAT problem we are given a CNF Φ. The goal is to find an assignment of the variables of Φ that satisfies maximum number of clauses. The MAX-SAT problem restricted to k-CNFs is called the MAX-k-SAT problem.

The model of random CNF we use is the random 3-CNFs with fixed density. Thus $\Phi(n, m)$ denotes a random CNF with m 3-clauses over n Boolean variables, where clauses are chosen uniformly and independently among $8\binom{n}{3}$ clauses without repetitions of variables. Repetitions of clauses are allowed.

We consider the following version of LS. Given a CNF, choose a random assignment π and repeat the following steps while this is possible: Find the set of variables W such that flipping a variable X from W, that is setting $\pi'(X) = \neg\pi(X)$ keeping all other values, decreases the number of unsatisfied constraints; choose a variable X from W uniformly at random; flip X.

The standard setting of the LS algorithm involves on the first step two random objects — a formula and an assignment. It is easy to see that for the purpose of studying performance one of them can be eliminated. We are going to fix an assignment, it will be $\pi(X) = 1$ for all X and change the formula. By $A_\Phi(X), B_\Phi(X)$ we denote the number of unsatisfied clauses in Φ containing $\neg X$ (this means that all literals in such clauses are negative), and the number of clauses in Φ containing X, and therefore satisfied, such that the two other literals are negative, respectively. As Φ is always clear from the context, we shall omit it. A formal description of LS is given in Fig. 1.

Along with the standard LS algorithm we will study its weakened version that we call the One-Pass Local Search algorithm (OLS), see Fig. 2. Obviously, it does not make any difference if one chooses a predefined order on the variable set in the very beginning of the OLS algorithm, as it is done in Fig.2, or a variable to flip is chosen at random on every step without repetitions.

INPUT: a random 3-CNF Φ with n variables

OUTPUT: the number of unsatisfied assignments in the resulting formula Φ'

Step 1   **for** $t = 1$ **to** n **do**
Step 1.1   **calculate** $A(X_t)$ and $B(X_t)$
Step 1.2   **if** $A(X_t) > B(X_t)$ **then do**
Step 1.2.1   **for every** clause $C \in \Phi$ containing X_t or $\neg X_t$ **replace** X_t with $\neg X_t$
                and $\neg X_t$ with X_t
            endif
        endfor
Step 2   **output** the number of clauses in Φ with 3 negative literals

Fig. 2. The One-Pass Local Search algorithm

2.2   Wormald's Theorem

The key stage in our analysis is the theorem by Wormald [9] that allows one to replace probabilistic analysis of combinatorial algorithm with analysis of a deterministic system of differential equations.

All random processes are discrete time random processes. Such a process is a probability space Ω denoted by $(Q_0, Q_1, \ldots)$, where each Q_i takes values in some set S. Consider a sequence Ω_n, $n = 1, 2, \ldots$, of random processes. The elements of Ω_n are sequences $(q_0(n), q_1(n), \ldots)$ where each $q_i(n) \in S$. For convenience the dependence of n will usually be dropped from the notation. Asymptotics, denoted by the notation o and O, are for $n \to \infty$, but uniform over all other variables. For a random X, we say $X = o(f(n))$ *always* if $\max\{x | \mathbf{P}(X = x) \neq 0\} = o(f(n))$. An event occurs *almost surely* (a.s.) if its probability in Ω_n is $1 - o(1)$. We denote by S^+ the set of all $h_t = (q_0, \ldots, q_t)$, each $q_t \in S$ for $t = 0, 1 \ldots$. By H_t we denote the *history* of the processes, that is the $n \times (t + 1)$-matrix with entries $Q_i(j)$, $0 \leq i \leq t, 1 \leq j \leq n$.

A function $f(u_1, \ldots, u_j)$ satisfies a *Lipschitz condition* on $D \subseteq \mathbb{R}^j$ if a constant $L > 0$ exists with the property that

$$|f(u_1, \ldots, u_j) - f(v_1, \ldots, v_j)| \leq L \sum_{i=1}^{j} |u_j - v_i|$$

for all $(u_1, \ldots, u_j)$ and $(v_1, \ldots, v_j)$ in D.

Theorem 1 (Wormald, [9]). *Let k be fixed. For $1 \leq \ell \leq k$, let $y^{(\ell)} : S^+ \to \mathbb{R}$ and $f_\ell : \mathbb{R}^{k+1} \to \mathbb{R}$, such that for some constant C and all ℓ, $|y^{(\ell)}| < Cn$ for all $h_t \in S^+$ for all n. Suppose also that for some function $m = m(n)$:*
  (i) for all ℓ and uniformly over all $t < m$, $\mathbf{P}(|Y_{t+1}^{(\ell)} - Y_t^{(\ell)}| > n^{1/5} \mid H_t) = o(n^{-3})$ always;
  (ii) for all ℓ and uniformly over all $t < m$,
    $\mathbf{E}(Y_{t+1}^{(\ell)} - Y_t^{(\ell)} \mid H_t) = f_\ell(t/n, Y_t^{(1)}/n, \ldots, y_t^{(k)}/n) + o(1)$ *always;*
  (iii) for each ℓ the function f_ℓ is continuous and satisfies a Lipschitz condition on D, where D is some bounded connected open set containing the intersection of $\{(t, z^{(1)}, \ldots, z^{(k)}) \mid t \geq 0\}$ with some neighbourhood of $\{(0, z^{(1)}, \ldots, z^{(k)}) \mid \mathbf{P}(Y_0^{(\ell)} = z^{(\ell)}n, 1 \leq \ell \leq k) \neq 0$ for some $n\}$.

Then:
(a) *For* $(0, \hat{z}^{(1)}, \ldots, \hat{z}^{(k)}) \in D$ *the system of differential equations*
$$\frac{dz_\ell}{ds} = f_\ell(s, z_1, \ldots, z_k), \ \ell = 1, \ldots, k, \ \text{has a unique solution in } D \text{ for } z_\ell \colon \mathbb{R} \to$$
$\mathbb{R}$ *passing through* $z_\ell(0) = \hat{z}^{(\ell)}$, $1 \leq \ell \leq k$, *and which extends to points arbitrarily close to the boundary of* D.
(b) *Almost surely* $Y_t^{(\ell)} = nz_\ell(t/n) + o(n)$ *uniformly for* $0 \leq t \leq \min\{\sigma n, m\}$ *and for each* ℓ, *where* $z_\ell(s)$ *is the solution in (a) with* $\hat{z}^{(\ell)} = Y_0^{(\ell)}/n$, *and* $\sigma = \sigma(n)$ *is the supremum of those* s *to which the solution can be extended.*

3   One Pass Local Search

3.1   Model

To analyze the OLS algorithm we use an extended version of the 'card game' framework, see [1]. Every clause of CNF Φ is represented by three cards. At step t the intermediate opens all cards with X_t or $\neg X_t$ and also tells us the 'polarity' of the remaining literals in the clauses containing $X_t, \neg X_t$ (that is how many of them are negative). Then we compare the numbers $a(X_t)$ of clauses containing $\neg X_t$, the remaining literals of which are negative, and $b(X_t)$ of clauses containing X_t, the remaining literals of which are negative. If $a(X_t) > b(X_t)$ then we flip X_t replacing everywhere X_t with $\neg X_t$ and $\neg X_t$ with X_t. Finally we remove clauses containing X_t and remove $\neg X_t$. If in the latter case a clause becomes empty we count it as unsatisfied. Note that in contrast to the card games used in [1], in the described game we have some information on the unopened cards, and therefore the formula obtained on each step is not quite random. Thus a more thorough analysis is required.

Such an analysis can be done by monitoring the dynamics of eight sets of clauses that we define at each step of the algorithm. Let Φ_t denote the formula obtained by step t. Variables (and the corresponding literals) from the set $\{X_1, \ldots, X_{t-1}\}$ will be called *processed* (they cannot change anymore), the remainining variables will be called *unprocessed*. We define the following 8 sets:

- $E_\varnothing$ is the set of all clauses in Φ_t that do no contain processed literals;
- E_1 is the set of all clauses in Φ_t containing a positive processed literal;
- E_0 is the set of all clauses in Φ_t that contain three negated processed literals;
- E_{++}, E_{--}, E_{+-} are the sets of all clauses in Φ_t that contain one negated processed literal and two positive, two negative, or a positive and negative unprocessed literals, respectively;
- E_+, E_- are the sets of all clauses in Φ_t containing two processed negative literals, and a positive, or a negative unprocessed literal, respectively.

We will denote sizes of these sets by $e_\varnothing, e_1, e_0, e_{++}, e_{+-}, e_{--}, e_+, e_-$ respectively, and the vector $(e_\varnothing, e_1, e_0, e_{++}, e_{+-}, e_{--}, e_+, e_-)$ by $\mathbf{e}$. These numbers will be our random variables from Wormald's theorem. All these values depend on t, but we always refer to them at the current step t, and so drop t from the notation. We also use v to denote $n - t + 1$.

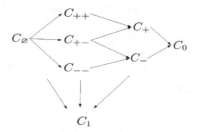

Fig. 3. Flow diagram

It is easy to see that clauses that once get to E_0 or E_1 never leave these sets, and that at each step for each clause that doesn't belong to $E_0 \cup E_1$ there is a chance to get to E_1. The other possible transitions of clauses between the sets are shown on Figure 3.

If $E_\diamond$ and $E_\star$ are some of the eight sets, then we will denote conditional probability for a clause to move from set $E_\diamond$ to $E_\star$ by $\mathbf{P}(E_\diamond \to E_\star)$, assuming a certain particular value of vector $\mathbf{e}$.

We will compute the probability that variable X_t is flipped at step t. This event happens when there are more unsatisfied clauses containing this variable (we denoted the set of such clauses by $A(X_t)$) than clauses that are satisfied only by X_t (we denoted this set by $B(x_t)$). Clauses from sets E_-, E_{--} and $E_\varnothing$ can fall into set $A(X_t)$, while clauses from sets E_+, E_{+-} and $E_\varnothing$ can fall into $B(X_t)$. Probability that a clause from E_- belongs to $A(X_t)$ equals $\frac{1}{v}$, this is the probability that X_t is written on the only card currently unrevealed in the clause. In a similar way we compute such probabilities for clauses from E_{--} and $E_\varnothing$, it is easy to see that they are equal to $\frac{2}{v}$ and $\frac{3}{8v}$ respectively. The probabilities that a clause from E_+, E_{+-}, and $E_\varnothing$ belongs to $B(X_t)$ equal $\frac{1}{v}, \frac{1}{v}$, and $\frac{3}{8v}$ respectively. Note that for different clauses the considered events are independent.

Now let $F(n_1, n_2, n_3, p_1, p_2, p_3)$ denote the event that exactly $n_1, n_2,$ and n_3 clauses from E_-, E_{--}, and $E_\varnothing$ respectively belong to $A(X_t)$, and exactly $p_1, p_2,$ and p_3 clauses from E_+, E_{+-} and $E_\varnothing$ belong to $B(X_t)$. By Bernoulli formula we have

$$
\mathbf{P}(F(n_1, n_2, n_3, p_1, p_2, p_3)) = \binom{e_+}{p_1} \left(\frac{1}{v}\right)^{p_1} \binom{e_{+-}}{p_2} \left(\frac{1}{v}\right)^{p_2} \binom{e_\varnothing}{p_3} \left(\frac{3}{8v}\right)^{p_3}
$$
$$
\times \binom{e_-}{n_1} \left(\frac{1}{v}\right)^{n_1} \binom{e_{--}}{n_2} \left(\frac{2}{v}\right)^{n_2} \binom{e_\varnothing}{n_3} \left(\frac{3}{8v}\right)^{n_3}
$$

As n tends to infinity, Bernoulli distribution tends to Poisson distribution and we have

$$
\mathbf{P}(F(n_1, n_2, n_3, p_1, p_2, p_3)) = \left(\frac{e_+}{v}\right)^{p_1} \left(\frac{e_{+-}}{v}\right)^{p_2} \left(\frac{3e_\varnothing}{8v}\right)^{p_3} \left(\frac{e_-}{v}\right)^{n_1}
$$
$$
\times \left(\frac{2e_{--}}{v}\right)^{n_2} \left(\frac{3e_\varnothing}{8v}\right)^{n_3} \frac{e^{\frac{e_+}{v} + \frac{e_{+-}}{v} + \frac{3e_\varnothing}{8v} + \frac{e_-}{v} + \frac{2e_{--}}{v} + \frac{3e_\varnothing}{8v}}}{n_1! n_2! n_3! p_1! p_2! p_3!} + O\left(\frac{1}{n}\right)
$$

The probability that X_t is flipped can then be calculated as follows:

$$\mathbf{P}(X_t \text{ is flipped}) = \mathbf{P}(|A(x_t)| > |B(x_t)|)$$

$$= \mathbf{P}\left(\bigvee_{p_1+p_2+p_3<n_1+n_2+n_3} |A(x_t)| = n_1 + n_2 + n_3 \ \& \ |B(t)| = p_1 + p_2 + p_3 \right)$$

$$= \sum_{n_1+n_2+n_3>p_1+p_2+p_3} \mathbf{P}(F(n_1, n_2, n_3, p_1, p_2, p_3))$$

It will be convenient for us to denote the sum similar to that appearing in the last line of the equation above, but over $n_1, n_2, n_3, p_1, p_2, p_3$ satisfying a certain condition Ξ, by $S(\Xi)$. Using this notation the probability that variable X_t is flipped can be expressed as

$$\mathbf{P}(X_t \text{ is flipped}) = S(n_1 + n_2 + n_3 > p_1 + p_2 + p_3), \tag{1}$$

when the parameters are clear from the context we denote this value by S.

Now we compute probability $\mathbf{P}(E_\varnothing \to E_{--})$. A clause goes from $E_\varnothing$ to E_{--} in two disjunctive cases. Firstly, if a clause has only negative literals, one of them is $\neg X_t$, and X_t is not flipped. Secondly, if a clause has two negative literals, a positive one, the positive literal is X_t, and X_t is flipped. The probability of the first event equals $\frac{3}{8v}$, and under this assumption the conditional probability that X_t is flipped equals $S(p_1+p_2+p_3 < n_1+n_2+n_3+1)$. The probability that a clause has two negative and one positive literal X_t equals $\frac{3}{8v}$ as well, and under this assumption the conditional probability that X_t flips equals $S(p_1 + p_2 + p_3 + 1 < n_1 + n_2 + n_3)$. We denote the two values specified in the last two sentences by S_+ and S_- respectively. Thus

$$\mathbf{P}(E_\varnothing \to E_{--}) = \frac{3}{8v}(S_+ + S_-) + o\left(\frac{1}{n}\right).$$

The other probabilities can be computed in a similar way:

$$\mathbf{P}(E_\varnothing \to E_{+-}) = \frac{6}{8v} + o\left(\frac{1}{n}\right), \quad \mathbf{P}(E_\varnothing \to E_{++}) = \frac{3}{8v} + o\left(\frac{1}{n}\right),$$

$$\mathbf{P}(E_{++} \to E_+) = \frac{2}{v}S + o\left(\frac{1}{n}\right), \quad \mathbf{P}(E_{+-} \to E_+) = \frac{1}{v}(1 - S) + o\left(\frac{1}{n}\right),$$

$$\mathbf{P}(E_{+-} \to E_-) = \frac{1}{v}S_- + o\left(\frac{1}{n}\right), \quad \mathbf{P}(E_{--} \to E_-) = \frac{2}{v}(1 - S_+) + o\left(\frac{1}{n}\right),$$

$$\mathbf{P}(E_- \to E_0) = \frac{1}{v}(1 - S_+) + o\left(\frac{1}{n}\right), \quad \mathbf{P}(E_+ \to E_0) = \frac{1}{v}S_- + o\left(\frac{1}{n}\right),$$

$$\mathbf{P}(E_\varnothing \to E_1) = \frac{6S}{8v} + \frac{7(1 - S)}{8v} + \frac{3S_+}{8v} + \frac{3(1 - S_-)}{8v} + o\left(\frac{1}{n}\right),$$

$$\mathbf{P}(E_{++} \to E_1) = \frac{2}{v}(1 - S) + o\left(\frac{1}{n}\right), \quad \mathbf{P}(E_{+-} \to E_1) = \frac{1}{v}(1 - S_- + S) + o\left(\frac{1}{n}\right),$$

$$\mathbf{P}(E_{--} \to E_1) = \frac{2}{v}(S_+ + o\left(\frac{1}{n}\right), \quad \mathbf{P}(E_+ \to E_1) = \frac{1}{v}(1 - S_-) + o\left(\frac{1}{n}\right),$$

$$\mathbf{P}(E_- \to E_1) = \frac{1}{v}S_+ + o\left(\frac{1}{n}\right).$$

The probabilities $P(E_\diamond \to E_\star)$ that are not mentioned above equal zero.

We are ready to check that random process $(\mathbf{e}(1), \mathbf{e}(2), \mathbf{e}(3), \ldots)$ satisfies conditions (i) - (iii) of Wormald's theorem.

(i) Let $e_\diamond$ be a component of $\mathbf{e}$. It is obvious that $|e_\diamond(t+1) - e_\diamond(t)|$ is less than the number of occurrences of X_t in Φ. The probability that X_t occurs in some clause equals $\frac{3}{n}$, therefore the probability that X_t occurs in k clauses equals $\binom{\varrho n}{k} \left(\frac{3}{n}\right)^k$. So assuming that n is large enough we have

$$\mathbf{P}(X_t \text{ occurres in more that } n^{1/5} \text{ clauses}) = \sum_{k=n^{1/5}}^{n} \binom{\varrho n}{k} \left(\frac{3}{n}\right)^k$$

$$= \sum_{k=n^{1/5}}^{n} \frac{\varrho n(\varrho n - 1) \ldots (\varrho n - k + 1) 3^k}{k! n^k} \leq \frac{(3\varrho)^{n^{1/5}} n}{n^{1/5}!} = o(n^{-3}).$$

(ii) Let $e_\star$ be a component of $\mathbf{e}$. Then we have

$$\mathbf{E}(e_\star(t+1) - e_\star(t) | H_t) =$$

$$\sum_{E_\diamond \neq E_\star} \left(\sum_{C \in E_\diamond(t)} \mathbf{P}(C \in E_\star(t+1)) - \sum_{C \in E_\star(t)} \mathbf{P}(C \in E_\diamond(t+1)) \right) =$$

$$\sum_{e_\diamond \neq e_\star} (e_\diamond \mathbf{P}(E_\diamond \to E_\star) - e_\star \mathbf{P}(E_\star \to E_\diamond)). \tag{2}$$

Thus we set $s = \frac{t}{n}$, $f_\star(s) = \frac{1}{n} e_\star(t)$ (as is easily seen, $v \approx n(1-s)$) and

$$p(s, n_1, n_2, n_3, p_1, p_2, p_3) = \left(\frac{3 f_\varnothing(s)}{8}\right)^{p_3 + n_3} \left(\frac{1}{1-s}\right)^{n_1 + n_2 + n_3 + p_1 + p_2 + p_3}$$

$$\times f_+^{p_1}(s) f_{+-}^{p_2}(s) f_-^{n_1}(s) (2f_{--}(s))^{n_2} \frac{e^{\frac{e_+(s) + e_{+-}(s) + e_-(s) + 2e_{--}(s) + 3/4 e_\varnothing(s)}{1-s}}}{n_1! n_2! n_3! p_1! p_2! p_3!}.$$

Then set

$$s_0(s) = \sum_{n_1 + n_2 + n_3 > p_1 + p_2 + p_3} p(s, n_1, n_2, n_3, p_1, p_2, p_3),$$

$$s_+(s) = \sum_{n_1 + n_2 + n_3 + 1 > p_1 + p_2 + p_3} p(s, n_1, n_2, n_3, p_1, p_2, p_3),$$

$$s_-(s) = \sum_{n_1 + n_2 + n_3 + 1 > p_1 + p_2 + p_3} p(s, n_1, n_2, n_3, p_1, p_2, p_3).$$

Note that this functions are represented by series. Later we show that this does not cause any problems. Finally, the required system of differential equations can be obtained from equations (2) using $s_0(s), s_+(s), s_-(s)$ to compute the probabilities instead of S_0, S_+, S_-.

(iii) The functions constructed above have two substantial deficiencies: they are not defined when $s = 1$, and the series used to define them do not converge uniformly in the naturally defined set D. However, this can be overcome using a standard trick, namely, for each $\epsilon > 0$, define set D such that it includes only points with $s \leq 1 - \epsilon$. It is not hard to see that, as the series above are non-negative and bounded with 1, they converge uniformly in any closed set. Then we find the required value as the limit when $\epsilon \to 0$.

Applying Wormald's theorem we prove the following

Theorem 2. *For any positive ϱ and there is a constant c such that for a random 3-CNF $\Phi(n, \varrho n)$ almost surely the OSL algorithm finds and assingment such that the number of satisfied clauses equals $cn + o(n)$.*

4   Local Search

4.1   Model

In this section we use similar techniques to analyze the Local Search algorithm. However, as every variable in this algorithm can be considered and flipped several times we cannot use the card game approach; instead we have to find quite a different set of random variables that represents properties of the problem crucial for the performance of the algorithm. Although we were unable to carry out a complete rigorous analysis, it turns out that such an analysis boils down to a certain simple assumption (see Assumption 1 below). This assumption looks very plausible, but we could not neither prove nor disprove it. Experiments show that our model is accurate enough, this is why we believe that either Assumption 1 is true, or it can be replaced with a property that gives rise to an equivalent model.

We need some notation. Let Φ be a 3-CNF and X a variable in Φ. By $Q_{i\alpha}(X)$, where $i \in \{0, 1, 2\}$ and $\alpha \in \{-, +\}$, we denote the set of clauses C such that $X \in C$ if $\alpha = +$, $\neg X \in C$ if $\alpha = -$, and among the other two literals there are exactly i positive. If $C \in Q_{i\alpha}$ we also say that C has *type* $i\alpha$ for X, and that variable X occupies *position* of type $i\alpha$ in the clause C. Let also $q_{i\alpha}(X)$ denote the size of $Q_{i\alpha}(X)$. By $E_{\bar{a}}$, $\bar{a} = (a_{0-}, a_{0+}, a_{1-}, a_{1+}, a_{2-}, a_{2+})$ we denote the set of all variables X of Φ such that $q_{i\alpha}(X) = a_{i\alpha}$ for all i and α. By $e_{\bar{a}}$ we denote the size of $E_{\bar{a}}$. As Φ is changing over time all these sets and numbers are actually fuctions of the number of steps made. Thus, sometimes we use notation $E_{\bar{a}}(t)$, $e_{\bar{a}}(t)$. Functions $e_{\bar{a}}(t)$ will be the random variables required in Wormald's theorem. If $X \in E_{\bar{a}}$ then variable X is said to have type $\bar{a}$. Note that as n grows the number of different tuples $\bar{a}$ and therefore the number of random variables also grow. To overcome this problem we will consider only those variables that appear in at most M clauses for some fixed M. Clearly, this does affect the analysis, but in a certain controllable way, as we shall see.

Before checking conditions (i)–(iii) of Theorem 1 we make a simple observation.

Lemma 1. *If Φ is a random 3-CNF of density ϱ with n variables, then for a variable X*

$$\mathbf{P}(q_{i\alpha}(X) = a) = \frac{\left(\frac{3\varrho}{8}\right)^a e^{3\varrho/8}}{a!} + o(1) \quad \text{if } i = 0, 2,$$

$$\mathbf{P}(q_{i\alpha}(X) = a) = \frac{\left(\frac{3\varrho}{4}\right)^a e^{3\varrho/4}}{a!} + o(1) \quad \text{if } i = 1,$$

$$\mathbf{P}(X \in E_{\overline{a}}) = \prod_{i,\alpha} \mathbf{P}(q_{i\alpha}(X) = a), \quad \mathbf{E}(e_{\overline{a}}) = n \cdot \mathbf{P}(X \in E_{\overline{a}}).$$

Lemma 2. *If Φ is a random 3-CNF of density ϱ with n variables, then $\mathbf{P}(|e_{\overline{a}} - \mathbf{E}(e_{\overline{a}})| > n^{1/5}) = o(n^{-3})$.*

Lemma 2 provides the initial values for equations from Theorem 1. Now we are verifying conditions (i)–(iii).

(i) Possible variations of random variables $e_{\overline{a}}$ are bounded with $2K$ where K is the degree of the flipped variable. Therefore condition (i) can be proved in the same way as for the OLS algorithm.

(ii) Suppose that on the current step t of LS the variable to flip is X. Since X is a variable picked uniformly at random from the set $B(t) = \bigcup_{\substack{\overline{a} \\ a_{0-} > a_{0+}}} E_{\overline{a}}(t)$,

we have $\mathbf{P}(X \in E_{\overline{a}}(t)) = \frac{e_{\overline{a}}(t)}{b(t)}$, where $b(t) = |B(t)|$. Also we have

$$\mathbf{E}(q_{i\alpha}(X)) = \sum_{\substack{\overline{a} \\ a_{0-} > a_{0+}}} a_{i\alpha} \cdot \mathbf{P}(X \in E_{\overline{a}}(t)).$$

We say that tuples $\overline{a}, \overline{b}$ are *adjacent* if there are i, j, α such that $|j - i| = 1$, $a_{i\alpha} = b_{i\alpha} + 1, a_{j\alpha} = b_{j\alpha} - 1$, and $a_{i'\alpha'} = b_{i'\alpha'}$ in all other cases. Intuitively, adjacency means that if $X \in E_{\overline{b}}$ then it can be moved to $E_{\overline{a}}$ or vice versa by flipping one literal in one of the clauses containing X. Let also $\overline{a}'$ denote the tuple such that $a'_{i-} = a_{i+}$ and $a'_{i+} = a_{i-}$.

Set $E_{\overline{a}}$ changes in two ways. First, variable X can move to or from $E_{\overline{a}}$, in this case it moves from or to $E_{\overline{a}'}$. Second, X may happen to be in the same clause with some other variable, Y, and then Y can move to or from $E_{\overline{a}}$. Such a variable moves then from or to $E_{\overline{b}}$ for some $\overline{b}$ adjacent with $\overline{a}$.

Clearly, the expectation of change of the first type equals $\mathbf{P}(X \in E_{\overline{a}'}) - \mathbf{P}(X \in E_{\overline{a}})$. Further computation we carry out under the following assumption.

Assumption 1. *Assuming history H_t, for a random clause C of the current formula, any positions p, r, $p \neq r$, in C, any tuples $\overline{a}, \overline{b}$, and any variables $X \in E_{\overline{a}}, Y \in E_{\overline{b}}$, the events "$X$ is in position p of clause C" and "Y is in position r of clause C" are independent.*

Let us take a variable $Y \in E_{\overline{a}}$ and calculate the probability of an event $\mathcal{G}^-$:"variable Y moves from $E_{\overline{a}}$ to $E_{\overline{b}}$", where $\overline{b}$ is some tuple adjacent to $\overline{a}$

and $\bar{a}, \bar{b}$ differ in components $i\alpha$ and $j\alpha$. This happens if in some clause C containing both X and Y some position occupied by Y changes its type from $i\alpha$ to $j\alpha$. Obviously, depending on $i\alpha$ the type of the position occupied by X may vary. We use $\hat{i}\hat{\alpha}$ to denote the possible type of such a position. Simple case analysis shows that $\hat{i} = j$ if $j < i$ and $\alpha = -$, or if $j > i$ and $\alpha = +$, otherwise $\hat{i} = i$. Then $\hat{\alpha} = -$ if $j < i$ and $\hat{\alpha} = +$ if $j > i$.

Let $\theta_{\bar{a} \to \bar{b}}$ denote the number of positions of type $\hat{i}\hat{\alpha}$ in C except the one possibly occupied by Y. It is easy to see that $\theta_{\bar{a} \to \bar{b}} = 1$ if $i = 1$, $\theta_{\bar{b} \to \bar{a}} = 1$ if $j = 1$, and $\theta_{\bar{a} \to \bar{b}} = 2$, $\theta_{\bar{b} \to \bar{a}} = 2$ otherwise. Thus, the number of positions in the clauses of Φ such that if X in such a position then $\mathcal{G}$ happens to some variable Y equals $\theta_{\bar{a} \to \bar{b}} a_{i\alpha}$. Let also $k_{\hat{i}\hat{\alpha}}(t) = \sum_{\bar{a}} a_{\hat{i}\hat{\alpha}} \cdot e_{\bar{a}}(t)$ be the number of positions of type $\hat{i}\hat{\alpha}$ in the formula.

Suppose that variable X that is flipped belongs to $E_{\bar{c}}$. Then among all $k_{\hat{i}\hat{\alpha}}(t)$ positions of type $\hat{i}\hat{\alpha}$ we have $c_{\hat{i}\hat{\alpha}}$ positions occupied by X, and $\theta_{\bar{a} \to \bar{b}} a_{i\alpha}$ positions such that the presence of X in one of them makes the event $\mathcal{G}^-$ happen for some Y. By Assumption 1, we have $\mathbf{P}(\mathcal{G}^- | X \in E_{\bar{c}}) = \frac{c_{\hat{i}\hat{\alpha}} \theta_{\bar{a} \to \bar{b}} a_{i\alpha}}{k_{\hat{i}\hat{\alpha}}(t)}$. Therefore,

$$\mathbf{P}(\mathcal{G}^-) = \sum_{\substack{\bar{c} \\ c_0 - > c_{0+}}} P(X \in E_{\bar{c}}) \frac{c_{\hat{i}\hat{\alpha}} \theta_{\bar{a} \to \bar{b}} a_{i\alpha}}{k_{\hat{i}\hat{\alpha}}(t)}$$

$$= \mathbf{E}(q_{\hat{i}\hat{\alpha}}(X) | q_{0-}(X) > q_{0+}(X)) \frac{\theta_{\bar{a} \to \bar{b}} a_{i\alpha}}{k_{\hat{i}\hat{\alpha}}(t)}$$

Similarly, the probability of an event $\mathcal{G}^+$:"variable Y moves from $E_{\bar{b}}$ to $E_{\bar{a}}$", where $\bar{b}$ is some tuple adjacent to $\bar{a}$ and $\bar{a}, \bar{b}$ differ in components $i\alpha$ and $j\alpha$, equals

$$\mathbf{P}(\mathcal{G}^+) = \mathbf{E}(q_{\hat{j}\hat{\alpha}}(X) | q_{0-}(X) > q_{0+}(X)) \frac{\theta_{\bar{b} \to \bar{a}} b_{j\alpha}}{k_{\hat{j}\hat{\alpha}}(t)}$$

Observing that the expectations of the numbers of variables that move to and from $E_{\bar{a}}$ (excluding X) equal

$$e_{\bar{b}} \mathbf{P}(\mathcal{G}^+) \qquad \text{and} \qquad e_{\bar{a}} \mathbf{P}(\mathcal{G}^-),$$

respectively, we get

$$\mathbf{E}(e_{\bar{a}}(t+1) - e_{\bar{a}}(t) \mid H_t) \qquad\qquad\qquad (3)$$
$$= \mathbf{P}(X \in E_{\bar{a}'}(t) \mid q_{0-}(X) > q_{0+}(X)) - \mathbf{P}(X \in E_{\bar{a}}(t) \mid q_{0-}(X) > q_{0+}(X))$$
$$+ \sum_{\substack{\bar{b} \text{ adjacent to } \bar{a} \\ i,j,\alpha}} \left(\frac{\theta_{\bar{b} \to \bar{a}} b_{j\alpha} e_{\bar{b}}(t)}{k_{\hat{j}\hat{\alpha}}(t)} - \frac{\theta_{\bar{a} \to \bar{b}} a_{j\alpha} e_{\bar{a}}(t)}{k_{\hat{i}\hat{\alpha}}(t)} \mathbf{E}(q_{\hat{i}\hat{\alpha}}(X) \mid q_{0-}(X) > q_{0+}(X)) \right).$$

Denoting $s = \frac{t}{n}$, $z_{\bar{a}}(s) = e_{\bar{a}}(sn)$ and

$$u(s) = \sum_{\substack{\bar{a} \\ a_0 - > a_{0+}}} z_{\bar{a}}, \qquad g_{i\alpha} = \sum_{\bar{a}} a_{i\alpha} z_{\bar{a}}, \qquad h_{i\alpha} = \sum_{\substack{\bar{a} \\ a_0 - > a_{0+}}} a_{i\alpha} \frac{z_{\bar{a}}}{u}$$

we get

$$\frac{dz_{\bar{a}}}{ds} = \frac{z_{\bar{a}'} - z_{\bar{a}}}{u} + \sum_{\substack{\bar{b} \text{ adjacent to } \bar{a} \\ i,j,\alpha}} \left(\frac{\theta_{\bar{b} \to \bar{a}} b_{j\alpha} z_{\bar{b}}}{g_{\hat{j}\hat{\alpha}}} - \frac{\theta_{\bar{a} \to \bar{b}} a_{j\alpha} z_{\bar{a}}}{g_{\hat{i}\hat{\alpha}}} h_{\hat{i}\hat{\alpha}} \right).$$

(iii) We are interested in the value of g_{0-} when $u(s)$ becomes 0 for the first time. Thus D can be chosen to be any open set with positive elements satisfying the condition $u > \epsilon$ for some $\epsilon > 0$. As before we can find the required value as limit as $\epsilon \to 0$.

Theorem 3. *If Assumption 1 is true then for any positive ϱ there is a constant c such that for a random 3-CNF $\Phi(n, \varrho n)$ almost surely the LS algorithm finds and assingment such that the number of satisfied clauses equals $cn + o(n)$.*

Proof. Applying Wormald's theorem we get that, for any positive ϱ and any M there is a constant c' such that for a random 3-CNF $\Phi(n, \varrho n)$ almost surely the SL algorithm finds and assingment such that the number of satisfied clauses equals $c'n + o(n)$ not containing variables of degree higher than M. We estimate how many clauses may contain a variable (or its negation) of degree higher than M. It is not hard to see that almost surely the number of such clauses is no more than

$$e^{\varrho/2} \cdot \sum_{k > M} k \binom{\varrho n}{k} \left(\frac{1}{2n-1} \right)^k,$$

which is $o(1) \cdot n$ where o means asymptotics as $M \to \infty$.

4.2 Experiments

In this subsection we report on experiments aiming to estimate constant c from Theorem 3 for different values of ϱ. In order to do this we solve numerically the system of differential equations built in the previous subsection. Unfortunately, even for small M this system contains far too many equations. For example, if $M = 15$ then the number of equations exceeds one million. However, while conducting experiments we observed some properties of functions involved that allow us to decrease the number of equations without loss of accuracy. We state these properties later after proper definitions.

To simplify the system of equations we introduce new random variables

$$E_{ab}(t) = \bigcup_{a_{0-} = a, a_{0+} = b}^{\bar{a}} E_{\bar{a}}(t), \quad e_{ab}(t) = \sum_{a_{0-} = a, a_{0+} = b}^{\bar{a}} e_{\bar{a}}(t).$$

It is also clear that $\mathbf{E}(e_{ab}(t+1) - e_{ab}(t)) = \sum_{a_{0-} = a, a_{0+} = b}^{\bar{a}} \mathbf{E}(e_{\bar{a}}(t+1) - e_{\bar{a}}(t))$. Along with $e_{ab}(t)$ we shall use the following random variables: $A(t), B(t), C(t)$, and $D(t)$ that are equal to the number of clauses with 0,1,2, and 3 positive literals, respectively. It is not hard to see that $A(t) = 1/3 \sum_{c,d} c \cdot e_{cd}(t)$, $B(t) = \sum_{c,d} d \cdot e_{cd}(t)$, and $D(t) = \varrho n - (A(t) + B(t) + C(t))$. Thus, as a matter of fact, we need only one extra random variable, $C(t)$. Now we compute the sum in the right side of this equation accordingly to the three parts of the expression (3) for $\mathbf{E}(e_{\bar{a}}(t+1) - e_{\bar{a}}(t))$. The first part

$$\sum_{a_{0-}=a, a_{0+}=b}^{\bar{a}} (\mathbf{P}(X \in E_{\bar{a}'}(t) \mid q_{0-}(X) > q_{0+}(X)) - \mathbf{P}(X \in E_{\bar{a}}(t) \mid q_{0-}(X) > q_{0+}(X)))$$

can be converted into

$$\mathbf{P}(X \in E_{ba} \mid q_{0-}(X) > q_{0+}(X)) - \mathbf{P}(X \in E_{ab} \mid q_{0-}(X) > q_{0+}(X))$$
$$= \begin{cases} \frac{e_{ba}(t)}{G(t)}, & \text{if } a < b, \\ -\frac{e_{ab}(t)}{G(t)} & \text{if } a > b, \end{cases}$$

where $G(t) = \sum_{c>d} e_{cd}(t)$.

It is easier to compute the second and third parts from scratch. Compute first the third part. Function $e_{ab}(t)$ can be decreased if for some variable $Y \in E_{ab}$ either (a) a certain clause of type $0-$ for Y contains $\neg X$, or (b) a certain clause of type $1-$ contains X, or (c) a certain clause of type $0+$ contains $\neg X$, or (d) a certain clause of type $1+$ contains X. The probabilities of these events are

$$\mathbf{P}(\neg X \in C \mid C \text{ of type } 0- \text{ for } Y, q_{0-}(X) = K_1) = \frac{2K_1}{A(t)},$$

$$\mathbf{P}(X \in C \mid C \text{ of type } 1- \text{ for } Y, q_{0+}(X) = K_2) = \frac{2K_2}{B(t)},$$

$$\mathbf{P}(\neg X \in C \mid C \text{ of type } 0+ \text{ for } Y, q_{1-}(X) = K_3) = \frac{2K_3}{B(t)},$$

$$\mathbf{P}(X \in C \mid C \text{ of type } 1+ \text{ for } Y, q_{1+}(X) = K_4) = \frac{2K_4}{C(t)}.$$

By Assumption 1,

$$P_1 = \mathbf{P}(\neg X \in C \mid C \text{ of type } 0- \text{ for } Y) = \sum_{K_1} \mathbf{P}(q_{0-}(X) = K_1)\frac{2K_1}{A(t)} = 2\frac{\mathbf{E}(q_{0-}(X))}{A(t)},$$

$$P_2 = \mathbf{P}(X \in C \mid C \text{ of type } 1- \text{ for } Y) = \sum_{K_2} \mathbf{P}(q_{0+}(X) = K_2)\frac{2K_2}{B(t)} = 2\frac{\mathbf{E}(q_{0+}(X))}{B(t)},$$

$$P_3 = \mathbf{P}(\neg X \in C \mid C \text{ of type } 0+ \text{ for } Y) = \sum_{K_3} \mathbf{P}(q_{1-}(X) = K_3)\frac{2K_3}{B(t)} = 2\frac{\mathbf{E}(q_{1-}(X))}{B(t)},$$

$$P_4 = \mathbf{P}(X \in C \mid C \text{ of type } 1+ \text{ for } Y) = \sum_{K_4} \mathbf{P}(q_{1+}(X) = K_4)\frac{2K_4}{C(t)} = 2\frac{\mathbf{E}(q_{1+}(X))}{C(t)}.$$

The expectations $\mathbf{E}(q_{0-}(X)), \mathbf{E}(q_{0+}(X))$ can be easily found, since

$$\mathbf{P}(q_{0-}(X) = K_1) = \frac{\sum_b e_{K_1 b}(t)}{G(t)}, \quad \mathbf{P}(q_{0+}(X) = K_2) = \frac{\sum_a e_{a K_2}(t)}{G(t)}.$$

The expectations $\mathbf{E}(q_{1-}(X)), \mathbf{E}(q_{1+}(X))$ we find using the following empirical observation.

Observation 1. *For a randomly chosen Y and any i, α, $i \neq 0$, and a, b*

$$\mathbf{E}(q_{i\alpha}(Y) \mid Y \in E_{ab}) \approx \mathbf{E}(q_{i\alpha}(Y)).$$

Thus, easy computation shows that $\mathbf{E}(q_{1-}(X)) = \frac{B(t)}{n}$ and $\mathbf{E}(q_{1+}(X)) = \frac{C(t)}{n}$. Then the expectation for the third part equals

$$e_{ab}(t)(P_1\mathbf{E}(q_{0-}(Y)) + P_2\mathbf{E}(q_{1-}(Y)) + P_3\mathbf{E}(q_{0+}(Y)) + P_4\mathbf{E}(q_{1+}(Y)))$$

$$= 2e_{ab}(t)\left(\frac{a\mathbf{E}(q_{0-}(X))}{A(t)} + \frac{B(t)}{n} + \frac{b}{n} + \frac{C(t)}{n}\right).$$

The second part of the expectation equals

$$2\frac{\mathbf{E}(q_{0+}(X))}{n}e_{(a-1)b} + 2\frac{\mathbf{E}(q_{0-}(X))(a+1)}{A(t)}e_{(a+1)b} + 2\frac{C(t)}{n^2}e_{a(b-1)} + 2\frac{b+1}{n}e_{a(b+1)}.$$

Similarly we have

$$\mathbf{E}(C(t+1) - C(t)) = \mathbf{E}(q_{1-}(X)) + \mathbf{E}(q_{2+}(X)) - \mathbf{E}(q_{1+}(X)) - \mathbf{E}(q_{2-}(X)).$$

Denoting $s = \frac{t}{n}$, $z_{ab}(s) = \frac{e_{ab}(sn)}{n}$, $p(s) = \frac{A(sn)}{n}$, $q(s) = \frac{B(sn)}{n}$, $r(s) = \frac{C(sn)}{n}$, $u(s) = \frac{D(sn)}{n}$, $g(s) = \frac{G(sn)}{n}$, $g_{ab} = z_{ba}$ if $a < b$ and $g_{ab} = -z_{ab}$ if $a > b$, and $h_1(s) = \frac{1}{g}\sum_{\substack{a,b \\ a>b}}az_{ab}$, $h_2(s) = \frac{1}{g}\sum_{\substack{a,b \\ a>b}}bz_{ab}$ we get

$$\frac{dz_{ab}}{ds} = \frac{g_{ab}}{g} + 2\left(\frac{h_2}{g}z_{(a-1)b} + \frac{(a+1)h_1}{g}z_{(a+1)b} + rz_{a(b-1)} + (b+1)z_{a(b-1)}\right)$$

$$- 2z_{ab}\left(\frac{ah_1}{p} + q + b + r\right),$$

$$\frac{dr}{ds} = 2q + u - 3r. \tag{4}$$

As the graphs in Fig. 4 show, these equations give a very good approximation for empirical results. The graphs show the evolution of $p(s)$ that is the relative

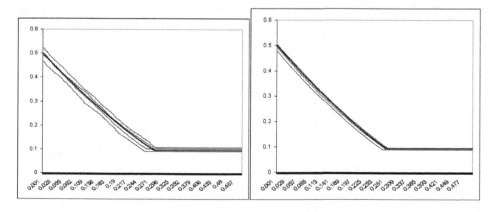

Fig. 4. Empirical performance of LS and its prediction

number of unsatisfied clauses. Thin lines are values observed when running LS for particular random problems, and the thick lines are computed from a numerical solution of the system above. In the examples shown $\varrho = 4$, $M = 30$, $n = 1000$ for the graph on the left and $n = 10000$ for the graph on the right.

The following table shows the dependance between ϱ and the constant c from Theorem 3 both empirical and predicted by the system (4). Experimental figures are average on 10 formulas with 1000 variables each.

ϱ	2	3	4	4.5	5	6	7	10	15	20	25
c (experiment)	1.98	2.95	3.91	4.39	4.86	5.80	6.74	9.53	14.11	18.69	23.23
c (system (4))	1.98	2.95	3.91	4.38	4.85	5.80	6.73	9.52	14.14	18.74	23.32

References

1. D. Achlioptas. Lower bounds for random 3-SAT via differential equations. *Theor. Comput. Sci.*, 265(1-2):159–185, 2001.
2. T. Asano and D. Williamson. Improved approximation algorithms for MAX SAT. *J. Algorithms*, 42(1):173–202, 2002.
3. M. Goemans and D. Williamson. New 3/4-approximation algorithms for the maximum satisfiability problem. *SIAM J. Discrete Math.*, 7(4):656–666, 1994.
4. J. Gu. Efficient local search for very large-scale satisfiability problem. *ACM SIGART Bulletin*, 3(1):8–12, 1992.
5. P. Hansen and B. Jaumard. Algorithms for the maximum satisfiability problem. *Computing*, 44:279–303, 1990.
6. H. Karloff and U. Zwick. A 7/8-approximation algorithm for MAX 3SAT? In *FOCS*, pages 406–415, 1997.
7. E. Koutsoupias and C. Papadimitriou. On the greedy algorithm for satisfiability. *Inf. Process. Lett.*, 43(1):53–55, 1992.
8. B. Selman, H. Levesque, and D. Mitchell. A new method for solving hard satisfiability problems. In *AAAI*, pages 440–446, 1992.
9. N. Wormald. Differential equations for random processes and random graphs. *The Annals of Applied Probability*, 5(4):1217–1235, 1995.

Implementing Survey Propagation on Graphics Processing Units

Panagiotis Manolios and Yimin Zhang

College of Computing
Georgia Institute of Technology
Atlanta, GA, 30318
{manolios, ymzhang}@cc.gatech.edu

Abstract. We show how to exploit the raw power of current graphics processing units (GPUs) to obtain implementations of SAT solving algorithms that surpass the performance of CPU-based algorithms. We have developed a GPU-based version of the survey propagation algorithm, an incomplete method capable of solving hard instances of random k-CNF problems close to the critical threshold with millions of propositional variables. Our experimental results show that our GPU-based algorithm attains about a nine-fold improvement over the fastest known CPU-based algorithms running on high-end processors.

1 Introduction

The Boolean satisfiability problem (SAT) has been intensely studied both from a theoretical and a practical point of view for about half a century. The interest in SAT arises, in part, from its wide applicability in domains ranging from hardware and software verification to AI planning. In the last decade several highly successful methods and algorithms have been developed that have yielded surprisingly effective SAT solvers such as Chaff [13], Siege [16], and BerkMin [8]. A major reason for the performance results of recent SAT solvers is that the algorithms and data structures used have been carefully designed to take full advantage of the underlying CPUs and their architecture, including the memory hierarchy and especially the caches [17].

We propose using graphics processing units (GPUs), to tackle the SAT problem. Our motivation stems from the observation that modern GPUs have peak performance numbers that are more than an order of magnitude larger than current CPUs. In addition, these chips are inexpensive commodity items, with the latest generation video cards costing around $500. Therefore, there is great potential for developing a new class of highly efficient GPU-based SAT algorithms. The challenge in doing this is that GPUs are specialized, domain-specific processors that are difficult to program and that were not designed for general-purpose computation.

In this paper, we show how the raw power of GPUs can be harnessed to obtain implementations of survey propagation and related algorithms that exhibit almost an order of magnitude increase over the performance of CPU-based algorithms. We believe that we are the first to develop a competitive SAT algorithm based on GPUs and the first to show that GPU-based algorithms can eclipse the performance of state-of-the-art CPU-based algorithms.

A. Biere and C.P. Gomes (Eds.): SAT 2006, LNCS 4121, pp. 311–324, 2006.
© Springer-Verlag Berlin Heidelberg 2006

Table 1. Performance comparison of NVIDIA's GTX7800 and Intel's Pentium Dual Core EE 840 processor

	Pentium EE 840 3.2GHz Dual Core	GeForce GTX7800
FLOPs	25.6 GFLOPs	313 GFLOPs
Memory bandwidth	19.2 GB/sec	54.4 GB/sec
Transistors	230M	302M
Process	90nm	110nm
Clock	3.2GHz	430Mz

The SAT algorithm we consider is survey propagation (SP), a recent algorithm for solving randomly generated k-CNF formulas that can handle hard instances that are too large for any previous method to handle [2,1]. By "hard" we mean instances whose ratio of clauses to variables is below the critical threshold separating SAT instances from UNSAT instances, but is close enough to the threshold for there to be a preponderance of metastable states. These metastable states make it difficult to find satisfying assignments, and it has been shown in previous work that instances clustered around the critical threshold are hard random k-SAT problems [12,3,5].

The rest of the paper is organized as follows. In Section 2, we describe GPUs, including their performance, architecture, and how they are programmed. In section 3, we provide an overview of the survey propagation algorithm. Our GPU-based parallel version of survey propagation is described in Section 4 and is evaluated in Section 5. We discuss issues arising in the development of GPU-based algorithms in Section 6 and conclude in Section 7.

2 Graphical Processing Units

A GPU is a specialized processor that is designed to render complex 3D scenes. GPUs are optimized to perform the kinds of operations needed to support real-time realistic animation, shading, and rendering. The performance of GPUs has grown at a remarkable pace during the last decade. This growth is fueled by the video game industry, a multi-billion dollar per year industry whose revenues exceed the total box office revenues of the movie industry.

The raw power of GPUs currently far exceeds the power of CPUs. For example, Table 1 compares a current Intel Pentium CPU and a current GPU, namely NVIDIA's GTX7800, in terms of floating point operations per second (FLOPs) and memory bandwidth. It is worth noting that the 313 GFLOPs number for the GTX 7800 corresponds to peak GFLOPs available in the GPU's shader, the part of the GPU that is programmable. The total GFLOPs of the GTX 7800 is about 1,300 GFLOPs.

In addition, the performance of GPUs is growing at a faster rate than CPU performance. Whereas CPU speed has been doubling every eighteen months, GPU performance has been doubling about every six months during the last decade, and estimates are that this trend will continue during the next five years. The rapid improvements

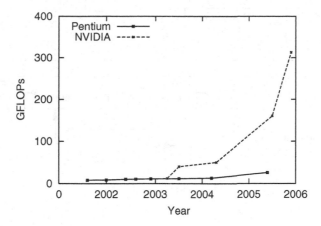

Fig. 1. Comparison of GPUs and CPUs with respect to peak floating point performance. The last two CPU numbers are for dual-core machines. The figure is partly based on data from Ian Buck, at Stanford University.

in GPU performance can been seen in Figure 1, which compares the peak floating point performance of GPUs with CPUs over the course of about five years.

The reason why GPUs have such high peak FLOPs and memory bandwidth is that they are architecturally quite different from CPUs; in fact, they should be thought of as parallel stream processors providing both MIMD (Multiple Instruction Multiple Data) and SIMD (Single Instruction Multiple Data) pipelines. For example, NVIDIA's GTX7800 has eight MIMD vertex processes and twenty four SIMD pixel processors. Each of the processors provides vector operations and is capable of executing four arithmetic operations concurrently.

There is increasing interest in using GPUs for general-purpose computation and many successful applications in domains that are not related to graphics have emerged. Examples include matrix computations [7], linear algebra [11], sorting [10], Bioinformatics [15], simulation [6], and so on. In fact, General Purpose computation on GPUs (GPGPU) is emerging as a new research field [9]. Owens et. al. have written a survey paper of the field that provides an overview of GPUs and a comprehensive survey of the general-purpose applications [14].

It is worth pointing out that there are significant challenges in harnessing the power of GPUs for applications. Since GPUs are targeted and optimized for video game development, the programming model is non-standard and requires an expert in computer graphics to understand and make effective use of these chips. For example, GPUs only support 32-bit floating point arithmetic (often not IEEE compliant). They do not provide support for any of the following: 64-bit floating point arithmetic, integers, shifting, and bitwise logical operations. The underlying architectures are largely secret and are rapidly changing. Therefore, developers do not have direct access to the hardware, but must instead use special-purpose languages and libraries to program GPUs. It takes a while for one to learn the many tricks, undocumented features (and bugs), perils, methods of debugging, etc. that are needed for effective GPU programming.

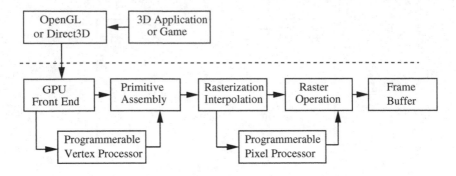

Fig. 2. Graphics pipeline

2.1 GPU Pipeline

Nearly all current GPUs use a graphics pipeline consisting of several stages, as outlined in Figure 2. The graphics pipeline takes advantage of the inherent parallelism found in the kinds of computations used in 3D graphics.

An application, say a game, communicates with a GPU through a 3D API such as OpenGL or Direct3D. The application uses the API to send commands and the vertex data modeling of the object to be rendered to the GPU. Objects are modeled by a sequence of triangles, each of which is a 3-tuple of vertices. After receiving this stream of vertices, the GPU processes each vertex using a vertex shader, an application-specific program that runs on the vertex processor and which computes vertex-specific values, such as position, color, normal vector, etc. These transformed vertices are then assembled into triangles (by the primitive assembly stage) and the vertex information is then used to perform interpolation and rasterization, producing a 2D raster image. Next, a pixel shader, an application-specific program running on the pixel processor is used to compute the color values of pixels. The pixel shader can access information stored on *textures*, memories organized as cubes. (In GPGPU applications, texture memory is used in lieu of main memory.) Finally, a color vector containing the four values R(red), G(green), B(blue), and A(alpha) is output to the frame buffer and displayed on the screen.

Notice that there are two kinds of programmable processors in the graphics pipeline, the vertex processor and pixel processor (also called the fragment processor). Both types of processors are capable of vector processing and can read from textures. However, the vertex processors are MIMD processors, whereas the pixel processors are SIMD processors.

One major difference between GPUs and CPUs is that GPUs are capable of "gathering" but not "scattering." Roughly speaking, gathering means being able to read any part of memory, while scattering means being able to write to any part of memory. The memories we are referring to are textures. Both vertex shaders and pixel shaders are capable of gathering, but as can be seen in Figure 2 vertex shaders cannot directly output data. Instead, they output a fixed number of values to the next stage in the pipeline. In contrast, pixel shaders are able to output data, by writing into the frame buffer. The output for each pixel is a 4-word value representing a color vector. Pixel shaders can also write to textures by utilizing OpenGL extensions. A major limiting factor for GPGPU applications is that the amount of information and its location are fixed before the pixel is processed. In graphics applications, this is not much of a limitation because one typically knows where and what to draw first.

For general purpose computation, pixel processors are usually preferable to vertex processors. There are several reasons for this. First, GPUs contain more pixel processors than vertex processors. Second, pixel processors can write to texture memory, whereas vertex processors cannot. Finally, pixel shader texturing is more highly optimized (and thus much faster) than vertex shader texturing.

2.2 OpenGL

Recall that the architectures of GPUs are closely guarded secrets. Therefore, developers do not have direct access to GPUs and instead have to access the chips via a software interface. One popular choice, which is what we use in this paper, is OpenGL (Open Graphics Library), a specification for a low-level, cross-platform, cross-language API for writing 3D computer graphics applications.

Currently, the OpenGL specification is managed by ARB, the OpenGL Architecture Review Board, which includes companies such as NVIDIA, ATI, Intel, HP, Apple, IBM, etc. OpenGL is an industry standard that is independent of the operating system and underlying hardware. Microsoft has its own API, DirectX, which is dedicated to the Window operating system.

OpenGL is very popular for general purpose computing with GPUs, in part due to its ability to quickly extend the specification with extensions in response to GPU hardware developments. These extension enable developers to more fully take advantage of the new functionality appearing in graphics chips. One example of this is the Frame Buffer Object (FBO), an essential component for general purpose computing with GPUs. The FBO allows shader programs to write to a specified texture, instead of writing to the frame buffer. This is quite useful because in the graphics pipeline, no matter what value is written to the frame buffer, it is turned into a value in the interval [0..1], which makes writing non-graphics applications quite difficult. A further benefit of using an FBO is that we can write to a texture and then use this texture as input in the next pass of the rendering process.

2.3 The Cg Programming Language

Cg (C for Graphics) is a high-level language for programming vertex and pixel shaders, developed by NVIDIA. Cg is based on C and has essentially the same syntax. However, Cg contains several features that make it suitable for programming graphics chips. Cg supports most of the operators in C, such as the Boolean operators, the arithmetic operators, etc., but also includes supports for vector data types and operations. For example, it supports float4, a vector of four floating point numbers, and it supports MAD, a vector multiply and add operator. Cg also supports several other graphic-based operations, *e.g.*, it provides functions to access the texture, as shown in the Cg example in Figure 3.

```
float4 main(uniform samplerRECT exampletexture, float4 pos : WPOS) {
        float4 color;
        color = texRECT(exampletexture, pos.xy);
        return color;
}
```

Fig. 3. This is an example of a pixel shader using Cg

```
void draw() {
        cgGLBindProgram(UpdateEta);
        glDrawBuffer(GL_COLOR_ATTACHMENT3_EXT);
        cgGLSetTextureParameter(etavarParam, varTex);
        cgGLEnableTextureParameter(etavarParam);
        glBegin(GL_QUADS);
            glVertex2f(0.0, 0.0);
            glVertex2f(100, 0.0);
            glVertex2f(100, 100);
            glVertex2f(0.0, 100);
        glEnd();
}
```

Fig. 4. This is an OpenGL code snippet

In addition to the features appearing in Cg that do not appear in C, there are also limitations in Cg that do not appear in C. For example, while user-defined functions are supported, recursion is not allowed. Arrays are supported, but array indices must be compile-time constants. Pointers are not supported; however, by using texture memory, which is 2-dimensional, they can be simulated by storing the 16 high-level bits and the 16 low-level bits in the in the x and y coordinates, respectively. Loops are allowed only when the number of loop iterations is fixed. In addition, the `switch`, `continue`, and `break` statements of C are not supported.

Figure 3 gives an example of a Cg program. The main entry of a Cg program can have any name. In the above example, we have a function `main` that is a pixel shader whose output is a float4 representing a 4-channel color vector. Our simple pixel shader takes a single texture, `exampletexture`, and a single float4, `pos`, as input and samples a color value (using the Cg function `texRECT`) from position `pos` in the texture.

In Figure 4, we provide an OpenGL example that simply draws a rectangle of size 100x100 pixels on the screen. It first installs the pixel shader program, `UpdateEta` (from Figure 3); then it chooses the texture to write to, GL_COLOR_ATTACHMENT3_EXT; then it selects the texture to read from, `varTex`; and finally it sends the rendering command (starting at `glBegin(GL_QUADS)`). Executing the rendering command results in running 10,000 pixel shader programs, one per pixel in the 100x100 area. Each pixel shader program outputs a color, as described above. Notice, that to utilize the GPU, we have to "draw" something, and this means that the position of the output has to be fixed.

3   Survey Propagation

Survey Propagation is a relatively new incomplete method based on ideas from statistical physics and spin glass theory in particular [1]. SP is remarkably powerful, able to solve very hard k-CNF problems, *e.g.*, it can solve 3-CNF problems near threshold with over 10^7 propositional variables.

In this section, we provide a brief overview of the algorithm. We start by recalling that a factor graph can be used to represent a SAT problem. It is bipartite graph whose nodes are the propositional variables and clauses appearing in the SAT instance. There is an edge between a variable and a clause iff the variable appears in the clause. If the clause contains a positive occurrence of variable, the edge is drawn with solid line; otherwise, it is drawn with a dotted line. An example is shown in Figure 5.

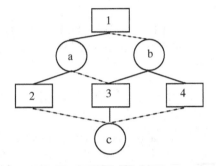

Fig. 5. The factor graph for $(x_1 \vee x_2 \vee \neg x_3) \wedge (\neg x_1 \vee x_3 \vee x_4) \wedge (\neg x_2 \vee x_3 \vee \neg x_4)$

Let C, V represent sets of clauses and variables, respectively. We will use $a, b, c, \ldots$ to denote clauses and $i, j, k, \ldots$ to denote variables. We define $V(a)$ to be the set of variables appearing in clause a and similarly $C(i)$ denotes the set of clauses containing variable i. $C^+(i)$ is the subset of $C(i)$ containing the clauses in which i appears positively; and $C^-(i)$ is the subset of $C(i)$ containing clauses in which i appears negatively (negated).

$$(3.1) \quad \begin{array}{lll} \text{if } a \in C^+(i) & C_a^u(i) = C^-(i); & C_a^s(i) = C^+(i) \setminus \{a\} \\ \text{if } a \in C^-(i) & C_a^u(i) = C^+(i); & C_a^s(i) = C^-(i) \setminus \{a\} \end{array}$$

The SP algorithm is an iterative message-passing algorithm that for every edge $\langle a, i \rangle$ in the factor graph passes messages consisting of a floating point number, $\eta_{a \to i}$, from clause a to variable i and passes messages consisting a 3-tuple of floating point numbers, $\Pi_{i \to a} = \langle \Pi_{i \to a}^u, \Pi_{i \to a}^s, \Pi_{i \to a}^0 \rangle$, from variable i to clause a. This process is initialized by randomly assigning values to $\eta_{a \to i}$ from the interval $(0..1)$ for all edges $\langle a, i \rangle$ in the factor graph. The process is then repeated, where each iteration is called a *cycle*, as described below. As is discussed in [1], the messages can be thought of as corresponding to probabilities of warnings. The value $\eta_{a \to i}$, sent from a to i, corresponds to the probability that clause a sends a warning to variable i, which it will do if it receives a "u" symbol from all of its other variables. In the other direction, the triple

$\Pi_{i \to a} = \langle \Pi^u_{i \to a}, \Pi^s_{i \to a}, \Pi^0_{i \to a} \rangle$ sent from i to a corresponds to the probability that i sends the "u" symbol (indicating that it cannot satisfy a) or sends the "s" symbol (indicating that it can satisfy a) or sends the "0" symbol (indicating that it is indifferent). The formal definitions follow.

$$(3.2) \qquad \eta_{a \to i} = \prod_{j \in V(a) \setminus \{i\}} \left[\frac{\Pi^u_{j \to a}}{\Pi^u_{j \to a} + \Pi^s_{j \to a} + \Pi^0_{j \to a}} \right]$$

$$(3.3) \qquad \prod_{j \to a}^{u} = \left[1 - \prod_{b \in C^u_a(j)} (1 - \eta_{b \to j}) \right] \prod_{b \in C^s_a(j)} (1 - \eta_{b \to j})$$

$$(3.4) \qquad \prod_{j \to a}^{s} = \left[1 - \prod_{b \in C^s_a(j)} (1 - \eta_{b \to j}) \right] \prod_{b \in C^u_a(j)} (1 - \eta_{b \to j})$$

$$(3.5) \qquad \prod_{j \to a}^{0} = \prod_{b \in C(j) \setminus \{a\}} (1 - \eta_{b \to j})$$

The message passing described above is iterated until we attain convergence, which occurs when each of the η values changes less than some predetermined value. Such a sequence of cycles is called a *round*. At the end of a round, we identify a predetermined fraction[1] of variables that the above process has identified as having the largest bias and assign them their preferred values. Having fixed the values of the variables just identified, we perform Boolean Constraint Propagation (BCP) to reduce current SAT problem to simpler one. If the ratio of clauses to variables becomes small, then the problem is under-constrained and we can use Walk-SAT or some other SAT algorithm to quickly find a solution. Otherwise, we again apply the SP algorithm to the reduced SAT problem. Of course, it is possible that either BCP encounters a contradiction or that SP fails to converge, in which case the algorithm fails.[2]

Most of running time of SP is spent trying to converge. Notice that this part of the algorithm requires performing a large number of memory reads and writes and also requires a large number of floating point operations. This is exactly what GPUs excel at doing, which is why we have chosen to develop a GPU-based SP algorithm.

4   Parallel SP on GPU

The basic idea for how to parallelize the SP algorithm is rather straightforward, because the order in which edges in the factor graph are updated does not matter. Therefore, we

---

[1] We use 1 percent, the same percentage used in the publicly available implementation by the authors of survey propagation.

[2] In our implementation, we say that SP fails to converge if it takes more that 1,000 cycles, which is the same parameter used in the code by the authors of survey propagation.

can implement the SP algorithm by running a program per edge in the factor graph, whose sole purpose is to update the messages on that edge. We can then update the messages concurrently. That is the basic idea of the GPU algorithm. In more detail, we ask the GPU to "draw" a quad on the screen where there is one pixel per edge. This allows us to use a pixel shader per edge to compute and update the edge messages, and to store the result in the texture memory. Of course, the CPU and GPU have to communicate after every round of the SP algorithm, so that the GPU can inform the CPU of what variables to fix, so the CPU can perform the BCP pass, and so the CPU can then update the GPU's data structures to reflect the implied literals. The CPU also determines when the clause to variable ratio is such that Walk-SAT should be used.

Given the irregular architecture of GPUs and the difficulty in programming them, one must carefully consider the data structures used and the details of the algorithm, something we now do.

4.1 Data Structures

When using GPUs, the textures are the only place where we can store large amounts of data. Therefore, all of the data used by our algorithm is encoded into textures, which are rectangular areas of memory. One read-only texture is used to represent the factor graph. In it we store, for each clause, pointers to all the literals appearing in that clause (*i.e.*, pointers to all the edges in the factor graph). The pointers for a particular clause are layed out sequentially, which make it easy for us to traverse the edges of a given clause.

We also have three read-write textures, which are used to store information about the variables, clauses, and edges. The variable texture has an entry per variable; similarly the clause and edge textures have entries per clause and edge, respectively.

The main components of the variable texture for entry j include $\prod_{b \in C^+(j)}(1 - \eta_{b \to j})$, $\prod_{b \in C^-(j)}(1 - \eta_{b \to j})$, and a pointer to the edge texture. In the edge texture, the edges with the same variable are layed out sequentially, so the pointer is to the first such edge and we also store the number of edges containing variable j.

The main components of the clause texture for entry a include a pointer into the read-only texture (which points to the first pointer in the read-only texture for that clause) and the value $\prod_{j \in V(a)} \left[\dfrac{\Pi_{j \to a}^u}{\Pi_{j \to a}^u + \Pi_{j \to a}^s + \Pi_{j \to a}^0} \right]$.

The main components of the edge texture for entry $\langle a, i \rangle$ are a pointer to the variable i (in the variable texture), a pointer to the clause a (in the clause texture), and $\eta_{a \to i} = \prod_{j \in V(a) \setminus \{i\}} \left[\dfrac{\Pi_{j \to a}^u}{\Pi_{j \to a}^u + \Pi_{j \to a}^s + \Pi_{j \to a}^0} \right]$.

In current version of OpenGL, the maximum texture size is limited to 256MB, and this is a major restriction because it limits the size of the problems we can consider. We note that the amount of memory available on GPUs is constantly increasing (already GPUs with 1GB memory are available) and that it is possible to use multiple GPUs together. Also, the OpenGL size constraints on textures will eventually be relaxed, but for now, one can distribute the data across multiple textures.

4.2 Algorithm

The algorithm is given as input a factor graph encoded in the textures as described above. Also, the η values are initialized with randomly generated numbers from the interval $(0..1)$. If successful, the algorithm returns a satisfying assignment. As previously described, the algorithm consists of a sequence of rounds, the purpose of which is to converge on the η values. A single round of the algorithm consists of a sequence of cycles, each of which includes four GPU passes, where we assume that we start with the correct η values and show how to compute the η values for the next cycle. Recall that computing on a GPU means we have to draw quads using OpenGL, which in turn means that the computation is being performed by pixel shaders. We omit many of the details and focus on the main ideas below.

1. For each variable j, compute $\prod_{b \in C^+(j)}(1 - \eta_{b \rightarrow j})$ and $\prod_{b \in C^-(j)}(1 - \eta_{b \rightarrow j})$ by iterating over all of the edges that variable appears in. Recall that we have a pointer to the first such edge in the variable texture and that we know what the number of such edges is.

2. For each clause a, compute $\prod_{j \in V(a)} \left[\frac{\Pi^u_{j \rightarrow a}}{\Pi^u_{j \rightarrow a} + \Pi^s_{j \rightarrow a} + \Pi^0_{j \rightarrow a}} \right]$ by iterating over all the edges this clause appears in. Recall that that we have a pointer to the read-only texture and we know the number of such edges. The pointer to the read-only memory points to the first such edge in the edge texture and the next pointer points to the next edge, and so on. By iterating and following the variable pointers in the edge texture, we can compute the above value. This is because we can use the values stored in the variable texture to compute $\Pi^u_{j \rightarrow a}$, $\Pi^s_{j \rightarrow a}$, and $\Pi^0_{j \rightarrow a}$ for each variable j occurring in a.

3. For each edge $\langle a, i \rangle$, compute $\eta_{a \rightarrow i} = \prod_{j \in V(a) \setminus \{i\}} \left[\frac{\Pi^u_{j \rightarrow a}}{\Pi^u_{j \rightarrow a} + \Pi^s_{j \rightarrow a} + \Pi^0_{j \rightarrow a}} \right]$. This can be done by iterating over the elements in the edge texture and using the pointers to the variable and clause of the edge. All that is required is a simple division, given the information already stored in the textures (and after recomputing $\Pi^u_{j \rightarrow a}$, $\Pi^s_{j \rightarrow a}$, and $\Pi^0_{j \rightarrow a}$).

4. Use an occlusion query to test for convergence. If so, this round is over and the GPU and CPU communicate as described previously. Otherwise, goto step 1. An occlusion query is a way for the GPU to determine how many of the pixel shaders have updated the frame buffer. In our case, if the difference between consecutive η values is below a certain threshold, the pixel shader does not update the frame buffer. If the occlusion query returns 0, that means that all of the pixel shaders were killed, *i.e.*, the algorithm has converged.

We note that a GPU's inherent limitations with respect to the support of dynamic data structures can lead to inefficiencies. For example, after BCP, the length of clause may be reduced. Unfortunately, due to the restrictions imposed by Cg, GPU-based programs cannot take advantage of reduced clause sizes and will still have to scan for k literals. Fortunately, if we lay out the literals in a clause in a sequential fashion (which we do), then there is a negligible effect on performance.

5   Experimental Results

We implemented our GPU-based survey propagation algorithm using Cg1.4, C++, and OpenGL2.0. The experiments were run on an AMD 3800+ 2.4GHz machine with an NVIDIA GTX 7900 GPU. The operating system we use is 32-bit WindowsXP. We note that this is a 64-bit machine and we expect to get better performance numbers if we use it for 64-bit computation, but since the NVIDIA GTX 7900 is a rather new GPU, the only drivers we could find were for 32-bit Windows. We also note that using NVIDIA's SLI (Scalable Link Interface) technology, we can use two NVIDIA GPUs, which should essentially double our performance numbers.

The CPU-based survey propagation program we used is from the authors of the survey propagation algorithm [1] and is the fastest implementation of the algorithm we know of. We ran the survey propagation algorithm on the fastest machine we had access to, which is an Intel(R) Xeon(TM) CPU 3.06GHz with 512 KB of cache, running Linux Redhat. (We did not use the same machine we ran the GPU experiments on because the Intel machine is faster.) The experimental data we used is available upon request.

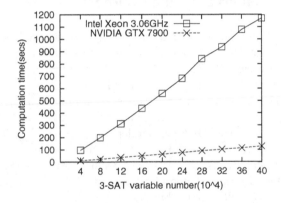

Fig. 6. A comparison between our GPU-based algorithm and the fastest CPU-based algorithm for survey propagation on random 3-SAT instances, with a clause to variable ratio of 4.2.

In Figure 6, we compare the two algorithms on a range of 3-SAT instances, where the clause to variable ration is 4.2; this means that the problems are hard as they are close to the threshold separating satisfiable problems from unsatisfiable problems [12]. The number of variables ranges from 40,000 to 400,000 and each data point corresponds to the average running time for three problems of that size. As is evident in Figure 6, our algorithm is over nine times as fast as the CPU based algorithm.

In Figure 7, we compare the two algorithms on a range of hard 4-SAT instances, where the clause to variable ration is 9.5. The results for the 4-SAT instances are similar to the results we obtained in the 3-SAT case. That is, for hard 4-SAT instances, our GPU-based algorithm attains about a nine-fold improvement in running times over the best known CPU-based algorithm.

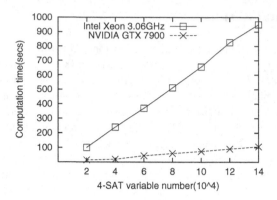

Fig. 7. A comparison between our GPU-based algorithm and the fastest CPU-based algorithm for survey propagation on random 4-SAT instances, with a clause to variable ratio of 9.5

6   Observations on Programming With GPUs

In this section, we outline some observations about programming with GPUs that we think are relevant to the SAT community. The currently available information on GPU programming is mostly geared to the graphics community, and our experience has been that it takes a while for non-specialists to understand. Hopefully, our observations will help to speed up the process for researchers interested in applying GPUs to SAT and similar problems. A good source of information on this topic is the GPGPU Website [9].

When considering using GPUs for general purpose computing, it is important to choose or develop a parallel algorithm. Recall that these processors are best thought of as parallel stream processors and all algorithms that have been successfully implemented on these chips are parallel algorithms. In fact, GPUs are a poor choice for performing reductions, *e.g.*, selecting the biggest element in an integer array turns out to be very difficult to implement efficiently on a GPU.

It is also important to be aware GPUs do not currently support integer operations. You may have noticed that Cg does have integer operations, but these are compiled away and are in fact emulated by floating point operations. Another important difference between CPU and graphics processors is that GPUs do not perform well in the presence of branch instructions, as they do not support branch prediction. Also, reading data from the GPU to the CPU is often a bottleneck. Finally, a major limitation of GPUs is that the per pixel output is restricted to be a four-word vector (extensions allowing sixteen four-word vectors are also currently available), which effectively rules out the use of GPUs for algorithms that do not fit this framework.

Since many optimization algorithms are iterative in nature, they may well be good candidates for implementing on graphics processors. When doing this, we suggest that one carefully encodes the problem into the texture. It is important to do this in a way that attains as much locality as possible because GPUs have very small caches, which means that it is crucial to read memory sequentially, as random access to memory will have a detrimental effect on performance.

It is also often necessary to divide algorithms into several passes. For example, recall that each pixel shader only outputs one four-word vector; if more than four words are needed, then multiple passes have to be used. The general idea is to partition algorithms into several steps, each of which performs a specific function and saves intermediate results to a texture. Subsequent passes can then use the result of the previous passes.

One optimization that is quite useful is to test convergence by using an occlusion query. Without this, one has to use the CPU to test for convergence, which will greatly affect performance. In contrast, an occlusion query gives precise information and can be pipelined, so it has negligible impact on the performance of GPUs.

7 Conclusions and Future Work

In this paper, we have shown how to harness the raw power of GPUs to obtain an implementation of survey propagation, an incomplete method capable of solving hard instances of random k-CNF problems close to the critical threshold. Our algorithm exhibits about an order of magnitude increase over the performance of the fastest CPU-based algorithms. As far as we know, we are the first to develop a competitive SAT algorithm based on graphics processors and the first to show that GPU-based algorithms can eclipse the performance of state-of-the-art CPU-based algorithms running on high-end processors.

We foresee many opportunities to exploit the power of GPUs in the context of SAT solving and verification algorithms in general. Graphics processors are undergoing rapid development and will almost certainly incorporate many new features that make them even more suitable for general purpose computation in a few years. Consider that programmable GPUs were first introduced in 2002 and now they support a rich instruction set and surpass the most powerful currently available CPUs both in terms of memory bandwidth and peak floating point performance.

For future work, we plan to add further improvements to our algorithm and want to explore using GPUs to help speed up complete SAT algorithms such as those based on DPLL [4]. One simple idea is to use GPUs as coprocessors which are used to compute better heuristics that the DPLL algorithm running on the CPU can take advantage of. Another idea we are exploring is the use of other non-standard processors such as the Cell processor.

References

1. A. Braunstein, M. Mezard, and R. Zecchina. Survey propagation: an algorithm for satisfiability. *Random Structures and Algorithms*, 27:201–226, 2005.
2. A. Braunstein and R. Zecchina. Survey and belief propagation on random k-SAT. In *6th International Conference on Theory and Applications of Satisfiability Testing, Santa Margherita Ligure, Italy (2003)*, volume 2919, pages 519–528, 2003.
3. S. A. Cook and D. G. Mitchell. Finding hard instances of the satisfiability problem: A survey. In Du, Gu, and Pardalos, editors, *Satisfiability Problem: Theory and Applications*, volume 35, pages 1–17. American Mathematical Society, 1997.
4. M. Davis, G. Logemann, and D. Loveland. A machine program for theorem proving. *Communications of the ACM*, 5(7):394–397, 1962.

5. O. Dubois, R. Monasson, B. Selman, and R. Zecchina. Statistical mechanics methods and phase transitions in optimization problems. *Theoretical Computer Science*, 265(3–67), 2001.
6. Z. Fan, F. Qiu, A. Kaufman, and S. Yoakum-Stover. GPU cluster for high performance computing. In *SC '04: Proceedings of the 2004 ACM/IEEE conference on Supercomputing*, pages 47–47, 2004.
7. K. Fatahalian, J. Sugerman, and P. Hanrahan. Understanding the efficiency of GPU algorithms for matrix-matrix multiplication. In *HWWS '04: Proceedings of the ACM SIGGRAPH/EUROGRAPHICS conference on Graphics hardware*, pages 133–138, 2004.
8. E. Goldberg and Y. Novikov. BerkMin: A fast and robust SAT-solver. In *Design, Automation, and Test in Europe (DATE '02)*, pages 142–149, Mar 2002.
9. GPGPU. General-Purpose Computation using GPUs, 2006. http://www.gpgpu.org.
10. P. Kipfer and R. Westermann. Improved GPU sorting. In M. Pharr and R. Fernando, editors, *GPU Gems 2: Programming Techniques for High-Performance Graphics and General-Purpose Computation*, pages 733–746. Addison Wesley, Mar 2005.
11. J. Kruger and R. Westermann. Linear algebra operators for GPU implementation of numerical algorithms. *ACM Transactions on Graphics*, 22(3):908–916, 2003.
12. M. Mezard and R. Zecchina. The random k-satisfiability problem: from an analytic solution to an efficient algorithm. *Physical Review E*, 66:056126, 2002.
13. M. W. Moskewicz, C. F. Madigan, Y. Zhao, L. Zhang, and S. Malik. Chaff: Engineering an efficient SAT solver. In *Design Automation Conference (DAC'01)*, pages 530–535, 2001.
14. J. D. Owens, D. Luebke, N. Govindaraju, M. Harris, J. Krger, Aaron, E.Lefohn, and T. J. Purcell. A survey of general-purpose computation on graphics hardware. In *Eurographics 2005, State of the Art Reports*, pages 21–51, 2005.
15. B. R. Payne. *Accelerating Scientific Computation in Bioinformatics by Using Graphics Processing Units as Parallel Vector Processors*. PhD thesis, Georgia State University, Nov. 2004.
16. L. Ryan. Siege homepage. See URL http://www.cs.sfu.ca/ ~loryan/personal.
17. L. Zhang and S. Malik. Cache performance of SAT solvers: A case study for efficient implementation of algorithms. In S. M. Ligure, editor, *Sixth International Conference on Theory and Applications of Satisfiability Testing (SAT2003)*, pages 287–298, 2003.

Characterizing Propagation Methods for Boolean Satisfiability

Eric I. Hsu and Sheila A. McIlraith

University of Toronto, Canada
{eihsu, sheila}@cs.toronto.edu

Abstract. Iterative algorithms such as Belief Propagation and Survey Propagation can handle some of the largest randomly-generated satisfiability problems (SAT) created to this point. But they can make inaccurate estimates or fail to converge on instances whose underlying constraint graphs contain small loops–a particularly strong concern with structured problems. More generally, their behavior is only well-understood in terms of statistical physics on a specific underlying model. Our alternative characterization of propagation algorithms presents them as value and variable ordering heuristics whose operation can be codified in terms of the Expectation Maximization (EM) method. Besides explaining failure to converge in the general case, understanding the equivalence between Propagation and EM yields new versions of such algorithms. When these are applied to SAT, such an understanding even yields a slight modification that guarantees convergence.

1 Introduction

The Survey Propagation (SP) algorithm [1] is one of the most exciting current approaches to the Boolean Satisfiability problem, rapidly solving problems with millions of variables under the most critically constrained settings of the clauses-to-variables ratio. Other successful applications of SP include coding and learning [2,3,4], while the older Belief Propagation (BP) framework that SP extends has been applied to Constraint Satisfaction Problems (CSPs) [5,6,2]. Nonetheless, both SP and BP are subject to some shortcomings that make them best-suited to large, randomly generated SAT instances.

In particular, these propagation algorithms do not always converge, or if they do, can converge to inaccurate estimates that eliminate valid solutions–especially on smaller or structured problems that contain short feedback cycles in their underlying constraint graphs. In such cases SP and BP cannot provide useful information to a surrounding search framework, necessitating a random restart. More generally, their behavior on loopy graphs is not been well-understood outside of a statistical physics interpretation [7,8] that is founded on Markov Random Fields. Within the constraint-based reasoning field, their behavior has been partially explained in terms of discrete inference for the special case of extreme values [9], but no more general understanding has emerged.

The contribution of this paper is to supplement existing mathematical presentations of propagation methods with informal intuition, elucidating connections

A. Biere and C.P. Gomes (Eds.): SAT 2006, LNCS 4121, pp. 325–338, 2006.

to related concepts. The most significant product is an alternative derivation of BP and SP in terms of the Expectation Maximization (EM) framework, one that is not dependent on the groundwork of existing physical explanations. This allows a new version of the BP and SP algorithms that always converges.

These results are demonstrated in BP for clearer exposition, but can be expressed in SP via the transformation shown in [1]. To that end, Section 2 provides background to our approach to characterizing BP and SP, and Section 3 presents necessary notation. In Section 4 we explain the two algorithms based on the background provided in Section 2, supplementing the equations provided in [1]. Finally, in Section 5 we present the EM algorithm in standard form, and then consider a formulation for SAT in Section 6. This produces a convergent update rule for solving SAT instances. The results of implementing this process appear in Section 7, followed by discussion in Section 8.

2   Approach

In this paper, we characterize SP in alternative terms, translating existing descriptions into familiar concepts, founded on insights into its behavior in the context of SAT. Specifically, we contend that SP works best as a variable and value ordering heuristic within a simple search framework. It can be roughly understood as an extension of BP into ternary space, where variables are positive in some fraction of the solutions, negative in another fraction, and recognized as unconstrained in a third proportion of solutions [1]. (The usefulness of such logics has been independently noted in completely different approaches [10].) Efforts in statistical physics to battle loops by clustering nodes together tend to parallel AI techniques for finding join-graphs, cluster-trees, etc. [11,12,13].

Familiar local search techniques find probabilistic variable settings Θ to maximize the probability $P(SAT|\Theta)$ that all clauses are satisfied, just as MAXSAT approximations target an expectation $E[SAT|\Theta]$ on the number of satisfied clauses [14,15,16]. In contrast, the BP and SP propagation algorithms can be viewed as estimators of $P(\Theta|SAT)$, the probability that the variables are configured a certain way given that all clauses are satisfied. Thus, SP can help detect the most prescient "backdoor" variables whose correct assignments trivialize the remaining problem, while BP settles for "backbone variables," which are constrained to be always positive or always negative in the majority of solutions [17,18]. So, propagation methods serve as heuristics that guide the search; if they were always right (and always converged) the search would be backtrack-free.

Such insights enable a final conclusion: that propagation methods actually perform a slightly altered version of the well-known Expectation Maximization (EM) algorithm, which seeks out posterior likelihoods complicated by hidden interactions [19]. This understanding engenders further comparison with Lagrangian optimization approaches to SAT, through known connections to EM [20]. Similarly, it provides for the development of specialized propagation methods based on EM variants for sparse problems and partial optimizations [21,22]. Perhaps the most interesting observation, though, is that EM always converges.

3   Problem and Notation

Definition 1. *A SAT instance is a set C of m clauses, constraining a set V of n Boolean variables. Each clause $c \in C$ is a disjunction of literals built from the variables in V. An assignment $X \in \{0,1\}^n$ to the variables satisfies the instance if it makes at least one literal true in each clause. The sets V_c^+ and V_c^- contain the indexes of the variables appearing positively and negatively in a clause c, respectively. The sets C_v^+ and C_v^- contain the indexes of the clauses that contain positive and negative literals for variable v, respectively.*

Definition 2. *A variable bias $\theta_v \in \mathbb{R}$, $0 \leq \theta_v \leq 1$, represents the probability that variable v will be positive rather than negative. $\Theta \in \mathbb{R}^n$ denotes a vector of biases for all n variables.*

In general, capitalized variables will correspond to vectors, and lower-case variables with subscript indexes will be their components.

4   The BP and SP Algorithms

BP and SP are message-passing algorithms that attempt to sample from the space of satisfying assignments. Here we explain the algorithms at an intuitive level, to supplement the formulas in [1]. The only concrete artifact that is necessary for our purposes is the BP update rule appearing in Equation (1).

Imagine a listing of all solutions to a given SAT problem. Clearly the chance to simply read from that list is wishful thinking for a polynomial algorithm, as this ability would instantly provide a solution if one existed. But what if it were possible to compile statistics over the contents of that list, despite being blind to the list itself? This would be very useful for guiding search, and comprises the goal of propagation methods as applied to SAT. BP and SP attempt this by repeatedly updating estimated biases, hopefully until convergence to a local maximum in likelihood. For BP this means a bias θ_v for each variable, indicating the estimated proportion of solutions in which the variable must be positive rather than negative. SP extends this space with a third state where a variable is not constrained to take either value; with BP the mass for this case ends up proportionately distributed between the positive and negative.

4.1   Sample Space for Determining Bias

In this section, we differentiate the sample space for determining the variable biases that BP and SP are estimating. To borrow terminology from Section 6.2, BP codifies a simplifying assumption that in any satisfying assignment, every variable is the "sole support" of some clause. That is, each variable believes that it is satisfying at least one clause that would otherwise be left totally unsatisfied by all of its other variables.

Figure 1(a) is a Venn diagram depicting the bias space for a given variable, as the shaded region. The area of the diagram as a whole spans the space of

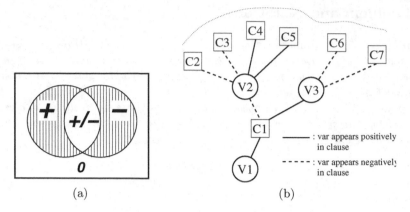

Fig. 1. (a) Probability Space for Variable Bias under BP; (b) Factor Graph Fragment

all satisfying assignments to the variables. The left circle, labeled "+", denotes those assignments where there exists some clause that is wholly dependent on the variable being positive. That is, the considered variable appears positively in some clause whose other literals all hold unsatisfactory values under the current assignment. Likewise, the right circle indicates that some clause requires the variable to be negative. Their intersection, labeled "+/−", is eliminated from the probability space, as no satisfying assignment could require a single variable to be both positive and negative.

Thus BP's goal is to determine, for each variable, the proportion of solutions in which it lies in the positive half of the shaded area, versus the negative half. In comparison, the power of SP lies in additionally considering region "0", where all clauses are already satisfied by variables other than the one under consideration. Stated differently, BP determines the bias of each variable, in terms of the chances that it would appear positively or negatively if a satisfying assignment were randomly drawn from the otherwise inaccessible list of solutions.

4.2   Algorithmic Framework

At a high level, the BP and SP algorithms accomplish the described task by passing messages over a given SAT problem's factor graph representation, as depicted in Figure 1(b).

Nodes representing variables connect to nodes representing clauses in which they appear. Edges can be distinguished, conceptually, by whether the variables appear as positive or negative literals in the clauses. The edges carry clause-to-variable messages in one direction, and variable-to clause messages in the other.

Each variable is randomly seeded with an initial bias, and informs all of its clauses by passing variable-to-clause messages along the edges. The clauses compile such reports and determine whether they are poorly supported–that is, they calculate the probability that their variables will jointly end up failing to satisfy them. From here they signal each variable as to whether they need their support by passing messages back along the edges, in the opposite direction. The variables weigh such requests, and begin a new iteration by updating and reporting their new biases. Equations (13-16) in [1] represent this process. A crucial detail is that a variable tells a clause what its bias would be *in the absence of that clause*. Likewise, clauses do not broadcast identical distress messages along all their outgoing edges. Rather, along each edge they report whether they are unlikely to be satisfied *in the absence of the corresponding variable*. This is what makes the algorithm exactly the same as Pearl's original BP, also known as the Sum-Product algorithm [5,6].

Thus, consider a state for Figure 1(b) where c_6 and c_7 have both informed v_3 that they are unlikely to be supported by their other variables. Consequently, v_3 sees that they need v_3's help, and tells c_1 that in its absence, v_3 would probably have to be negative. Likewise, if v_2 is getting stronger messages from c_4 and c_5 than from c_2 and c_3, then v_2 can report to c_1 that v_2 will also not be of much help. Thus, c_1 will send a strong message to v_1, *whether or not v_1 is already positively biased*. This message could be interpreted as either "I need you to come support me" or "don't listen to your other clauses, I'm highly dependent on you" depending on the strength of v_1's existing bias toward the positive.

The graph and messages are conceptual, though. After much derivation, the entire dynamics can be operationalized as a single update rule. Clause-to-variable messages can be bypassed by expressing variable-to-clause messages in terms of other variable-to-clause messages, two edges removed. (In fact, the original SP derivation and code happen to employ the opposite clause-to-variable representation, shown as Equations (17) and (18) [1,23].) The variables' incoming messages can themselves be represented as changes to the bias, culminating in the update rule for Θ shown below as Equation (1).

Recall that Θ is a vector containing biases for each of the variables, and undergoes this update once for each individual bias θ_v. The entire process is repeated in hopes of eventual convergence. The products expressed in terms of i's and j's represent the probability that a clause c will be left unsatisfied by all of its variables outside of v. Subtracting from 1 creates the negation of this proposition. So the numerator represents the chance that none of the variable's negative clauses require it. In other words, it is the inverse of the proposition that some negative clause requires the variable. Thus the rule can be understood in terms of the diagram in Figure 1(a). Here the numerator represents the inverse of the entire circle labeled "−." Because the "+/−" region is excluded by sampling only satisfying assignments, and the "0" region is excluded by the BP assumption, the inverse yields the left moon of the shaded sample space. Thus, the numerator of the update rule denotes the area of the positive shaded region, while the sum in the denominator constructs the complete sample space.

$$\theta'_v \leftarrow \frac{\prod_{c \in C_v^-} \left[1 - \prod_{i \in V_c^+} (1 - \theta_i) \prod_{j \in V_c^- \backslash v} \theta_j \right]}{\prod_{c \in C_v^-} \left[1 - \prod_{i \in V_c^+} (1 - \theta_i) \prod_{j \in V_c^- \backslash v} \theta_j \right] + \prod_{c \in C+v} \left[1 - \prod_{i \in V_c^+} (1 - \theta_i) \prod_{j \in V_c^- \backslash v} \theta_j \right]} \tag{1}$$

So we represent two states with one parameter; the probability that variable v must be negative is just $1 - \theta_v$. In the case of SP, we use two equations to represent three states. The rule for positive variables is identical to (1), but with an added term in the denominator for including "0" in the outcome space, and extra factors in the numerator for excluding it from the sample space. A second rule will represent a variable's negative bias, flipping the products for $c \in C^+$ for those over $c \in C^-$. Finally, the "0" state where a variable is unbiased is left to marginalization; it is just one minus the first two probabilities.

4.3 Applying BP/SP Via Unit Decimation

Propagation algorithms cannot solve SAT instances on their own. Rather, they can be embedded within a simple search framework that consults them when deciding which variable to fix next. Originally this was tied to a decimation algorithm, where blocks of several variables are fixed all at once. This is risky because the probabilities are not conditional: perhaps v_1 and v_2 both are positive in most satisfying assignments, but often are not positive at the same time.

Conceptually and in practice, one expedient is to consider a more extreme version of this methodology, where only one variable is fixed at a time (at a greater cost in terms of number of surveys). The conditional probabilities are essentially produced by re-running the survey on simplified problems where the previous choices have already been fixed. More concretely, the rest of this paper will consider BP and SP within the following framework:

Algorithm 1. BP/SP with Unit Decimation

BP/SP(SAT-$instance$, $t_{timeout}$)
  repeat
    $survey \leftarrow$ SP(SAT-$instance$, $t_{timeout}$) or BP(SAT-$instance$, $t_{timeout}$).
    $assignment \leftarrow$ CHOOSE-ASSIGNMENT(SAT-$instance$, $survey$).
    SAT-$instance \leftarrow$ FIX(SAT-$instance$, $assignment$)
    If all are clauses satisfied, return solution.
  until all variables fixed
  Return failure.

On each iteration, BP or SP produces a survey estimating the biases of the variables, in either binary or ternary space respectively. If the $t_{timeout}$ parameter is reached before convergence, the entire algorithm fails. With a survey in hand, the algorithm uses CHOOSE-ASSIGNMENT to identify a single variable to fix, and whether to fix it to true or false. One straightforward rule is to choose the variable with the most extreme positive or negative bias, and fix it in that direction. With SP, the extra "0" space allows more choices, such as fixing the variable with the smallest such bias. Next, the SAT instance is simplified to reflect the assignment and the process repeats. This process can incorporate unit-propagation or other such inference processes. The algorithm terminates if enough assignments have been made to satisfy the problem, or if all variables are assigned yet there is still an unsatisfied clause.

Viewed as such, the surveys serve as simultaneous variable and value ordering heuristics. If we only require a single solution, and the heuristics are always correct, then we have a complete reasoning process. For if a survey returns any bias whatsoever for a variable, then some percentage of the solutions features the variable at that value. Thus there is at least one solution remaining. If a variable must not be fixed a certain way, it must have zero bias in that direction.

In practice, the algorithm fixes the most important variables in the first few iterations, and then the maximum bias drops below a certain level. At that point the simplified problem can be passed to a regular solver like WalkSAT for improved speed [14]. Also, completeness is not guaranteed; the algorithms converge to local maxima in terms of survey correctness. Finally, BP and SP may simply fail to converge. For tree networks like the excerpt in Figure 1(b) it is clear that the algorithms converge. But for factor graphs with cycles, it is easy to visualize the algorithms' incompleteness, as feedback loops of messages being passed around and around. Alternatively, such structure can cause the algorithm to converge, but to the wrong answer: randomly initializing biases via a uniform distribution can tilt the optimization process away from endless loops, but only by immediately jumping into local maxima. Without a uniform understanding of the algorithms, such behavior has historically been difficult to characterize.

5 The EM Algorithm

In this section we consider the general EM algorithm [19], so that we can later exploit its transformation into BP and derive an improved way to calculate surveys, one that always converges. At a high level, EM accepts a vector of observations Y, and determines the model parameters Θ that maximize the likelihood of having seen Y. Maximizing $log\,P(Y|\Theta)$ would ordinarily be straightforward, but for the additional complication that we posit some latent variables Z that contributed to the generation of Y, but that we did not get to observe. That is, we want to set Θ to maximize $log\,P(Y,Z|\Theta)$, but cannot marginalize on Z.

So, we bootstrap by constructing $\tilde{P}(Z)$ to estimate $P(Z|Y,\Theta)$ and then use this distribution to maximize the expected likelihood $P(Y,Z|\Theta)$ with respect to Θ. The first step is called the E-Step, and the second is the M-Step. The two are repeated until convergence, which is guaranteed.

6 Transformation from BP to EM Approaches for SAT

Operationally, the BP and the EM algorithm appear to share nothing more than their dualized iterative dynamics. Yet even here there are differences: BP can actually be expressed in terms of just one set of messages (either function-to-variable or variable-to-function) while EM cannot (unless it is used stochastically by updating one variable at a time.) On convergence, neither algorithm can promise more than a local maximum, but EM is guaranteed to converge, while BP is not in the case of graphs with cycles.

6.1 Free-Energy Characterizations of BP and EM

It turns out, though, that the two approaches are actually derived from equivalent energy minimization equations, called "Variational Free Energy" and "Gibbs Free Energy" in the EM and BP literature, respectively. Such expressions arise from trying to minimize the Kullback-Leibler distance between two probability distributions, meaning that the distributions are made to give similar predictions across a common domain of outcomes. The following is a high-level overview of this equivalence, while less germane details can be found in [24].

In the case of BP, we want our belief $b(x)$ to match a factorized approximation of the truth $p(x)$, across all values of x [8]. With EM, there are two distinct steps in getting $\tilde{P}(Z)$, our estimated distribution on Z, to match the true probability $P(Z|Y, \Theta)$. During E, we adjust $\tilde{P}(Z)$, and during M, we adjust Θ [21].

However, the proof of equivalence, based on [8]'s Markov Random Fields representation of factor graphs, is not necessarily constructive. In particular, pushing a straightforward SAT formulation from BP through to EM is liable to produce an inoperable restatement of the problem. Typical interpretations of such formulas produce English phrasings like "bias all variables toward 0 or 1 (by avoiding entropy) while still satisfying all clauses (by avoiding free energy.)" Furthermore, the generic Random Field structure relies on an approximation that breaks any guarantee of convergence. Despite this, a BP-inspired, yet convergent, SAT solution method can be reverse-engineered into the EM framework, by essentially lying to the algorithm.

6.2 SAT Formulation for EM

The trick is to tell EM that we have seen that all the clauses are satisfied, but not how exactly they went about choosing satisfying variables for support. We ask the algorithm to find the variable biases that will best support our claimed sighting, via some hypothesized support configuration. This produces the desired $P(\Theta|SAT)$. In this section explicitly derive this formulation from first principles, resulting in a modification of the update rule (1), reflected in (9).

First we set the EM variables as in Table 1. Y will be a vector of all 1's, meaning that all clauses are satisfied, while each $\theta_v \in \Theta$ represents variable v's probability

Table 1. SAT Formulation for EM

Vector	Status	Interpretation	Domain
Y	Observed	whether clauses are SAT	$\{0, 1\}^m$
Z	Unobserved	support configurations for clauses	$\{s_{c,v}\}^m\ c \in C, v \in V$
Θ	Parameters	variable biases	$(0, 1)^n$

of being positive. Finally, Z contains some value $s_{c,v}$ for each clause c, denoting that "clause c is solely-supported by variable v".

Definition 3. *A variable v is the* sole support *to a clause c when its current assignment satisfies the clause, and none of the clause's other variables satisfy it under the current assignment.*

The Z terms invite elaboration on the space of possible configurations for a given clause. Under a particular variable assignment, a clause is either satisfied by multiple variables, or else by exactly one, or else by none–in which case it is unsatisfied. In order to sample only from the space of satisfying assignments (or more presciently, because $Y = [1]^m$) we eliminate the last case from the probability space. Further, by the simplifying assumption of BP, we eliminate the first case: all constraints are tight in that all supporting variables think that they are the only supports. Reinstating this case yields SP, and a more unreadable derivation. In short, v is the sole support of c when it satisfies c and all of c's other variables do not. For each c, exactly one $s_{c,v}$ must hold–these are the hidden values that EM will weight with its artificial distribution.

6.3   Deriving a SAT Algorithm from EM

For the E-Step we derive said distribution $\tilde{P}(Z)$ for the latent variables, decomposing it into a single $\tilde{p}_c(z_c) = p(z_c|y_c, \Theta)$ for each clause:

$$\tilde{p}_c(s_{c,v}) = \prod_{w \in V_c^+ \setminus v} (1 - \theta_w) \prod_{w \in V_c^- \setminus v} \theta_w \tag{2}$$

The equation states that for $s_{c,v}$ ("v is the sole support of c") to hold, all variables outside of v that were supposed to be positive turned out negative, and vice versa–so v is the sole support. It might seem that v's own support (θ_v if $v \in V_c^+$, $1 - \theta_v$ if $v \in V_c^-$) should appear as a factor above. But it is precisely its exclusion that guarantees that we sample from a space of satisfied clauses, maintaining consistency with the conditioned $y_c = 1$.

In the M-Step we use this distribution to get a lower bound on the logarithm of the expected probability of Y. The log is crucial for maintaining convergence via Jensen's Inequality: $log\,E[p(x)] \geq E[log\,p(x)]$. It also allows us to decompose the set of data (clauses) into terms in a sum. So in short we will set $\Theta \leftarrow argmax_\Theta$ $F(\Theta)$, where:

$$F(\Theta) = E_{\tilde{p}}[log\,P(Y, Z|\Theta)] \tag{3}$$

$$= \sum_c E_{\tilde{p}_c}[log\,p(z_c|\Theta)] \tag{4}$$

This is effected by making logs of products into sums of logs, and by observing that any valid z_c, i.e. one that is given any weight by the corresponding $\tilde{p}_c$, already signifies that its clause is satisfied by definition. In other words, z_c implies y_c, enabling its removal from the joint probability.

By similar reasoning to that in (2) we derive, after some simplification,

$$F(\Theta) = \sum_c \sum_{s_{c,v}} \tilde{p}_c(s_{c,v}) \left[\sum_{i \in V_c^+ \setminus v} log\,(1 - \theta_i) + \sum_{j \in V_c^- \setminus v} log\,\theta_j \right] \tag{5}$$

To optimize, we take the first derivative with respect to each variable v:

$$\frac{dF}{d\Theta_v} = \sum_{c \in C_v} \sum_{\substack{s_{c,w} \\ w \neq v}} \tilde{p}_c(s_{c,w}) \left[\begin{cases} \frac{1}{\theta_v - 1} & \text{if } v \in V_c^+ \\ \frac{1}{\theta_v} & \text{if } v \in V_c^- \end{cases} \right] \tag{6}$$

Because the various support profiles $s_{c,v}$ partition the space of possible configurations for c, we can marginalize out the probability that the clause is solely supported by some variable other than v:

$$\sum_{\substack{s_{c,w} \\ w \neq v}} \tilde{p}_c(s_{c,w}) = 1 - \tilde{p}_c(s_{c,v}) \tag{7}$$

By substituting into (6), and splitting v's clauses into those where it appears positively and negatively, we obtain:

$$\begin{aligned} \frac{dF}{d\Theta_v} &= \alpha \cdot \frac{1}{\theta_v} + \beta \cdot \frac{1}{\theta_v - 1} \\ \text{where} \quad \alpha &= \sum_{c \in C_v^-} (1 - \tilde{p}_c(s_{c,v})) \text{ and} \\ \beta &= \sum_{c \in C_v^+} (1 - \tilde{p}_c(s_{c,v})) \end{aligned} \tag{8}$$

Finally, by setting the derivative to zero, and substituting (2) for $\tilde{p}_c(s_{c,v})$, we arrive at an EM update rule for the θ's:

$$\frac{dF}{d\Theta_v} = 0 \quad \Rightarrow \quad \alpha \cdot (\theta_v - 1) = -\beta \cdot (\theta_v) \quad \Rightarrow \quad \theta_v = \frac{\alpha}{\alpha + \beta} \quad \Rightarrow$$

$$\theta'_v \leftarrow \frac{\displaystyle\sum_{c \in C_v^-} \left[1 - \prod_{i \in V_c^+} (1 - \theta_i) \prod_{j \in V_c^- \setminus v} \theta_j \right]}{\displaystyle\sum_{c \in C_v^-} \left[1 - \prod_{i \in V_c^+} (1 - \theta_i) \prod_{j \in V_c^- \setminus v} \theta_j \right] + \sum_{c \in C+v} \left[1 - \prod_{i \in V_c^+} (1 - \theta_i) \prod_{j \in V_c^- \setminus v} \theta_j \right]} \tag{9}$$

6.4   Comparison with BP

Thus we exhibit a transformation between this EM formulation for SAT, and previous approaches based on BP; the above is almost identical to (1). The sole difference, the replacement of products by sums, is the crux of ensuring convergence. A high-level syntactic understanding is that the logarithms allow us to treat each clause as a separate term in a sum, making the update rule into a (log) odds expression rather than a standard probability. A high-level operational understanding is that when walking toward local maxima, we want to avoid large steps that can overshoot a given peak, resulting in a sort of orbiting nonconvergence. Ratios of sums produce gentler steps than those of products.

In fact, the steps here are bounded in such a way that they are guaranteed never to increase our distance from the nearest local maximum. This is a general property of EM, and a more detailed (and rigorous) explanation can be found in most introductions to the algorithm, or to related variational methods [22]. In essence, the use of Jensen's Inequality in formulating (3) ensures that we have a lower bound on the local maximum, i.e. an admissible heuristic. Further, the fact that the E step fully optimizes the energy equation ensures tightness, meaning that we can only raise this lower bound between alternations of E and M.

7   Implementation and Empirical Results

We have examined such theoretical claims by implementing the EM-derived version of BP ("EMBP") within existing SP code [23]. The code also implements regular BP, allowing comparison of the three approaches within a common infrastructure. (An EM version of SP is also possible, but was not implemented.) As expected, EMBP always proceeded directly to single local maximum in likelihood, and thus always converged. A second question of interest, though, concerns the quality of such maxima. Although the EM formulation always converges, it can still fail to find a solution when one exists. Indeed, even on convergence, all three algorithms arrive at only a *local* maximizer for the log likelihood of $P(\Theta|SAT)$; this peak might not correctly reflect the truth. Thus, even under unit decimation it is still possible to make an incorrect decision that eliminates all remaining solutions. Though they were not implemented in the proof of concept, backtracking or restarts would be necessary at this point.

Figure 2 addresses such issues over both random and structured problems. Across one hundred trials per test suite, the three approaches either solved (satisfiable) instances, failed to converge at some point during execution, or else aborted upon fixing all variables without finding an assignment (either because of inaccurate local maxima, or because an instance was indeed unsatisfiable.) The three graphs represent the relative proportion of these cases on random problems as the clause-to-variable ratio α crosses the phase transition area, while the table represents these percentages over the entirely satisfiable "Inductive Inference," "Logistics," "Parity," and "Quasigroup Completion" test suites of the Satlib benchmark library. In short, EMBP always converges, but it appears to give worse answers than BP and SP on random problems, and better answers on the selected structured problems.

As the random problems become more constrained, the traditional propagation techniques encounter increasing risk of non-convergence, essentially on unsatisfiable instances. Further on, they begin to recapture the ability to converge, and only abort due to incorrect maxima. (The maxima must be incorrect, as these are unsatisfiable instances.) While EMBP always converges, it will begin aborting with an earlier threshold than BP and SP. This is consistent with the hypothesis that by overshooting their targets, traditional propagation methods are able

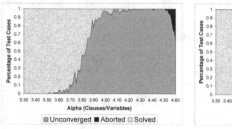

(a) BP: Belief Propagation

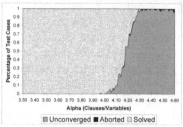

(b) SP: Survey Propagation

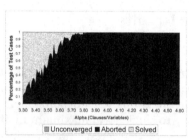

(c) EMBP: EM version of BP

Suite	Method	Unconverged	Aborted	Solved
ii	BP	2.7	73.5	23.8
	SP	0	68.0	32.0
	EMBP	0	60.5	39.5
logistics	BP	0	100	0
	SP	0	100	0
	EMBP	0	100	0
parity	BP	0	99.8	0.2
	SP	0	99.9	0.1
	EMBP	0	97.0	3.0
qg	BP	0	100	0
	SP	0	100	0
	EMBP	0	95.4	4.6

(d) Structured Problems

Fig. 2. Outcomes for (a) BP, (b) SP, and (c) BP-EM Propagation on Random Problems; Percentage Outcomes on (d) Structured Problems

to sample a larger space of local maxima than EM methods, but at the risk of failing to converge. All three approaches remain practical, though, with the use of restarts–so long as there is a non-negligible probability of success, repeated attempts will eventually cure single-mindedness and wanderlust alike. Similarly, with the exception of logistics, the structured results are positive despite relatively low success rates. Recall that the framework is backtrack-free: each run is first randomly initialized and then continues on to success only by making an entirely correct string of decisions for fixing variables. (It is this initialization that makes BP and SP fail by abortion rather than non-convergence.) Still, within the restart framework, EMBP is superior to BP/SP on these structured problems–it is significantly more likely to find a solution for inductive inference and parity problems, and is the only approach with any chance of solving a quasi-group completion problem.

8   Summary and Discussion

The main contribution of this paper is to provide a clearer understanding of the BP and SP algorithms by relating them to the EM algorithm. This exposition provides deeper insight into the differing performance of these algorithms on structured and unstructured problems. It also enables development of variants of these algorithms that were guaranteed to converge. A secondary contribution of this paper is to provide an intuitive but nonstandard explanation of BP and SP by characterizing unit decimation as a variable/value heuristic and relating these algorithms' purpose to finding backbones and backdoors.

We hope that relating BP and SP to EM will allow more tangible gains through the application of related ideas. EM is widely used in statistically-inclined research communities and features many variants suggested by theoretical works like [21]. There are incremental versions that converge more quickly and enable online processing of new clauses. Sparse versions can handle near-zero probabilities symbolically, another expedient that has been used in similar form [23] with propagation algorithms, but not systematically. Other variants alter the artificial distribution $\tilde{P}(Z|Y, \Theta)$, for instance by encouraging it to give more mass to fewer possibilities; the commonly-used K-Means algorithm is an extreme example of this idea. Finally, variational methods can be crudely viewed as developing less exact or less convergent techniques to more efficiently operate on the Markov Random Fields underlying BP's energy equations [22].

Another avenue for future work is to consider $P(SAT|\Theta)$ in lieu of $P(\Theta|SAT)$. While there are no clear semantics for the priors $P(\Theta)$ and $P(SAT)$, the two conditional probabilities are proportional via Bayes' rule. The unit decimation framework suggests an alternate employment of local search for finding solutions to SAT and MAXSAT. Instead of using searches as walks to maxima, each walk can be considered a sample to use as a variable and value ordering heuristic.

References

1. Braunstein, A., Mezard, M., Zecchina, R.: Survey propagation: An algorithm for satisfiability. Random Structures and Algorithms **27** (2005) 201–226
2. Kask, K., Dechter, R., Gogate, V.: Counting-based look-ahead schemes for constraint satisfaction. In: Proc. of 10th International Conference on Constraint Programming (CP '04), Toronto, Canada. (2004)
3. Wang, Y., Zhang, J., Fossorier, M., Yedidia, J.: Reduced latency iterative decoding of LDPC codes. In: IEEE Conference on Global Telecommunications (GLOBECOM). (2005)
4. Braunstein, A., Zecchina, R.: Learning by message passing in networks of discrete synapses. Physics Review Letters **96**(5) (2006)
5. Pearl, J.: Probabilistic Reasoning in Intelligent Systems. Morgan Kaufmann, San Mateo (1988)
6. Kschischang, F.R., Frey, B.J., Loeliger, H.A.: Factor graphs and the sum-product algorithm. IEEE Transactions on Information Theory **47**(2) (2001)   ∙
7. Braunstein, A., Zecchina, R.: Survey propagation as local equilibrium equations. Journal of Statistical Mechanics: Theory and Experiments **PO6007** (2004)
8. Yedidia, J., Freeman, W., Weiss, Y.: Understanding belief propagation and its generalizations. In Nebel, B., Lakemeyer, G., eds.: Exploring Artificial Intelligence in the New Millennium. Morgan Kaufmann (2003) 239–256
9. Dechter, R., Mateescu, R.: A simple insight into properties of iterative belief propagation. In: Proc. of 19th International Conference on Uncertainty in Artificial Intelligence (UAI '03), Acapulco, Mexico. (2003)

10. Lardeux, F., Saubion, F., Hao, J.K.: Three truth values for the SAT and MAX-SAT problems. In: Proc. of the Nineteenth International Joint Conference on Artificial Intelligence (IJCAI '05), Edinburgh, Scotland. (2005)
11. Yedidia, J., Freeman, W., Weiss, Y.: Constructing free-energy approximations and generalized belief propagation algorithms. IEEE Transactions on Information Theory **51**(7) (2005) 2282–2312
12. Dechter, R., Kask, K., Mateescu, R.: Iterative join-graph propagation. In: Proc. of 18th International Conference on Uncertainty in Artificial Intelligence (UAI '02), Edmonton, Canada. (2002) 128–136
13. Kask, K., Dechter, R., Larrosa, J., Pfeffer, A.: Cluster-tree decompostitions for reasoning in graphical models. Artificial Intelligence **166**(1-2) (2005)
14. Selman, B., Kautz, H., Cohen, B.: Local search strategies for satisfiability testing. DIMACS Series in Discrete Mathematics and Theoretical Computer Science **26** (1996)
15. Goemans, M., Williamson, D.: New 3/4-approximation algorithms for the maximum satisfiability problem. SIAM Journal on Discrete Mathematics **7** (1994) 656–666
16. Goemans, M., Williamson, D.: Improved approximation algorithms for maximum cut and satisfiability problems using semidefinite programming. Journal of the ACM (42) (1995) 1115–1145
17. Williams, R., Gomes, C., Selman, B.: Backdoors to typical case complexity. In: Proc. of 18th International Joint Conference on Artificial Intelligence (IJCAI '03), Acapulco, Mexico. (2003)
18. Gomes, C., Selman, B., Crato, N., Kautz, H.: Heavy-tailed phenomena in satisfiability and constraint satisfaction problems. Journal of Automated Reasoning **24**(1-2) (2000) 67–100
19. Dempster, A., Laird, N., Rubin, D.: Maximum likelihood from incomplete data via the EM algorithm. Journal of the Royal Statistical Society **39**(1) (1977) 1–39
20. Shang, Y., Wah, B.: A discrete Lagrangian-based global-search method for solving satisfiability problems. Journal of Global Optimization **12**(1) (1998) 61–99
21. Neal, R., Hinton, G.: A view of the EM algorithm that justifies incremental, sparse, and other variants. In Jordan, M., ed.: Learning in Graphical Models. Kluwer Academic Publishers (1998) 355–368
22. Jordan, M., Ghahramani, Z., Jaakkola, T., Saul, L.: An introduction to variational methods for graphical models. In Jordan, M., ed.: Learning in Graphical Models. MIT Press (1998)
23. Braunstein, A., Leone, M., Mezard, M., Weigt, M., Zecchina, R.: Sp-1.3 survey propagatrion implementation. (http://www.ictp.trieste.it/~zecchina/SP/)
24. Hsu, E., McIlraith, S.: Characterizing loopy belief propagation as expectation maximization (2006) Manuscript in preparation.

Minimal False Quantified Boolean Formulas*

Hans Kleine Büning[1] and Xishun Zhao[2],**

[1] Department of Computer Science,
Universität Paderborn
33095 Paderborn Germany
kbcsl@upb.de
[2] Institute of Logic and Cognition,
Sun Yat-sen University
510275 Guangzhou, P.R. China
hsszxs@mail.sysu.edu.cn

Abstract. This paper is concerned with the minimal falsity problem MF for quantified Boolean formulas. A QCNF formula (i.e., with CNF-matrix) is called minimal false, if the formula is false and any proper subformula is true. It is shown that the minimal falsity problem is PSPACE-complete. Then the deficiency of a QCNF formula is defined as the difference between the number of clauses and the number of existentially quantified variables. For quantified Boolean formulas with deficiency one, MF is solvable in polynomial time.

1 Introduction

A propositional formula in CNF is called minimal unsatisfiable (MU) if the formula is unsatisfiable and any proper subformula is satisfiable. Minimal unsatisfiable formulas have been studied, not only because of their theoretical interests, but also because of their applications for example in formal verification. Now quantified Boolean formulas QBF are gaining their importance since they are suitable for the representation of many problems such as planning, abduction, non-monotonic reasoning, and games. This has motivated much research activity in the QBF area. In this paper we shall introduce and investigate the minimal falsity problem for quantified Boolean formulas with CNF matrices and without free variables. A QCNF formula is said to be minimal false if the formula is false and deleting any clause results in a true formula. Since every false QCNF formula contains a minimal false subformula, it is natural to believe that deep understanding of minimal false formulas might be helpful for designing more efficient QBF solvers.

For propositional CNF formulas the so called deficiency — the difference between the number of clauses and the number of variables — has been used

---

* Some previous results of this of paper has been presented in the Workshop on Theory and Application of Quantified Boolean Formulas [3].
** This research was partially supported by NSFC projects under grant numbers: 60573011, 10410638 and a MOE project under grant number 05JJD72040122.

A. Biere and C.P. Gomes (Eds.): SAT 2006, LNCS 4121, pp. 339–352, 2006.

successfully for the characterization of classes of MU formulas. The problem whether a formula is minimal unsatisfiable (MU) is known to be D^P–complete [11]. D^P is the class of problems which can be described as the difference of two NP-problems. But the problem whether for fixed k a formula with deficiency k is minimal unsatisfiable is solvable in polynomial time [4,8,9]. For MU(1) and MU(2) the class of minimal unsatisfiable formulas with deficiency $1, 2$, natural characterizations can be given [2,5].

The deficiency for quantified Boolean formulas with CNF matrices will be defined as the difference between the number of clauses and the number of existential variables. It will be proved that for an arbitrary minimal false formula Φ, any proper subformula has less deficiency than Φ and the deficiency is always greater than or equal to 1. This gives a chance to reduce the complexity of the minimal falsity problem by restricting the deficiency. Indeed, in section 3, it will be shown that the minimal falsity problem is PSPACE-complete, but if we restrict to the QCNF formulas with deficiency one, the minimal falsity problem becomes solvable in polynomial time (see Section 4). It remains open, whether for fixed deficiency greater than 1 the minimal falsity problem remains solvable in polynomial time.

The algorithms for testing minimal falsity of QCNF formulas with deficiency 1 depends essentially on the structure of MU(1) formulas. Although, formulas HIT-MU (the class of MU formulas in which any two clauses hit each other) fail to have these structural properties, we find that for QEHIT formulas in which any two different clauses contain a pair of complementary existential literals, the minimal falsity depends mainly on the position of the occurrences of universal literals. Hence, we can show that the minimal falsity problem for QEHIT is solvable in polynomial time. However, the minimal falsity problem for QHIT formulas whose matrices are hitting is not solvable in polynomial time unless P equals PSPACE.

2   Notations

A literal is a variable or a negated variable. Clauses are disjunctions of literals. Clauses are also considered as sets of literals. A propositional formula in conjunctive normal form (CNF) is a conjunction of clauses. Usually, CNF formulas are considered as multi-sets. Thus, they may contain multiple occurrences of clauses. A CNF formula φ' is called a subformula of φ if $\varphi' \subseteq \varphi$, i.e., φ' is a subset of φ. The set of all variables occurring in a formula φ is denoted as $var(\varphi)$. For a set X of variables, $\overline{X} := \{\neg x \mid x \in X\}$.

Suppose φ is a propositional formula, v is a partial truth assignment, i.e., $dom(v) \subseteq var(\varphi)$. Then $\varphi[v]$ is the formula obtained from φ by deleting all true clauses and removing all false literals under v. If $dom(v) = \{x\}$ and $v(x) = \epsilon$, then we just write $\varphi[x/\epsilon]$ instead of $\varphi[v]$.

A CNF formula is termed minimal unsatisfiable, if the formula is unsatisfiable and removing an arbitrary clause leads to a satisfiable formula. The set of minimal unsatisfiable formulas is denoted as MU.

Any quantified Boolean formula Φ in QCNF has the form $\Phi = Q_1 x_1 \cdots Q_n x_n \varphi$, where $Q_i \in \{\exists, \forall\}$ and φ is a CNF formula. $Q_1 x_1 \cdots Q_n x_n$ is the prefix of Φ, and φ is called the matrix of Φ. Sometimes we use an abbreviation and write $\Phi = Q\varphi$. If $X = \{x_1, \cdots, x_n\}$, we often write $\exists X$ (resp. $\forall X$) for $\exists x_1 \cdots \exists x_n$ (resp. $\forall x_1 \cdots \forall x_n$).

Let $\Phi = Q_1 x_1 \cdots Q_n x_n \varphi$, $\Phi' = Q_1 x_1 \cdots Q_n x_n \varphi'$ be two QCNF formulas. We say Φ' is a subformula of Φ, denoted as $\Phi' \subseteq \Phi$, if φ' is a subformula of φ, i.e., every clause of φ' is also a clause of φ.

A literal x or $\neg x$ is called an universal resp. existential literal, if the variable x is bound by an universal quantifier resp. existential quantifier. Sometimes universally resp. existentially quantified variables are denoted as $\forall$– resp. $\exists$–variables. A clause without existential resp. universal variables is called an universal resp. existential clause. All the formulas in QCNF are closed, that means any variable is bound by a quantifier.

Suppose we have a formula $Q\varphi \in$ QCNF with $\exists$–variables $x_1, \cdots, x_n$. Then $\varphi_{|\exists}$ is the formula obtained from φ by removing all occurrences of universal literals. Please note, that the propositional formula $\varphi_{|\exists}$ may contain multiple occurrence of clauses. If a clause in φ contains only universal literals then the result is the empty clause $\sqcup$.

3   Minimal Falsity MF and MF(k)

In this section we shall introduce and investigate the minimal falsity of QCNF formulas. The minimal falsity problem will be shown PSPACE-complete. The notion of deficiency will be generalized to QCNF formulas. It will be shown that any minimal false formula has deficiency greater or equal than 1.

Definition 1. *A quantified Boolean formula $\Phi \in QCNF$ is termed* minimal false, *if Φ is false and after removing an arbitrary clause the resulting formula is true. The set of minimal false formulas is denoted as MF.*

Lemma 1. *(Simplification)   Suppose $\Phi = \forall y Q\varphi$ is in MF . Then the variable y occurs either only positively or only negatively in Φ and $Q\varphi[y/\epsilon] \in MF$, where $\epsilon = 0$, if y occurs positively and $\epsilon = 1$, otherwise.*

Proof. Because of the falsity of Φ, the formula is false w.l.o.g. for $y = 0$. In that case no clause with occurrence of $\neg y$ is required for the falsity. Since Φ is in MF, $\neg y$ does not occur in any clause.   □

For any subset $X \subseteq$ QCNF, we can regard X as the decision problem of determining whether an arbitrary given QCNF formula belongs to X.

Theorem 1. *The minimal falsity problem MF is PSPACE-complete.*

Proof. Obviously, the minimal falsity problem is in PSPACE. Let QCNF$_{3\exists}$ be the class of formulas in QCNF for which every clause contains three existential literals besides the universal literals. It is well-known that the evaluation problem for

QCNF$_{3\exists}$ remains PSPACE–complete [6]. The PSPACE-hardness of the minimal falsity problem will be shown by a polynomial reduction $\omega : \text{QCNF}_{3\exists} \to \text{QCNF}$, for which Φ is true if and only if $\omega(\Phi)$ is minimal false.

Let $\Phi = Q(\alpha_1 \wedge \cdots \wedge \alpha_m)$ be a QCNF$_{3\exists}$ formula with clauses $\alpha_i = u_i \vee L_{i,1} \vee L_{i,2} \vee L_{i,3}$, here u_i is the clause consisting of all universal literals in α_i. For new variables $\{x_1, x_2, \cdots, x_m\}$ let π_i $(1 \leq i \leq m)$ be the clause

$$x_1 \vee \cdots \vee x_{i-1} \vee x_{i+1} \vee \cdots \vee x_m.$$

Let $\omega(\Phi)$ be the following formula

$$\exists x_1 \cdots \exists x_m Q$$
$$\begin{pmatrix} (\alpha_1 \vee \pi_1) \wedge (\alpha_2 \vee \pi_2) \wedge \cdots \wedge (\alpha_m \vee \pi_m) & \wedge \\ \bigwedge_{1 \leq i \leq m}((\neg L_{i,1} \vee \pi_i \vee \neg x_i) \wedge (\neg L_{i,2} \vee \pi_i \vee \neg x_i) \wedge (\neg L_{i,3} \vee \pi_i \vee \neg x_i)) \wedge \\ \bigwedge_{1 \leq i < j \leq m}(\neg x_i \vee \neg x_j) & \wedge \\ (x_1 \vee x_2 \vee \cdots \vee x_m) \end{pmatrix}$$

and $\lambda(\Phi) = \omega(\Phi) - \{(x_1 \vee \cdots \vee x_m)\}$. At first we show

(1) If Φ is true then $\omega(\Phi)$ is minimal false.

Suppose Φ is true. We then have to prove (a) $\omega(\Phi)$ is false and (b) every proper subformula is true.

(a) Suppose on the contrary $\omega(\Phi)$ is true. Then there is a truth assignment v defined on $x_1, \cdots, x_m$ such that $\omega(\Phi)[v]$ is true. Because of the clause $(x_1 \vee \cdots \vee x_m)$ and the clauses $(\neg x_i \vee \neg x_j)$, $1 \leq i < j \leq m$, there is exactly one x_i with $v(x_i) = 1$. Then we have

$$\omega(\Phi)[v] = Q(\alpha_i \wedge \neg L_{i,1} \wedge \neg L_{i,2} \wedge \neg L_{i,3}),$$

The formula is false in contradiction to our assumption. Hence, $\omega(\Phi)$ is false.

(b) Let f be an arbitrary clause in $\omega(\Phi)$. We shall prove that $\omega(\Phi) - \{f\}$ is true by a case distinction on the clause f.

Case 1. $f = (x_1 \vee \cdots \vee x_m)$. For the truth assignment v which assigns 0 to each x_i, $(\omega(\Phi) - \{f\})[v] = \Phi$. Thus, $\omega(\Phi) - \{f\}$ is true.

Case 2. $f = (\neg x_i \vee \neg x_j)$ for some $i < j$. Let v be the truth assignment which sets $x_i = x_j = 1$ and other $x_k = 0$. Then v satisfies the matrix of $\omega(\Phi) - \{f\}$.

Case 3. $f = (\alpha_i \vee \pi_i)$ for some $i = 1, \cdots, m$. For the assignment v which sets $x_i = 1$ and other $x_k = 0$, we have $(\omega(\Phi) - \{f\})[v] = Q(\neg L_{i,1} \wedge \neg L_{i,2} \wedge \neg L_{i,3})$, which is true.

Case 4. $f = (\neg L_{i,j} \vee \pi_i \vee \neg x_i)$ for some $i = 1, \cdots, m$, $j = 1, 2, 3$. W.l.o.g we assume $j = 1$. For the assignment v which sets $x_i = 1$ and other variables $x_k = 0$, we have $(\omega(\Phi) - \{f\})[v] = Q(\alpha_i \wedge \neg L_{i,2} \wedge \neg L_{i,3})$. The formula is true for example for $L_{i,1} = 1, L_{i,2} = L_{i,3} = 0$.

It remains to prove: (2) If $\omega(\Phi)$ is minimal false then Φ is true.

Suppose $\omega(\Phi)$ is minimal false. Then $\lambda(\Phi)$ is true, because the formula is a proper subformula. Let v be a truth assignment defined on $x_1, \cdots, x_m$ such that $\lambda(\Phi)[v]$ is true. Because of the clauses $(\neg x_i \vee \neg x_j)$, $1 \leq i < j \leq m$, we have two cases.

Case 1. $v(x_i) = 0$ for all $i = 1, \cdots, m$. Then Φ is true, because of $\lambda(\Phi)[v] = \Phi$.
Case 2. There exists exactly one x_i with $v(x_i) = 1$. The assignment v leads to $\lambda(\Phi)[v] = Q(\alpha_i \wedge \neg L_{i,1} \wedge \neg L_{i,2} \wedge \neg L_{i,3})$, but this formula is false. Hence, case 2 cannot occur. Altogether, Φ must be true. $\qquad \square$

For propositional formulas in CNF, the deficiency is defined as the difference between the number of clauses and the number of variables. It has been shown that $MU(k)$, the set of minimal unsatisfiable formulas with deficiency k, is solvable in polynomial time [4] and any minimal unsatisfiable formula has deficiency greater than 0. For the definition of the deficiency for QCNF only the number of existential variables will be taken into account.

Definition 2. *For a formula $\Phi = Q\varphi \in QCNF$ with t clauses, the* deficiency *is defined as $d(\Phi) := t - |var(\varphi_{|\exists})|$. The* maximum deficiency *of Φ is defined as $d^*(\Phi) := \max\{d(\Phi') \mid \Phi' \subseteq \Phi\} = \max\{d(\varphi'_{|\exists}) \mid \varphi' \subseteq \varphi\} := d^*(\varphi_{|\exists})$. For fixed k, MF(k) is the set of MF formulas with deficiency k.*

For a formula Φ in MF the existential part $\varphi_{|\exists}$ of the matrix φ is unsatisfiable. But $\varphi_{|\exists}$ is not necessarily minimal unsatisfiable.

Example 1. The formula $\Phi = \exists x \forall z \exists a \exists b \exists c \ \varphi$ with
$\varphi = (z \vee a \vee b) \wedge (x \vee z \vee \neg a) \wedge (\neg x \vee \neg z \vee \neg a) \wedge (\neg z \vee a \vee c) \wedge (\neg b) \wedge (\neg c)$ is in MF, but the existential part $(a \vee b) \wedge (x \vee \neg a) \wedge (\neg x \vee \neg a) \wedge (a \vee c) \wedge (\neg b) \wedge (\neg c)$ of the matrix is not minimal unsatisfiable.

Theorem 2. *Let Φ be a formula in QCNF without universal clauses.*

1. If Φ is false then $d^*(\Phi) \geq 1$.
2. If $\Phi \in MF$ then $d(\Phi) \geq 1$ and $d(\Phi') < d(\Phi)$ for any proper subformula Φ'.

Proof. (1) Suppose, Φ is false and has the form $Q\varphi$. Then the existential part of the matrix $\varphi_{|\exists}$ is false and contains a minimal unsatisfiable subformula. Since any minimal unsatisfiable formula has deficiency greater than or equal to 1 [2], there is a subformula $\varphi' \subseteq \varphi$ with $d(Q\varphi') \geq 1$ and therefore $d^*(\phi) \geq 1$.

(2) Suppose $\Phi := Q\varphi$ is in MF. If $d(\Phi') < d(\Phi)$ for any proper subformula then $d^*(\Phi) = d(\Phi)$ and $d(\Phi) \geq 1$, because of (1). Now we suppose: There exists a proper subformula $\Phi' \subset \Phi$ with $d(\Phi') \geq d(\Phi)$. We choose a proper subformula with $d(\Phi') = d^*(\Phi)$. Let $\Phi' := Q\varphi'$, and $\Phi'' := Q(\varphi - \varphi')$. Moreover, let $\Phi''_{-\exists var(\varphi')}$ be the formula obtained from Φ'' by removing all positive and negative occurrences of existential variables occurring in φ'. We claim that $\Phi''_{-\exists var(\varphi')}$ is true. Otherwise, $\Phi''_{-\exists var(\varphi')}$ would have a subformula with deficiency at least 1. Combining this formula with Φ' (and recovering the removed literals) we would have a subformula with deficiency greater than $d^*(\Phi)$, a contradiction.

Φ'' is true, because $\Phi''_{-\exists var(\varphi')}$ is true. Since Φ' and $\Phi''_{-\exists var(\varphi')}$ have distinct existential variables, Φ'' is true independently of Φ'. Consequently, Φ' must be false. That contradicts the assumption that Φ is in MF. $\qquad \square$

We conclude this section by explaining why we do not adopt the alternative definition of the deficiency which takes all variables into account.

Definition 3. *For a formula* $\Phi = Q(\varphi_1 \wedge \cdots \wedge \varphi_t) \in QCNF$ *with* s *variables we define* $d_{all}(\Phi) = t - s$ *and for any fixed integer* k, $MF_{all}(k) := \{\Phi \in MF \mid d_{all}(\Phi) = k\}$.

Lemma 2. *For any fixed integer* k, $MF_{all}(k)$ *is PSPACE-complete.*

Proof. For any fixed integer k, $MF_{all}(k)$ is in PSPACE, because MF is in PSPACE. The PSPACE-hardness will be shown by a poly-time reduction F_k : QCNF $\longrightarrow$ QCNF for which Φ is in MF if and only if $F_k(\Phi) \in MF_{all}(k)$.

For $\Phi = Q(\varphi_1 \wedge \cdots \wedge \varphi_t)$ with s variables we have $d_{all}(\Phi) = t - s$. We shall proceed by a case distinction on $r := t - s$.

Case 1. $r = k$. Then define $F_k(\Phi) := \Phi$.

Case 2. $r > k$. Pick new variables $y, x_1, \cdots, x_{r-k}$ and define $F_k(\Phi)$ to be the following formula

$$Q\forall x_1 \cdots \forall x_{r-k} \exists y \, (\varphi_1 \vee x_1 \vee \cdots \vee x_{r-k} \vee y) \wedge (\varphi_2 \vee \neg x_1 \vee \cdots \vee \neg x_{r-k} \vee y)$$
$$\wedge (\neg y) \wedge (\varphi_3 \wedge \cdots \wedge \varphi_t).$$

Obviously, Φ is in MF if and only if $F_k(\Phi)$ is in MF and $d_{all}(F_k(\Phi)) = t + 1 - (s + (r - k) + 1) = k$.

Case 3. $r < k$. Let $m := k - r$. One can easily construct a MF formula $\Sigma_{m+1} := \exists X(\sigma_1 \wedge \cdots \wedge \sigma_{2(m+1)})$ such that $X \cap var(\Phi) = \emptyset$ and X has $m+1$ variables, here, each σ_i is a clause. Then $d_{all}(\Sigma_{m+1}) = m + 1$. For a new variable y we define

$$F_k(\Phi) := \exists y \exists X Q((\sigma_1 \vee y) \wedge \sigma_2 \wedge \cdots \wedge \sigma_{2(m+1)} \wedge (\varphi_1 \vee \neg y) \wedge \varphi_2 \wedge \cdots \wedge \varphi_t).$$

It is not hard to see that Φ is in MF if and only if $F_k(\Phi)$ is in MF and $d_{all}(F_k(\Phi)) = 2(m + 1) + t - (m + 1 + 1 + s) = k$. $\qquad\square$

4   MF(1) Is Solvable in Polynomial Time

In this section we demand w.l.o.g. that QCNF formulas contain neither tautological nor universal clauses. Moreover, the inner-most and the outermost quantifier are existential quantifiers.

It is known that MU(1), the set of minimal unsatisfiable CNF formulas with deficiency 1, can be solved in polynomial time. There are QCNF formulas $\Phi = Q\varphi$ for which $\varphi_{|\exists}$ is in MU(1), but Φ is not minimal false. Take for example $\Phi = \forall y \exists x (y \vee x) \wedge (\neg y \vee \neg x)$.

Lemma 3. *(MF(1) versus MU(1)) Let* $\Phi = Q\varphi$ *be a formula in QCNF. If* $Q\varphi \in MF(1)$ *then* $\varphi_{|\exists} \in MU(1)$.

Proof. Suppose $\Phi = Q\varphi \in MF(1)$ with n $\exists$–variables. Then $\varphi_{|\exists}$ is unsatisfiable, and $d(\varphi_{|\exists}) = 1$. Using Theorem 2 we see that any proper subformula Φ' of Φ has deficiency $d(\Phi') < 1$. That means any proper subformula of $\varphi_{|\exists}$ has a deficiency less than 1. That implies the satisfiability of the proper subformulas of $\varphi_{|\exists}$, because any unsatisfiable propositional formula contains a minimal unsatisfiable formula with deficiency greater than 0 [2]. Altogether we have shown $\varphi_{|\exists} \in$ MU(1). $\qquad\square$

In our polynomial time algorithm for deciding MF(1) we make use of the following definition.

Definition 4. *Let $\alpha \in MU(1)$, $X \subseteq var(\alpha)$, $f, g \in \alpha$. We say f and g are directly connected without X if $f = g$, or if there is some $L \notin X \cup \overline{X}$ such that $L \in f$ and $\neg L \in g$. We say f and g are connected without X if there are in α clauses $f = f_1, f_2, \cdots, f_n = g$ such that f_i and f_{i+1} are directly connected without X.*

Please notice, that if X is empty, then any two clauses in α are connected without X, because $\alpha \in MU(1)$.

Whether two clauses are connected without X can be decided in polynomial time as follows: For every clause we have a node labeled with the clause, two clauses are connected by an edge if there is $x \notin X$ such that x occurs in one of the clauses and $\neg x$ in the other clause. Then two clauses are connected without X if and only if there is a path between the clauses. That can be decided in polynomial time.

Theorem 3. *Let $\Phi = \exists X_1 \forall Y_1 \cdots \exists X_m \forall Y_m \exists X_{m+1} \varphi$ be a formula with $\varphi_{|\exists}$ in $MU(1)$. Then, Φ is false $\Leftrightarrow$ for all variables $y \in Y_i, 1 \le i \le m$, for all $f, g \in \varphi_{red} : (y \in f, \neg y \in g) \Rightarrow f$ and g are not connected without $X_1 \cup \cdots \cup X_i$ in $\varphi_{|\exists}$. Here, φ_{red} is the result of removing all universal literals from φ except y and $\neg y$.*

By Theorem 3, to see whether a formula $\Phi = \exists X_1 \forall Y_1 \cdots \exists X_m \forall Y_m \exists X_{m+1} \varphi$ is in MF(1), first check that $\varphi_{|\exists} \in MU(1)$. If this condition holds, check for each $y \in Y_i, i = 1, \cdots, m$, and for any two clauses f, g containing y and $\neg y$ respectively, that f, g are not connected without $X_1 \cup \cdots \cup X_i$. Since there at most quadratic many such pairs and connectivity can be tested in polynomial time, MF(1) can be solved in polynomial time.

Example 2. Let $\Phi := \exists x_1 \forall y_1 \forall y_2 \exists x_2 \exists x_3 \exists x_4 \exists x_5 \ \varphi$, where φ is the following formula (the columns are the clauses)

$$\varphi = \left\{ \begin{array}{cccccc} x_2 & y_1 & \neg y_1 & \neg x_3 & y_2 & \neg x_1 \\ x_4 & x_1 & x_3 & \neg x_4 & \neg x_1 & \neg x_5 \\ \neg x_2 & & & x_5 & & \end{array} \right\}$$

Clearly, $\varphi_{|\exists}$ is in MU(1). In φ, the second clause is directly connected without $\{x_1\}$ to the first one which is directly connected without $\{x_1\}$ to the fourth clause, and the fourth clause is directly connected without $\{x_1\}$ to the third one. Thus, Φ is true. In fact for $x_1 = 0, x_2 = y_1, x_3 = y_1$, and $x_4 = \neg y_1$ the formula is true.

Example 3. Let $\Phi := \exists x_1 \exists x_3 \forall y \exists x_2 \exists x_4 \ \varphi$, where φ is the following formula.

$$\varphi = \left\{ \begin{array}{ccccc} x_1 & \neg x_1 & \neg x_3 & x_3 & x_4 \\ \neg x_2 & y & x_2 & \neg y & \\ & \neg x_2 & & & \neg x_4 \end{array} \right\}$$

Clearly, $\varphi_{|\exists}$ is in MU(1). In φ, only the second clause and the fourth clause contain a pair of complementary universal literals, and they are not connected without $\{x_1, x_3\}$. Thus, the formula is false.

The proof of Theorem 3 can be divided into two parts (Lemma 4 and Lemma 5).

Lemma 4. *Let* $\Phi = \exists X_1 \forall Y_1 \cdots \exists X_m \forall Y_m \exists X_{m+1} \varphi$ *be a formula with* $\varphi_{|\exists}$ *in* MU(1). *Then,* Φ *is false* $\Leftrightarrow$ *for all variables* $y \in Y_i, 1 \leq i \leq m$, *the formula* $\exists X_1 \cdots \exists X_i \forall y \exists X_{i+1} \cdots \exists X_{m+1} \varphi_{red}$ *is false. Here,* φ_{red} *is the result of removing all universal literals from* φ *except* y *and* $\neg y$.

Lemma 5. *Let* $\Phi = \exists X \forall y \exists Z \varphi$ *be a formula with* $\varphi_{|\exists}$ *in* MU(1). *Then,* Φ *is false* $\Leftrightarrow$ $\forall f, g \in \varphi : (y \in f, \neg y \in g) \Rightarrow f$ *and* g *are not connected without* X *in* $\varphi_{|\exists}$.

Our remaining task is to prove Lemma 4 and Lemma 5. Let us first recall some properties of MU(1) formulas. For any formula $\varphi \in$ MU(1) and $x \in var(\varphi)$, $\varphi[x/0]$ (resp. $\varphi[x/1]$) contains a unique minimal unsatisfiable subformula which is also in MU(1) [2], denoted as φ_x (resp. $\varphi_{\neg x}$). We call $(\varphi_x, \varphi_{\neg x})$ a splitting of φ over x. In general, the splitting formulas φ_x and $\varphi_{\neg x}$ may have common clauses. However, we have the following nice structural property.

Proposition 1. *([2]) For any formula* $\varphi \in MU(1)$, *there is a variable* $x \in var(\varphi)$ *such that* φ_x *and* $\varphi_{\neg x}$ *have distinct variables, that is,* $var(\varphi_x) \cap var(\varphi_{\neg x}) = \emptyset$. *Hence, they contain no common clause.*

Let φ, x, $\varphi_{\neg x}$, φ_x be as in Proposition 1, then we call $(\varphi_x, \varphi_{\neg x})$ a disjunctive splitting of φ over x.

Suppose φ_L is a splitting formula, $y \in var(\varphi_L)$, then we can split φ_L further and get formulas $(\varphi_L)_K$ with $K \in \{y, \neg y\}$. For simplicity, we write $(\varphi_L)_K$ as φ_{LK}. Generally, we have $\varphi_{L_1 \cdots L_k}$ (which we still call a splitting formula) after several steps of splitting. Please notice, that when performing splitting, we remove the occurrences of splitting literals. During the proof of the polynomial-time solvability of MF(1) we have to recover some of the removed occurrences of literals. Suppose θ is a splitting formula, L a literal, θ^L denotes the formula obtained from θ by recovering the occurrences of L properly. Please notice, that if the original clauses from which θ is obtained (by deleting some splitting literals) does not contain L, then L will not occur in θ^L.

Example 4. Let φ be the following formula:

$$\varphi := \begin{pmatrix} x_1 & \neg x_1 & & & & \\ & & x_2 & \neg x_2 & & & \\ & & & x_3 & \neg x_3 & & \\ & & & y & & \neg y & \neg y \\ & & & & & x_4 & \neg x_4 \\ & & & & & x_5 & x_5 & \neg x_5 \end{pmatrix}$$

We can see

$$\varphi_y = \begin{pmatrix} x_1 & \neg x_1 \\ & x_2 & \neg x_2 \\ & & x_3 & \neg x_3 \end{pmatrix}, \quad \varphi_{yx_2} = (x_1, \neg x_1), \quad \varphi_{y\neg x_2} = (x_3, \neg x_3).$$

Then

$$\varphi_y^y = \begin{pmatrix} x_1 & \neg x_1 \\ & x_2 & \neg x_2 \\ & & x_3 & \neg x_3 \\ & & & y \end{pmatrix}, \quad \varphi_{yx_2}^y = \varphi_{yx_2}, \quad \varphi_{y\neg x_2}^y = \begin{pmatrix} x_3 & \neg x_3 \\ & y \end{pmatrix}.$$

Proposition 2. *Suppose $\varphi \in MU(1)$, $y \in var(\varphi)$ with a disjunctive splitting $(\varphi_y, \varphi_{\neg y})$. Suppose further $x \in \varphi_L$ for some $L \in \{y, \neg y\}$. Then for $K \in \{x, \neg x\}$, we have the following:*

If φ_{LK}^L does not contain L then $\varphi_K = \varphi_{LK}$.
If φ_{LK}^L contains L then $\varphi_K = \varphi_{LK}^L + \varphi_{\neg L}^{\neg L}$, moreover, $\varphi_{LK} = \varphi_{KL}$.

Proof. If φ_{LK}^L does contain L then $\varphi_{LK} = \varphi_{LK}^L$. That means $\varphi_{LK} \subseteq \varphi[K/0]$. Thus $\varphi_K = \varphi_{LK}$. Suppose φ_{LK}^L contains L then $\varphi_{LK}^L + \varphi_{\neg L}^{\neg L}$ is in MU(1), and a subformula of $\varphi[K/0]$, therefore, it must be φ_K. □

In Example 4, we can see that $\varphi_{x_2} = \varphi_{yx_2}$, while $\varphi_{\neg x_2} = \varphi_{y\neg x_2}^y + \varphi_{\neg y}^{\neg y}$.

Lemma 6. *Let φ be a MU(1) formula, $x \in var(\varphi)$. Then $\varphi[L/0] - \varphi_L$ can be satisfied by a partial truth assignment defined on $var(\varphi[L/0] - \varphi_L) - var(\varphi_L)$, for $L \in \{x, \neg x\}$.*

Proof. We prove the lemma by induction on the number of variables in φ. If φ has only one variable, the lemma clearly holds. Suppose φ has more than one variable. If $(\varphi_x, \varphi_{\neg x})$ is a disjunctive splitting, then the assertion follows. So, we assume $(\varphi_x, \varphi_{\neg x})$ is non-disjunctive. By Proposition 1, there is some $y \neq x$ such that $(\varphi_y, \varphi_{\neg y})$ is a disjunctive splitting. W.o.l.g., we assume that $x \in var(\varphi_y)$. There are two possibilities.

Case 1. y occurs in φ_{yx}^y. Then we have $\varphi_x = \varphi_{yx}^y + \varphi_{\neg y}^{\neg y}$. Hence

$$\varphi[x/0] - \varphi_x = (\varphi_y[x/0] - \varphi_{yx})^y.$$

Now the assertion follows from the induction hypothesis.
Case 2. y does not occur in φ_{yx}^y. Then $\varphi_x = \varphi_{yx}$. Thus

$$\varphi[x/0] - \varphi_x = (\varphi_y[x/0] - \varphi_{yx})^y + \varphi_{\neg y}^{\neg y}.$$

By the induction hypothesis, $\varphi_y[x/0] - \varphi_{yx}$ is satisfied by a truth assignment t defined on $var(\varphi_y[x/0] - \varphi_{yx}) - var(\varphi_{yx})$. Please note that $y \notin var(\varphi_x)$ and $var(\varphi_{\neg y}) \cap var(\varphi_x) = \emptyset$. Thus we can extend t to t' such that $t'(y) = 0$ and t' satisfies $\varphi_{\neg y}^{\neg y}$. Hence, the assertion is valid.

By the same argument we can show the assertion for $\varphi[x/1] - \varphi_{\neg x}$. □

Suppose $\Phi = Q\varphi$ is a QCNF formula such that $\varphi_{|\exists} \in MU(1)$. From a splitting $((\varphi_{|\exists})_x, (\varphi_{|\exists})_{\neg x})$ we recover the occurrences of all universal literals, the result is denoted as $(\varphi_x, \varphi_{\neg x})$ which we still call a splitting of φ over x.

Corollary 1. *Let $\Phi = \exists x Q\varphi$ be a QCNF formula with $\varphi_{|\exists} \in MU(1)$. Then $\Phi[x/0]$ (resp. $\Phi[x/1]$) and $Q\varphi_x$ (resp. $Q\varphi_{\neg x}$) have the same truth. Moreover, Φ is false if and only if both $Q\varphi_x$ and $Q\varphi_{\neg x}$ are false.*

Proof of Lemma 4. We first show the direction from left to right. If Φ is false then $\exists X_1 \cdots \exists X_i \forall y \exists X_{i+1} \cdots X_{m+1} \varphi_{red}$ is false, because removing some universal literals preserves the falsity. Here, φ_{red} is the result of removing all universal literals from φ except y and $\neg y$.

For the other direction, we suppose Φ is true. Pick $x \in X_1$, let $X_1' := X_1 - \{x\}$. Then w.o.l.g we assume $\Phi[x/0] = \exists X_1' \forall Y_1 \cdots \exists X_m \forall Y_m \exists X_{m+1} \varphi[x/0]$ is true.

Then by Corollary 1, the formula $\Psi := \exists X_1' \forall Y_1 \cdots \exists X_m \forall Y_m \exists X_{m+1} \varphi_x$ is true.

If X_1' is empty and $m = 1$, then $\Psi = \forall Y_1 \exists X_2 \varphi_x$. Since Ψ is true and the existential part of φ_x is in $MU(1)$, there must be some $y \in Y_1$ occurring positively and negatively. Then $\forall y \exists X_2 (\varphi_x)_{red}$ is true. By Lemma 6, then $\forall y \exists X_2 (\varphi[x/0])_{red}$ is true, hence $\exists X_1 \forall y \exists X_2 \varphi_{red}$ is true.

Suppose X_1' is empty but $m > 1$. Then $\Psi = \forall Y_1 \exists X_2 \forall Y_2 \cdots \exists X_m \forall Y_m \exists X_{m+1} \varphi_x$. Suppose some $y \in Y_1$ occurs both negatively and positively in φ_x, then it is easy to see that $\forall y \exists X_2 \cdots \exists X_{m+1} \varphi_x$ is true. By Lemma 6, $\forall y \exists X_2 \cdots \exists X_{m+1} \varphi[x/0]$ is true. Therefore, $\exists X_1 \forall y \exists X_2 \cdots \exists X_{m+1} \varphi$ is true. Suppose no variable $y \in Y_1$ occurs both negatively and positively, then $\exists X_2 \forall Y_2 \cdots \exists X_m \forall Y_m \exists X_{m+1} (\varphi_x)'$ is also true, here $(\varphi_x)'$ is obtained from φ_x by removing all occurrences of y or $\neg y$ for all $y \in Y_1$. Now Lemma 4 follows from the induction hypothesis and Lemma 6.

If X_1' is non-empty, Lemma 4 follows from the induction hypothesis and Lemma 6. □

Lemma 7. *Let $\Phi = \exists X \forall y \exists Z \varphi$ be in QCNF with $\varphi_{|\exists} \in MU(1)$. Then, Φ is true if and only if there exist $L_1, \cdots, L_s \in X \cup \overline{X}$ such that $var(\varphi_{L_1 \cdots L_s}) \cap X = \emptyset$ and $\forall y \exists Z \varphi_{L_1 \cdots L_s}$ is true.*

Proof. For $X = \{x_1, \cdots, x_n\}$, we proceed by an induction on n. For $n = 1$ the claim follows from Corollary 1. Suppose $n > 1$. Again by Corollary 1, Φ is true if and only if either $\exists x_2 \cdots \exists x_n Q\varphi_{x_1}$ is true or $\exists x_2 \cdots \exists x_n Q\varphi_{\neg x_1}$ is true. Now the lemma follows from the induction hypothesis. □

Before proving Lemma 5 we will show some propositions on splittings.

Proposition 3. *Suppose $\varphi \in MU(1)$, $y \in var(\varphi)$ with $(\varphi_y, \varphi_{\neg y})$ a disjunctive splitting. We consider $\varphi_{LL_1 \cdots L_s}$, here $L \in \{y, \neg y\}$.*

(1) Suppose $\varphi^L_{LL_1 \cdots L_s}$ does not contain L, then $\varphi_{L_1 \cdots L_s} = \varphi_{LL_1 \cdots L_s}$.

(2) Suppose $\varphi^L_{LL_1 \cdots L_s}$ contains L, then $\varphi_{L_1 \cdots L_s} = \varphi^L_{LL_1 \cdots L_s} + \varphi^{\neg L}_{\neg L}$.

(3) Suppose $\varphi^y_{yL_1 \cdots L_s}$ contains y, and $\varphi^{\neg y}_{\neg y L_{s+1} \cdots L_m}$ contains $\neg y$, then
$$\varphi_{L_1 \cdots L_m} = \varphi^y_{yL_1 \cdots L_s} + \varphi^{\neg y}_{\neg y L_{s+1} \cdots L_m}.$$

Proof. (1) This can be seen by induction on s. When $s = 1$ the assertion follows from Proposition 2. Suppose $s > 1$. If $\varphi_{LL_1}^L$ does not contain L then $\varphi_{L_1} = \varphi_{LL_1}$, and the assertion follows. So, we assume $\varphi_{LL_1}^L$ contains L. Then by Proposition 2, $\varphi_{LL_1} = \varphi_{L_1L}$. Thus, $\varphi_{LL_1\cdots L_s} = \varphi_{L_1LL_2\cdots L_s}$. Then by the induction hypothesis, $\varphi_{L_1L_2\cdots L_s} = \varphi_{L_1LL_2\cdots L_s}$, and the assertion follows.

(2) If $s = 1$, the claim follows from Proposition 2. Suppose $s > 1$. From Proposition 2, $\varphi_{LL_1} = \varphi_{L_1L}$, then $\varphi_{LL_1\cdots L_s} = \varphi_{L_1LL_2\cdots L_s}$. Thus by the induction hypothesis, $\varphi_{L_1L_2\cdots L_s} = \varphi_{L_1LL_2\cdots L_s}^L + \varphi_{L_1\neg L}^{\neg L}$. Since $\varphi_{L_1} = \varphi_{LL_1}^L + \varphi_{\neg L}^{\neg L}$, $\varphi_{L_1\neg L} = \varphi_{\neg L}$. The claim follows.

(3) Please notice, that for each $i = 1, \cdots s$, $\varphi_{L_1\cdots L_i}^y$ contains y. Thus from Proposition 2, $\varphi_{yL_1L_2\cdots L_s} = \varphi_{L_1yL_2\cdots L_s} = \cdots = \varphi_{L_1\cdots L_s y}$. Then by (2) we have

$$\varphi_{L_1\cdots L_s} = \varphi_{yL_1\cdots L_s}^y + \varphi_{\neg y}^{\neg y} = \varphi_{L_1\cdots L_s y}^y + \varphi_{\neg y}^{\neg y}.$$

Again by (2) we obtain $\varphi_{L_1\cdots L_m} = \varphi_{yL_1\cdots L_s}^y + \varphi_{\neg yL_{s+1}\cdots L_m}^{\neg y}$. □

Lemma 8. *Let $\varphi \in MU(1)$, $X \subseteq var(\varphi)$, $f, g \in \varphi$. f and g are connected without X if and only if there is a splitting formula $\varphi_{L_1\cdots L_m}$ such that $(f - \{L_1, \cdots, L_m\})$ and $(g - \{L_1, \cdots, L_m\}) \in \varphi_{L_1\cdots L_m}$, $L_i \in X \cup \overline{X}$, $i = 1, \cdots, m$, and $\varphi_{L_1\cdots L_m}$ contains no variable in X.*

Proof. ($\Leftarrow$) Let the splitting formula $\varphi_{L_1\cdots L_m}$ contain the two clauses $(f - \{L_1, \cdots, L_m\})$ and $(g - \{L_1, \cdots, L_m\})$, but contain no variable in X. Then, $(f - \{L_1, \cdots, L_m\})$ and $(g - \{L_1, \cdots, L_m\})$ are connected without X, hence f, g must be connected without X.

($\Rightarrow$) Let f and g be connected without X. We proceed by induction on the number of variables in φ. Suppose φ has only one variable, say x. If X is empty, clearly the assertion is true since φ itself contains no variable in X. Suppose $X = \{x\}$, then there is no variable outside X. That means, $f = g$, say the unit clause x, then the empty clause $f - \{x\}$ is in φ_x which contains no variable. Hence, the assertion is true.

Suppose, φ has more than one variable. Let y be a variable such that $(\varphi_y, \varphi_{\neg y})$ is a disjunctive splitting.

Case 1. $y \in X$. Since f and g are connected without X, f and g must be in the same part. Say for example $(f - \{y\}), (g - \{y\}) \in \varphi_y$. Obviously, $(f - \{y\}), (g - \{y\})$ are connected without $X \cap var(\varphi_y)$. Then by the induction hypothesis, there is a splitting formula $\varphi_{yL_1\cdots L_m}$ which do not contain any variable in X, but contains both $(f - \{y, L_1, \cdots, L_m\})$ and $(g - \{y, L_1, \cdots, L_m\})$. The lemma follows.

Case 2. $y \notin X$.
Subcase 2.1. f and g lie in the same part. W.o.l.g., we assume that $(f - \{y\}), (g - \{y\}) \in \varphi_y$. Now $(f - \{y\}), (g - \{y\})$ are still connected without $X \cap var(\varphi_y)$. Then by the induction hypothesis there is a splitting formula $\varphi_{yL_1\cdots L_m}$ which do not contain any variable in X, but contains both $(f - \{y, L_1, \cdots, L_s\})$ and $(g - \{y, L_1, \cdots, L_s\})$. If $\varphi_{yL_1\cdots L_s}^y$ contains no y, then $\varphi_{L_1\cdots L_s} = \varphi_{yL_1\cdots L_s}$, and the lemma follows. So, suppose $\varphi_{yL_1\cdots L_s}^y$ contains y. Let $\neg y \vee h \in \varphi$ which is

connected to itself. Then by the induction hypothesis there is a splitting formula $\varphi_{\neg y L_{s+1}\cdots L_m}$ in which no variable of X occurs but the clause $(h - \{L_{s+1}\cdots L_m\})$ appears. Now we can see that $\varphi_{\neg y L_{s+1}\cdots L_m}^{\neg y}$ contains $\neg y$. Then

$$\varphi_{y L_1 \cdots L_s}^{y} + \varphi_{\neg y L_{s+1}\cdots L_m}^{\neg y}$$

is in MU(1), which is in fact $\varphi_{L_1\cdots L_m}$. The lemma follows.

Subcase 2.2. f and g lie in different part. Say, $(f - \{y\}) \in \varphi_y$ while $(g - \{\neg y\}) \in \varphi_{\neg y}$. Please notice that f and g are connected without X. Thus, f must be connected without X to a clause f' containing y, and g must be connected without X to clause g' containing $\neg y$. Then by the induction hypothesis, there are $\varphi_{y L_1 \cdots L_s}$ and $\varphi_{\neg y L_{s+1}\cdots L_m}$ such that they do not contains variables in X, but $(f - \{y, L_1, \cdots, L_s\}), (f' - \{y, L_1, \cdots, L_s\}) \in \varphi_{y L_1 \cdots L_s}$, and $(g - \{\neg y, L_{s+1}, \cdots, L_m\}), (g' - \{\neg y, L_{s+1}, \cdots L_m\}) \in \varphi_{\neg y L_{s+1}\cdots L_m}$. Then we can see that $\varphi_{y L_1 \cdots L_s}^{y}$ contains y and $\varphi_{\neg y L_{s+1}\cdots L_m}^{\neg y}$ contains $\neg y$. Thus $\varphi_{y L_1 \cdots L_s}^{y} + \varphi_{\neg y L_{s+1}\cdots L_m}^{\neg y}$ is in MU(1), which is in fact $\varphi_{L_1\cdots L_m}$. The assertion follows.   □

Proof of Lemma 5. For $\Phi = \exists X \forall y \exists Z \varphi$ with $\varphi_{|\exists}$ in MU(1) we have

Φ is true   (by Lemma 7)

$\Leftrightarrow$ there exist $L_1, \cdots, L_s \in X \cup \overline{X}$ such that in $\varphi_{L_1 \cdots L_s}$ no variable of X occurs and $\forall y \exists z_1 \cdots \forall z_k \varphi_{L_1 \cdots L_s}$ is true.

$\Leftrightarrow$ there exist $L_1, \cdots, L_s \in X \cup \overline{X}$ such that in $\varphi_{L_1 \cdots L_s}$ no variable of X occurs and there are $f, g \in \varphi$ such that $y \in f, \neg y \in g$ and $f - \{L_1, \cdots, L_m\},\ g - \{L_1, \cdots, L_m\} \in \varphi_{L_1 \cdots L_m}$.   (reordering)

$\Leftrightarrow$ there are $f, g \in \varphi$ with $y \in f, \neg y \in g$ and there exist $L_1, \cdots, L_s \in X \cup \overline{X}$ such that in $\varphi_{L_1 \cdots L_s}$ no variable of X occurs and $f - \{L_1, \cdots, L_m\},\ g - \{L_1, \cdots, L_m\} \in \varphi_{L_1 \cdots L_m}$.   (by Lemma 8)

$\Leftrightarrow$ there are f, g such that $y \in f, \neg y \in g$ and f, g are connected without X.   □

5   Minimal Falsity and Quantified Hitting Formulas

For propositional formulas the class of hitting formulas is defined as HIT := $\{\alpha_1 \wedge \cdots \wedge \alpha_t \in \text{CNF} \mid$ for all $i \neq j$, α_i and α_j contain a complementary pair of literals, $t \geq 2\}$. Let $f = x \vee f'$ and $g = \neg x \vee g'$ be two clauses, then x is called a hitting variable for f and g. If a hitting formula α is unsatisfiable then α is minimal unsatisfiable and for every partial truth assignment v the formula $\alpha[v]$ is either the formula with only the empty clause or in HIT. Moreover, the satisfiability problem for HIT is solvable in polynomial time [7].

Now we consider quantified hitting formulas. Let QHIT be the class of QCNF formulas with matrix in HIT.

Theorem 4.   (1) The evaluation problem and the minimal falsity problem for QHIT formulas are PSPACE-complete.
(2) For fixed k, $MF(k) \cap QCNF$ and $MF(k) \cap QHIT$ are poly-time reducible.

Proof. At first we introduce a poly-time reduction $F : \text{QCNF} \rightarrow \text{QHIT}$ and show for $\Phi \in \text{CNF}$: (Φ is (minimal) false $\Leftrightarrow$ $F(\Phi)$ is (minimal) false) and (Φ is in $\text{MF}(k) \Leftrightarrow F(\Phi)$ is in $\text{MF}(k)$).

Let $Q\varphi$ be a QCNF formula with $\varphi = (\alpha_1 \wedge \cdots \wedge \alpha_m)$. For new variables $\boldsymbol{x} = x_1, \cdots, x_{m-1}, y$ we define $F(\Phi) := Q \forall x_1 \cdots \forall x_{m-1} \exists y (\alpha_1' \wedge \cdots \wedge \alpha_{m+1}')$, where

$$
\begin{aligned}
\alpha_1' &:= \alpha_1 && \vee\ x_{m-1} \vee \cdots\cdots \vee x_3 \vee\ x_2\ \vee\ x_1 \vee y, \\
\alpha_2' &:= \alpha_2 && \vee\ x_{m-1} \vee \cdots\cdots \vee x_3 \vee\ x_2\ \vee \neg x_1 \vee y, \\
\alpha_3' &:= \alpha_3 && \vee\ x_{m-1} \vee \cdots\cdots \vee x_3 \vee \neg x_2 \vee y, \\
&\cdots\cdots \\
\alpha_{m-1}' &:= \alpha_{m-1} && \vee\ x_{m-1} \vee \neg x_{m-2} \vee y, \\
\alpha_m' &:= \alpha_m && \vee \neg x_{m-1} \vee y \\
\alpha_{m+1}' &:= \neg y
\end{aligned}
$$

The formula $F(\Phi)$ is in QHIT and obviously Φ is (minimal) false if and only if the formula $F(\Phi)$ is (minimal) false. Moreover, both formulas have the same deficiency. The problems are in PSPACE-complete, because the evaluation problem and the minimal falsity problem for QCNF are PSPACE-complete. □

Now we investigate a weaker notion of quantified hitting formulas for which every pair of clauses has an existential hitting variable. Let QEHIT be the class of QCNF formulas $Q\varphi$ for which the existential part $\varphi_{|\exists}$ is in HIT. A QEHIT formula is false if and only it is minimal false, because for $\varphi_{|\exists}$ in HIT, $\varphi_{|\exists}$ is false if and only if $\varphi_{|\exists}$ is minimal unsatisfiable. We will show that the minimal falsity and the evaluation problem for QEHIT are solvable in polynomial time.

Lemma 9. *For* $\Phi = \exists X_1 \forall Y_1 \cdots \exists X_m \forall Y_m \exists X_{m+1} \varphi \in QEHIT$, $\Phi \in MF$ *if and only if* $\varphi_{|\exists} \in MU$ *and for all* $1 \leq i \leq m$, *for all* $y \in Y_i$, *any pair of clauses* $(y \vee f)$ *and* $(\neg y \vee g)$ *has a hitting variable* $x \in X_1 \cup \cdots \cup X_i$.

Proof. ($\Rightarrow$) Suppose, for a variable $y \in Y_i$ and clauses $(y \vee f)$ and $(\neg y \vee g)$ in φ, f and g have no hitting variable x with $x \in X_1 \cup \cdots \cup X_i$. We define a partial truth assignment v such that $v(L) = 0$ for any literal L such that its variable is in $X_1 \cup \cdots \cup X_i$ and L occurs in f or g. Since under v the two clause $y \vee f$ and $\neg y \vee g$ are not true, we see that in $\Phi' := \forall y Q' \varphi[v]$ both y and $\neg y$ occur.

Suppose Φ is in MF, then Φ' must be false. Further, Φ' is minimal false because $(\varphi[v])_{|\exists} = \varphi_{|\exists}[v] \in \text{HIT}$. This contradicts Lemma 1 which states that y cannot occur both positively and negatively in Φ'.

($\Leftarrow$) For each $i = 1, \cdots, m$, and for each truth assignment v on $X_1 \cup \cdots \cup X_i$, no $y \in Y_i$ occurs both positively and negatively in $\varphi[v]$. Thus, Φ and $\exists X_1 \cdots \exists X_i \exists X_{i+1} \forall Y_{i+1} \cdots \exists X_m \forall Y_m \exists X_{m+1} \varphi'$ have the same truth, here φ' is obtained from φ by deleting all occurrences of literals whose variables are in $Y_1 \cup \cdots \cup Y_i$. It follows that Φ is false since $\varphi_{|\exists}$ is in MU. □

Theorem 5. *QEHIT-MF can be solved in polynomial time.*

Proof. Given a formula Φ in QEHIT, if the inner-most quantifies is $\forall y$ then remove the quantifier and remove all occurrences of y and $\neg y$. Iteratively applying this procedure, we get a formula such that the inner-most quantifier is existential. If the outer-most quantifier is also $\forall y$, then check whether y has no complementary occurrences. If this condition holds, then remove the occurrences of y. Applying this procedure we obtain a formula whose prefix begins with an existential quantifier. Next we test the condition in Lemma 9. Whether $\varphi_{|\exists}$ is in MU, is decidable in polynomial time, since $\varphi_{|\exists} \in$HIT [7]. Whether the required hitting $\exists$-variable exists, can be solved in cubic time. $\square$

6   Conclusions and Future Work

In this paper, we have generalized the notion of minimal unsatisfiable CNF formulas to that of minimal false QCNF formulas. The minimal falsity problem is PSPACE-complete even for quantified hitting formulas. The deficiency for CNF formulas has been also extended to QCNF formulas. The minimal falsity for QCNF formulas with deficiency one can be solved in polynomial time. In a forthcoming paper we will prove that $MF(k)$ is in D^P for any fixed k. However, whether $MF(k)$ is decidable in polynomial time remains open.

References

1. B. Aspvall, M. F. Plass, R. E. Tarjan: A Linear–Time Algorithm for Testing the Truth of Certain Quantified Boolean Formulas, *Information Processing Letters*, **8** (1979), pp. 121-123
2. G. Davydov, I. Davydova, H. Kleine Büning: An efficient algorithm for the minimal unsatisfiability problem for a subclass of CNF, *Annals of Mathematics and Artificial Intelligence*, **23** (1998), pp. 229–245
3. D. Ding, H. Kleine Büning and X. Zhao: Minimal Falsity for QBF with Fixed Deficiency, Proceedings of the Workshop on Theory and Applications of Quantified Boolean Formulas, IJCAR 2001, Siena, (2001), pp. 21-28.
4. H. Fleischner, O. Kullmann, S.Szeider, Polynomial-time recognition of minimal unsatisfiable formulas with fixed clause-variable deficiency, *Theoretical Computer Scicence*, **289** (2002), pp. 503-516.
5. H. Kleine Büning: On Some Subclasses of Minimal Unsatisfiable Formulas, *Discrete Applied Mathematics*, **107** (2000), pp. 83–98
6. H. Kleine Büning, T. Lettmann: *Propositional Logic: Deduction and Algorithms*, Cambridge University Press, 1999.
7. H. Kleine Büning, X. Zhao : On the structure of some classes of minimal unsatisfiable formulas, *Discrete Applied Mathematics*, **130** (2003), No.2, pp. 185-207.
8. O. Kullmann: An Application of Matroid Theory to the ASAT Problem, *Electronic Colloquium on Computational Complexity*, Report 18, 2000.
9. O. Kullmann: Lean-sets: Generalizations of minimal unsatisfiable clause-sets, *Discrete Applied Mathematics*, **130** (2003), 209-249.
10. O. Kullmann: The Combinatorics of Conflicts between Clauses, In: *Lecture Notes in Computer Science* , volume 2919, pages 426-440, Springer.
11. C.H. Papadimitriou, D. Wolfe: The complexity of facets resolved, *Journal of Computer and System Science*, **37** (1988), 2-13.

Binary Clause Reasoning in QBF[*]

Horst Samulowitz and Fahiem Bacchus

Department of Computer Science, University Of Toronto,
Toronto, Ontario, Canada
{horst, fbacchus}@cs.toronto.edu

Abstract. Binary clause reasoning has found some successful applications in
SAT, and it is natural to investigate its use in various extensions of SAT. In this
paper we investigate the use of binary clause reasoning in the context of solving
Quantified Boolean Formulas (QBF). We develop a DPLL based QBF solver that
employs extended binary clause reasoning (hyper-binary resolution) to infer new
binary clauses both before and during search. These binary clauses are used to
discover additional forced literals, as well as to perform equality reduction. Both
of these transformations simplify the theory by removing one of its variables.
When applied during DPLL search this stronger inference can offer significant
decreases in the size of the search tree, but it can also be costly to apply. We are
able to show empirically that despite the extra costs, binary clause reasoning can
improve our ability to solve QBF.

1 Introduction

DPLL based SAT solvers standardly employ only unit propagation during search. Unit
propagation has the advantage that it can be very efficiently implemented, but at the
same time it is relatively limited in its inferential power. The more powerful inferential
mechanism of reasoning with binary clauses has been investigated in [1,2]. In particular,
Bacchus [2] demonstrated that by using a rule of hyper-binary resolution, which allows
the binary clause subtheory to be clashed against its non-binary counterpart, binary
clause reasoning can be very effective in pruning the size of the search space. It can
also be dramatically effective in decreasing the time required to solve SAT problems,
but not always.

The difficulty arises from the extra time required to perform binary clause reasoning,
which tends to scale non-linearly with the size of the SAT theory. Hence, on very large
SAT formulas, binary clause reasoning is often not cost effective. QBF instances, on
the other hand, are generally much smaller than SAT instances. First, QBF allows a
much more compact representation of many problems, so problems that would be very
large in SAT can be quite small when represented in QBF. Second, QBF is in practice a
much harder problem than SAT, so it is unlikely that "solvable" instances will ever be as
large as solvable SAT instances. This makes the application of extensive binary clause
reasoning more attractive on QBF instances, since such reasoning is more efficient on
smaller theories.

---

[*] This research was supported by the Canadian Government through their NSERC program.

A. Biere and C.P. Gomes (Eds.): SAT 2006, LNCS 4121, pp. 353–367, 2006.

In this paper we investigate using binary clause reasoning with QBF. We find that our intuition that such reasoning might be useful for QBF to be empirically true. However, we also find that there are a number of issues arising from the use of such reasoning. First, there are some issues involved in employing such reasoning soundly in a QBF setting. We describe these issues and show how they can be resolved. Second we have found that such reasoning does not universally yield an improvement. Instead one has to be careful about when and where one employs such reasoning.

We have found that binary clause reasoning to be almost universally useful prior to search when used in a QBF preprocessor (akin to the SAT preprocessor of [3]), and we present a more detailed description of preprocessing in [4]. To study the dynamic use of binary clause reasoning during search we have implemented a QBF solver that performs binary clause reasoning at every node of the search tree. Our empirical results indicate that binary clause reasoning can be effective when used dynamically. However, it is not as uniformly effective as it is in a preprocessor context. We provide some insights as to when it can be most useful applied dynamically.

In the rest of the paper we first provide some background, then we discuss how binary clause reasoning can be soundly employed in QBF. We then demonstrate that binary clause reasoning is effective in improving our ability to solve QBF instances. Part of that improvement actually occurs prior to search, and we briefly discuss our findings on this point. These empirical observations lead to the development of a preprocessor for QBF that we describe in [4]. Then we investigate the dynamic use of binary clause reasoning, and show that it also can be effective in the dynamically, but not universally so. Our overall conclusion is that binary clause reasoning does have an important role to play in solving QBF but that further investigation is required to isolate more precisely where it can be most effectively applied.

2 Background

2.1 QBF

A quantified boolean formula (QBF) has the form $Q.F$, where F is a propositional formula expressed in CNF and Q is a sequence of quantified variables ($\forall x$ or $\exists x$). We require that no variable appear twice in Q, that F contains no free variables, and that Q contains no extra or redundant variables.

A **quantifier block** qb of Q is a maximal contiguous subsequence of Q where every variable in qb has the same quantifier type. We order the quantifier blocks by their sequence of appearance in Q: $qb_1 \leq qb_2$ iff qb_1 is equal to or appears before qb_2 in Q. Each variable x in F appears in some quantifier block $qb(x)$, and the ordering of the quantifier blocks imposes a partial order on the variables. For two variables x and y we say that $x \leq_q y$ iff $qb(x) \leq qb(y)$. The variables in the same quantifier block are unordered. We also say that x is **universal** (**existential**) if its quantifier in Q is $\forall$ ($\exists$).

For example, $\exists e_1 e_2.\forall u_1 u_2.\exists e_3 e_4.(e_1, \neg e_2, u_2, e_4) \wedge (\neg u_1, \neg e_3)$ is a QBF with $Q = \exists e_1 e_2.\forall u_1 u_2.\exists e_3 e_4$ and $F = (e_1, \neg e_2, u_2, e_4) \wedge (\neg u_1, \neg e_3)$. The quantifier blocks in order are $\exists e_1 e_2$, $\forall u_1 u_2$, and $\exists e_3 e_4$, the u_i variables are universal while the e_i variables are existential, and $e_1 \leq_q e_2 <_q u_1 \leq_q u_2 <_q e_3 \leq_q e_4$.

The **restriction** of a formula $Q.F$ by a literal ℓ (denoted by $Q.F|_\ell$) is the new formula $Q'.F'$ where F' is F with all clauses containing ℓ removed and $\bar{\ell}$, the negation of ℓ, removed from all remaining clauses, and Q' is Q with the variable of ℓ and its quantifier removed. For example, $\left(\forall xz.\exists y.(\bar{y}, x, z) \wedge (\bar{x}, y)\right)|_{\bar{x}} = \forall z.\exists y(\bar{y}, z)$.

Semantics. A SAT model $\mathcal{M}_s$ of a CNF formula F is a truth assignment π to the variables of F that satisfies every clause in F. In contrast a QBF model (**Q-model**) $\mathcal{M}_q$ of a quantified formula $Q.F$ is a **tree** of truth assignments in which the root is the empty truth assignment, and every node n assigns a truth value to a variable of F not yet assigned by one of n's ancestors. The tree $\mathcal{M}_q$ is subject to the following conditions:

1. For every node n in $\mathcal{M}_q$, n has a sibling if and only if it assigns a truth value to a universal variable x. In this case it has exactly one sibling that assigns the opposite truth value to x. Nodes assigning existentials have no siblings.
2. Every path π in $\mathcal{M}_q$ (π is the sequence of truth assignments made from the root to a leaf of $\mathcal{M}_q$) must assign the variables in an order that respects $<_q$. That is, if n assigns x and one of n's ancestors assigns y then we must have that $y \leq_q x$.
3. Every path π in $\mathcal{M}_q$ must be a SAT model of F.

Thus a Q-model has a path for every possible setting of the universal variables of Q, and each of these paths is a $\leq_q$ ordered SAT model of F. We say that $Q.F$ is QSAT iff it has a Q-model. The QBF problem is to determine whether or not $Q.F$ is QSAT.

A more standard way of defining QSAT is the recursive definition: (1) $\forall x Q.F$ is QSAT iff both $Q.F|_x$ and $Q.F|_{\bar{x}}$ are, and (2) $\exists x Q.F$ is QSAT iff at least one of $Q.F|_x$ and $Q.F|_{\bar{x}}$ is. By removing the quantified variables one by one, in $\leq_q$ order, we arrive at either a QBF with an empty clause in its body F (which is not QSAT) or a QBF with an empty body F (which is QSAT). These two definitions are provably equivalent.

The advantage of our "tree-of-models" definition is that it makes the following observations more apparent.

A. If F' has the same satisfying assignments (SAT models) as F then $Q.F$ will have the same satisfying models (Q-models) as $Q.F'$. **Proof**: $\mathcal{M}_q$ is a Q-model of $Q.F$ iff each path in $\mathcal{M}_q$ is a SAT model of F iff each path is a SAT model of F' iff $\mathcal{M}_q$ is a Q-model of $Q.F'$. This observation allows us to transform F with any model preserving SAT transformation. Note that the transformation must be model preserving, i.e., it must preserve all SAT models of F. Simply preserving whether or not F is SAT is not sufficient.

B. A Q-model preserving (but not SAT model preserving) transformation that can be performed on $Q.F$ is **universal reduction** (UR) [5]. A universal variable u is called a *tailing universal* in a clause c if for every existential variable $e \in c$ we have that $e <_q u$. The universal reduction of a clause c is the process of removing all tailing universals from c. UR preserves the set of Q-models. This can be seen by observing that any path in a Q-model must satisfy the universally reduction of every clause in the theory: if it doesn't then another path of the Q-model will falsify c contradicting the fact that it is a Q-model.

2.2   Hyper-binary Resolution and Equality Reduction

Now we recall the techniques for binary clause reasoning in SAT first presented in [2,3]. We first define the hyper-binary resolution (*HypBinRes*) rule of inference that generates new binary unit clauses.

Definition 1 (*HypBinRes*). *Given a single n-ary clause* $c = (l_1, l_2, ..., l_n)$, *D a subset of c, and the set of binary clauses* $\{(\ell, \bar{l}) | l \in D\}$, *infer the new clause* $b = (c - D) \cup \{\ell\}$ *if b is either **binary** or **unary***.

For example, from (a, b, c, d), $(h, \bar{a})$, $(h, \bar{c})$ and $(h, \bar{d})$, we infer the new binary clause (h, b), similarly from (a, b, c) and $(b, \bar{a})$ the rule generates (b, c). The rule also covers the standard case of resolving two binary clauses (from (l_1, l_2) and $(\bar{l}_1, \ell)$ infer (ℓ, l_2)) and it can generate unit clauses (e.g., from $\{(l_1, \ell), (\bar{l}_1, \ell)\}$ we infer $(\ell, \ell) \equiv (\ell)$). *HypBinRes* is a hyper-resolution step because it collapses in one step a sequence of ordinary resolution steps.

The advantage of *HypBinRes* inference is that it does not blow up the theory (it can only add binary or unary clauses to the theory) and it can discover a lot of new unit clauses. These unit clauses can then be used to simplify the formula by doing unit propagation which in turn might allow more applications of *HypBinRes*. Applying *HypBinRes* and unit propagation until closure (i.e., until nothing new can be inferred) uncovers *all* failed literals. That is, in the resulting reduced theory there will be no literal ℓ such that forcing ℓ to be true followed by unit propagation results in a contradiction. This and other results about *HypBinRes* are proved in the above references.

In addition to uncovering unit clauses we can use the binary clauses to perform equality reductions. In particular, if we have two clauses $(\bar{x}, y)$ and $(x, \bar{y})$ we can replace all instances of y in the formula by x (and $\bar{y}$ by $\bar{x}$) or all instances of x by y. This might result in some tautological clauses which can be removed, and some clauses which are reduced in length because of duplicate literals. Such reductions might enable further *HypBinRes* inferences.

Taken together *HypBinRes* and equality reduction (*HypBinRes+eq*) can significantly reduce a SAT formula removing many of its variables and clauses. Such inference can be applied prior to search in a preprocessor, and as shown in [3] this can yield significant reductions in the number of variables and clauses in a theory. One can also incrementally maintain *HypBinRes+eq* closure during search as it is done in [2].

To maintain *HypBinRes+eq* closure during search we must trigger the *HypBinRes* inference step incrementally. It would be too expensive to continually search exhaustively for possible new applications of *HypBinRes*. During search the formula is restricted by literals that we choose to make true, or that are forced by unit propagation. This gives rise to only two different opportunities for additional applications of *HypBinRes*. First, if a k-ary clause is reduced to a binary clause the new binary clause might enable new *HypBinRes* steps. Second, when k-ary clauses are reduced in size it is possible that a previously existing set of binary clauses can generate an *HypBinRes* inference that was not available on the longer clause. For example, if we have the n-ary clauses $(h, \bar{d}, x)$ (a, b, c, d) and the binary clauses $(h, \bar{a})$, $(h, \bar{c})$ no *HypBinRes* inference is possible. A new *HypBinRes* inference could be applied if either we make x false generating a new binary clause $(h, \bar{d})$, or if we make d false reducing the clause (a, b, c, d) to (a, b, c)

against which the existing binary clauses can be resolved. The dynamic *HypBinRes* solver described in [2] kept track of these two types of situations, testing only these situations for new possible *HypBinRes* steps.

2.3   Hyper-binary Resolution and Equality Reduction in QBF

There are two problems with employing *HypBinRes+eq* in the context of QBF. First, it is not sound for QBF unless some additional restrictions are applied. Second, it misses out on some important additional inferences that can be achieved through universal reduction. We elaborate on these two issues.

Given a QBF $Q.F$ applying *HypBinRes+eq* and unit propagation to F results in a formula F'. However, the new QBF formula $Q'.F'$ might not be Q-equivalent to $Q.F$ (where Q' is Q with all variables not in F' removed), so this straightforward approach to using *HypBinRes* is not sound. The problem here is that F' does not have exactly the same SAT models as F so condition **A** above does not apply. In particular, the models of F' do not make assignments to variables that have been removed by unit propagation and equivalence reduction. Hence, a Q-model of $Q'.F'$ might not be extendable to a Q-model of $Q.F$. For example, if a universal variable in F was forced, then $Q'.F'$ might be QSAT, but $Q.F$ is not—no Q-model of $Q.F$ can exist since no path that sets the forced universal to its opposite value can be a SAT model of F.

Making unit propagation sound for QBF is quite simple. In particular, unit propagation only causes a problem when a universal variable is forced. We can deal with this by regarding the unit propagation of a universal variable as the derivation of a failure (i.e., the derivation of an empty clause).

Making equality reduction sound for QBF is a bit more subtle. Consider a formula F in which we have the two clauses $(x, \bar{y})$ and $(\bar{x}, y)$. Since every path in any Q-model satisfies F, this means that along any path x and y must have the same truth value. However, in order to soundly replace all instances of one of these variables by the other in F, we must respect the quantifier ordering. In particular, if $x <_q y$ then we must replace y by x. We call this $<_q$-*preferred equality reduction*. It would be unsound to do the replacement in the other direction. For example, say that x appears in quantifier block 3 while y appears in quantifier block 5 with both x and y being existential. The binary clauses above will enforce the constraint that along any path of any Q-model once x is assigned y must get the same value. In particular, y will be invariant as we change the assignments to the universal variables in quantifier block 4. This constraint will continue to hold if we replace y by x in all of the clauses of F. However, if we perform the opposite replacement, we would be able to make y vary as we vary the assignments to the universal variables in block 4: i.e., the opposite replacement would weaken the theory perhaps changing the formula's Q-SAT status. The same reasoning holds if x is universal and y is existential. However, if y is universal the two binary clauses imply that we will never have the freedom to assign y both of its values irrespective of the assignment of x. That is, in this case the QBF is UNQSAT, and we can again treat this case as if the empty clause has been derived.

These considerations suffice to make *HypBinRes+eq* sound for QBF. However, they remain weaker than they should be. To achieve more powerful inference we must take

2clsQ($Q.F$, $Level$)

    if F contains an [empty clause/is empty]
        Compute a new [clause/cube] and backtrack level btL by [conflict/solution] analysis
        return([FALSE/TRUE], btL)
    Pick a variable v from the outermost quantifier block
    for $\ell \in \{v, \bar{v}\}$
        $Q'.F' = Q.F|_\ell$ reduced by *HypBinRes+UR*, equality reduction, unit propagation,
                    and universal reduction
        $(Succ, btL) = $ 2clsQ($Q'.F'$, $Level + 1$)
        if $btL <$ Level **return**($Succ, btL$)
    if v is [universal/existential]
        Compute new [cube/clause] from the [cubes/clauses] learned from v and $\bar{v}$ by resolution
        Compute backtrack level btL from new [cube/clause]
        return([TRUE/FALSE], btL)

Fig. 1. 2clsQ Algorithm. Invoked with original QBF and *Level=1*. Returns (TRUE, 0) indicating QSAT or (FALSE, 0) indicating UNQSAT.

into account universal reduction. In particular, we can apply the following modification of *HypBinRes* that "folds" UR into the inference rule.

Definition 2 (*HypBinRes+UR*). *Given a single n-ary clause $c = (l_1, l_2, ..., l_n)$, D a subset of c, and the set of binary clauses $\{(\ell, \bar{l})|l \in D\}$, infer the universal reduction of the clause $(c - D) \cup \{\ell\}$ if this reduction is either binary or unary.*

For example, from $(u_1, e_3, u_4, e_5, u_6, e_7)$, $(e_2, \bar{e}_7)$, $(e_2, \bar{e}_5)$ and $(e_2, \bar{e}_3)$ we infer the new binary clause (u_1, e_2) when $u_1 \leq_q e_2 \leq_q e_3 \leq_q u_4 \leq_q e_5 \leq_q u_6 \leq_q e_7$. This example also shows that *HypBinRes+UR* is able to derive clauses that *HypBinRes* cannot. Since clearly *HypBinRes+UR* can derive anything *HypBinRes* can, *HypBinRes+UR* is a more powerful rule of inference. It should be noted that UR cannot be applied after *HypBinRes* as *HypBinRes* can only generate binary clauses. Instead UR must be folded into the *HypBinRes* rule as we have specified here.

Interestingly, once we add UR to *HypBinRes* many of the issues we had with soundness automatically resolve themselves, and we obtain the following result:

PROPOSITION 1. *Let F' be the result of applying HypBinRes+UR, unit propagation, UR (i.e., UR outside of HypBinRes as well as inside), and $<_q$ preferred equality reduction to F until closure. Then the Q-models of $Q'.F'$ are in 1-1 correspondence with the Q-models of $Q.F$.*

The only further constraint is that UR must be applied prior to unit propagation. In particular, if we have a unit clause containing a single universal variable, we should not unit propagate that universal. Rather we should immediately apply UR to obtain the empty clause.

As an example of how applying UR resolves some of the soundness issues mentioned above, consider the case where we have the two binary clauses $(x, \bar{y})$ and $(\bar{x}, y)$ with $x <_q y$. As pointed out above, when y is universal we have an immediate failure. In fact, applying universal reduction detects this failure: after UR we obtain the two clauses (x) and $(\bar{x})$ which immediately resolve to the empty clause. Hence, this proposition tells us that we can apply *HypBinRes+UR+eq* in QBF quite cleanly: we simply have to restrict equality reduction to respect the quantifier ordering and give precedence to UR over unit propagation.

3   2clsQ

We have implemented *HypBinRes+UR+eq* in a DPLL based QBF solver by modifying the 2clsEq SAT solver [2]. The resulting QBF solver, 2clsQ, performs *HypBin-Res+UR+eq* reasoning at every node of the search tree. An abstract outline of its algorithm is shown in Figure 1. The following changes were made to the 2clsEq SAT solver to make it into a QBF solver. First, branching had to be constrained so that the quantifier ordering is respected. Second, equality reduction had to be modified so that it respects the quantifier ordering. In the 2clsEq implementation an entire set of variables could be detected to be equivalent at once, so we must pick a variable v from the outermost quantifier block among that set and then replace all of the other variables with v.

Third, we had to modify the code that tested for possible new applications of *Hyp-BinRes* to account for universal reduction. When a new binary clause (x, y) is generated we can continue to test all clauses containing $\bar{x}$ as well as all clauses containing $\bar{y}$ to see if this new binary clause triggers any new applications of *HypBinRes+UR*. For example, if $\bar{x} \in c$, we determine the set S of other literals $\ell \in c$ that can be resolved away from c by binary clauses of the form $(y, \bar{\ell})$. Then we check if $c - S$ can be universally reduced to a clause of length 2 or less. The other trigger for new applications of *HypBinRes* occurs when a k-ary clause has been reduced in size, as discussed above. Unfortunately, this situation is relatively expensive to extend to *HypBinRes+UR*. With just *HypBinRes* when a clause c has just been reduced in size to length i, we only need to look for a literal x such that there are $i - 1$ binary clauses $(x, \bar{\ell})$ with $\ell \in c$. From these clauses we can then infer a new binary clause (x, y), where $y \in c$ is the single literal not covered in the set of clauses $(x, \bar{\ell})$. This can be accomplished relatively efficiently by first taking any two literals of c, l_1 and l_2 and examining the set of literals $L = \{y|$ either $(y, \bar{l}_1)$ or $(y, \bar{l}_2)$ exists$\}$. We then know that any literal x satisfying the above condition must be in L—any such literal must have a binary clause with one of $\bar{l}_1$ or $\bar{l}_2$—and we can restrict our attention to the literals in L.

Unfortunately, this strategy for limiting the set of literals to examine for potential new *HypBinRes* steps against a clause breaks down when we move to *HypBinRes+UR*. For example, consider the clauses $c = (e_1, u_1, u_2, u_3, e_2, u_4, u_5, e_3)$, $(e, \bar{e}_2)$, $(e, \bar{e}_3)$ with $e <_q e_1 <_q u_1, u_2, u_3 <_q e_2 <_q u_4, u_5 <_q e_3$. We can infer the new binary clause (e_1, e) by applying *HypBinRes+UR*. In this case, the literal e has only two binary clauses that can resolve against c, and so it does not fall into the set L defined above. Hence, it is not possible to limit our attention to the literals in L. It is still possible to detect all possible *HypBinRes+UR* inferences available from c in polynomial time, but it becomes more expensive to do so. Hence, in our implementation we do only a partial, and cheaper, test for new *HypBinRes+UR* inference on k-ary clauses that have been reduced in size. That is, we do not achieve *HypBinRes+UR* closure in 2clsQ.

Fourth, the algorithm employs both conflict and solution analysis for learning new clauses and solution cubes. Since literals can be forced from an extensive combination of binary clause reasoning and equality reduction, it was very difficult to implement 1-UIP clause learning. Instead, 2clsQ learns 'all decision clauses' [6]. The learned clauses are used to enhance unit propagation. However, we do not perform *HypBinRes+UR* or equality reduction against them as this appears to be too expensive. Solution analysis (cube learning) is done in the manner introduced in [6,7]. The learned cubes are also used to prune branches in the search. In particular, when a universal variable is set this might trigger a cube making search below that setting unnecessary.

Finally, we modified the original 2clsEq branching heuristics to take into account the varying nature of QBF search. In our implementation we combined two branching heuristics in the following way. Whenever 2clsQ encounters a conflict we try to generate more conflicts by branching on variables that cause the largest number of unit propagations (under *HypBinRes* this number is equal to the number of binary clauses the variable appears in). On the other hand when 2clsQ finds a solution we try to generate more solutions by branching on variables that will satisfy the most clauses. Thus the branching heuristic switches dependent on what "mode" the search is in.

4   Empirical Results

To evaluate the empirical effect of binary clause reasoning we considered all of the non-random benchmark instances from QBFLib (2005) [8] (508 instances in total). We discarded the instances from the benchmark families von Neumann and Z since these can all be solved very quickly by any state of the art QBF solver (less than 10 sec. for the entire suite of instances). We also discarded the instances in benchmark families Uclid, Jmc, and Jmc-squaring. None of these instances can be solved within a time bound of 5,000 seconds by any of the QBF solvers we tested. This left us with 465 instances from 18 different benchmark families. We tested all of these instances.

We tested 2clsQ [9] along with five other state of the art QBF solvers **Quaffle** [7] (version as of Feb. 2005), **Quantor** [10] (version as of 2004), **Qube** (release 1.3) [11], **Skizzo** [12] (release 0.82) and **SQBF** [13]. Quaffle, Qube and SQBF are based on search, whereas Quantor is based on variable elimination. Skizzo uses mainly a combination of variable elimination and search, but it also applies a variety of other kinds of reasoning on the symbolic and the ground representations of the instances.[1] All tests were run on a Pentium 4 3.60GHz CPU with 6GB of memory. The time limit for each run of any of the solvers was set to 5,000 seconds.

Table 1 shows the performance of 2clsQ and the other five solvers on the 465 problem instances we tested. The table is broken down by benchmark family as the structural properties of the families can be quite distinct. This structural distinctions are reflected in fact that the "best" solver for each family varies widely, where we measure best by the success rate of the solver on that families' instances breaking ties by CPU time consumed. By this measurement 2clsQ is best on 3 families, which is better than any other search based solver (Quaffle, Qube, and SQBF), but not as good as Skizzo which is best on 8 families. Another comparison is to examine the average success rate over

---

[1] Skizzo also employs some binary clause reasoning and equality reduction. But hyper binary resolution is not used, non-binary clauses are not involved in the inference steps ([12] incorrectly claims that hyper-binary resolution is used).

Table 1. Percentage of each Benchmark family solved and time taken for solved instances in CPU seconds (5,000 sec. consumed by each unsolved instances is not counted). For each family the solver with highest success rate is show in bold, where ties are broken by time required to solve these instances. The summary line shows the average success rate over all benchmark families and the total time taken (on solved instances only).

Benchmark Families (# instances)	2clsQ		Quaffle		Qube		SQBF		Quantor		Skizzo	
	Succ. %	time	Succ. %	time	Succ. %	time	Succ. %	time	Succ. %	time	Succ. %	time
ADDER (16)	44%	5,267	13%	1	19%	72	13%	3	25%	25	**50%**	955
adder (16)	19%	0	44%	5	**44%**	0	38%	2,677	25%	30	44%	454
Blocks (16)	50%	46	75%	1,284	69%	1,774	75%	2,043	**100%**	308	69%	2,068
C (24)	21%	16	21%	5,356	8%	4	17%	4,741	21%	140	**25%**	1,070
Chain (12)	**100%**	0	67%	6,075	83%	4,990	58%	4,192	100%	0	100%	1
Connect (60)	**100%**	7	70%	254	75%	7,013	67%	0	67%	14	68%	802
Counter (24)	33%	4,319	38%	5	33%	2	38%	9	50%	217	**54%**	1,035
EV-Pursuer(38)	26%	2,836	26%	1,963	18%	4,401	**32%**	4,759	3%	74	29%	1,450
FlipFlop (10)	100%	4	**100%**	0	100%	1	80%	5,027	100%	3,260	100%	6
K (107)	35%	20,575	35%	18,451	37%	25,397	33%	5,563	64%	3,855	**88%**	2,081
Lut (5)	100%	19	**100%**	1	100%	3	100%	1,246	100%	3	100%	9
Mutex (7)	43%	22	29%	43	43%	64	43%	1	43%	0	**100%**	1
Qshifter (6)	33%	59	17%	0	33%	29	33%	1,108	100%	26	**100%**	8
S (52)	8%	9	2%	0	4%	401	2%	1	25%	910	**27%**	643
Szymanski (12)	**67%**	2,741	0%	0	8%	0	8%	1,203	25%	7	41%	1,147
TOILET (8)	75%	528	75%	61	63%	496	100%	1,308	100%	4,135	**100%**	1
toilet (38)	84%	47	97%	115	**100%**	58	97%	395	100%	684	100%	84
Tree (14)	100%	296	100%	37	**100%**	0	93%	1,051	**100%**	0	**100%**	0
Summary	58%	36,793	50%	33,653	52%	44,708	51%	35,326	64%	10,432	71%	11,817

all benchmark families, shown in the final row of the table. A high average displays fairly robust performance across structurally distinct instances. On this measure 2clsQ is again superior to the other search based solvers with an average success rate of 58%, higher than any of the other search based solvers, but again not as good as Skizzo or Quantor. In terms of CPU time, the search based solvers are roughly comparable over their solvable instances, but both Quantor and Skizzo are notably faster.

Our first results lead to the following conclusions. Binary clause reasoning improves search based solvers, but the non-search solver Quantor and the mixture of search and variable elimination employed in Skizzo often have superior performance. The superior performance of Skizzo indicates that mixing search and variable elimination (as done by Skizzo) is very effective. We also observe that both Quantor and Skizzo are still inferior to *some* search based solver on 43% of the families. Furthermore, if we examine those cases where a solver is able to achieve a strictly higher success rate than any other solver (indicating that it can solve some instances not solvable by any of the other solvers), we see that 2clsQ achieves this on 2 families, Quaffle on zero, Qube on zero, SQBF on one, Quantor on one, and Skizzo on 6 families. Thus we conclude that binary clause reasoning as embodied in 2clsQ has some potential in increasing our ability to solve QBF (as to the techniques embedded in SQBF, Quantor, and Skizzo).

Table 2. Experiments from Table 1 repeated except that the other solvers are supplied with instances preprocessed by binary clause reasoning. Again unsolved instances consumed 5,000 sec., and for each family the solver with highest success rate is show in bold, where ties are broken by time required to solve these instances. The summary line shows the average success rate over all benchmark families and the total time taken (on solved instances only).

Benchmark Families (# instances)	2clsQ Succ. %	time	Quaffle Succ. %	time	Qube Succ. %	time	SQBF Succ. %	time	Quantor Succ. %	time	Skizzo Succ. %	time
ADDER (16)	44%	5,267	13%	1	19%	26	13%	1	25%	26	**50%**	792
adder (16)	19%	0	44%	4	**44%**	1	38%	1,546	25%	27	44%	550
Blocks (16)	50%	46	88%	1,025	69%	242	82%	3,434	**100%**	79	88%	11
C (24)	21%	16	25%	4,947	21%	683	25%	20	29%	5,189	**29%**	1,483
Chain (12)	100%	0	100%	0	100%	0	100%	0	100%	0	100%	0
Connect (60)	100%	7	100%	7	100%	7	100%	7	100%	7	100%	7
Counter (24)	33%	4,319	38%	5	33%	1	38%	20	50%	141	**54%**	731
EV-Pursuer(38)	26%	2,836	26%	1,961	18%	2,537	32%	4,508	5%	4,809	**39%**	5,753
FlipFlop (10)	100%	4	100%	4	100%	4	100%	4	100%	4	100%	4
K (107)	35%	20,575	36%	21,446	42%	30,606	35%	12,859	83%	6,898	**91%**	5,333
Lut (5)	100%	19	**100%**	1	100%	6	100%	66	100%	3	100%	9
Mutex (7)	43%	22	29%	49	43%	71	43%	6	43%	1	**100%**	100
Qshifter (6)	33%	59	17%	0	33%	29	33%	2,103	100%	29	**100%**	8
S (52)	8%	9	8%	9	10%	452	8%	9	31%	1,538	**37%**	1,538
Szymanski (12)	67%	2,741	0%	0	25%	199	0%	0	25%	109	**75%**	4,680
TOILET (8)	75%	528	75%	84	63%	325	100%	621	**100%**	3	**100%**	3
toilet (38)	84%	47	97%	221	100%	90	97%	3,061	100%	243	**100%**	50
Tree (14)	100%	296	100%	8	100%	1	93%	1,251	**100%**	0	**100%**	0
Summary	58%	36,793	55%	29,772	56%	35,281	57%	29,518	69%	19,108	81%	23,895

4.1 Dynamic Binary Clause Reasoning

In SAT it was observed that binary clause reasoning could be very beneficial even when done prior to search, in a preprocessing phase [3]. Hence, a natural question was to investigate the difference between dynamic and static (i.e., before search) application of binary clause reasoning. As part of that investigation we constructed a QBF preprocessor that applies *HypBinRes+UR+eq* to simplify a QBF instance. We found that this yielded a very consistent speedup for all of the other QBF solvers, and we describe those results in more detail in [4].

Without getting into the details of our preprocessor results, we can still use our preprocessor to throw light on the effect of dynamic binary clause reasoning. In particular, we are interested in the question of how much of 2clsQ's benefits accrue from the dynamic application of binary clause reasoning. Is utilizing binary clause reasoning solely in a preprocessor sufficient, or is it also useful to use such reasoning dynamically during search? To answer this question we compare the performance of 2clsQ with the other solvers on *preprocessed* instances. By using the preprocessed instances, 2clsQ's only "advantage" over the other solvers is its dynamic application of binary clause reasoning. Our results are shown in Table 2.

These results show that a significant part of the gains achieved from binary clause reasoning occurs statically prior to search. In terms of average success rate, 2clsQ still at 58% is now closer in performance to the other search based solvers all of which have gained, and still inferior to Quantor and Skizzo which have gained significantly from binary clause preprocessing. We also see that two of the families where 2clsQ was achieving superior performance, Chain and Connect, have been so reduced by preprocessing that all solvers now achieve similar performance on them. In fact, all instances of Connect are completely solved by preprocessing, and all instances of Chain are reduced to simple SAT problems by preprocessing.

Nevertheless, the results do show that dynamic binary clause reasoning improves the efficiency of search in QBF solvers. In particular, 2clsQ remains more effective than other other purely search based solvers even when the effect of inference prior to search is factored out. The question now is whether or not these improvements to search are useful, given the effectiveness of variable elimination used by Quantor and Skizzo.

4.2   Filtering Out Instances Best Solved by Variable Elimination

To address this question we look more closely at how effective dynamic binary clause reasoning is on instances that are more suitably solved by search. In particular, it does not really matter much if (dynamic) binary clause reasoning improves the efficiency of solving by search instances that are more easily solved by variable elimination.

We examined those instances that would be solved very quickly by variable elimination, and to factor out the effect of binary clause reasoning prior to search we first preprocessed these instances. In particular, we found that a large number of instances (approximately 285) could be solved by Quantor after preprocessing in 25 seconds or less. In fact Quantor and Skizzo are obtaining a significant head start in their average success rate over the search base solvers from these "easy" instances.

After filtering out these instances a number of benchmark families were completely eliminated. That is, all of their instances were best suited for variable elimination after preprocessing. This left us with the benchmark families Adder, adder, C, Connect, Counter, EV-Pursue, K, Mutex, S, Toilet and Szymanski. However, even among these families several instances were eliminated as being easy. In this analysis we also eliminated all instances that could not be solved by any of the solvers as such instances are not useful when comparing solvers. In total we ended up with 72 instances remaining in 10 different benchmark families.

In Table 3 we show the results of the solvers on these remaining preprocessed instances. In the table a '-' is used to indicate that the particular solver could not solve the instances within a 5,000 CPU second time bound. These results show that dynamic binary clause reasoning as performed in 2clsQ is effective on these harder instances. 2clsQ solves more of these instances than any other solver (27) except Skizzo. We also see that Quantor (i.e., pure variable elimination with 20 solved instances is less effective on these remaining instances than the improved search achieved by dynamic binary clause reasoning in 2clsQ. We also see that Skizzo, with its combination of search and variable elimination remains by far the most effective approach on these remaining instances with 57 instances solved.

Table 3. Solver performance on "non-easy" preprocessed instances (i.e., instances that could not be solved in 25 seconds by Quantor after preprocessing. Uniquely solved instances shown in bold.

Family	Instance	2clsQ	Quaffle	Qube	SQBF	Quantor	Skizzo
ADDER	Adder2-4-c	0	-	26	-	-	111
	Adder2-6-c	7	-	-	-	-	-
	Adder2-8-s	-	-	-	-	-	12
	Adder2-8-c	16	-	-	-	-	-
	Adder2-10-s	-	-	-	-	-	437
	Adder2-10-c	3,812	-	-	-	-	-
	Adder2-12-s	-	-	-	-	-	230
	Adder2-12-c	1,432	-	-	-	-	-
adder	**adder-8-sat**	-	-	-	-	-	12
	adder-8-unsat	-	0	0	0	-	-
	adder-10-unsat	-	0	0	935	-	-
	adder-12-sat	-	-	-	-	-	314
	adder-12-unsat	-	0	0	191	-	-
	adder-14-unsat	-	0	0	419	-	-
	adder-16-unsat	0	2	0	-	-	-
C	**C6288-10-1-1-out**	-	-	-	-	-	1,436
	C880-10-1-1-inp	1	4	3	3	905	23
Counter	**counter-16**	-	-	-	-	-	721
	counter-r-8	-	-	-	-	60	1
	counter-re-8	-	-	-	-	79	3
EV-Pursue	ev-pr-4x4-5-3-1-lg	1	1	0	1	82	24
	ev-pr-4x4-5-3-1-s	-	-	-	-	-	6
	ev-pr-4x4-7-3-1-lg	17	3	16	1	-	1,469
	ev-pr-4x4-7-3-1-s	-	-	-	-	-	973
	ev-pr-4x4-9-3-1-lg	180	65	2,174	2	-	-
	ev-pr-4x4-9-3-1-s	-	-	-	-	-	1,679
	ev-pr-4x4-11-3-1-lg	390	990	-	3	-	-
	ev-pr-4x4-13-3-1-lg	-	-	-	4	-	-
	ev-pr-4x4-15-3-1-lg	-	-	-	5	-	-
	ev-pr-4x4-17-3-1-lg	-	-	-	7	-	-
	ev-pr-6x6-5-5-1-2-lg	4	5	2	24	-	2
	ev-pr-6x6-5-5-1-2-s	-	-	-	-	-	258
	ev-pr-6x6-7-5-1-2-lg	60	67	44	172	-	2
	ev-pr-6x6-7-5-1-2-s	-	-	-	-	-	462
	ev-pr-6x6-9-5-1-2-lg	823	784	-	3,708	-	235
	ev-pr-6x6-11-5-1-2-s	-	-	-	-	-	606
	ev-pr-8x8-5-7-1-2-lg	3	2	3	2	4,727	3
	ev-pr-8x8-7-7-1-2-lg	68	9	298	578	-	8
	ev-pr-8x8-9-7-1-2-lg	1,292	34	-	-	-	12

Table 3. (*continued*)

Family	Instance	2clsQ	Quaffle	Qube	SQBF	Quantor	Skizzo
	ev-pr-8x8-11-7-1-2-lg	-	-	-	-	-	18
K	k-branch-n-4	141	-	93	1,190	-	12
	k-branch-n-8	-	-	-	-	-	40
	k-branch-p-4	1,858	389	20	147	32	0
	k-branch-p-8	-	-	-	-	-	0
	k-branch-p-12	-	-	-	-	-	52
	k-d4-n-8	-	-	-	-	-	0
	k-d4-n-12	-	-	·	-	-	0
	k-d4-n-16	-	-	-	-	-	0
	k-d4-n-20	-	-	-	-	-	1
	k-d4-n-21	-	-	-	-	-	1
	k-lin-n-20	1,493	-	1,370	-	66	74
	k-lin-n-21	1,511	-	1,593	-	82	87
	k-ph-n-16	287	261	4,729	4,334	198	198
	k-ph-n-20	2,636	2,204	-	-	1,790	1,806
	k-ph-n-21	4,254	3,668	-	-	2,950	2,977
	k-ph-p-12	-	-	-	-	1,689	-
Mutex	*mutex-16s*	-	-	-	-	-	1
	mutex-32s	-	-	-	-	-	9
	mutex-64s	-	-	-	-	-	22
	mutex-128s	-	-	-	-	-	70
S	s499-d4-s	-	-	-	-	228	107
	s499-d8-s	-	-	-	-	-	1,878
	s641-d2-s	-	-	-	-	294	18
	s713-d2-s	-	-	-	-	448	29
	s820-d2-s	-	-	-	-	429	33
	s3330-d2-s	-	-	-	-	107	11
Szymanski	szymanski-12-s	221	-	-	-	105	1,183
	szymanski-14-s	677	-	-	-	-	954
	szymanski-16-s	1,780	-	-	-	-	1,992
	szymanski-18-s	-	-	-	-	-	373
Toilet	toilet-a-10-01.16	-	59	37	-	103	32
	toilet-c-10-01.16	-	72	46	655	110	9
Solved Instances		27	21	22	20	20	57
Total time on solved instances		23,238	11,884	10,628	13,019	14,534	21,252
Number of Uniquely solved Instances		3	0	0	3	1	22

Finally, if we look at the number of uniquely solved instances we see that both 2clsQ and SQBF can solve 3 instances not solvable by any other solver. These include instances that to the best our knowledge have never been solved before, e.g., 'Adder2-10-c' and 'Adder2-12-c'. These two solver embed techniques for improving search, and we see that these techniques can be useful for improving our ability to solve QBF. Skizzo can 20 instances not solvable by any other solver, so we see that combining variable elimination and search appears to be the most powerful current technique for solving QBF. However, the search employed by Skizzo does not include the innovations of SQBF or 2clsQ. Hence, our results point to at least one direction for building a QBF solver superior to any that currently exists.

It is also worth noting that 2clsQ and SQBF implement many of the techniques of Quaffle and Qube, so it is hardly surprising that all of the instances solved by these two solvers are also solved by some other search based solver. This does not detract from the techniques pioneered in these solvers, like clause and cube learning, which are essential for search based solvers. The uniquely solved instances speaks instead to the value of the new techniques utilized in other solvers: variable elimination in Quantor, binary clause reasoning in 2clsQ, SAT solving lookahead in SQBF, and the mixture of variable elimination and search in Skizzo. The data indicates that these new techniques all have some value in improving our ability to solve QBF.

5   Conclusion

Our main conclusion is that extended binary clause reasoning is effective for QBF. If used prior to search in a preprocessor it is able to speed up both search based and variable elimination based solvers, as shown in [4]. Our empirical results also show that such reasoning can also be useful in a dynamic context, and that certain problem instances can be solved with such reasoning that do not seem to be otherwise solvable.

However, although our empirical results identify binary clause reasoning as being useful techniques for solving QBF, understanding more clearly how to best to combine this reasoning with other kinds of inference, especially variable elimination, remains an open question. In future work we plan to investigate this question more fully to see if we can find ways of applying binary clause reasoning in a more focused manner that can cooperate with other kinds of inference.

References

1. Van Gelder, A., Tsuji, Y.K.: Satisfiability testing with more reasoning and less guessing. In Johnson, D., Trick, M., eds.: Cliques, Coloring and Satisfiability. Volume 26 of DIMACS Series. American Mathematical Society (1996) 559–586
2. Bacchus, F.: Enhancing davis putnam with extended binary clause reasoning. In: Eighteenth national conference on Artificial intelligence. (2002) 613–619
3. Bacchus, F., Winter, J.: Effective preprocessing with hyper-resolution and equality reduction. In: Sixth International Conference on Theory and Applications of Satisfiability Testing (SAT 2003), Lecture Notes in Computer Science 2919. (2003) 341–355
4. Samulowitz, H., Davies, J., Bacchus, F.: Preprocessing QBF. Submitted to CP (2006)
5. Büning, H.K., Karpinski, M., Flügel, A.: Resolution for quantified boolean formulas. Inf. Comput. 117 (1995) 12–18
6. Zhang, L., Madigan, C.F., Moskewicz, M.W., Malik, S.: Efficient conflict driven learning in a Boolean satisfiability solver. In: International Conference on Computer-Aided Design (ICCAD'01). (2001) 279–285

7. Zhang, L., Malik, S.: Towards symmetric treatment of conflicts and satisfaction in quantified boolean satisfiability solver. In: Principles and Practice of Constraint Programming (CP2002). (2002) 185–199
8. Giunchiglia, E., Narizzano, M., Tacchella, A.: Quantified Boolean Formulas satisfiability library (QBFLIB) (2001) http://www.qbflib.org/.
9. Samulowitz, H., Bacchus, F.: QBF Solver 2clsQ (2006) available at http://www.cs.toronto.edu/~fbacchus/sat.html.
10. Biere, A.: Resolve and expand. In: Seventh International Conference on Theory and Applications of Satisfiability Testing (SAT). (2004) 238–246
11. Giunchiglia, E., Narizzano, M., Tacchella, A.: QUBE: A system for deciding quantified boolean formulas satisfiability. In: International Joint Conference on Automated Reasoning (IJCAR). (2001) 364–369
12. Benedetti, M.: skizzo: a QBF decision procedure based on propositional skolemization and symbolic reasoning. Technical Report TR04-11-03 (2004)
13. Samulowitz, H., Bacchus, F.: Using SAT in QBF. In: Principles and Practice of Constraint Programming, Springer-Verlag, New York (2005) 578–592 available at http://www.cs.toronto.edu/~fbacchus/sat.html.

Solving Quantified Boolean Formulas with Circuit Observability Don't Cares

Daijue Tang and Sharad Malik

Princeton University, Princeton, NJ 08544, USA
{dtang, sharad}@princeton.edu

Abstract. Traditionally the propositional part of a Quantified Boolean Formula (QBF) instance has been represented using a conjunctive normal form (CNF). As with propositional satisfiability (SAT), this is motivated by the efficiency of this data structure. However, in many cases, part of or the entire propositional part of a QBF instance can often be represented as a combinational logic circuit. In a logic circuit, the limited observability of the internal signals at the circuit outputs may make their assignments irrelevant for specific assignments of values to other signals in the circuit. This circuit observability don't care (ODC) information has been used to advantage in circuit based SAT solvers. A CNF encoding of the circuit, however, does not capture the signal direction and this limited observability, and thus cannot directly take advantage of this. However, recently it has been shown that this don't care information can be encoded in the CNF description and taken advantage of in a DPLL based SAT solver by modifying the decision heuristics/Boolean constraint propagation/conflict-driven-learning to account for these don't cares. Thus far, however, the use of these don't cares in the CNF encoding has not been explored for QBF solvers. In this paper, we examine how this can be done for QBF solvers as well as evaluate its practical benefits through experimentation. We have developed and implemented the usage of circuit ODCs in various parts of the DPLL-based procedure of the Quaffle QBF solver. We show that DPLL search based QBF solvers can use circuit ODC information to detect satisfying branches earlier during search and make satisfiability directed learning more effective. Our experiments demonstrate that significant performance gain can be obtained by considering circuit ODCs in checking the satisfiability of QBFs.

1   Introduction

Checking the satisfiability of QBFs has been an important research topic in recent years. Many problems in real-world applications such as AI planning [1], formal verification [2,3] and games [4,5,6] can be formulated as QBF instances. A QBF is a Boolean formula with its variables quantified by either universal or existential quantifiers. Evaluating QBF belongs to the class of P-SPACE complete problems, widely considered harder than NP-complete problems like SAT. Since QBF is more expressive and able to offer more compact problem encodings, in

A. Biere and C.P. Gomes (Eds.): SAT 2006, LNCS 4121, pp. 368–381, 2006.

recent years, there has been an increasing interest in using QBF as an alternative to SAT and Binary Decision Diagram (BDD) based techniques for automated Boolean reasoning. As a result, many QBF solvers have been developed. These QBF solvers take CNF as the standard input format for the propositional part of the QBF formulas since CNF has been proved to be an efficient data structure for implementing various techniques used in most contemporary SAT solvers.

Many QBF instances are derived from circuit application domains such as equivalence checking [7] and bounded model checking [8]. Translating a combinational circuit into its corresponding CNF is done by introducing one Boolean variable for each circuit node and conjuncting the logic consistency conditions of all the gates in that circuit [9]. However, since any Boolean function can be implemented using a multi-level circuit structure consisting of simple logic gates, it may be preferable to use a circuit representation for the logic formula. Structural information of circuits has been proved to be quite useful for solving SAT instances [10,11]. In particular, circuit ODC information can accelerate SAT evaluation through either improved encodings of SAT instances or modifying SAT solver algorithms to utilize this information [12,13,14,15,16]. Since QBF is a natural extension of SAT, techniques that are helpful for solving SAT instances are often adapted in QBF evaluations. However, we are not aware of any published research that deals with circuit ODCs in QBF evaluations. In this paper, we address the issue of using the circuit ODC information in solving QBF instances. Our approach of handling circuit ODCs extends the basic framework of considering circuit ODCs in CNF-based SAT solvers [16] where the don't care information is propagated during the learning process. Further utilization of circuit ODCs in the context of solving QBF instances adds additional possibilities due to the existence of universal variables. We focus on managing circuit ODCs during satisfiability directed backtrack and learning which are the main element that distinguishes search based QBF algorithms and SAT algorithms. We also discuss the impact of different encodings of QBF instances on the performance of QBF solvers and usage of don't care information. We demonstrate that exploiting circuit ODCs can provide significant speedup in search based QBF solvers.

2   Encoding Circuit ODC Conditions in CNFs

In a combinational circuit, a signal is said to be unobservable under a partial/total assignments to the inputs if its value does not affect the outcome of any primary output. Consider the circuit in Figure 1(a). If a and b are both assigned to 1, then the value of e is 1, which further implies that g is 1. This is true regardless of the assignment to c, d, and f, which means c, d and f are all unobservable in this case. To convert a logic circuit into its CNF representation, each circuit node is represented by a Boolean variable and each logic gate is encoded as a set of clauses consistent with the logic function of that gate. Additional constraints on the circuit can also be translated into clauses which are conjuncted with the original CNF representation for that circuit to form a new

CNF. For example, for the circuit in Figure 1(a), if the output signal g is fixed to 1, then the CNF describing the circuit consistency condition and the output constraint is shown in Figure 1(b). Note that we use the QDIMACS format for CNFs in the figure. The QDIMACS format is the standard input format of CNF based SAT and QBF solvers. If at some point e is assigned to be 1, then g is 1 independent of the value of f. In another words, f and all the signals that fan-out only to the logic cone of f, namely c and d, is not observable at the output g. Therefore, there is no need to consider clauses corresponding to gate G_2, which are the shaded clauses in Figure 1(b). To incorporate the unobservability condition of c, d and f in the circuit CNF representation, we encode the assignment $e = 1$ as a don't care literal e and add it to the clauses corresponding to the gates whose input and output signals are made unobservable due to this assignment. The resulting augmented CNF is shown in Figure 1(c). Note that for an arbitrary gate, a total ordering of the gate inputs must be maintained such that only a lower order signal can appear in the ODC condition of a higher order one. This issue is discussed in the context of using ODCs in logic synthesis [17]. We will consider those unobservability conditions of a circuit signal s that are individual signal assignments, and represent them as a set of don't care literals and add them to the corresponding clauses [16]. A clause can be ignored at a certain branch in the search when at least one of its don't care literals is assigned to be 1. An alternative approach introduces unobservability variables to account for the don't care conditions in which case these new variables appear as normal literals in clauses [13]. However, this approach is unable to completely exploit the unobservability in decision making and learning.

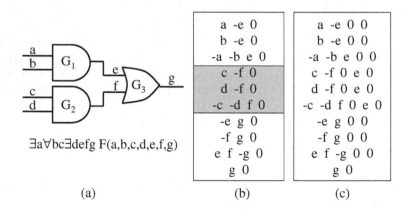

(a)   (b)   (c)

Fig. 1. Circuit ODC example

An arbitrary Boolean function f can be implemented using a multi-level circuit structure consisting of simple logic gates. The primary inputs of the circuit are the variables of f. Boolean logic functions are represented by simple logic gates such as AND, OR, and NOT. For example, a conjunction function is represented by an AND gate, a disjunction function is represented by an OR gate, an equivalence function is represented by an XNOR gate (an XNOR gate can be decomposed as combinations of simple gates), etc. A propositional formula is true if and only if the output of its corresponding circuit evaluates to 1. As an example, the propositional formula $(a \wedge b) \vee (c \wedge d)$ is represented by the circuit shown in Figure 1(a). Since translating any combinational circuit into its CNF is straightforward and requires only a single traversal of the circuit, a Boolean function can be translated into its CNF representation by first converting it to a logic circuit and then translating this circuit into a CNF. A Boolean formula can also be translated into CNF forms by using the distributive law, the De-Morgan's law and other rules of Boolean logic. For example, the propositional formula $(a \wedge b) \vee (c \wedge d)$ is equivalent to $(a \vee c) \wedge (b \vee c) \wedge (a \vee d) \wedge (b \vee d)$ in CNF by the distributive law. In general, the conjunction of two CNFs is still a CNF; the disjunction of two CNFs can be converted to CNF using the distributive law; the negation of a CNF is a DNF which can also be converted to a CNF by applying the distributive law.

Direct CNF translation using Boolean algebra properties can eliminate the need to consider circuit ODCs. For example, for the formula $(a \wedge b) \vee (c \wedge d)$ described previously, when a and b are both assigned to 1, all clauses in the CNF $(a \vee c) \wedge (b \vee c) \wedge (a \vee d) \wedge (b \vee d)$ are satisfied. Yet for the CNF in Figure 1(b), all the possible implications from the assignment $a = 1$ and $b = 1$ is $e = 1$ and $g = 1$, which makes the clauses for G_2 still undecided. Only when the don't care literal e is added to the undecided clauses to indicate that these clauses can be ignored, can we declare that all the clauses are satisfied. In general, suppose we have two CNFs $f_A = C_{1A}C_{2A} \cdots C_{mA}$ and $f_B = C_{1B}C_{2B} \cdots C_{nB}$, where $C_{iA}(i = 1, 2, \cdots, m)$ are clauses of f_A and $C_{jB}(j = 1, 2, \cdots, n)$ are clauses of f_B. For $f_A \wedge f_B$, if we consider f_A and f_B as inputs to an AND gate, then either $f_A = 0$ or $f_B = 0$ makes the other CNF unobservable. Since a CNF is 0 if and only if at least one of its clauses is 0, either f_A or f_B being 0 makes $f_A \wedge f_B$ be 0. This means that the don't care conditions for conjunctions are taken care of naturally in CNFs. Similarly, for $f_A \vee f_B$, if we consider f_A and f_B as inputs to an OR gate, then either $f_A = 1$ or $f_B = 1$ makes the other CNF unobservable. By the distributive law, the CNF for $f_A \vee f_B$ is $f_{AorB} = \prod_{i,j}(C_{iA} + C_{jB})$ where $i = 1, 2, \cdots, m$ and $j = 1, 2, \cdots, n$. If $f_A = 1$, then every clause C_{iA} is 1, which implies that every clause in f_{AorB} is 1. We can derive the same for $f_B = 1$ similarly. Thus, the distributive law takes care of the don't care conditions for disjunctions of CNFs. Therefore, for CNFs obtained from Boolean functions by iteratively applying the Boolean algebra rules, we do not need to consider circuit ODCs. However, this approach of converting Boolean functions to CNFs can possibly result in an exponential growth of the formula size. Although the approach of converting Boolean functions to circuit implementations first and then translating the circuits to CNFs involves introduction of additional variables corresponding to the internal nodes of the circuits, it is easy to implement and guarantees linear complexity for translation. In this case, to take circuit ODC conditions present in Boolean functions into account during the satisfiability search, we encode these conditions as don't care literals in CNFs as described previously. Note that we assume a conventional circuit to CNF translation. That is, every circuit node is a variable in the resulting CNF. Adjacent logic gates in a circuit can be merged to eliminate some of the internal circuit node variables and be represented by a unified set of clauses, which can be viewed as a local CNF translation by Boolean algebra rules, thus the local don't care conditions are already taken into account in the unified set of clauses [14]. It is worth emphasizing that the circuit encoding approach works not only for Boolean functions derived from applications with circuit structures, but also for the Boolean functions where auxiliary variables are used during the CNF translations to avoid the exponential growth of the CNF formulas. For the latter case, the auxiliary variables introduced in the CNF translations are essentially internal nodes of circuits and thus possible sources for don't care literals.

3   SAT Evaluation with Circuit ODCs

Several research efforts on SAT evaluation have exploited the fact that the search procedure does not need to branch on the variables that appear only in the don't care regions and can ignore the clauses corresponding to the don't care regions of circuits [12,15,16]. Since contemporary SAT solvers spend most of their time on the Boolean Constraint Propagation (BCP) procedure which is used to deduce new assignments of variables from existing assignments, reducing the number of variables and clauses in consideration means decreasing the time spent on BCP and therefore less time for SAT solving. Moreover, by considering circuit ODCs, the search is guided away from the don't care regions, which implies fewer assignments before conflicts.

Encoding circuit ODC conditions in CNFs, as described in Section 2, enables CNF-based SAT solvers to avail of the don't care information without keeping the circuit structure information during search. If the ODC information is kept as don't care literals which are separated from normal literals in clauses, a slightly modified CNF format is used. To handle this modified CNF format, several changes were made to the ZChaff SAT solver [18]:

1. The two literal watching BCP strategy needs to be modified to handle the don't care literals. When a don't care literal is found to be true, the clause where this literal is present is satisfied. Unlike normal literals, a don't care literal cannot be watched since it cannot be implied a value by the BCP process.

2. The set of don't care literals of a conflict driven learned clause is the union of the sets of don't care literals of the clauses involved in generating the learned clause by resolution. In this way, don't care information is propagated during the conflict driven learning process and used for future search space pruning.

3. The VSIDS decision heuristic can also accommodate the presence of don't care literals such that a variable is more likely to be branched on if it appears more often as don't care literals as well as normal literals.

4   Exploiting Circuit Oberservability Don't Cares in Quaffle

4.1   Quaffle's Basic Framework

Quaffle is based on the DPLL algorithm [19], which is the most widely used search based algorithm for QBF solvers. Quaffle evaluates the QBF by branching on the variables according to their quantification order. This decision strategy is in natural correspondence with the semantics of the formula. There is no restriction of the decision order for variables within the same quantification level. The deduction engine propagates further assignments to unassigned variables through implications. Implication rules of QBF solvers are more complicated than those for SAT solvers due to the mixing of existential and universal variables and the different quantification levels of the literals in a clause [20]. Decisions and implications on a subset of variables generate a partial assignment. Let f denote the propositional part of a QBF. For a partial assignment of the variables, the valuation of f can be undetermined, satisfied (f is true) or conflicting (f is false). Similar to most DPLL based SAT algorithms, if f is undetermined, Quaffle continues the search by deciding and implying values for unassigned variables; if f is false, Quaffle analyzes the conflict, derives learned clauses and backtracks from the current branch. Our treatment of don't care literals in these two cases are the same as that used in the ZChaff SAT solver. Thus the benefits gained by considering circuit ODCs in CNF-based SAT solvers can also potentially be obtained by exploiting circuit ODCs in Quaffle. In the case of f being true under a partial assignment, Quaffle has to backtrack to search for other satisfying assignments until all combinations of universal variable assignments have been covered. This is an important difference between solving SAT and QBF instances. In SAT, once a complete or partial assignment is found to make f true, the solving process is completed. Yet a satisfying assignment in QBF does not mean the end of the search. Similar to backtracking from a conflict, backtracking from a satisfying leaf can also be non-chronological. Satisfying assignments can be recorded as cubes in Quaffle. Satisfiability directed learning derives learned cubes by performing selected consensus on existing cubes in the same way that conflict driven learning derives learned clauses by performing selected resolution on existing clauses[20,21,22].

4.2   Using ODCs for Early Detection of Satisfying Branches

Due to the existence of universally quantified variables, a potentially exponential number of satisfying assignments need to be enumerated by QBF solvers in the search tree. The earlier the satisfying branches are detected, the better the satisfying space is pruned. A major advantage of considering circuit ODCs in search based QBF solvers is the possibility of early detection of satisfying branches. A conventional CNF formula is satisfied when all the clauses in this CNF are satisfied. In a CNF augmented with circuit ODC encoding by adding don't care literals, a clause can be satisfied with either a normal literal or a don't care literal. If a clause is satisfied by a don't care literal, then there is no need to consider assignments to normal literals of that clause. Thus fewer assignments may be needed to satisfy all the clauses in such augmented CNFs, which means more satisfying leafs are pruned by the current branch. Once a satisfying branch is detected, the solver can backtrack and explore new search space. Consider the example in Figure 1. Suppose we have a QBF where the propositional part is implemented by the combinational circuit in Figure 1(a) with its output constrained to one. The quantifications of the variables in this QBF is shown at the bottom of Figure 1(a). Suppose that the first two decisions of the QBF procedure are $a = 1$ and $b = 1$, then $e = 1$ and $g = 1$ by deduction. With these assignments, all the clauses in Figure 1(c) are satisfied due to the addition of don't care literals. The solver can then flip the assignment of variable b and explore new search space. This is not the case for the conventional CNF in Figure 1(b). In this case, the solver has to continue branching on variable c, d and f until all clauses are satisfied, and then the value of c is flipped while b's value remains the same if chronological backtracking is used. After both values of c are tried, can we backtrack to flip the value of b. In practice, it is not uncommon for QBF instances to have several levels of both universal and existential quantifications and a large number of variables. Thus, it is particularly important to derive and backtrack from the satisfying branch early during the search to avoid exploring the sub-trees of this branch. Suppose at a certain point during the search, all unassigned variables are in the don't care regions of the circuit, then those clauses corresponding to the don't care regions can be satisfied by the don't care literals. If there are n unassigned universal variables at this branch, then the QBF solver could potentially save the effort of exploring 2^n satisfying leafs by exploiting the circuit ODC encoding.

Note that for any two inputs of the same gate, although it is possible that each signal can have a controlling value that makes the other signal unobservable, only one signal can appear as the don't care condition of the other signal in the circuit ODC enhanced CNF encoding. This is because the encoding of the circuit ODC condition is static and at least one input with a controlling value of a gate should be responsible for the unobservability of other inputs of that gate [17]. Therefore, a total ordering of all the gate inputs is used so that only a lower ordered signal can appear in the circuit ODC condition of a higher ordered one. Note also that QBF variables are decided in accordance with their quantification orders and assigned variables can satisfy clauses via both normal true literals and don't care true literals. Thus, we propose the ordering heuristic of using quantification levels to order variables. The intuition behind this heuristic is to satisfy as many clauses as early during search as possible. Specifically, for two variables x and y, if y or primary input variables that fanin to y have higher quantification order than x or primary input variables that fanin to x, then it is very likely that x is assigned prior to y since decision order is restricted by quantification order. Thus there is a greater chance that x is assigned a controlling value that makes y unobservable than vice versa. If x is assigned a controlling value prior to y, then giving x a higher priority as don't care literals in the total ordering, we satisfy the clauses representing the unobservable region that fans into y.

4.3   Using ODCs for Deriving Larger Satisfying Cubes

Conflict driven learning is an important technique for both SAT and QBF solvers [23,24]. Learned clauses generated by conflict driven learning are very useful for preventing certain conflicting search space to be revisited. Combinations of conflicting assignments that have been seen previously during search will make corresponding learned clauses unsatisfying and thus cannot appear again. Similar to conflict driven learning, satisfiability directed learning is the technique used in QBF solvers to prevent revisiting an already satisfied space. Combinations of satisfying assignments are recorded as satisfying cubes. Satisfiability directed learning is the process of selected consensus on cubes, just as conflict driven learning is the process of selected resolution on clauses. When a satisfying leaf is reached, at least one true literal from each clause in a CNF representation is chosen to form a cover set of the satisfying assignment. A satisfying cube is the conjunction of the literals in a cover set. For augmented CNFs with don't care literals, since there may exist clauses that are satisfied by don't care literals, we may potentially get cover sets consisting of fewer literals. which means the satisfying cubes are bigger and a larger satisfying space is pruned. Let us again consider the example in Figure 1. With the augmented CNF with don't care literals as shown in Figure 1(c), a partial assignment $a = 1, b = 1, e = 1, g = 1$ satisfies all the clauses, therefore the satisfying cube is $a \wedge b \wedge e \wedge g$. On the contrary, suppose that no circuit information is used and we only have the CNF in Figure 1(b). After the partial assignment $a = 1, b = 1, e = 1, g = 1$, we need to continue the search since not all clauses are satisfied. If we continue to assign $c = 1, d = 1, f = 1$, then the whole CNF is satisfied. The satisfying cube in this case is $a \wedge b \wedge c \wedge d \wedge e \wedge f \wedge g$, which is larger (has more literals) than the cube we get from considering ODCs in CNF. Note that we maintain a total ordering of the circuit nodes when generating the don't care literals. The higher the priority of one circuit node is in the total ordering, the more likely literals corresponding to that circuit node appear as don't care literals in other clauses. This implies that the total ordering can be a good criterion for choosing literals in cover sets. Among all the true literals in a clause, the literal that has the highest priority in the total ordering is chosen to be included in the cover set first. In fact, if we always choose the true literal that has the highest priority in the total ordering in a clause to be included in the cover set, then we only need to consider to cover all the clauses that have no true don't care literal while ignoring the clauses that contain true don't care literals. The following is an informal proof for this claim. A cover set is a set of variable assignments that makes the propositional part of a QBF satisfiable. Clauses that are satisfied by the don't care literals represent the logic condition of the unobservable part of the circuit, where a satisfying assignment A_S always exist. Moreover, since the unobservable part of the circuit does not share variables with other parts of the circuit, A_S will not conflict with the cover set variable assignments. The observable part of the circuit corresponds to clauses where no don't care literal is true. Since we always choose among the true normal literals in a clause the one literal that has the highest priority in the total ordering, we make sure that the controlling variable is in the cover set. Therefore, our selection of the cover set makes sure that the observable part of the circuit is satisfied.

4.4   Comparison with the Conditional QBF Solver

A recent work explored QBF encodings of adversarial games and proposed conditional QBF solvers to solve such problems [6]. In the QBF encodings of adversarial games between two players, one player corresponds to the existential variables and the other player corresponds to the universal variables. The interleaving of the existential and universal quantifications corresponds to the interleaved actions of the two players. The resulting QBF formula is true if and only if the existential player has a winning strategy. For this problem, many combinations of universal variable assignments correspond to "illegal" actions that violate the rules of the game. These combinations of universal variable assignments make the propositional formula f of the QBF true. Once a partial assignment is identified as an illegal action of the universal player, f must be satisfied by this partial assignment. Therefore, a set of indicator variables is used to indicate such occurrences of actions that are known to make f true. The conditional QBF solver keeps tracking the values of the indicator variables. If at least one of the indicator variables is true, the solver backtracks from current branch and searches for other satisfying assignments. This can potentially save a lot of search effort since an indicator variable being true can happen much earlier during search than the whole propositional CNF being satisfied.

In the above two-player adversarial games, if we use A_U and A_E to represent the characteristic functions of the legal actions of the universal and existential players respectively and G to represent the condition that the existential player wins, then the propositional formula of the game QBF is $A_U \to A_E \wedge G$ which is equivalent to $\neg A_U \vee (A_E \wedge G)$. It is easy to see that if this formula is implemented using a circuit, then the circuit node for $\neg A_U$ is an input of the OR gate whose output is the primary output of the circuit. Therefore, if A_U evaluates to 0, then $\neg A_U$ is the controlling value of the OR gate, which makes the value of $(A_E \wedge G)$ unobservable. Let v_{AU} denote the variable representing the output of the function A_U in the circuit. If $v_{AU} = 0$, then the rest of the variable assignments become don't cares. Thus we can add $\neg v_{AU}$ as don't care literals for clauses representing $(A_E \wedge G)$. In general, such illegal combinations of universal variable assignments is disjuncted with the rest of the propositional formula in QBF. Therefore, the indicator variables are essentially don't care literals for clauses of the other part of the disjunction. An indicator variable being 1 means that the corresponding don't care literal is true. If the circuit ODC condition is encoded in CNF, then the QBF solver that handles don't care enhanced CNF can also detect that the CNF is true when one of the indicator variables is 1. Thus solving QBF with circuit ODCs can potentially achieve the same performance gain as that obtained by the conditional QBF solver. Our experiments prove that manually adding indicator variables as don't care literals for the formulas makes Quaffle perform exactly the same as the conditional QBF solver.

Conditional QBF solver requires the identification of the indicator variables before the actual solving begins. In practice, identifying such variables requires familiarity with the problem at hand and its encoding. Moreover, if a variable is falsely identified as an indicator variable, the result returned by the conditional QBF solver might be wrong. On the other hand, QBF solvers that handle circuit ODCs require the circuit ODC augmented CNFs. This means that the circuit structure information needs to be properly identified, which is facilitated by a recent development of the CNF to circuit converter [25]. Given a CNF where part of it or the entire formula is constructed from combinational circuits, this converter can generate the original structure of the circuits. Therefore, don't care literals can be extracted automatically from CNFs originating from circuits and then added to the original CNF formula. Compared to the conditional QBF solver, this approach is less error prone and more robust. In addition, it is more general in the sense that indicator variables are global don't care conditions while don't care literals can capture local circuit ODC information as well.

5   Experimental Results

We implemented a Quaffle-based QBF solver that is able to handle the circuit ODC augmented CNF. For our experiments, we implemented generators for three categories of QBF instances. The first category is a two-player game where each player takes turns to mark an element in a one dimensional array at each time step. A marked element cannot be marked again. A player wins if it marked n consecutive elements in the array, where n is an integer that is greater than one. The generated QBF instance is true if and only if the first player has a winning strategy against the second player for all its possible moves. We call this game consecutive-n. The second category is a simplified Evader/Pursuer chess problem [6]. The third category is the sequential circuit state space diameter problem [26,27]. For all these problems, we first generate the circuit implementation of the propositional formula, then convert the formula to its CNF representation. We use the CNF to circuit converter to reconstruct the circuit structure information from the CNF and automatically identify the ODC information and then generate the circuit ODC augmented CNF. The reason that we chose to write our own generators for these QBF instances is that most QBF benchmarks available are not built from circuit structures. Thus it is hard to extract circuit ODC information from these benchmarks. Note that our encodings are not the most optimal encodings for these problems. In practice, deriving an optimal encoding requires a lot of effort and an understanding of the underlying algorithms and implementations of the QBF solvers. Moreover, our purpose is to use these testcases to evaluate the QBF solver performance in the presence of circuit ODC information instead of providing an optimal solution for a certain class of problems.

All of our experiments were run on a workstation with Intel Pentium IV 2.8 GHz Processor and 1GB physical memory running Linux Fedora Core 1. We

Table 1. Performance summary

Testcase	T/F	Quaffle Time(s)	#SAT	Quaffle-DC-A Time(s)	#SAT	Quaffle-DC-M Time(s)	#SAT
s27-3	T	-	-	-	-	1.59	78748
s27-4	T	-	-	-	-	41.79	1423282
s27-5	T	-	-	-	-	1054.37	26450250
chess-6-5-0	F	149.52	51394	2.55	44594	0.14	394
chess-6-5-1	T	184.88	65536	4.64	58736	4	14536
chess-8-5-0	F	1081.67	156399	550.75	156399	0.34	544
chess-8-5-1	T	1118.98	262144	836.21	262144	33	61832
chess-8-7-0	F	-	-	-	-	0.79	915
chess-8-7-1	T	-	-	-	-	60.89	61832
cont-4-4-0	F	3	610530	0.7	146050	0.03	2323
cont-4-4-1	T	5.1	1048934	1.49	328143	0.09	5514
cont-4-6-1	T	-	-	-	-	1.1	45467
cont-6-6-1	T	-	-	-	-	232.43	7013225

imposed a one hour time limit on each QBF solving run. We call the modified Quaffle that handles the circuit ODC augmented CNFs Quaffle-DC. Table 1 shows the performance comparison of using Quaffle on the original testcases (columns 3 and 4), using Quaffle-DC for the testcases with automatically generated don't care literals (columns 5 and 6) and using Quaffle-DC for the testcases with manually generated don't care literals (columns 7 and 8). Manually generated don't care literals are exactly the indicator variables for the conditional QBF solver. These literals are identified through understanding the QBF encoded problems. If these literals are true, then the propositional part of the QBF is true. We use "-" to indicate timeout of the one hour time limit. The label #SAT indicates the number of satisfying solutions enumerated for each QBF solving. For these experiments, we turned off the satisfiability directed non-chronological backtracking and learning which slowed both Quaffle and Quaffle-DC down significantly for these testcases. The result shows that exploiting the circuit ODC information greatly improves the performance of the Quaffle QBF solver. Moreover, Quaffle-DC runs faster on instances with manually generated ODC information than those with automatically extracted ODC information. The number of SAT solutions enumerated is far less in the case of manually extracted ODC information. But automatically extracted ODC information does not necessarily reduce the number of SAT leaves due to the chronological nature of satisfiability backtracking. Yet the run time is reduced significantly with automatically extracted ODC information. The reason for this is that early detection of satisfiability in a certain branch saves a huge amount of deduction effort for the QBF solver. The better performance of Quaffle-DC on manually extracted ODCs than automatically generated ODCs is likely because of the following two reasons: automatic extraction of the circuit ODC information generates far more don't care literals than actually needed for early satisfaction of the CNF and these extra don't care literals slow down the deduction procedure;

Table 2. Performance Comparison with Satisfiability Directed Learning

Testcase	T/F	Quaffle			Quaffle-DC-A		
		Time(s)	#SAT	Cube Length	Time(s)	#SAT	Cube Length
chess-4-5-0	F	47.48	11226	4285	14.37	6066	882
chess-4-5-1	T	31.45	7577	4284	17.95	7805	1002
chess-6-5-0	F	350.18	28943	10462	150.51	28970	2018
chess-6-5-1	T	626.42	51904	10462	293.3	53462	2467
cont-4-4-0	F	168.54	315190	495	0.36	1234	265
cont-4-4-1	T	564.09	1049230	495	2.59	8101	268

the automatically generated don't care literals may not capture the global don't care condition which result in more satisfied clauses than the local don't care literals, while the global don't care condition can be recognized by humans during the encoding process. Since we used chronological backtracking and no learning when a satisfying assignment is reached, the performance gain obtained by considering circuit ODCs is mostly the result of early detection of the satisfying partial assignments.

To evaluate how the presence of the circuit ODC information influences the satisfiability directed learning, we conducted another set of experiments with the satisfiability directed non-chronological backtracking and learning in Quaffle and Quaffle-DC turned on. We measured the average length (in terms of the number of literals) of the satisfying induced cubes. The manually generated global don't care literals are added to every clause, they do not correspond to the actual controlling signals of the entire circuit. Thus we cannot use these don't care literals for satisfiability directed learning. Therefore, we only use the automatically generated ODCs for this set of experiments. Table 2 shows that with don't care literals, the average length of the satisfying induced cubes is shorter, which means that satisfiability directed learning is more effective when considering circuit ODCs.

6   Conclusions

We have considered exploiting circuit ODC information in solving QBF instances in CNFs. We discussed the effects of the CNF encoding approaches on the valuation of a CNF under a partial assignment. The circuit ODC information needs to be explicitly encoded in the CNF if auxiliary variables are introduced while translating a Boolean formula into its CNF. Search based QBF solvers can use this information to backtrack early from a satisfying branch. If satisfiability directed learning is incorporated, then using the circuit ODC information may result in shorter satisfying cubes which better prune the satisfying search space. Our experimental results show that the circuit ODC information is very helpful for solving QBF instances where many combinations of small subsets of variables lead to satisfying assignments for the propositional parts of the formulas.

Acknowledgments

This work is inspired by the work of Carlos Ansotegui, Carla P. Gomes, and Bart Selman [6]. We especially thank Carlos Ansotegui for providing us the encoding of the Evader/Pursuer chess problem, the conditional Quaffle QBF solver and the insightful discussions. We would also like to thank Zhaohui Fu for providing the CNF to circuit translator for our experiments [25].

References

1. Rintanen, J.: Constructing conditional plans by a theorem prover. Journal of Artificial Intelligence Research **10** (1999) 323–352
2. Biere, A., Cimatti, A., Clarke, E.M., Zhu, Y.: Symbolic model checking without BDDs. In: Proceedings of Tools and Algorithms for the Analysis and Construction of Systems (TACAS'99). (1999)
3. Sheeran, M., Singh, S., Stålmark, G.: Checking safety properties using induction and a SAT-solver. In: Proceedings of the Third International Conference on Formal Methods in Computer-Aided Design (FMCAD 2000). (2000)
4. Gent, I.P., Rowley, A.G.D.: Encoding connect-4 using quantified boolean formulae. In: Modelling and Reformulating CSP. (2003) 78–93
5. Alur, R., Madhusudan, P., Nam, W.: Symbolic computational techniques for solving games. In: International Journal on Software Tools for Technology Transfer (STTT). (2005) 118–128
6. Ansótegui, C., Gomes, C.P., Selman, B.: The achilles' heel of QBF. In: Proceedings of the 21th National (US) Conference on Artificial Intelligence (AAAI). (2005) 275–281
7. Scholl, C., Becker, B.: Checking equivalence for partial implementations. In: Proceedings of the Design Automation Conference (DAC). (2001)
8. Katz, J., Hanna, Z., Dershowitz, N.: Space-efficient bounded model checking. In: Proceedings of the IEEE/ACM Design, Automation, and Test in Europe (DATE). (2005) 686–687
9. Tseitin, G.S.: On the Complexity of Derivation in Propositional Calculus. In: Studies in Constructive Mathematics and Mathematical Logic. (1968) 115–125
10. Ganai, M.K., Ashar, P., Gupta, A., Zhang, L., Malik, S.: Combining strengths of circuit-based and CNF-based algorithms for a high-performance SAT solver. In: Proceedings of the Design Automation Conference (DAC). (2002) 747–750
11. Lu, F., Wang, L., Cheng, K., Huang, R.: A circuit SAT solver with signal correlation guided learning. In: Design, Automation, and Test in Europe (DATE '03), Munich, Germany. (2003) 892–897
12. Gupta, A., Gupta, A., Yang, Z., Ashar, P.: Dynamic detection and removal of inactive clauses in SAT with application in image computation. In: Design Automation Conference. (2001) 536–541
13. Velev, M.N.: Encoding global unobservability for efficient translation to sat. In: Proc. of The Seventh International Conference on Theory and Applications of Satisfiability Testing (SAT2004). (2004)
14. Velev, M.N.: Comparison of schemes for encoding unobservability in translation to sat. In: Asia and South Pacific Design Automation Conference (ASP-DAC). (2005) 1056–1059

15. Safarpour, S., Veneris, A.G., Drechsler, R., Lee, J.: Managing don't cares in boolean satisfiability. In: Proceedings of the IEEE/ACM Design, Automation, and Test in Europe (DATE). (2004) 260–265
16. Fu, Z., Yu, Y., Malik, S.: Considering circuit observability don't cares in CNF satisfiability. In: Proceedings of the IEEE/ACM Design, Automation, and Test in Europe (DATE). (2005) 1108–1113
17. Bartlett, K., Brayton, R.K., Hachtel, G.D., Jacoby, R.M., Morrison, C.R., Rudell, R.L., Sangiovanni-Vincentelli, A., Wang, A.R.: Multi-level logic minimization using implicit don't cares. IEEE Transactions on Computer-Aided Design of Integrated Circuits and Systems **7** (1988) 723–740
18. Moskewicz, M.W., Madigan, C.F., Zhao, Y., Zhang, L., Malik, S.: Chaff: Engineering an efficient SAT solver. In: Proceedings of the Design Automation Conference (DAC). (2001)
19. Davis, M., Logemann, G., Loveland, D.: A machine program for theorem proving. Communications of the ACM **5** (1962) 394–397
20. Zhang, L., Malik, S.: Towards a symmetric treatment of satisfaction and conflicts in quantified Boolean formula evaluation. In: Proceedings of 8th International Conference on Principles and Practice of Constraint Programming (CP2002). (2002)
21. Letz, R.: Lemma, model caching in decision procedures for quantified Boolean formulas. In: International Conference on Automated Reasoning with Analytic Tableaux and Related Methods (Tableaux2002). (2002)
22. Giunchiglia, E., Narizzano, M., Tacchella, A.: Learning for quantified Boolean logic satisfiability. In: Proceedings of the 18th National (US) Conference on Artificial Intelligence (AAAI). (2002)
23. Zhang, L., Madigan, C.F., Moskewicz, M.W., Malik, S.: Efficient conflict driven learning in a Boolean satisfiability solver. In: Proceedings of the IEEE/ACM International Conference on Computer-Aided Design (ICCAD). (2001)
24. Zhang, L., Malik, S.: Conflict driven learning in a quantified Boolean satisfiability solver. In: Proceedings of the IEEE/ACM International Conference on Computer-Aided Design (ICCAD). (2002)
25. Fu, Z., Malik, S.: Extracting logic circuit structure from conjunctive normal form descriptions. In: submission to the Design Automation Conference (DAC). (2006)
26. Mneimneh, M., Sakallah, K.: Computing vertex eccentricity in exponentially large graphs: QBF formulation and solution. In: Sixth Intermational Conference on Theory and Application of Satisfiability Testing. (2003)
27. Tang, D., Yu, Y., Ranjan, D., Malik, S.: Analysis of DPLL Based Search Algorithms for Satisfiability of Quantified Boolean Formulas arising from Circuits. In: Proc. of The Seventh International Conference on Theory and Applications of Satisfiability Testing (SAT2004). (2004)

QBF Modeling: Exploiting Player Symmetry for Simplicity and Efficiency

Ashish Sabharwal[1], Carlos Ansotegui[2], Carla P. Gomes[1],
Justin W. Hart[1], and Bart Selman[1]

[1] Dept. of Computer Science, Cornell University, Ithaca, NY 14853-7501, U.S.A.
{sabhar, gomes, jwh38, selman}@cs.cornell.edu
[2] Dept. of Computer Science, Universitat de Lleida, E-25001 Lleida, Spain
carlos@diei.udl.es

Abstract. Quantified Boolean Formulas (QBFs) present the next big challenge for automated propositional reasoning. Not surprisingly, most of the present day QBF solvers are extensions of successful propositional satisfiability algorithms (SAT solvers). They directly integrate the lessons learned from SAT research, thus avoiding re-inventing the wheel. In particular, they use the standard conjunctive normal form (CNF) augmented with layers of variable quantification for modeling tasks as QBF. We argue that while CNF is well suited to "existential reasoning" as demonstrated by the success of modern SAT solvers, it is far from ideal for "universal reasoning" needed by QBF. The CNF restriction imposes an inherent asymmetry in QBF and artificially creates issues that have led to complex solutions, which, in retrospect, were unnecessary and sub-optimal. We take a step back and propose a new approach to QBF modeling based on a game-theoretic view of problems and on a dual CNF-DNF (disjunctive normal form) representation that treats the existential and universal parts of a problem symmetrically. It has several advantages: (1) it is generic, compact, and simpler, (2) unlike fully non-clausal encodings, it preserves the benefits of pure CNF and leverages the support for DNF already present in many QBF solvers, (3) it doesn't use the so-called indicator variables for conversion into CNF, thus circumventing the associated illegal search space issue, and (4) our QBF solver based on the dual encoding (**Duaffle**) consistently outperforms the best solvers by two orders of magnitude on a hard class of benchmarks, even without using standard learning techniques.

1 Introduction

The automated propositional reasoning community has come a long way since the development of the first practical propositional satisfiability algorithms (SAT solvers) nearly a decade ago. SAT solvers have been successfully used on real-world problems from a variety of areas like hardware and software verification, planning, and scheduling. Quantified Boolean Formula (QBF) reasoning extends the scope of SAT to domains requiring adversarial analysis, like conditional planning [17], unbounded model checking [16, 3], and discrete games [7]. In the simplest case, consider a two-player game. Here a winning strategy is a partial game

A. Biere and C.P. Gomes (Eds.): SAT 2006, LNCS 4121, pp. 382–395, 2006.

tree that, for every possible game play of the opponent, indicates how to proceed so as to guarantee a win. This is more complex than the single-agent reasoning SAT solvers offer, and requires modeling and analyzing adversarial actions of another agent with competing interests. The QBF approach thus supports a much richer setting. However, it also poses new and sometimes unforeseen challenges. In terms of worst-case complexity, deciding the truth of a QBF is PSPACE-complete [18] whereas SAT is "only" NP-complete.[1] Even with very few quantification levels, the explosion in the search space is tremendous in practice. Further, as the winning strategy example indicates, even a solution to a QBF may require exponential space to describe, causing practical difficulties [2].

Nonetheless, several tools for deciding the truth of a given QBF (QBF solvers) have been developed, such as Quaffle [20], sKizzo [3], Quantor [4], QuBE [8], Semprop [10], Evaluate [5], Decide [15], and QRSat [13]. Most of these tools extend the concepts underlying many successful SAT solvers, which use the DPLL procedure [6] as their backbone. As a result, they inherit conjunctive normal form (CNF) as the input representation, which has been the standard for SAT solvers for over a decade. Internally, many solvers also employ disjunctive normal form (DNF) in order to cache partial solutions for efficiency [21].

While the performance of QBF solvers has been promising, translating a QBF into a (much larger) SAT specification and using a good SAT solver is often faster in practice — a fact well-recognized and occasionally exploited [4, 3]. This motivates the need for further investigation into the design of QBF solvers and possible fundamental weaknesses in the modeling methods used.

The main contribution of this paper is a new generic QBF modeling technique that uses a dual CNF-DNF representation and, with a fairly straightforward adaptation of a modern QBF solver, improves the state of the art by two orders of magnitude on a set of computationally challenging benchmarks. The dual representation splits problem constraints into a CNF and a DNF part in a natural manner based on a game-theoretic view. Note that we do not go to fully non-clauses encodings, which also have promise but are unable to directly exploit rapid advances in CNF-based SAT solvers. We also differ from an independent dual CNF-DNF approach recently proposed [19] in that we do not convert a full CNF encoding into a logically equivalent full DNF encoding and provide both to the solver. Our approach exploits the representational power of DNF to simplify the model while addressing the issues associated with pure CNF representations.

We think of a problem $\mathcal{P}$ as a two-player game $\mathcal{G}$ with a bounded number of turns. This is different from the standard interpretation of a QBF as a game [14]; in our approach, one must formulate the higher level problem $\mathcal{P}$ as a game $\mathcal{G}$ *before* modeling it as a QBF. The sets of "rules" to which the players of $\mathcal{G}$ are bound may differ from one player to the other. In general, any QBF reasoning task has a natural game playing interpretation at a high level, which we exploit. We illustrate this correspondence with a circuit minimization problem [cf. 14]

---

[1] Assuming P $\neq$ NP, PSPACE-complete problems are significantly harder than NP-complete problems; cf. [14].

that underlies practical QBF benchmarks involving adder circuits and sorting networks [12], a graph coloring problem, and a chess-like problem [11, 1].

The key idea underlying our approach is to exploit a dichotomy between the players: *we model rules for the existential player as CNF clauses, (the negations of) rules for the universal player as DNF terms, and split game state information equally into clauses and terms.* This symmetric dual format places "equal responsibility" on the two players, in stark contrast with current QBF encodings which tend to leave most work for the existential player. We are able to avoid many pitfalls of current techniques while increasing the reasoning efficiency. In particular, we bring to QBF solvers *unit propagation across quantifiers* which has been a stumbling block so far. We are also able to completely avoid the use of the so-called auxiliary indicator variables and the associated illegal search space issue inherent in the translation of QBF problems into pure CNF form [1].[2]

We evaluate our approach with `Duaffle` (short for `dual-Quaffle`), our QBF solver for the dual encoding. It is an adaptation of the solver `Quaffle`, which already supports DNF terms for solution learning. Our empirical evaluation on computationally difficult chess-based instances shows that `Duaffle` consistently outperforms the best solvers by several orders of magnitude. More generally, this paper demonstrates that by taking a step back and re-thinking basic modeling techniques, one can significantly extend the reach of QBF reasoning systems.

2   Preliminaries

We begin by discussing how adversarial tasks can be treated as games, and then describe our QBF notation and a systematic way of encoding games as QBF.

2.1   Treating Adversarial Tasks as Games

Most discrete adversarial tasks have a natural albeit somewhat non-traditional game playing interpretation with an existential and a universal player. Interestingly, the rules for the existential player are often different from those of the universal player. We illustrate this with two simple but concrete examples.

Example 1. **The Circuit Minimization Problem:** Given a Boolean circuit C, is there a smaller circuit that computes the same function as C? Observe that the answer is yes iff there *exists* a circuit C_E such that $size(C_E) < size(C)$ and *for all* inputs ρ, $C_E(\rho) = C(\rho)$. This problem lies in the complexity class Σ_2^P, which is believed to be beyond NP and is characterized by QBFs with exactly two levels of quantification beginning with the existential [cf. 14].

We can think of circuit minimization as a game with two turns. First, the existential player E commits to a circuit C_E by specifying the type its gates, their connections, and the output line. The rules for E are that C_E must be a legal circuit with $size(C_E) < size(C)$. Second, the universal player U produces

---

[2] While this can also be handled by providing semantic information about auxiliary variables as additional input [1], this has the undesirable effect of mixing the declarative nature of problem specification with the procedural nature of solution technique.

an input ρ and the polynomial-size computations of C_E and C on ρ. The rule for U is that it must correctly compute $C_E(\rho)$ and $C(\rho)$. The goal of E is to ensure that $C_E(\rho) = C(\rho)$ no matter how ρ is chosen.   □

Example 2. **The Chromatic Number Problem:** Given a graph $\mathcal{H}$ and a positive integer k, does $\mathcal{H}$ have chromatic number k? The chromatic number of a graph is the minimum number of colors needed to color its vertices so that no two adjacent vertices have the same color [cf. 14]. Observe that the answer is yes iff $\mathcal{H}$ has a legal coloring with k colors but no legal coloring with $k - 1$ colors.

We can again think of this as a game between E and U. First, E produces a coloring σ of the vertices of $\mathcal{H}$. The rule for E is that σ must be a legal k-coloring respecting the edges of $\mathcal{H}$. Second, U produces a second coloring τ of the vertices of $\mathcal{H}$. The rule for U is that τ must be a legal $(k - 1)$-coloring of $\mathcal{H}$. E wins iff she is able to produce a valid σ and U is not able to produce a valid τ.   □

2.2   Quantified Boolean Formulas

Let $V = \{x_1, \ldots, x_n\}$ be a set of n propositional (Boolean, TRUE-FALSE, 1-0) variables. A conjunctive normal form or CNF formula over V is a conjunction of clauses, where each clause is a disjunction of literals, and a literal is a variable or its negation. A disjunctive normal form or DNF formula is a disjunction of terms (sometimes called cubes), where each term is a conjunction of literals.

A Quantified Boolean Formula (QBF) is a Boolean formula in which variables are quantified as existential ($\exists$) or universal ($\forall$). We will use the term QBF for *totally quantified Boolean formulas in prenex form* beginning with $\exists$:

$$F \;=\; \exists x_1^1 \ldots \exists x_1^{t(1)} \; \forall x_2^1 \ldots \forall x_2^{t(2)} \; \ldots \; Q x_k^1 \ldots Q x_k^{t(k)} \;\; M$$

where M is a Boolean formula referred to as the *matrix* of F, x_i^j above are distinct and include all variables appearing in M, and Q is $\exists$ if k is odd and $\forall$ if k is even. Defining $V_i = \left\{ x_i^1, \ldots x_i^{t(i)} \right\}$ and using associativity within each level of quantification, we can simplify the notation to $F = \exists V_1 \forall V_2 \exists V_3 \ldots Q V_k \, M$. A QBF solver is an algorithm that determines the truth value of such formulas F, i.e., whether there exist values of variables in V_1 such that for every assignment of values to variables in V_2, and so on, M is satisfied (set to TRUE).

For two Boolean formulas G and G', $G = G'$ will denote *syntactic equality* (they "look" the same) and $G \equiv G'$ will denote semantic equality (they evaluate to the same truth value for every variable assignment). For two QBFs F and F', $F = F'$ will denote *syntactic equality*, while $F \equiv F'$ will denote semantic equality between the matrices (i.e., the Boolean parts) of F and F'.

2.3   QBF and Two-Player Games

A QBF $F = \exists V_1 \forall V_2 \ldots Q V_k \, M$ has a natural interpretation as a two-player game $\mathcal{G}$ (see standard texts, e.g. [14]). The idea is to have an existential player E and a universal player U, who take turns setting variables in $V_1, V_2, \ldots, V_k$ in order. If M is satisfied after all variables are set, E wins. Otherwise, U wins.

Our interest in this work, however, is in going the other direction, that is, treating arbitrary adversarial tasks as discrete games and modeling them as QBF. Given a discrete two-player game $\mathcal{G}$ with players E and U, a bound k on the total number of turns, and the guarantee that after k turns either E or U will be declared a winner (i.e., there is no "draw"), we can construct a QBF $F = \exists V_1 \forall V_2 \ldots Q V_k\ M$ that models $\mathcal{G}$ in the following manner.[3]

We will follow the systematic framework described by Ansotegui et al. [1]. It is based on a highly successful technique used in SAT-based planning [9] and can be applied to any well-defined discrete game $\mathcal{G}$ without draws. The variables of F model the possible *moves* of E and U as well as global *state* information maintained about the game as it is played. The possible moves in the ith turn naturally correspond to variables in V_i. The rules and goal of $\mathcal{G}$ are formulated as follows: (1) *precondition* and *effect* axioms for each move in relation to the game state before and after the move, (2) *mutual exclusion* axioms restricting a player to one move per turn, (3) *frame* axioms ensuring that parts of the game state not affected by the current move stay unchanged, (4) *initial state* axioms, and (5) *goal* axioms stating the winning conditions for one of the players chosen arbitrarily. With no draws, it clearly suffices to describe one player's goals.

The transition axioms for the ith turn are the conjunction of the precondition, effect, mutual exclusion, and frame axioms for that turn, denoted by $Tr^i = Pr^i \wedge Mf^i \wedge Me^i \wedge Fr^i$. With a bound k on the total number of turns in $\mathcal{G}$, all transition axioms for the existential player E and the universal player U can be grouped together as $Tr_E = Tr^1 \wedge Tr^3 \wedge \ldots \wedge Tr^{odd(k)}$ and $Tr_U = Tr^2 \wedge Tr^4 \wedge \ldots \wedge Tr^{even(k)}$, where $odd(k)$ and $even(k)$ denote the largest odd and even integers up to k, respectively. Let I denote the initial state axioms and G_E the goal axioms for E. The following Boolean formulas represent two alternative formulations of $\mathcal{G}$:

$$M_1 = I \wedge Tr_E \wedge (Tr_U \to G_E) \qquad M_2 = Tr_U \to (I \wedge Tr_E \wedge G_E) \qquad (1)$$

In general, the choice of the formulation is dictated by the requirements of the game being modeled. Formulation M_1 has the property that it evaluates to TRUE on a variable assignment iff (a) E adheres to all her rules *and* (b) either E achieves her goal or U violates his rules. This fits the game interpretations of the circuit minimization and graph coloring examples we saw in Sect. 2.1. In graph coloring, for instance, E *must* adhere to her rules of producing a valid k-coloring of $\mathcal{H}$ irrespective of whether U is able to produce a $(k-1)$-coloring. On the other hand, M_2 evaluates to TRUE iff *either* (a) E adheres to all her rules and achieves her goal, *or* (b) U violates his rules. This relieves E of all responsibility if U violates a rule. This formulation fits games like chess where E doesn't even need to continue playing the game according to her rules if U

---

[3] Interestingly, without the possibility of a draw, exactly one of E and U is guaranteed to have a winning strategy *even before they start playing the game*. This is because if E does not have a choice of moves that will make her win irrespective of the moves of U, then U's winning strategy is simply the "witness" of this fact. This corresponds to the only two possible evaluations of the QBF F, namely, TRUE and FALSE.

makes an illegal move; she is immediately declared the winner. While chess may also be formulated as M_1, using M_2 increases the reasoning efficiency.

Let S^i denote the state variables for $\mathcal{G}$ during the ith turn, A^i the move or action variables, and I^i a set of auxiliary "indicator" variables [1] used to detect when the formula may be declared satisfiable. Assuming k is odd, the complete CNF-based QBF formulation of $\mathcal{G}$ is given by:

$$\underbrace{\exists S^1 A^1 S^2}\ \ \underbrace{\forall A^2}\ \ \underbrace{\exists I^2 S^3 A^3 S^4}\ \ \ldots\ \ \underbrace{\forall A^{k-1}}\ \ \underbrace{\exists I^{k-1} S^k A^k S^{k+1}}\ \ M_i \qquad (2)$$

where $i \in \{1, 2\}$ is chosen based on the requirements of $\mathcal{G}$.

3   A New QBF Modeling Technique

In this section, we present a new QBF modeling technique based on a game-theoretic view of the underlying problem and a dual CNF-DNF representation. We also describe a QBF solver that uses this dual representation. We begin with the motivation behind using DNF.

CNF is the generally accepted input format for SAT solvers, and for two good reasons. First, many problems of interest are naturally expressed as a conjunction of several constraints. Second, before SAT solvers reach their goal of finding any *one* satisfying assignment, they typically encounter many falsifying assignments. It is therefore extremely beneficial for them to be able to deduce *locally* from a single CNF clause that all extensions of the current partial assignment will be falsifying. This forms the basis of DPLL-based backtrack search as well as heuristics for local search. On the other hand, due to universal quantification, a QBF solver must continue its search even after one satisfying assignment is found. It must therefore also detect *satisfiability* quickly. While the satisfaction of a CNF formula is a *global* property (all clauses must be satisfied), the satisfaction of a DNF formula can be guaranteed *locally* by evaluating an individual term.

This fact is exploited by QBF solvers that implement "solution learning" [21]. We take this observation a step further, using a combination of CNF and DNF as part of the input formula itself. Interestingly, adding DNF-based solution learning to the solver Quaffle, while theoretically natural and desirable, has limited practical impact on many problem instances over and above what "conflict clause" learning already achieves. In fact, the "conditional" variant of Quaffle called QuaffleC [1], which outperforms all state-of-the-art QBF solvers on our benchmarks, doesn't even use solution learning and DNF because of technical reasons. On the other hand, using DNF as part of the problem specification itself, as we will see, can be extremely effective.

Our modeling technique is based on the interpretation of adversarial tasks as games as discussed in Sect. 2.1. For modeling games as QBF, recall the generic framework of Sect. 2.3 and, in particular, the matrices $M_1 = I \wedge Tr_E \wedge (Tr_U \to G_E)$, $M_2 = Tr_U \to (I \wedge Tr_E \wedge G_E)$ in Eqn. (1) and the variable quantification in Eqn. (2). Two crucial observations about this representation of games motivate our modeling approach. (A) The implications $Tr_U \to \ldots$ in M_1 and

M_2 must be translated into a CNF formula by either expanding it out, which is typically costly, or adding new auxiliary variables, which cause problems with unit propagation and lead to the illegal search space issue. This is discussed in detail by Ansotegui et al. [1] and is handled using a fairly intricate machinery of individual and grouped "indicator" variables that flag the violation of any rule by U and "propagate" this information *globally* to all clauses. This makes the model undesirably complex. (B) The variable quantification in Eqn. (2) clearly depicts the "unequal treatment" of E and U. While U only decides actions at even-numbered turns, E is left with the responsibility of deciding actions at odd-numbered turns, maintaining the correct game state at every turn, and setting and propagating appropriate indicator variables when U violates a rule.

3.1   Modeling Games in a Dual CNF-DNF Form

Representing games as QBF in the framework of Sect. 2.3 boils down to specifying the initial state, the rules of the game, and the goal for a player as a Boolean formula, and quantifying appropriately over its variables. In our approach, we *model the rules for the existential player E as a CNF formula G and, unlike existing encoding techniques, model (the negations of) the rules for the universal player U as a DNF formula H*, respecting the following behavior: violation of a rule by E should directly falsify a clause of G and violation of a rule by U should directly *satisfy* a term of H. The dual formula will encode the winning conditions for E.

Before going into the details for the general setting, we illustrate the complete dual encoding for the chromatic number problem described earlier.

Example 3. **Dual Encoding of the Chromatic Number Problem:** Let $(\mathcal{H}, k)$ be the problem input. Let $n = |V(\mathcal{H})|$ and $[m]$ denote $\{1, 2, \ldots, m\}$. Recall the game playing interpretation of this problem from Sect. 2.1. The corresponding dual QBF encoding has nk existential variables $x_{i,j}$ with $i \in [n], j \in [k]$ for the rules of the existential player E, and $n(k-1)$ universal variables $y_{i,j}$ with $i \in [n], j \in [k-1]$ for the rules of the universal player U. Semantically, $x_{i,j}$ (or $y_{i,j}$) is TRUE iff E (or U, respectively) assigns color j to vertex i.

We construct a CNF formula F_{CNF} such that it is satisfied by a variable assignment iff the x variables form a legal k-coloring of $\mathcal{H}$. The first set of clauses in F_{CNF} will say that every vertex must be assigned some color by x, the second set will say that a vertex can get only one color, and the third set will say that if two vertices share an edge, then they do not get the same color. Formally,

$$F_{\text{CNF}} = \bigwedge_{i \in [n]} (x_{i,1} \vee \ldots \vee x_{i,k}) \;\wedge\; \bigwedge_{\substack{i \in [n] \\ j \neq j' \in [k]}} (\overline{x}_{i,j} \vee \overline{x}_{i,j'}) \;\wedge\; \bigwedge_{\substack{(i,i') \in E(\mathcal{H}) \\ j \in [k]}} (\overline{x}_{i,j} \vee \overline{x}_{i',j})$$

We now construct a DNF formula F_{DNF} which is satisfied by an assignment iff the y variables do *not* form a legal $(k-1)$-coloring of $\mathcal{H}$. The first set of terms in F_{DNF} will say that some vertex is not assigned any color by y, the second set will say that two different colors are assigned to a single vertex, and the third set will say that two adjacent vertices are assigned the same color. Formally,

$$F_{\mathrm{DNF}} = \bigvee_{i \in [n]} (\overline{y}_{i,1} \wedge \ldots \wedge \overline{y}_{i,k-1}) \quad \vee \quad \bigvee_{\substack{i \in [n] \\ j \neq j' \in [k-1]}} (y_{i,j} \wedge y_{i,j'}) \quad \vee \quad \bigvee_{\substack{(i,i') \in E(\mathcal{H}) \\ j \in [k-1]}} (y_{i,j} \wedge y_{i',j})$$

Finally, the dual QBF encoding of the chromatic number problem is given by

$$F_{\mathrm{chr\text{-}num}}(\mathcal{H}, k) \;=\; \exists x_{1,1} x_{1,2} \ldots x_{n,k} \; \forall y_{i,1} y_{i,2} \ldots y_{n,k-1} \; F_{\mathrm{CNF}} \wedge F_{\mathrm{DNF}}$$

The game playing interpretation implies that $F_{\mathrm{chr\text{-}num}}(\mathcal{H}, k)$ is TRUE iff the chromatic number of $\mathcal{H}$ is k. □

More generally, we begin by thinking of the rules for E and U as standard clauses encoding various axioms like preconditions and effects for each turn, as defined in Sect. 2.3. For E, these directly become part of the CNF portion. For U, we negate each of these clauses to obtain DNF terms, which directly become part of the DNF portion. The overall QBF encoding is created from the perspective of E by encoding conditions under which E would win. We illustrate the translation of rules into clauses and terms with a simple example.

Example 4. **The Game of Chess:** We use standard chess notation, with board columns a-g and rows 1-8. A typical set of precondition axioms would be: if the white player moves a rook from square b2 to square b4 at step s, then (a) that rook must be at b2 to begin with, (b) b3 must be empty, and (c) there must not be a white piece at b4. Treated as clauses, these translate into:

$$
\begin{aligned}
C_1 &= (\text{NOT move-wRook-b2-b4-s OR at-wRook-b2-s}) \\
C_2 &= (\text{NOT move-wRook-b2-b4-s OR empty-b3-s}) \\
C_3 &= (\text{NOT move-wRook-b2-b4-s OR NOT at-wPiece1-b4-s}) \\
C_4 &= (\text{NOT move-wRook-b2-b4-s OR NOT at-wPiece2-b4-s})
\end{aligned}
$$

The clause C_1, for instance, says that the CNF formula is immediately falsified if a white rook tries to move from square b2 to b4 without actually being there at step s. When modeling the white player as the existential player E, we use the above set of clauses. The axioms for the black player modeled as the universal player U state the converse, i.e., the conditions under which it violates a rule or fails to reach its goal, causing E to win. These are the *negations* of the standard axiom clauses, and are modeled as DNF terms of the form:

$$
\begin{aligned}
D_1 &= (\text{move-bRook-b2-b4-s AND NOT at-bRook-b2-s}) \\
D_2 &= (\text{move-bRook-b2-b4-s AND NOT empty-b3-s}) \\
D_3 &= (\text{move-bRook-b2-b4-s AND at-bPiece1-b4-s}) \\
D_4 &= (\text{move-bRook-b2-b4-s AND at-bPiece2-b4-s})
\end{aligned}
$$

The term D_2, e.g., says that the DNF formula is satisfied if a black rook attempts to move from b2 to b4 and the intermediate square b3 is non-empty. □

Given this symmetric way of encoding the rules for E and (the negations of) the rules for U as a collection of clauses and terms, respectively, we are ready to state the complete new encoding in the generic framework of Sect. 2.3. Recall

Eqn. (1) describing two possible matrices M_1 and M_2 of the QBF formulation of a game $\mathcal{G}$. Note that since there is no draw, $G_U \equiv \neg G_E$. We rewrite M_1 and M_2 in the following manner, which immediately suggests a natural split into CNF and DNF parts and how to logically combine them. We use M_i' to emphasize the syntactic difference with $M_i, i \in \{1, 2\}$; semantically $M_i' \equiv M_i$.

$$M_1' = \underbrace{(I \wedge Tr_E)}_{\text{CNF}} \wedge \underbrace{(\neg Tr_U \vee \neg G_U)}_{\text{DNF}} \qquad M_2' = \underbrace{(I \wedge Tr_E \wedge G_E)}_{\text{CNF}} \vee \underbrace{\neg Tr_U}_{\text{DNF}} \qquad (3)$$

We see that while M_1' combines the CNF and DNF parts with the AND operator, M_2' uses the OR operator. Which one of M_1' and M_2' is chosen for a particular game $\mathcal{G}$ at hand is dictated by the requirements of $\mathcal{G}$ as discussed in Sect. 2.3. Particularly, if the game stops as soon as U violates a rule, M_2' is preferred.

Recall that Tr_U is the conjunction of transition clauses for even-numbered turns, so that $\neg Tr_U$ is naturally expressed as a DNF formula with terms corresponding to negated original clauses:

$$\left. \begin{array}{rcl} \neg Tr_U &=& \neg Tr^2 \vee \neg Tr^4 \vee \ldots \vee \neg Tr^{even(k)} \\ \neg Tr^i &=& \neg Pr^i \vee \neg Mf^i \vee \neg Me^i \vee \neg Fr^i \end{array} \right\} \text{ DNF}$$

Similarly for $\neg G_U$. Equation (3) is the heart of our dual representation. All that remains to be specified is variable quantification. As in Sect. 2.3, we use S^i for state variables and A^i for move or action variables during the ith turn. (Indicator variables I^i are not used.) The complete dual CNF-DNF encoding of $\mathcal{G}$ is:

$$\exists S^1 \; \exists A^1 S^2 \; \forall A^2 S^3 \; \exists A^3 S^4 \; \forall A^4 S^5 \; \ldots \; Q A^k S^{k+1} \quad M_i' \qquad (4)$$

where $i \in \{1, 2\}$. Intuitively, this quantification says that given the initial state, E makes her move A^1 and brings $\mathcal{G}$ to state S^2 while obeying her rules, U then makes his move A^2 and brings $\mathcal{G}$ to state S^3 while obeying his rules, and so on, for k turns. Contrasting this with the original quantification in Eqn. (2) immediately highlights our symmetric treatment of the two players.

3.2 Duaffle: A QBF Solver Using the Dual Encoding

We adapted the QBF solver Quaffle to create a new solver Duaffle (short for dual-Quaffle) that determines the truth value of QBF formulas in the dual CNF-DNF form described above. The input format for Duaffle is a straightforward extension of the standard QDIMACS format [cf. 12]. Specifically, the formula is specified as a collection of CNF clauses and DNF terms along with variable quantification, as defined in Eqns. (3)-(4) and illustrated in Example 3. In addition, Duaffle takes as input a parameter specifying which of M_1' and M_2' in Eqn. (3) is used in the problem formulation. We identify these two formulations with the Boolean operator that is used to combine the corresponding CNF and DNF parts, namely, AND and OR.

In general, the behavior of a QBF solver with a mix of CNF and DNF as input is defined by what we call its *solver policy*: the actions it takes when it encounters any of the nine combinations of the CNF and DNF parts being

		DNF part		
		U	F	T
	U	BRN	UNS	BRN
CNF part	F	UNS	UNS	UNS
	T	BRN	UNS	SAT

(a) **Duaffle** with AND

		DNF part		
		U	F	T
	U	BRN	BRN	SAT
	F	BRN	UNS	SAT
	T	SAT	SAT	SAT

(b) **Duaffle** with OR

		DNF part		
		U	F	T
	U	BRN	BRN	SAT
	F	(UNS)	UNS	SAT
	T	SAT	SAT	SAT

(c) **Duaffle** with OR optimized for pure games

Fig. 1. Solver policies of **Duaffle** and the optimization for pure games

undetermined (denoted U), falsified (F), or satisfied (T) by a partial variable assignment. The possible actions include declaring the current branch unsatisfiable (UNS), declaring it satisfiable (SAT), or continuing to branch further by setting more variables (BRN). **Duaffle** implements two policies that correspond to the AND and OR dual formulations. These are given in Figure 1(a)-(b).

Implementation: Modern QBF solvers such as **Quaffle** already have the data structures and reasoning methods to support the DNF format we need. These are used for solution learning. The input format of **Quaffle** is pure CNF with quantification. **Duaffle** is created by adapting **Quaffle** so as to receive a dual CNF-DNF input, follow the solver policies in Fig. 1(a)-(b), and use a modified constraint propagation mechanism necessary for our dual formulation.

Quaffle assumes certain restrictions on the CNF and DNF formulas it operates on, most notably that the DNF part logically implies the CNF part (because DNF terms are added only through solution learning). Besides resulting in a different solver policy than what we need, this also makes **Quaffle**'s constraint propagation mechanism unsuitable for **Duaffle**. Consider a simple quantified DNF term: $\forall x \exists y \ (x \wedge y)$. Let $F = F_{CNF} \wedge F_{DNF}$ be the complete formula. In the game-playing interpretation, the goal of the universal player U is to make F FALSE. If U sets $x = $ TRUE, the existential player E can set $y = $ TRUE, so that $F_{DNF} = $ TRUE. When $F_{DNF} \rightarrow F_{CNF}$ (the working assumption of **Quaffle**), this implies $F_{CNF} = $ TRUE, so that F itself is satisfied and U loses. Therefore, U can safely infer from the DNF term $(x \wedge y)$ that x must be set to FALSE. In general, **Quaffle** can ignore variables with deeper existential (universal) quantification when performing standard unit propagation on a universal (existential, resp.) variable in a term (clause, resp.), achieving faster propagation.

In **Duaffle**, where $F_{DNF} \not\rightarrow F_{CNF}$, such inference by U would be incorrect. When $x = $ TRUE and E sets $y = $ TRUE to satisfy the DNF term $(x \wedge y)$, this could make a clause in F_{CNF} FALSE, so that F is falsified and U still wins. One must therefore ignore quantification levels and revert back to a simpler SAT-type notion of unit propagation: a universal (or existential) variable is implied by a term (or clause, resp.) iff all other literals in it are TRUE (or FALSE, resp.). Fortunately, the cost incurred by the removal of quantifier-sensitive unit propagation is more than paid off by the benefits of the dual model, such as propagation across quantifiers (see Sect. 4). Partly due to these reasons, the experimental results we report are based on `Duaffle⁻`, a restricted version of `Duaffle` with no conflict learning or solution learning. If today's SAT and QBF solvers are any indication, the performance of `Duaffle⁻` can only improve by re-integrating learning.

Optimization: Figure 1(c) depicts an optimization to `Duaffle` when using the OR formulation (i.e., matrix M_2') on "pure" games. Recall that M_2' can be used for any game in which E immediately wins as soon as U violates a rule. Such games are typically *pure* in the sense that they also follow the converse: U immediately wins if E violates a rule. This converse is not captured by the OR connective in M_2'. The optimization for the solver policy is the following: if the DNF part is still undetermined but the CNF part is FALSE, declare the branch to be UNS and backtrack. The correctness of this relies on the top-down structure of `Quaffle`, which sets variables respecting the quantification order. As a result, the DNF part being undetermined and the CNF part being FALSE imply that the game has indeed already been played according to the rules till the current turn.

4   Experimental Results

We evaluated our approach on a challenging set of QBF formulas encoding a rich variant of the game of chess. This game fits well in the M_2' dual formulation using the OR connective.

The Game xChess: xChess is based on Evader-Pursuer, a chess-like game introduced as a QBF benchmark by Madhusudan et al. [11] and later extended to several pieces [1]. We generalize it further by introducing more refined movements of various pieces. The input is an $n \times n$ chess board with an initial configuration consisting of some white and black pieces, the rules defining legal moves of each piece, the maximum number k of turns, and the goal square g. The players take alternating turns as usual, starting with white. The white player wins iff the white king, K_w, is placed at g at or before step k. K_w is always part of the initial board configuration. We assume that k is odd.

The rules for the moves, which are part of the problem input for xChess, are defined as follows. The sets of legal moves for pawns and knights are defined as an arbitrary subset of their possible moves in standard chess. The set of legal moves for every other piece is defined by an 8-tuple, which denotes the maximum number of squares the piece can move in each of the eight directions (horizontal, vertical, and diagonal). Thus, one can *create* new kinds of pieces by appropriately defining the rules for their moves, yielding a fairly rich setting.

Table 1 summarizes the results obtained on several xChess instances on a 550 MHz 8 processor Intel Pentium III Linux machine with 4 GB shared memory. The first set of instances encode an unreachability argument based on the number of moves (details in Sect. 5). The second and third sets have a mix of wins for white and black, and range in hardness from being solved in a few seconds to several minutes to hours. These instances have an average of 7 quantifier alternations. We compare the performance of five state-of-the-art QBF solvers on a pure CNF encoding against `Duaffle⁻` (`Duaffle` without solution- or conflict-learning) with the pure games optimization on the dual encoding with the OR formalism. The

Table 1. QBF solvers on xChess instances. T/F indicates formula is TRUE (white wins) or FALSE (black wins). Run-time is in seconds. — denotes time-out after 1 hour, -m- denotes out of memory, and -e- denotes runtime error related to stack overflow.

xChess instance name	T/F	Pure CNF Encoding vars cls (×10³)		Quantor	Semprop	sKizzo	Quaffle	QuaffleC	New Dual Encoding vars cls trms (×10³)			Duaffle⁻
conf-r1	F	5	42	—	12	4.0	15	1.3	3	22	14	0.01
conf-r2	F	7	60	—	25	5.8	33	2.5	5	29	22	0.02
conf-r3	F	10	77	—	55	9.3	62	4.1	6	36	29	0.03
conf-r4	F	12	94	—	85	26	124	6.4	7	43	36	0.04
conf-r5	F	23	207	—	985	84	676	34	13	88	75	0.08
conf-r6	F	27	239	—	2042	73	713	49	15	101	88	0.10
conf1a	T	13	155	—	627	83	—	161	7	55	63	1.8
conf1b	F	13	155	—	682	176	2939	124	7	55	63	1.3
conf1c	T	13	155	-e-	659	804	—	156	7	55	63	2.1
conf1d	F	13	155	—	706	1930	1473	148	7	55	63	2.2
conf2a	T	9	83	—	—	—	—	438	4	24	35	65.9
conf2b	F	9	83	—	—	—	—	275	4	24	35	56.9
conf3a	T	17	176	—	—	-m-	—	653	12	94	62	5.2
conf3b	F	16	162	—	—	—	2128	327	11	79	62	2.2
conf4	F	17	163	—	—	—	—	274	11	73	74	32.0
conf5	F	8	77	—	1018	427	142	11	5	41	26	0.1
conf01	F	19	210	-e-	1225	492	—	539	9	61	99	6.4
conf02	F	12	100	-e-	93	30	6.0	1.0	7	12	69	0.0
conf03	T	9	88	—	—	1532	—	83	6	47	31	1.4
conf04	T	10	92	—	—	-e-	2352	100	7	47	37	3.5
conf05	F	15	181	-e-	3290	448	510	196	9	94	66	0.1
conf06	F	12	123	—	—	-m-	—	633	7	47	54	30.6
conf07	F	10	84	-e-	261	42	78	3.5	6	12	48	0.0
conf08	T	13	142	—	—	1509	—	1088	8	59	64	31.2

solvers used are the conditional solver QuaffleC [1], Quaffle [20], sKizzo version 0.8.1 [3], Semprop version 010604 [10], and Quantor version 2004.01.25 [4]. These were among the top five solvers in QBF Evaluation 2005 [12].

The results clearly show that the benchmark suite of xChess instances is challenging for the best available QBF solvers. While Semprop, sKizzo, and Quaffle solve many of the instances in a few minutes, QuaffleC performs the best on the pure CNF encoding. Surprisingly, Quantor was unable to solve any of the instances of xChess we considered. As the last column of the table shows, by using the dual encoding along with Duaffle⁻ optimized for pure games, we consistently achieve two orders of magnitude improvement even over QuaffleC.

The first set of xChess instances, conf-r1 to conf-r6, highlight an important benefit of the dual encoding, namely, *fast unit propagation across quantifiers*, which previous approaches did not achieve. The net effect is that while QuaffleC needs thousands of branching decisions and conflict-learning to solve these instances,

`Duaffle` solves them during its preprocessing stage by simple constraint propagation without even a single explicit branch. This is explained as follows. These instances are based on an "unreachability" argument, namely, the white player simply has one too few steps to make the white king, K_w, reach the goal square g, and therefore must lose. In our framework, this can be inferred by constraint propagation across quantifiers: if the distance between K_w and g after the white player's turn t is d (denoted $dist(K_w, g, t) = d$), then $dist(K_w, g, t+1) = d, dist(K_w, g, t+2) \geq d - 1, dist(K_w, g, t+3) \geq d - 1, dist(K_2, g, t+4) \geq d - 2$, and so on, till $dist(K_w, g, k) \geq 1$, where k is the total number of allowed turns. These distance inequalities manifest themselves in the sets of falsified location variables capturing squares at which K_w *cannot* be after t turns.

For the above inference to work, state information from turn t to $t + 2$ to $t + 4$, and so on, must be carried across intermediate turns of the black player through frame axioms (Sect. 2.3), which involve universal variables. Technically, a CNF clause can never imply and fix the value of *universal* variables at steps $t+1, t+3$, etc., hindering the process of determining the locations not reachable by K_w. With pure CNF, a solver must branch on intermediate universal variables and later learn that this was irrelevant. In the dual encoding, universal state variables for K_w are instead implied and set by DNF terms encoding frame axioms, bridging state information between consecutive existential layers.

Note also that the number of variables in the dual encodings of xChess instances is roughly a half of pure CNF encodings because auxiliary variables are not needed. Variables in the dual encoding correspond precisely to the set of possible moves and locations for each piece, making the QBF model very clean. The "rules" are split into CNF clauses and DNF terms in proportion to the richness of the sets of pieces the two players have in each instance.

5   Conclusion

This paper demonstrates that by using a well-designed combination of CNF and DNF formulas as the input for QBF solvers, one can avoid many issues traditionally associated with QBF reasoning. Most tasks one intends to model as QBF have natural interpretations as generalized two-player games. Such tasks fit well into our game-theoretic formalism and translate into our dual representation. In addition to being simpler and avoiding the illegal search space issue, the dual model enhances in QBF solvers an essential technique that has made SAT solvers highly successful, namely, constraint propagation, which is now achieved across quantifiers. Our solver `Duaffle` outperforms state-of-the-art solvers by orders of magnitude. Finally, we believe that the full potential of solution learning techniques, which were inhibited by a pure CNF input highly biased towards conflict learning, will be unveiled once learning is re-integrated into `Duaffle⁻`.

Acknowledgments

We thank the anonymous reviewers for helpful comments. This work was supported by the Intelligent Information Systems Institute, Cornell University

(AFOSR grant F49620-01-1-0076) and DARPA (REAL grant FA8750-04-2-0216). The work of Carlos Ansotegui was also partially supported by the *Ministerio de Educación y Ciencia*, Spain (projects TIN2004-07933-C03-03 and TIC2003-00950).

References

[1] C. Ansotegui, C. P. Gomes, and B. Selman. The Achilles' heel of QBF. In *20th AAAI*, pages 275–281, Pittsburgh, PA, July 2005.

[2] M. Benedetti. Extracting certificates from quantified Boolean formulas. In *19th IJCAI*, pages 47–53, Edinburgh, Scotland, July 2005.

[3] M. Benedetti. sKizzo: a suite to evaluate and certify QBFs. In *20th CADE*, volume 3632 of *LNCS*, pages 369–376, Tallinn, Estonia, July 2005.

[4] A. Biere. Resolve and expand. In *7th SAT*, volume 3542 of *LNCS*, pages 59–70, Vancouver, BC, Canada, May 2004. Selected papers.

[5] M. Cadoli, M. Schaerf, A. Giovanardi, and M. Giovanardi. An algorithm to evaluate QBFs and its experimental evaluation. *J. Auto. Reas.*, 28(2):101–142, 2002.

[6] M. Davis, G. Logemann, and D. Loveland. A machine program for theorem proving. *CACM*, 5:394–397, 1962.

[7] I. P. Gent and A. G. Rowley. Encoding Connect-4 using quantified Boolean formulae. In *Work. Modelling and Reform. CSP*, pages 78–93, Ireland, Sept. 2003.

[8] E. Giunchiglia, M. Narizzano, and A. Tacchella. QUBE: A symtem for deciding QBFs satisfiability. In *IJCAR*, vol. 2083 of *LNCS*, pg. 364–369, Italy, June 2001.

[9] H. A. Kautz and B. Selman. Planning as satisfiability. In *Proc., 10th Euro. Conf. on AI*, pages 359–363, Vienna, Austria, Aug. 1992.

[10] R. Letz. Lemma and model caching in decision procedures for quantified Boolean formulas. In *TABLEAUX*, vol. 2381 of *LNCS*, pg. 160–175, Denmark, July 2002.

[11] P. Madhusudan, W. Nam, and R. Alur. Symbolic computation techniques for solving games. *Elec. Notes TCS*, 89(4), 2003.

[12] M. Narizzano and A. Tacchella (Organizers). QBF 2005 evaluation, June 2005. URL http://www.qbflib.org/qbfeval/2005.

[13] C. Otwell, A. Remshagen, and K. Truemper. An effective QBF solver for planning problems. In *Proc. MSV/AMCS*, pages 311–316, Las Vegas, NV, June 2004.

[14] C. H. Papadimitriou. *Computational Complexity*. Addison-Wesley, 1994.

[15] J. Rintanen. Improvements to the evaluation of quantified Boolean formulae. In *16th IJCAI*, pages 1192–1197, Stockholm, Sweden, July 1999.

[16] J. Rintanen. Partial implicit unfolding in the Davis-Putnam procedure for quantified Boolean formulae. In *8th Intl. Conf. Logic for Prog., AI, and Reason.*, volume 2250 of *LNCS*, pages 362–376, Havana, Cuba, Dec. 2001.

[17] J. Rintanen. Constructing conditional plans by a theorem prover. *JAIR*, 10: 323–352, 1999.

[18] L. J. Stockmeyer and A. R. Meyer. Word problems requiring exponential time. In *Conf. Record of 5th STOC*, pages 1–9, Austin, TX, Apr.-May 1973.

[19] L. Zhang. Solving QBF by combining conjunctive and disjunctive normal forms. In *21th AAAI*, Boston, MA, July 2006. To appear.

[20] L. Zhang and S. Malik. Conflict driven learning in a quantified Boolean satisfiability solver. In *ICCAD*, pages 442–449, San Jose, CA, Nov. 2002.

[21] L. Zhang and S. Malik. Towards a symmetric treatment of satisfaction and conflicts in QBF evaluation. In *8th CP*, pages 200–215, Ithaca, NY, Sept. 2002.

Solving #SAT Using Vertex Covers

Naomi Nishimura[1,*], Prabhakar Ragde[1,*], and Stefan Szeider[2,**]

[1] School of Computer Science, University of Waterloo,
Waterloo, Ontario, N2L 3G1, Canada
{nishi, plragde}@uwaterloo.ca
[2] Department of Computer Science, Durham University,
Durham DH1 3LE, England, United Kingdom
stefan.szeider@durham.ac.uk

Abstract. We propose an exact algorithm for counting the models of propositional formulas in conjunctive normal form (CNF). Our algorithm is based on the detection of strong backdoor sets of bounded size; each instantiation of the variables of a strong backdoor set puts the given formula into a class of formulas for which models can be counted in polynomial time. For the backdoor set detection we utilize an efficient vertex cover algorithm applied to a certain "obstruction graph" that we associate with the given formula. This approach gives rise to a new hardness index for formulas, the clustering-width. Our algorithm runs in uniform polynomial time on formulas with bounded clustering-width.

It is known that the number of models of formulas with bounded clique-width, bounded treewidth, or bounded branchwidth can be computed in polynomial time; these graph parameters are applied to formulas via certain (hyper)graphs associated with formulas. We show that clustering-width and the other parameters mentioned are incomparable: there are formulas with bounded clustering-width and arbitrarily large clique-width, treewidth, and branchwidth. Conversely, there are formulas with arbitrarily large clustering-width and bounded clique-width, treewidth, and branchwidth.

1 Introduction

#SAT is the problem of determining the number of satisfying truth assignments or models of a given propositional formula in conjunctive normal form (CNF). This problem is computationally equivalent to several problems that arise in automatic reasoning and artificial intelligence. However, since the problem is #P-complete (Valiant [27]), it is very unlikely that it can be solved in polynomial time. #SAT remains #P-hard even for monotone 2CNF formulas and Horn 2CNF formulas, and it is NP-hard to approximate the number of models of a formula with n variables within $2^{n^{1-\epsilon}}$ for $\epsilon > 0$. This approximation hardness holds also for monotone 2CNF formulas and Horn 2CNF formulas [23].

---

[*] Supported by the Natural Science and Engineering Research Council of Canada.
[**] Research partially supported by the Nuffield Foundation (NAL/01012/G).

A. Biere and C.P. Gomes (Eds.): SAT 2006, LNCS 4121, pp. 396–409, 2006.

An alternative to restricting the language of formulas is to impose *structural restrictions* in terms of certain (hyper)graphs associated with formulas. In particular, graph parameters that restrict the structure of associated primal graphs, incidence graphs, and formula hypergraphs have been considered; see Sect. 8 for definitions of the various graphs and graph parameters. Bacchus, Dalmao, and Pitassi [1] propose an algorithm that solves #SAT in time $n^{O(1)}2^{O(k)}$ for formulas with n variables whose formula hypergraphs have *branchwidth* k. The algorithm is based on the DPLL procedure and uses caching techniques for an efficient reuse of solutions for subproblems. A similar time complexity can be achieved by restricting the *treewidth* of primal graphs and by dynamic programming on tree-decompositions; this approach is described by Gottlob, Scarcello, and Sideri [12] for SAT and can be extended to #SAT in a straight-forward way. Bounding the *clique-width* of directed incidence graphs yields larger classes of formulas for which #SAT is tractable: Fisher, Makowsky, and Ravve [8] obtain an algorithm for #SAT by combining Oum and Seymour's approximation algorithm for clique-width [21] with a general result of Courcelle, Makowsky, and Rotics [4] on counting problems expressible in a certain fragment of Monadic Second Order Logic. The algorithm solves #SAT in time $n^{O(1)}O(f(k))$ for formulas with n variables whose directed incidence graphs have clique-width k; here f denotes a simply exponential function. The latter result is more general than the results for bounded treewidth and branchwidth in the sense that every class of formulas with bounded treewidth or bounded branchwidth also has bounded clique-width; however, there are classes of formulas with bounded clique-width but unbounded treewidth and unbounded branchwidth, see Sect. 8. Practical application of the clique-width based algorithm is, however, very limited due to a huge hidden constant in the estimation of its running time.

Note that the algorithms considered above are so-called *fixed-parameter algorithms*, since the bound on the running time is, although exponential in the parameter k, uniformly polynomial in n. The main advantage of fixed-parameter algorithms is that the running time increases moderately when n becomes large, in contrast to algorithms with running time $n^{O(k)}$. We will review the basic concepts of parameterized complexity in Sect. 2.2.

1.1   Our Approach: Backdoor Sets

The concept of strong backdoor sets with respect to a base class $\mathcal{C}$ of formulas was introduced by Williams, Gomes, and Selman [28] as a tool for analyzing the performance of local search SAT algorithms. Backdoor sets have recently received a lot of attention in satisfiability research [14,16,18,20,24,26].

A set B of variables of a formula F is a *strong $\mathcal{C}$-backdoor set* if for all truth assignments $\tau : B \to \{0,1\}$, the restriction $F[\tau]$ of F to τ belongs to the base class $\mathcal{C}$. Note that if a strong $\mathcal{C}$-backdoor set of size k is found, then we can decide the satisfiability of the given formula by deciding the satisfiability of 2^k formulas that belong to the base class $\mathcal{C}$. Based on this concept, Nishimura, Ragde, and Szeider [20] propose algorithms for SAT that search for strong backdoor sets of bounded size with respect to the base classes HORN and 2CNF. The detection of strong backdoor sets is based on the fact that a set B of variables is a strong HORN-backdoor set (strong 2CNF-backdoor set) of a formula F if and only if $F - B$ is a Horn formula (2CNF formula, respectively); here $F - B$ denotes the formula obtained from F by removing all the literals $x, \overline{x}$ for $x \in B$ from the clauses of F. We also say that B is a *deletion C-backdoor set* if $F - B \in C$. In general, deletion C-backdoor sets are not necessarily strong C-backdoor sets. However, if all subsets of a formula in C also belong to C (C is *clause-induced*), then indeed deletion C-backdoor sets are strong C-backdoor sets.

In this paper we extend the algorithmic use of backdoor sets for SAT to the counting problem #SAT. It is easy to see that the number of models of a formula F equals the sum over the number of models of the restrictions $F[\tau]$ for all truth assignments $\tau : B \rightarrow \{0, 1\}$ for a set B of variables of F. Hence, if we can solve #SAT for the elements of a base class C in polynomial time, then we can solve #SAT for a formula F in time $O(2^k n^{O(1)})$ provided that we know a strong C-backdoor set of F of size at most k. Hence, to convert the above considerations into an algorithm for #SAT, we need to identify a base class C for which the following holds:

1. #SAT can be solved in polynomial time for formulas in C, and
2. for a given formula F we can find strong C-backdoor sets of bounded size efficiently.

The second condition can be relaxed to deletion C-backdoor sets if C is clause-induced.

To this end, we introduce the clause-induced class CLU of *cluster formulas*. A cluster formula is a variable-disjoint union of so-called *hitting formulas*; any two clauses of a hitting formula clash in at least one literal. The known polynomial-time algorithm for computing the number of models of a hitting formula can be extended in a straight-forward way to compute the number of models of a cluster formula.

A strong CLU-backdoor set of size k of a formula F with n variables can obviously be found by exhaustive search, considering all $O(n^k)$ sets of k variables. This approach does not yield a fixed-parameter algorithm and becomes inefficient for large n even if k is small. We show in Sect. 5 that under a certain complexity theoretic assumption, there is no algorithm that is significantly faster than exhaustive search. We overcome this limitation by restricting by k the size of a smallest deletion CLU-backdoor set. We propose a fixed-parameter algorithm that either finds for a given formula a *strong* CLU-backdoor set of size at most k or decides that the given formula has no *deletion* CLU-backdoor set of size at most k.

To develop such an algorithm, we proceed as follows. We associate with every formula F a certain graph $G(F)$, the *obstruction graph* of F, which can be obtained in polynomial time. The vertex set of $G(F)$ is the set of variables of F. We show that every *vertex cover* of $G(F)$ is a strong CLU-backdoor set of F; recall that a vertex cover is a set S of vertices such that every edge is incident with a vertex in S. Now we can apply known vertex cover algorithms, e.g., the algorithm of Chen, Kanj, and Xia [3] for the detection of strong CLU-backdoor sets. Of related interest is Gramm et al.'s work [11] on a graph editing problem involving *cluster graphs* (i.e., disjoint unions of cliques).

1.2 Clustering-Width

We define the *clustering-width* of a formula F as the size of a smallest vertex cover of the obstruction graph of F. It follows from our results that the clustering-width of a formula F is a lower bound on the size of a smallest deletion CLU-backdoor set of F and an upper bound on the size of a smallest strong CLU-backdoor set of F.

Finally, we exhibit a class of formulas of bounded clustering-width for which all the parameters clique-width, branchwidth, and treewidth are unbounded. We also exhibit a class of formulas with unbounded clustering-width for which all the parameters clique-width, branchwidth, and treewidth are bounded. In other words, there are formulas that are easy for our algorithm and arbitrarily hard for the known algorithms, and formulas where the converse prevails.

It would be interesting to complement our theoretical results with empirical evidence on the significance of our new parameter. In particular, it would be interesting to know the clustering-width of CNF formulas that encode real-world instances from different domains. However, one must choose the encoding carefully in order to avoid a large clustering-width caused by the gadgets of the encoding itself. On the other hand, as indicated above, it can be checked very efficiently whether a CNF formula has small clustering-width. Hence, any other #SAT algorithm can be extended by a subroutine that checks the clustering-width and performs our algorithm if the clustering-width is small.

2 Preliminaries

2.1 SAT and #SAT

We consider propositional formulas in conjunctive normal form (CNF), represented as sets of clauses. That is, a *literal* is a (propositional) variable x or a negated variable $\overline{x}$; a *clause* is a finite set of literals not containing a complementary pair x and $\overline{x}$; a *formula* is a finite set clauses. For a literal $\ell = \overline{x}$ we write $\overline{\ell} = x$; for a clause C we put $\overline{C} = \{\,\overline{\ell} : \ell \in C\,\}$. For a clause C, $\mathrm{var}(C)$ denotes the set of variables x with $x \in C$ or $\overline{x} \in C$. Similarly, for a formula F we write $\mathrm{var}(F) = \bigcup_{C \in F} \mathrm{var}(C)$.

We say that two clauses C, D *overlap* if $C \cap D \neq \emptyset$; we say that C and D *clash* if C and $\overline{D}$ overlap. Note that two clauses can clash and overlap at the same time.

A *truth assignment* (or *assignment*, for short) is a mapping $\tau : X \rightarrow \{0,1\}$ defined on some set X of variables. We extend τ to literals by setting $\tau(\overline{x}) = 1 - \tau(x)$ for $x \in X$. $F[\tau]$ denotes the formula obtained from F by removing all clauses that contain a literal x with $\tau(x) = 1$ and by removing from the remaining clauses all literals y with $\tau(y) = 0$; $F[\tau]$ is the *restriction* of F to τ. Note that $\mathrm{var}(F[\tau]) \cap X = \emptyset$ holds for every assignment $\tau : X \rightarrow \{0,1\}$ and every formula F. A truth assignment $\tau : X \to \{0,1\}$ *satisfies* a formula F if $F[\tau] = \emptyset$. A truth assignment $\tau : \mathrm{var}(F) \to \{0,1\}$ that satisfies F is a *model* of F. We denote by $\#(F)$ the number of models of F. A formula F is *satisfiable* if $\#(F) > 0$. The satisfiability problem SAT is the problem of deciding whether a given formula is satisfiable. #SAT, the counting version of SAT, is the problem of determining $\#(F)$ for a given formula F. SAT and #SAT are complete problems for the complexity classes NP and #P, respectively.

The following concept of connectedness of formulas will be useful below. We call a formula F *connected* if for any two clauses $C, D \in F$ there exists a sequence of clauses $C_1, \ldots, C_r \in F$ such that $C_1 = C$, $C_r = D$, and $\mathrm{var}(C_i) \cap \mathrm{var}(C_{i+1}) \neq \emptyset$ holds for all $i \in \{1, \ldots, r-1\}$. A maximal connected subset of a formula is a *connected component*.

2.2 Parameterized Complexity

Next we give a brief and rather informal review of the most important concepts of parameterized complexity. For an in-depth treatment of the subject we refer the reader to other sources [7,19].

The instances of a parameterized problem can be considered as pairs (I, k) where I is the *main part* of the instance and k is the *parameter* of the instance; the latter is usually a non-negative integer. A parameterized problem is *fixed-parameter tractable* if instances (I, k) of size n can be solved in time $O(f(k)n^c)$ where f is a computable function and c is a constant independent of k.

The framework of parameterized complexity offers a *completeness theory*, similar to the theory of NP-completeness, that allows the accumulation of strong theoretical evidence that a parameterized problem is *not* fixed-parameter tractable. This completeness theory is based on the *weft hierarchy* of equivalence classes $W[1], W[2], \ldots, W[P]$ of certain parameterized decision problems under *parameterized reductions*. A parameterized reduction is a straightforward extension of a polynomial-time many-one reduction that ensures a parameter for one problem maps into a parameter for another (see [7] for details).

Below we will refer to the following parameterized decision problem, which is known to be W[2]-complete [7].

HITTING SET
Instance: A family S of finite sets $S_1, \ldots, S_m$.
Parameter: An integer $k \geq 0$.
Question: Is there a subset $R \subseteq \bigcup_{i=1}^m S_i$ of size at most k such that $R \cap S_i \neq \emptyset$ for all $i = 1, \ldots, m$? (R is a *hitting set* of S)

3 Backdoor Sets

Consider a *base class* C of formulas for which the problems #SAT and recognition can be solved in polynomial time. Furthermore, consider a formula F and a set B of variables of F. A set $B \subseteq \mathrm{var}(F)$ is a *strong backdoor set* of F with respect to C (or *strong C-backdoor set*, for short) if $B \subseteq \mathrm{var}(F)$ and for every truth assignment $\tau : B \to \{0,1\}$ we have $F[\tau] \in \mathcal{C}$. For every formula F and every set $B \subseteq \mathrm{var}(F)$ we have $\#(F) = \sum_{\tau:B\to\{0,1\}} \#(F[\tau])$. Thus, if B is a strong $\mathcal{C}$-backdoor set of a formula F, then determining $\#(F)$ reduces to determining the number of satisfying assignments for $2^{|B|}$ formulas of the base class $\mathcal{C}$. Thus, when we have found a small strong $\mathcal{C}$-backdoor set of F, we can compute $\#(F)$ efficiently. A key question is whether we can find a small backdoor set if it exists. To study this question, we define for every base class $\mathcal{C}$ the following parameterized problem.

> STRONG $\mathcal{C}$-BACKDOOR
> *Input:* A formula F.
> *Parameter:* A positive integer k.
> *Question:* Does F have a strong $\mathcal{C}$-backdoor set of size at most k?

For base classes that have a certain property, we can relax the problem STRONG $\mathcal{C}$-BACKDOOR as follows. For a formula F and a set X of variables let $F - X$ denote the formula obtained from F by removing all literals x and $\overline{x}$ from the clauses of F. We call a set $B \subseteq \mathrm{var}(F)$ a *deletion backdoor set* with respect to a base class $\mathcal{C}$ (or *deletion $\mathcal{C}$-backdoor set*, for short) if $F - B \in \mathcal{C}$. Furthermore, we define a base class $\mathcal{C}$ to be *clause-induced* if for every $F \in \mathcal{C}$ and every $F' \subseteq F$, also $F' \in \mathcal{C}$.

Lemma 1. *Let F be a formula and $\mathcal{C}$ a clause-induced base class. Every deletion $\mathcal{C}$-backdoor set of F is also a strong $\mathcal{C}$-backdoor set.*

Proof. The result follows directly from the fact that $F[\tau] \subseteq F - X$ holds for every truth assignment $\tau : X \to \{0,1\}$. □

For a base class $\mathcal{C}$, deletion backdoor sets can be larger than strong backdoor sets. However, if the detection of strong $\mathcal{C}$-backdoor sets is fixed-parameter intractable, we can still hope that the detection of deletion $\mathcal{C}$-backdoor sets is fixed-parameter tractable. We state the corresponding parameterized problem:

> DELETION $\mathcal{C}$-BACKDOOR
> *Input:* A formula F.
> *Parameter:* A positive integer k.
> *Question:* Does F have a deletion $\mathcal{C}$-backdoor set of size at most k?

4   Hitting Formulas and Cluster Formulas

A formula is a *hitting formula* if any two of its clauses clash (see [17]). A *cluster formula* is the variable-disjoint union of hitting formulas. In other words, a formula is a cluster formula if and only if all its connected components are hitting formulas. We denote the class of all hitting formulas by HIT and the class of all cluster formulas by CLU.

The next lemma is due to an observation of Iwama [15].

Lemma 2. *A hitting formula F with n variables has exactly $2^n - \sum_{C \in F} 2^{n-|C|}$ models.*

Proof. Let F be a hitting formula with n variables. For a clause $C \in F$ let T_C denote the set of all truth assignments $\tau : \text{var}(F) \to \{0,1\}$ that *do not* satisfy C. Obviously $|T_C| = 2^{n-|C|}$ since T_C contains exactly those assignments that set all literals in C to 0. Since F is a hitting formula, the sets T_C and $T_{C'}$ are disjoint for any two distinct clauses $C, C' \in F$. Hence the lemma follows. □

Lemma 3. *#SAT can be solved in polynomial time for cluster formulas.*

Proof. If a formula F is the variable-disjoint union of formulas $F_1, \ldots, F_q$, then $\#(F) = \prod_{i=1}^{q} \#(F_i)$. Thus the result follows directly from Lemma 2. □

By means of the previous lemma we can consider CLU as the base class for a backdoor set approach to #SAT. Observe that CLU is clause-induced.

5   Finding Smallest Strong CLU-Backdoor Sets

In this section we show that the detection of strong CLU-backdoor sets is fixed-parameter intractable.

We shall use the following construction. Let D be a directed graph. We associate with D a formula F_D where every arc a of D corresponds to a variable x_a of F, and every vertex v of D corresponds to a clause C_v of F. The clause C_v contains the literals x_a for outgoing arcs a of v, and the literals $\overline{x_b}$ for incoming arcs b of v. Note that if D is the orientation of a complete graph, then F_D is a hitting formula.

Theorem 1. *The problem* STRONG CLU-BACKDOOR *is W[2]-hard.*

Proof. (Sketch.) We give a parameterized reduction from the W[2]-complete problem HITTING SET as defined in Sect. 2.2. Let $S = S_1, \ldots, S_m$ be an instance of HITTING SET; $\bigcup_{i=1}^{m} S_i = \{x_1, \ldots, x_n\}$. Let D be an orientation of a complete graph with $r = (m+1)(k+1)$ vertices. Consider the hitting formula F_D. We partition F_D into formulas $F_1, \ldots, F_m, H$ such that each of the partite sets contains exactly $k+1$ clauses. For $i = 1, \ldots, m$ we put

$$F_i' = \{ C \cup S_i : C \in F_i \}.$$

Finally, we put $C^* = \{\overline{x_1}, \ldots, \overline{x_n}\}$ and

$$F = \{C^*\} \cup \bigcup_{i=1}^{m} F_i' \cup H.$$

We can show that S has a hitting set of size at most k if and only if F has a strong CLU-backdoor set of size at most k. □

The NP-hardness of the non-parameterized version of STRONG CLU-BACKDOOR (where the parameter is taken as part of the input) follows from the proof of Theorem 1.

We will show in sections below that the concept of deletion backdoor sets can be used to find small strong backdoor sets with respect to CLU. Next we give an example that shows that for the base class CLU, smallest deletion backdoor sets can be larger that smallest strong backdoor sets.

Consider the formula

$$F = \{\{x_1, \ldots, x_n\}, \{\overline{x_1}, \ldots, \overline{x_n}, y_1, \ldots, y_n\}, \{\overline{y_1}, \ldots, \overline{y_n}\}\}.$$

Note that each of the variables of F forms a strong CLU-backdoor set of F; e.g., $B = \{x_1\}$ is a strong CLU-backdoor set. However, we need to delete at least n variables in order to obtain a cluster formula. Thus a smallest strong CLU-backdoor set of F has size 1, but every deletion CLU-backdoor set of F has size at least n.

6   Obstructions

In the following results, it is helpful to characterize cluster formulas in terms of obstructions. An *overlap obstruction* is a formula $\{C_1, C_2\}$ consisting of two clauses that overlap but do not clash. With an overlap obstruction we associate the following pair of sets of variables:

$$\{\mathrm{var}(C_1 \cap C_2), \ \mathrm{var}(C_1 \bigtriangleup C_2)\}.$$

Here $C_1 \bigtriangleup C_2$ denotes the symmetric difference $(C_1 \setminus C_2) \cup (C_2 \setminus C_1)$ of C_1 and C_2. A *clash obstruction* is a formula $\{C_1, C_2, C_3\}$ where C_1 and C_2 clash such that $(C_1 \setminus C_3) \cap \overline{C_2} \neq \emptyset$, C_2 and C_3 clash such that $(C_3 \setminus C_1) \cap \overline{C_2} \neq \emptyset$, and C_1 and C_3 do not clash. (Any two of the three clauses may overlap.) With a clash obstruction we associate the following pair of sets of variables:

$$\{\mathrm{var}((C_1 \setminus C_3) \cap \overline{C_2}), \ \mathrm{var}((C_3 \setminus C_1) \cap \overline{C_2})\}.$$

We say that an overlap or clash obstruction F' is an *obstruction of* a formula F if F' is a subset of F. A pair $\{X, Y\}$ of sets of variables is a *deletion pair* of F if the pair is associated with an overlap or clash obstruction of F. It follows from the definitions of overlap and clash obstructions that the two sets in a deletion pair are nonempty and disjoint.

Lemma 4. *A formula is a cluster formula if and only if it has no overlap or clash obstruction.*

Proof. If a formula F contains an overlap or clash obstruction, then there are two clauses $C, D \in F$ that belong to the same connected component of F but do not clash. Hence F is not a cluster formula.

Conversely, consider a formula F that does not contain any overlap or clash obstructions. We show that F is a cluster formula. Consider a connected component F' of F. If $|F| = 1$ then F' is a hitting formula; hence assume $|F| > 1$. We show that any two clauses of F' clash. Choose two arbitrary clauses $C, D \in F'$.

Since F' is connected, there is a sequence of clauses $C_1, \ldots, C_r \in F$ such that $C_1 = C$, $C_r = D$, and $\mathrm{var}(C_i) \cap \mathrm{var}(C_{i+1}) \neq \emptyset$ holds for all $i \in \{1, \ldots, r-1\}$. We observe that C_i and C_{i+1} clash for all and $i \in \{1, \ldots, r-1\}$ since otherwise C_i and C_{i+1} would form an overlap obstruction. It now follows inductively that the clauses C_1 and C_i clash for all $i \in \{3, \ldots, r\}$ since otherwise C_1, C_{i-1}, and C_i would form a clash obstruction. Thus, indeed, C and D clash. Whence F' is a hitting formula. □

The next result is a consequence of Lemma 4. We omit the proof due to space limitations.

Lemma 5. *Let F be a formula and $B \subseteq \mathrm{var}(F)$. If $F - B$ is a cluster formula, then $X \subseteq B$ or $Y \subseteq B$ holds for every deletion pair $\{X, Y\}$ of F.*

7   Finding Backdoor Sets Using Vertex Covers

For a formula F let G_F denote the graph with vertex set $\mathrm{var}(F)$; two variables x and y are joined in G_F by an edge if and only if there is a deletion pair $\{X, Y\}$ of F with $x \in X$ and $y \in Y$. We call G_F the *obstruction graph* of F. Note that the obstruction graph of a formula can be constructed in polynomial time.

We consider vertex covers of obstruction graphs. Recall that a *vertex cover* of a graph is a set of vertices that contains at least one end of every edge of the graph. It is NP-hard to determine, given a graph and an integer k, whether the graph has a vertex cover of size at most k. Parameterized by the size of the vertex cover, however, the problem is fixed-parameter tractable. In fact, vertex cover is the best studied problem in parameterized complexity with a long history of improvements. The current best worst-case time complexity for the parameterized vertex cover problem is due to Chen, Kanj, and Xia [3]:

Theorem 2. *Given a graph G on n vertices, one can find in time $O(1.273^k + nk)$ (and in polynomial space) a vertex cover of G of size at most k, or determine that no such vertex cover exists.*

The next two lemmas relate backdoor sets and vertex covers of obstruction graphs. The first is a direct consequence of Lemma 5.

Lemma 6. *Every deletion CLU-backdoor set of a formula F is a vertex cover of the obstruction graph of F.*

Lemma 7. *Every vertex cover of the obstruction graph of a formula F is a strong CLU-backdoor set of F.*

Proof. (Sketch.) Let B be a vertex cover of the obstruction graph of a formula F. Assume to the contrary that B is not a strong CLU-backdoor set of F. Thus, there is an assignment $\tau : B \to \{0, 1\}$ such that $F[\tau] \notin \mathrm{CLU}$. Let $B_0 = \{y \in B \cup \overline{B} : \tau(y) = 0\}$; i.e., B_0 is the set of all literals over variables of B that are mapped to 0 under τ. By Lemma 4, $F[\tau]$ contains overlap or clash obstructions.

We assume that $F[\tau]$ contains an overlap obstruction; for clash obstructions the argument is similar. Let C_1, C_2 be two clauses of $F[\tau]$ that overlap but do not clash. For the associated obstruction pair $\{X, Y\}$ with $X = \text{var}(C_1 \cap C_2)$ and $Y = \text{var}(C_1 \triangle C_2)$ choose $x \in X$ and $y \in Y$. By definition of $F[\tau]$ it follows that F contains clauses C_1', C_2' with $C_1 = C_1' \setminus B_0$ and $C_2 = C_2' \setminus B_0$. It follows that C_1' and C_2' overlap but do not clash, thus $\{C_1', C_2'\}$ is an overlap obstruction of F. We have $x \in X \subseteq \text{var}(C_1' \cap C_2')$ and $y \in Y \subseteq \text{var}(C_1' \triangle C_2')$. Thus xy is an edge of G_F. Since B is a vertex cover of G_F, either x or y must belong to B. This contradicts the fact that $\text{var}(F[\tau]) \cap B = \emptyset$. Whence it follows that B^* is indeed a strong CLU-backdoor set of F. $\qquad\square$

From Theorem 2 and the previous two lemmas we get immediately the main result of this section.

Theorem 3. *Given a formula with n variables together with its obstruction graph and an integer k, in time $O(1.273^k + nk)$ we can find a strong CLU-backdoor set of F of size at most k, or decide that the size of every deletion CLU-backdoor set of F exceeds k.*

8   Comparison with Other Parameters

In this section we introduce a general framework for comparing parameters that allow fixed-parameter algorithms for #SAT. Here we consider as a parameter any computable function p that assigns to each formula F a non-negative integer $p(F)$. We assume that the parameter is invariant under changing the names of variables.

The following three parameters arise from the considerations of this paper. We denote by $\text{str}_{\text{CLU}}(F)$ the size of a smallest strong backdoor set of a formula F with respect to CLU, and we denote by $\text{del}_{\text{CLU}}(F)$ the size of a smallest deletion backdoor set of F with respect to CLU. The *clustering-width* $\text{clu}(F)$ of F is the size of a smallest vertex cover of the obstruction graph of F. Consequently, HIT is the class of formulas with clustering-width 0. From Lemmas 1 and 6 we know that for every formula F the following holds:

$$\text{str}_{\text{CLU}}(F) \leq \text{clu}(F) \leq \text{del}_{\text{CLU}}(F). \tag{1}$$

For a parameter p we consider the following generic parameterized problem.

#SAT(p)
Instance: A formula F and a non-negative integer k such that $p(F) \leq k$.
Parameter: The integer k.
Question: What is the total number of models of F? (I.e., what is the number $\#(F)$?)

The definition of fixed-parameter tractability carries over from decision problems to counting problems in a natural way. Flum and Grohe [9] provide a framework of intractability of parameterized counting problems.

Note that the above formulation of #SAT(p) is a "promise problem" in the sense that we only need to consider instances (F, k) for which we can take as granted that $p(F) \leq k$ holds. However, for most parameters p considered in the sequel for which #SAT(p) is fixed-parameter tractable, deciding whether $p(F) \leq k$ actually holds is also fixed-parameter tractable with respect to the parameter k. An exception is the parameter del$_{\mathrm{CLU}}$; however, also in that case we do not depend on the promise as will be discussed below.

By Theorem 2, deciding whether clu(F) $\leq k$ is fixed-parameter tractable; if clu(F) $\leq k$, then it is also fixed-parameter tractable to produce a strong CLU-backdoor set B of F of size at most k. We then compute #(F) as the sum of #($F[\tau]$) over all truth assignments $\tau : B \rightarrow \{0, 1\}$. Whence we have the following corollary to Theorem 2.

Corollary 1. *The problem #SAT(clu) is fixed-parameter tractable.*

Note that the algorithm outlined above also checks whether the promise clu(F) $\leq k$ is true. Furthermore, from (1) it follows that every instance (F, k) of #SAT(del$_{\mathrm{CLU}}$) is also an instance of #SAT(clu). Whence Corollary 1 also implies fixed-parameter tractability of #SAT(del$_{\mathrm{CLU}}$).

Corollary 2. *The problem #SAT(del$_{\mathrm{CLU}}$) is fixed-parameter tractable.*

Although we do not know whether DELETION $\mathcal{C}$-BACKDOOR is fixed-parameter tractable, we emphasize that the algorithm for Corollary 2 will not produce an incorrect solution, even if the promise del$_{\mathrm{CLU}}(F) \leq k$ does not hold. Consider F and k with del$_{\mathrm{CLU}}(F) > k$. The algorithm checks whether clu(F) $\leq k$. If clu(F) $\leq k$, then the algorithm outputs the correct solution #SAT(F). If, however, clu(F) $> k$, then we know by (1) that also del$_{\mathrm{CLU}}(F) > k$, hence the algorithm can reject the input.

8.1   Treewidth, Branchwidth, and Clique- Width

Several parameters are defined in terms of the following directed and undirected graphs associated with a formula F. The *primal graph* $P(F)$ is the graph whose vertices are the variables of F, and where two variables x and y are joined by an edge if and only if F contains a clause C with $x, y \in \mathrm{var}(C)$. The *incidence graph* $I(F)$ is the bipartite graph where one vertex class consists of the variables of F, the other vertex class consists of the clauses of F; a variable x and a clause C are joined by an edge if and only if $x \in \mathrm{var}(C)$. The *directed* or *signed incidence graph* $I_d(F)$ arises from $I(F)$ by orienting edges from C to x if $x \in C$, and from x to C if $\bar{x} \in C$. The *underlying graph* G_D of a directed graph D is the undirected graph obtained from D by "forgetting" the orientation of edges and by identifying possible parallel edges. Thus $I(F)$ is the underlying graph of $I_d(F)$. For an undirected graph G we consider its treewidth tw(G), its branchwidth bw(G), and its clique-width cwd(G); clique-width is also defined for directed graphs. For definitions of these graph parameters we refer the reader to related work [2,6,5,1,13,25]. By means of primal, incidence and directed incidence graphs, these graph parameters apply to formulas as follows: For a formula F we call $\mathrm{tw}(F) = \mathrm{tw}(P(F))$ the *primal treewidth* of F, $\mathrm{tw}^*(F) = \mathrm{tw}(I(F))$ the *incidence treewidth* of F, $\mathrm{bw}(F) = \mathrm{bw}(P(F))$ the *branchwidth* of F, $\mathrm{cwd}(F) = \mathrm{cwd}(I_d(F))$ the *clique-width* of F.

For two formula parameters p and q we say that p *dominates* q if there is a computable function f such that $p(F) \leq f(q(F))$ holds for all formulas F. We say that p and q are *incomparable* if neither p dominates q nor q dominates p. Note that if $\#\mathrm{SAT}(p)$ is fixed-parameter tractable and p dominates q, then also $\#\mathrm{SAT}(q)$ is fixed-parameter tractable. From known results it follows that clique-width dominates incidence treewidth, and that, in turn, incidence treewidth dominates primal treewidth and branchwidth [25]. Whence, clique-width can be considered as the most general parameter considered so far. Fischer, Makowsky, and Ravve [8] show that $\#\mathrm{SAT}(\mathrm{cwd})$ is fixed-parameter tractable, combining an earlier result of Courcelle, Makowsky, and Rotics [5] and a recent result of Oum and Seymour [21]. By the above relationships among the various parameters, this result also implies the fixed-parameter tractability of $\#\mathrm{SAT}(\mathrm{tw}^*)$, $\#\mathrm{SAT}(\mathrm{tw})$, and $\#\mathrm{SAT}(\mathrm{bw})$:

Theorem 4. *The problems* $\#\mathrm{SAT}(\mathrm{cwd})$, $\#\mathrm{SAT}(\mathrm{tw}^*)$, $\#\mathrm{SAT}(\mathrm{tw})$, *and* $\#\mathrm{SAT}(\mathrm{bw})$, *are fixed-parameter tractable.*

The question arises how our new parameter, the clustering-width, is related to the other parameters. Does any of the above parameters dominate clustering-width, or does clustering-width dominate any of the other parameters? We will show that the answer to both questions is 'no': clustering-width is *incomparable* with any of the other parameters.

Lemma 8. *The class* HIT *has unbounded clique-width.*

Proof. Let $n \geq 3$ be an integer and let G denote an $n \times n$ grid. That is, G is a bipartite graph with n^2 vertices $v_{i,j}$, $i, j \in \{1, \ldots, n\}$, where two vertices $v_{i,j}$ and $v_{i',j'}$ are joined by an edge if and only if either $i = i'$ and $|j - j'| = 1$, or $|i - i'| = 1$ and $j = j'$. Let V_1, V_2 be a bipartition of the vertex set of G. We obtain a formula F with $I(F) = G$ by considering vertices in V_1 as variables and putting $F = \{ N(v_{i,j}) : v_{i,j} \in V_2 \}$; here $N(v_{i,j})$ denotes the set of neighbors of $v_{i,j}$ in G.

Consider a directed graph D whose underlying graph is the complete graph K_m for $m = |V_2|$. We construct the hitting formula F_D as described at the beginning of Sect. 5; we assume that F and F_D do not share variables. Observe that $|F_D| = m$; thus we can write $F = \{C_1, \ldots, C_m\}$ and $F_D = \{C_{1,D}, \ldots, C_{m,D}\}$, ordering the clauses arbitrarily.

Let H be the formula $\{C_1 \cup C_{1,D}, \ldots, C_m \cup C_{m,D}\}$. Clearly H is a hitting formula since F_D is a hitting formula. Golumbic and Rotics [10] show that the clique-width of $n \times n$ grids, $n \geq 3$, is exactly $n+1$, hence $\mathrm{cwd}(G) = n+1$. Note that $I(F) = G$ is isomorphic to a vertex-induced subgraph of $I(H)$; this implies that $\mathrm{cwd}(H) \geq \mathrm{cwd}(G) = n+1$ (see Courcelle and Olariu [6]). Moreover, also noted by Courcelle and Olariu, the clique-width of a directed graph is at least as large as the clique-width of its underlying graph; hence we have $\mathrm{cwd}(I_d(H)) \geq \mathrm{cwd}(I(H)) \geq \mathrm{cwd}(I(F)) = \mathrm{cwd}(G) = n+1$. We conclude that for every positive integer n there exists a hitting formula H with $\mathrm{cwd}(H) > n$. $\square$

Lemma 9. *The class of formulas with primal treewidth* 1 *has unbounded clustering-width.*

Proof. Let $\mathcal{C}$ denote the class of formulas with primal treewidth 1. Let n be an even positive integer and consider the formula

$$F = \{\{x_0, x_1\}, \{x_1, x_2\}, \ldots, \{x_{n-1}, x_n\}\}.$$

The primal graph of F is a path. Since paths have treewidth 1, $F \in \mathcal{C}$ follows.

For every $i = 1, \ldots, n - 1$, the formula F contains the overlap obstruction $\{\{x_{i-1}, x_i\}, \{x_i, x_{i+1}\}\}$ with the corresponding deletion pair $\{\{x_i\}, \{x_{i-1}, x_{i+1}\}\}$. There are no clash obstructions. The obstruction graph is therefore a path P on the vertices $x_1, \ldots, x_n$. Any vertex cover of P contains at least $n/2$ vertices, hence $\mathrm{clu}(F) \geq n/2$ follows.

As we can choose arbitrarily large n, $\mathcal{C}$ has unbounded clustering-width. $\square$

In view of the relationships omong the parameters cwd, tw*, tw, and bw stated above, the last two lemmas imply the following result.

Theorem 5. *The parameters* cwd, tw*, tw, *and* bw, *are all incomparable with clustering-width.*

References

1. F. Bacchus, S. Dalmao, and T. Pitassi. Algorithms and complexity results for #SAT and Bayesian Inference. In *44th Annual IEEE Symposium on Foundations of Computer Science (FOCS'03)*, 340–351, 2003.
2. H. L. Bodlaender. A partial k-arboretum of graphs with bounded treewidth. *Theoret. Comput. Sci.*, 209(1-2):1–45, 1998.
3. J. Chen, I. A. Kanj, and G. Xia. Simplicity is beauty: Improved upper bounds for vertex cover. Technical Report TR05-008, DePaul University, Chicago IL, 2005.
4. B. Courcelle, J. A. Makowsky, and U. Rotics. Linear time solvable optimization problems on graphs of bounded clique-width. *Theory Comput. Syst.*, 33(2):125–150, 2000.
5. B. Courcelle, J. A. Makowsky, and U. Rotics. On the fixed parameter complexity of graph enumeration problems definable in monadic second-order logic. *Discr. Appl. Math.*, 108(1-2):23–52, 2001.
6. B. Courcelle and S. Olariu. Upper bounds to the clique width of graphs. *Discr. Appl. Math.*, 101(1-3):77–114, 2000.
7. R. G. Downey and M. R. Fellows. *Parameterized Complexity*. Monographs in Computer Science. Springer Verlag, 1999.
8. E. Fischer, J. A. Makowsky, and E. R. Ravve. Counting truth assignments of formulas of bounded tree-width or clique-width. *Discr. Appl. Math.* to appear.
9. J. Flum and M. Grohe. The parameterized complexity of counting problems. *SIAM J. Comput.*, 33(4):892–922, 2004.

10. M. C. Golumbic and U. Rotics. On the clique-width of some perfect graph classes. *Internat. J. Found. Comput. Sci.*, 11(3):423–443, 2000. Selected papers from the Workshop on Graph-Theoretical Aspects of Computer Science (WG 99), Part 1 (Ascona).

11. J. Gramm, J. Guo, F. Hüffner, and R. Niedermeier. Graph-modeled data clustering: fixed-parameter algorithms for clique generation. *Theory Comput. Syst.*, 38(4):373–392, 2005.

12. G. Gottlob, F. Scarcello, and M. Sideri. Fixed-parameter complexity in AI and nonmonotonic reasoning. *Artificial Intelligence*, 138(1-2):55–86, 2002.

13. G. Gottlob and S. Szeider. Fixed-parameter algorithms for artificial intelligence, constraint satisfaction, and database problems. Submitted, April 2006.

14. Y. Interian. Backdoor sets for random 3-SAT. In *Informal Proceedings of SAT 2003*, 231–238, 2003.

15. K. Iwama. CNF-satisfiability test by counting and polynomial average time. *SIAM J. Comput.*, 18(2):385–391, 1989.

16. P. Kilby, J. K. Slaney, S. Thiébaux, and T. Walsh. Backbones and backdoors in satisfiability. In *Proceedings, The Twentieth National Conference on Artificial Intelligence and the Seventeenth Innovative Applications of Artificial Intelligence Conference (AAAI 2005)*, 1368–1373, 2005.

17. H. Kleine Büning and X. Zhao. Satisfiable formulas closed under replacement. In *Proceedings for the Workshop on Theory and Applications of Satisfiability*, volume 9 of *Electronic Notes in Discrete Mathematics*. Elsevier Science Publishers, North-Holland, 2001.

18. I. Lynce and J. P. Marques-Silva. Hidden structure in unsatisfiable random 3-SAT: An empirical study. In *16th IEEE International Conference on Tools with Artificial Intelligence (ICTAI 2004)*, 246–251. IEEE Computer Society, 2004.

19. R. Niedermeier. *Invitation to Fixed-Parameter Algorithms*. Oxford Lecture Series in Mathematics and Its Applications. Oxford University Press, 2006.

20. N. Nishimura, P. Ragde, and S. Szeider. Detecting backdoor sets with respect to Horn and binary clauses. In *Informal Proceedings of SAT 2004*, 96–103, 2004.

21. S. Oum and P. Seymour. Approximating clique-width and branch-width. *J. Combin. Theory, Ser. B*, to appear.

22. N. Robertson and P. D. Seymour. Graph minors. X. Obstructions to tree-decomposition. *J. Combin. Theory Ser. B*, 52(2):153–190, 1991.

23. D. Roth. On the hardness of approximate reasoning. *Artificial Intelligence*, 82(1-2):273–302, 1996.

24. Y. Ruan, H. A. Kautz, and E. Horvitz. The backdoor key: A path to understanding problem hardness. In *Proceedings of the 19th National Conference on Artificial Intelligence, 16th Conference on Innovative Applications of Artificial Intelligence*, 124–130. AAAI Press / The MIT Press, 2004.

25. S. Szeider. On fixed-parameter tractable parameterizations of SAT. In *Theory and Applications of Satisfiability, 6th International Conference, SAT 2003, Selected and Revised Papers, LNCS*, vol. 2919, 188–202. Springer, 2004.

26. S. Szeider. Backdoor sets for DLL subsolvers. *Journal of Automated Reasoning*, 2005. In press.

27. L. G. Valiant. The complexity of computing the permanent. *Theoret. Comput. Sci.*, 8(2):189–201, 1979.

28. R. Williams, C. Gomes, and B. Selman. On the connections between backdoors, restarts, and heavy-tailedness in combinatorial search. In *Informal Proceedings of SAT 2003*, 222–230, 2003.

Counting Models in Integer Domains

António Morgado[1], Paulo Matos[1], Vasco Manquinho[1],
and João Marques-Silva[2]

[1] IST/INESC-ID, Technical University of Lisbon, Portugal
{ajrm, pocm, vmm}@sat.inesc-id.pt
[2] School of Electronics and Computer Science, University of Southampton, UK
jpms@ecs.soton.ac.uk

Abstract. This paper addresses the problem of counting models in integer linear programming (ILP) using Boolean Satisfiability (SAT) techniques, and proposes two approaches to solve this problem. The first approach consists of encoding ILP instances into pseudo-Boolean (PB) instances. Moreover, the paper introduces a model counter for PB constraints, which can be used for counting models in PB as well as in ILP. A second alternative approach consists of encoding instances of ILP into instances of SAT. A two-step procedure is proposed, consisting of first mapping the ILP instance into PB constraints and then encoding the PB constraints into SAT. One key observation is that not all existing PB to SAT encodings can be used for counting models. The paper provides conditions for PB to SAT encodings that can be safely used for model counting, and proves that some of the existing encodings are safe for model counting while others are not. Finally, the paper provides experimental results, comparing the PB and SAT approaches, as well as existing alternative solutions.

1 Introduction

Besides its well-known theoretical relevancy, the problem of counting models of Boolean Satisfiability (SAT) formulas (#SAT) has a large number of key application areas [2,18]. Recent years have seen significant improvements in algorithms for #SAT, which include the utilization of well-known SAT techniques as well as the identification of connected components and component caching, but also variable lifting and blocking clauses [6,12,17,18]. Nevertheless, model counting is also extremely important in non-Boolean domains, including Integer Linear Programming (ILP) [5,11] and Linear Integer Arithmetic (LIA) [7,16]. This paper focus on ILP, but the techniques proposed can be extended to LIA.

Existing algorithms for counting models in ILP [5,11] are extremely sensitive to the number of variables in the problem formulation, being able to solve instances with a very small number of variables. Hence, in many practical applications, existing algorithms are ineffective.

This paper proposes two alternative solutions to counting models in ILP, by considering the utilization of SAT-based techniques. The first approach consists of encoding instances of ILP into instances of pseudo-Boolean (PB) constraints.

A. Biere and C.P. Gomes (Eds.): SAT 2006, LNCS 4121, pp. 410–423, 2006.

Moreover, the paper introduces a model counter for PB constraints, which can be used for counting models in PB as well as in ILP. A second alternative approach consists of encoding instances of ILP into instances of SAT. A two-step procedure is proposed, consisting of first mapping the ILP instance into PB constraints and then encoding the PB constraints into SAT. One key concern is that not all existing PB to SAT encodings can be used for counting models. The paper provides conditions for encodings that can be safely used for model counting, and proves that some of the existing PB to SAT encodings are safe for model counting. Finally, the paper provides experimental results, comparing the PB and the SAT approaches, as well as existing alternative solutions. The results provide interesting insights into the problem of counting models in ILP. First, the PB counter, albeit a preliminary prototype, is competitive with SAT counters, which integrate more sophisticated techniques including the identification of connected components and component caching. Second, the very effective SAT-techniques used in Cachet [18] may not scale for integer domains.

The paper is organized as follows. Section 2 presents the notation used throughout the paper. Afterwards, the paper addresses the encoding of ILP into PB constraints, and describes a model counter for PB formulations. Section 5 details the second approach to model counting in ILP, based on encoding ILP into SAT. This section proves that some existing encodings will yield correct results, whereas others can overestimate the number of integer models. Section 6 compares the two approaches and also evaluates an alternative solution [11]. Section 7 surveys related work, and the paper concludes in Section 8.

2   Definitions

An *Integer Linear Programming* (ILP) problem with n variables and m constraints can be defined as follows [14]:

$$\sum_{j=1}^{n} a_{ij}x_j \le b_i,$$
$$x_j, a_{ij}, b_i \in \mathbb{Z} \tag{1}$$
$$j \in \{1, \ldots, n\}, i \in \{1, \ldots, m\}$$

where a_{ij} denote the coefficients of the problem variables x_j in the set of m linear constraints. Note that all ILP problem instances can be rewritten as defined in (1). ILP problem instances are usually defined with a cost function to minimize or maximize. However, for the purpose of counting the number of models, the cost function is irrelevant. Hence, for simplification, we focus solely on the constraints.

An ILP problem instance is said to be a *Linear Pseudo-Boolean* (PB) problem instance (also known as 0-1 ILP) if the variable domain of the ILP variables is Boolean. In this case, all constraints correspond to pseudo-Boolean constraints. A particular type of pseudo-Boolean constraints are propositional clauses and a problem instance where all constraints are propositional clauses is an instance of the *Propositional Satisfiability* (SAT) problem.

In a propositional formula, a literal l_j denotes either a variable x_j or its complement $\bar{x}_j$. If a literal $l_j = x_j$ and x_j is assigned value 1 or $l_j = \bar{x}_j$ and x_j is assigned value 0, then the literal is said to be true. Otherwise, the literal is said to be false.

A propositional clause is a disjunction of literals such as $l_1 \vee l_2 \vee \ldots \vee l_k$ where l_j is a literal representing either x_j or $\bar{x}_j$. We should observe that propositional clauses can also be represented as linear inequalities, e.g. $\sum_{j=1}^{k} l_j \geq 1$. One can obtain a linear inequality as in (1) if we replace literals $\bar{x}_j$ by $1 - x_j$. In the context of SAT we will represent propositional clauses as a disjunction of literals, instead of linear inequalities. However, in the context of ILP or pseudo-Boolean, we use the linear inequalities formalism.

Whenever an assignment to *all* problem variables is found such that all problem constraints become satisfied, we say that a model has been found. However, it may occur that a *partial assignment* (i.e. not all problem variables are assigned) is able to satisfy all problem constraints. In this case, the partial assignment represents a set of models for the problem instance.

We say that an ILP instance defines a *convex polytope* if the number of integer solutions (models) to the ILP constraints is finite. Note that all PB and SAT problem instances define rational convex polytopes since the value of the problem variables is bounded. Hence, the number of solutions is $O(2^n)$ for both PB and SAT instances, where n is the number of problem variables. However, not all ILP instances define convex polytopes. For example, the following ILP has an infinite number of solutions:

$$x_1 - x_2 \leq 10, \ x_1 - x_3 \leq 5$$
$$x_1, x_2, x_3 \in \mathbb{Z} \tag{2}$$

In section 3 we discuss how to determine if a set of ILP constraints define a convex polytope by finding lower and upper bounds on the value of all problem variables.

3   Encoding ILP into Pseudo-Boolean

This section presents a procedure to encode an Integer Linear Programming (ILP) problem instance into a Linear Pseudo-Boolean (PB) problem instance. The resulting PB instance can then be solved using specific Boolean techniques [1,8]. A key aspect of this encoding is to determine lower and upper bounds on the possible values of the integer valued variables in the ILP. We assume that the ILP instance defines a convex polytope; otherwise the number of integer solutions would be infinite. Hence, every integer variable is guaranteed to have a lower and an upper bound.

Given an ILP instance as presented in section 2, let $lower(x_j)$ and $upper(x_j)$ denote respectively the lower and upper bound on the value of variable x_j in the ILP. If specified in the problem instance, the values of $lower(x_j)$ and $upper(x_j)$ can be determined directly from the instance bounds or by constraints of the type $x_j \geq l$ and $x_j \leq u$.

In general, the lower and upper bounds of an integer valued variable x_j can be determined by solving a linear programming relaxation (LPR) as follows:

$$\text{minimize/maximize } x_j$$
$$\text{subject to} \quad \sum_{j=1}^{n} a_{ij}x_j \leq b_i, \tag{3}$$
$$x_j \in \mathbb{R}, a_{ij}, b_i \in \mathbb{Z}$$

where we use *minimize* or *maximize* depending on whether we are interested in obtaining a lower or a upper bound, respectively. Note that in this formulation all problem variables are no longer integer and there are known polynomial time algorithms for solving these formulations [14].

Let z_j^m and z_j^M denote respectively the solutions of (3) in the minimization and maximization formulations. By relaxing the variable integer constraints we can obtain a lower and upper bound on the value of x_j, since no integer solution to (1) can be obtained such that $x_j < z_j^m$ or $x_j > z_j^M$. Hence, we have $lower(x_j) = \lceil z_j^m \rceil$ and $upper(x_j) = \lfloor z_j^M \rfloor$, where $\lceil z_j^m \rceil$ denotes the smallest integer value not lower than z_j^m and $\lfloor z_j^M \rfloor$ denotes the largest integer value not higher than z_j^M.

Observe that in order to obtain lower bounds on the problem variables, the well-known replacement of each problem variable x_j with $x_j' - x_j''$, where $x_j' \geq 0$ and $x_j'' \geq 0$, cannot be used. For example, suppose we have the following constraints for variable x_1:

$$x_1 \geq -1, x_1 \leq 3 \tag{4}$$

In this formulation, x_1 is clearly bounded. However, if we replace x_1 with $x_1' - x_1''$, we would get:

$$x_1' - x_1'' \geq -1, \; x_1' - x_1'' \leq 3$$
$$x_1' \geq 0, x_1'' \geq 0 \tag{5}$$

For this new formulation, both variables x_1' and x_1'' are not bounded, since we can always find arbitrary large values for x_1' and x_1'' such that the constraints are satisfied.

One should also note that if (3) is unbounded for any given problem variable x_j, then the original ILP does not define a convex polytope. Otherwise, if (3) is bounded for all problem variables, then the ILP is a convex polytope and there is a finite number of integer solutions to (1).

Since all integer variables x_j of (1) are limited between $lower(x_j)$ and $upper(x_j)$ in a convex polytope, we can apply a substitution of all variables x_j with $y_j - lower(x_j)$ so that in the new ILP we have all new problem variables y_j bounded between 0 and $upper(y_j)$ where $upper(y_j) = upper(x_j) - lower(x_j)$. Afterwards, we can encode each integer variable y_j as a set of weighted bits as follows:

$$y_j = \sum_{i=0}^{b_j} 2^i y_j^i \tag{6}$$
$$y_j^i \in \{0, 1\}$$

where b_j is the number of bits necessary to represent $upper(y_j)$ and variables y_j^i are Boolean. Additionally, we can also add the following constraints:

$$\sum_{i=0}^{b_j} 2^i y_j^i \leq upper(y_j) \tag{7}$$

As a result of integer variable replacements from (6) and the addition of upper bound constraints from (7), we get a pseudo-Boolean instance that encodes the convex polytope defined by the original ILP.

It is important to ensure that the number of solutions of the PB instance is the same as in the original ILP. Indeed, for this encoding, every unique satisfiable assignment for the ILP instance corresponds to a unique satisfiable assignment for the PB instance, because the integer variables are encoded as a set of weighted bits as it is represented in the computer memory.

4   Model Counting in Pseudo-Boolean Formulations

One way for performing model counting in Pseudo-Boolean (PB) formulations is to implicitly enumerate all possible variable assignments using a backtrack search PB solver. Moreover, current state-of-the-art PB solvers are able to perform *conflict learning* [1,8] and thus prevent entering areas of the search space where no satisfiable assignment exists. This technique has been found particularly useful when the solver has to implicitly visit the complete search space.

It is possible to modify backtrack search PB solvers to count models in PB formulations. Basically, whenever a new solution is found, a propositional clause is added such that it prevents accounting for the same solution later in the search. In the context of model counting, these clauses are known as *blocking clauses* [12,17].

The most straightforward way of generating a new blocking clause is to consider the negation of the search path when a new satisfiable assignment is found. Therefore, if the search path corresponds to the decision assignments $\{x_1 = v_1, x_2 = v_2, \ldots, x_k = v_k\}$, then the blocking clause is defined by:

$$\sum_{j=1}^{k} l_j \geq 1 \tag{8}$$

where $l_j = x_j$ if $x_j = 0$ in the search path or $l_j = \bar{x}_j$ if $x_j = 1$. Observe that this blocking clause prevents the current search path to occur later in the search. Hence, the models corresponding to the partial solution will not be counted twice. Note that a partial assignment of k problem variables that satisfies all problem constraints implicitly represents a set of models. Therefore, whenever a PB solver finds a new solution considering only k variables, it has in fact found 2^{n-k} possible ways of satisfying the problem constraints.

Simplification of Satisfying Partial Assignments. It is well-known that the problem of computing the satisfying partial assignment with the smallest number of specified variables can be formulated as an integer linear program [15]. However, we are just interested in simplifying satisfying partial assignments computed by a PB solver.

Variable lifting denotes a number of techniques used for the elimination of assignments that can be declared redundant [12,17]. A simple variable lifting technique consists of removing from a satisfying partial assignment all variable assignments that are not used to satisfy any constraint. Moreover, these variable assignments cannot also be used in constraints that imply other variable assignments. When using this technique, we can immediately conclude that all implied assignments cannot be removed from the partial assignment since they are necessary to satisfy at least one constraint. Otherwise, these assignments would not be implied. Hence, we only have to check decision assignments.

Suppose we have the following decision assignment $x_1 = x_2 = x_3 = 0$ and that $x_5 = 0$ is an implied assignment. Consider also the following constraints:

$$(x_1 + x_2 + x_3 \leq 1) \wedge (x_2 + x_3 \leq 1) \wedge (x_2 + x_4 \leq 1) \wedge (-x_3 + x_5 \leq 0) \qquad (9)$$

Clearly, the assignment to x_1 is not relevant to satisfy the problem constraints. Note that x_3 cannot be considered irrelevant, since it is necessary to imply the value of x_5. Hence, the resulting blocking clause would be $x_2 + x_3 \geq 1$. Since there are two variables (x_1 and x_4) that are not relevant to satisfy the constraints in this partial assignment, then we conclude that 4 models have been found. In [12,17] other lifting techniques are presented. However, they require a significant computational overhead.

Additional SAT Techniques. The identification of connected components [6] and component caching [18] are among the most effective techniques for model counting instances of SAT. These techniques are not yet integrated in the PB model counter described above, since they will require significant re-implementation effort, and our objective is first to evaluate whether these techniques are effective for ILP and PB model counting.

5   Encoding Pseudo-Boolean Constraints as SAT

Several algorithms exist that are based on modifying a SAT solver in order to deal with pseudo-Boolean constraints [1,8], as well as generalizing other techniques like conflict analysis. A different approach is based on encoding all pseudo-Boolean constraints into propositional clauses and use a SAT solver directly [1,3,4,9,19]. Some of these encodings are polynomial whereas others are exponential in the worst case. The objective of encoding PB constraints into SAT is to take advantage of the powerful techniques of SAT solvers in dealing with propositional clauses. This section defines *counting safety*, and shows that not all encodings are counting safe. Moreover, this section also shows that some encodings are counting safe and so can be used for model counting.

Definition 1 (Counting Safety). *A PB formulation to SAT encoding is counting safe iff the number of models in the PB formulation and in the encoded SAT formulation are the same.*

5.1   Unsafe Encodings

The vast majority of PB to SAT encodings solely aim the discovery of one solution and may introduce auxiliary variables. These variables may lead to double counting of the same solution in pseudo-Boolean. For example, consider the following PB constraints:

$$2x_0 + 4x_1 + 8x_2 + 3y_0 + 6y_1 + 12y_2 \leq 18$$
$$-2x_0 - 4x_1 - 8x_2 + 1y_0 + 2y_1 + 4y_2 \geq -10$$

If *any* of the encodings proposed in [9] is used with this example, and the resulting CNF formula is given to model counter, e.g. *cachet* [18], the number of models reported will be at least 38. However, the correct number of models for this example is 31. Hence, the encodings proposed in [9] do not satisfy the counting safety property, and cannot be used for model counting. The next section addresses encodings which are counting safe.

5.2   Safe Encodings

Both the well-known Warners PB to SAT encoding [19] as well as the more recent arc-consistency encoding of Bailleux, Boufkhad and Roussel (BBR) [4] can be shown to be counting safe. Due to space constraints, this section addresses solely the BBR encoding; a detailed analysis of Warners encoding is available in [13]. Next, we provide a brief description of the BBR PB to SAT encoding [4].

Consider a pseudo-Boolean constraint ω with the constraint literals l_j sorted according to their coefficients a_j:

$$\omega = \sum_{j=1}^{n} a_j l_j \leq b, \tag{10}$$
$$\text{where } 0 < a_1 \leq a_2 \leq \ldots \leq a_n$$

Let $\omega_{i,k}$ represent the constraint ω considering only the first i literals ($0 \leq i \leq n$) with the right-hand side value k, i.e. $\omega_{i,k} : \sum_{j=1}^{i} a_j l_j \leq k$. Therefore, the original constraint ω corresponds to $\omega_{n,b}$.

In order to generate the CNF encoding for a given constraint ω, we need to introduce new Boolean variables $D_{i,k}$ which represent the satisfaction of constraints $\omega_{i,k}$ obtained from ω. Hence, we have $D_{i,k} = 1$ iff constraint $\omega_{i,k}$ is satisfied. As a result, $D_{n,b} = 1$ represents the satisfaction of the original pseudo-Boolean constraint ω in the CNF encoding.

When building the CNF encoding, variables $D_{i,k}$ are said to be *terminal* if $k \leq 0$ or if $\sum_{j=1}^{i} a_j \leq k$. Otherwise, variables $D_{i,k}$ are said to be *non-terminal*.

The CNF encoding for a pseudo-Boolean constraint ω proceeds as follows:

1. Start with a set of variables containing variables x_j in constraint ω, as well as variable $D_{n,b}$, and an empty set of propositional clauses. Mark all variables x_j.
2. Consider an unmarked variable $D_{i,k}$.
3. If $D_{i,k}$ is a non-terminal variable, add two new variables $D_{i-1,k}$ and $D_{i-1,k-a_i}$ to the set of variables, if they are not already in this set. Moreover, mark selected variable $D_{i,k}$ and add the following propositional clauses:

$$\bar{D}_{i-1,k-a_i} + D_{i,k} \geq 1 \tag{11}$$
$$\bar{D}_{i,k} + D_{i-1,k} \geq 1 \tag{12}$$
$$\bar{D}_{i,k} + \bar{l}_i + D_{i-1,k-a_i} \geq 1 \tag{13}$$
$$\bar{D}_{i-1,k} + l_i + D_{i,k} \geq 1 \tag{14}$$

4. If $D_{i,k}$ is a terminal variable, and if $k \neq 0$, then:

$$D_{i,k} = \begin{cases} 0 \text{ if } k < 0. \text{ Add } \bar{D}_{i,k} \geq 1 \text{ to the clause set.} \\ 1 \text{ if } \sum_{j=1}^{i} a_j \leq k. \text{ Add } D_{i,k} \geq 1 \text{ to the clause set.} \end{cases} \tag{15}$$

Otherwise, if $k = 0$, then add the following set of clauses:

$$\bar{D}_{i,k} + \bar{l}_j \geq 1, 1 \leq j \leq i \tag{16}$$
$$\sum_{j=1}^{i} l_j + D_{i,k} \geq 1 \tag{17}$$

5. If there are any unmarked variables, go to step 2. Otherwise, the propositional clause set contains the CNF encoding of constraint ω and the procedure terminates.

The following example illustrates how the proposed CNF encoding works:

$$\omega : 2\bar{x}_1 + 3x_2 + 3x_3 \leq 5 \tag{18}$$

Figure 1 presents the new variables created for the CNF encoding. For each non-terminal variable, two new additional variables are created whereas terminal variables are represented as leaf nodes. The full encoding for constraint ω as a set of propositional clauses is as follows:

$$\begin{array}{lll} D_{3,5} \geq 1 & \bar{D}_{1,-1} + D_{2,2} \geq 1 & \bar{D}_{1,-1} \geq 1 \\ \bar{D}_{2,2} + D_{3,5} \geq 1 & \bar{D}_{2,2} + D_{1,2} \geq 1 & D_{1,2} \geq 1 \\ \bar{D}_{3,5} + D_{2,5} \geq 1 & \bar{D}_{2,2} + \bar{x}_2 + D_{1,-1} \geq 1 & \\ \bar{D}_{3,5} + \bar{x}_3 + D_{2,2} \geq 1 & \bar{D}_{1,2} + x_2 + D_{2,2} \geq 1 & \\ \bar{D}_{2,5} + x_3 + D_{3,5} \geq 1 & D_{2,5} \geq 1 & \end{array} \tag{19}$$

One should note that in addition to the propositional clauses added by the encoding procedure, it is also necessary that variable $D_{3,5}$ be assigned value 1.

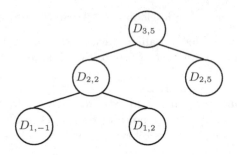

Fig. 1. Additional variables in CNF encoding

This is required since $D_{3,5}$ represents the satisfaction of the original pseudo-Boolean constraint.

Finally, we refer to [4] for details, namely proofs of correction of the encoding and of the maintenance of generalized arc consistency in the resulting CNF encoding, as well as examples of constraints for which this encoding provides exponential increase in the size of the CNF encoding.

Theorem 1. *The BBR arc-consistency encoding [4] is counting safe.*

Proof:
We want to prove that to each model for the PB constraints there is a corresponding unique model for the SAT encoding, and that to each model for the SAT encoding there is a unique corresponding model for the PB constraints.

($\leftarrow$) If we have a model for the SAT encoding then we can only have one model for the PB constraints, which is the model for the SAT encoding restricted to the variables for the PB constraints.

($\rightarrow$) Suppose we have a model for the PB constraints. We know that the encoding is correct [4], so we already know that there exists at least one model for the SAT encoding corresponding to the model for the PB constraints. What we want to show is that this model is unique. We also know that the model in SAT corresponds to the same assignments made to the variables of the PB constraints plus the assignments to the variables introduced by the encoding. In order to show that the model is unique all we have to prove is that these new variables $D_{i,b}$ can only have one possible assignment for satisfying the created instance. We are going to prove this claim by induction on n. Let us first consider a PB constraint $\sum_{j=1}^{n} a_j x_j \leq k$ and the corresponding variable node $D_{n,k}$.

Base: $n = 1$:
There can be two cases, depending on whether $D_{1,k}$ is terminal. If $D_{1,k}$ is terminal we may have $k \neq 0$ or $k = 0$. If $k \neq 0$, due to the definition of a terminal node, either $k < 0$ and the value of $D_{1,k}$ is 0, or $a_1 \leq k$ and the value of $D_{1,k}$ is 1. Otherwise, if $k = 0$, then the encoding adds clauses $(\bar{D}_{1,0} \lor x_1)$, $(D_{1,0} \lor \bar{x}_1)$. Since x_1 has a determined value it implies the unique value of $D_{1,k}$.

We now consider the case when $D_{1,k}$ is not terminal. In this case, the encoding adds the following clauses: $(\bar{D}_{0,k-a_1} \lor D_{1,k})$, $(D_{1,k} \lor D_{0,k})$, $(\bar{D}_{1,k} \lor \bar{x}_1 \lor D_{0,k-a_1})$, and $(\bar{D}_{0,k} \lor x_1 \lor D_{1,k})$. Since $D_{1,k}$ is not terminal, then if $k > 0$ we have $D_{0,k} = 1$ or if $k < a_1$ we have $D_{0,k-a_1} = 0$. The first and the second clauses added by the encoding are trivially satisfied. By removing the false literals of the third and fourth clauses we get $(\bar{D}_{1,k} \lor \bar{x}_1)$, $(x_1 \lor D_{1,k})$. Since x_1 has a determined value both clauses imply the unique value of $D_{1,k}$.

Step:

Hypothesis: For $i < n$, $D_{i,b}$ has only one possible assignment that satisfies the created instance.

We now consider node $D_{n,k}$. If the node is terminal with $k < 0$, then the encoding adds the clause $(\bar{D}_{n,k})$ and $D_{n,k}$ can only be assigned value false.

If the node is terminal with $k \neq 0$ and $k \geq \sum_{j=1}^{n} a_j$, then the encoding adds the clause $(D_{n,k})$ and $D_{n,k}$ can only be assigned value true.

If the node is terminal with $k = 0$, then the encoding adds the clauses $(\bar{D}_{n,0} \lor \bar{x}_j)$, $1 \leq j \leq n$ and $(x_1 \lor x_2 \lor \ldots \lor x_i \lor D_{n,0})$. Two situations may occur. If all the variables x_j, $1 \leq j \leq n$, are false, then (17) implies the value of $D_{n,0}$ to true. If at least one x_j, $1 \leq j \leq n$ is true, then (16) implies the value of $D_{n,0}$ to false and x_j satisfies (17).

If the node is non-terminal, then we apply the hypothesis to $D_{n-1,k}$ and to $D_{n-1,k-a_{n-1}}$. We get that these variable nodes have a fixed known value. We have four cases depending on the value of the variable nodes:

- If $D_{n-1,k}$ and $D_{n-1,k-a_{n-1}}$ are false, then from (12) $D_{n,k}$ can only be false and the other clauses are all satisfied.
- If $D_{n-1,k}$ is false and $D_{n-1,k-a_{n-1}}$ is true, then from (11) and (12) there is a contradiction, and the encoded instance is unsatisfiable. This situation is acceptable since we cannot have $(\sum_{j=1}^{n-1} a_j x_j \leq k - a_{n-1})$ being satisfiable and at the same time being unable to satisfy the same left hand side of the equation with a higher right hand side $(\sum_{j=1}^{n-1} a_j x_j \leq k)$.
- If $D_{n-1,k}$ is true and $D_{n-1,k-a_{n-1}}$ is false, then we can have two cases:
  x_n **is false** , then from (14) $D_{n,k}$ can only be true and the other clauses are all satisfied;
  x_n **is true** , then from (13) $D_{n,k}$ can only be false and the other clauses are all satisfied.
- If $D_{n-1,k} = 1$ and $D_{n-1,k-a_{n-1}} = 1$, then from (11) $D_{n,k} = 1$ and all the other clauses are satisfied.

Finally, the results holds for any constraint, and so necessarily holds for all constraints in an instance of PB. ∎

6   Experimental Results

This section presents experimental results for model counting in integer domains. Instances from different problems are considered. Moreover, different model counting approaches are evaluated. The existing implementation of Barvinok's algorithm, _LattE_ [11], is evaluated. A prototype PB model counter, described in Section 4 is evaluated. Finally, SAT model counters are evaluated. In order to use the SAT model counters, we use the counting safe encodings of Warners [19] as well as the arc-consistency encoding proposed by Bailleux, Boufkhad and Roussel (BBR) [4]. All CPU times presented are for a AMD Athlon 1.9GHz processor with 1GB of physical memory. The time limit for each instance was set to one hour. If the time limit was reached, we provide a partial solution when one is available, i.e. the number of models found when the search was stopped. This is preceded by the sign ≥ since the total number of models must be higher than or equal to the ones already found. If all models are found, we provide the total time in seconds. In cases where time limit was reached and the counter did not provide any count of the number of models, **TO** (Time Out) is shown. All processes were run with 900MB of allowed memory. **MO** (Memory Out) is shown for the cases where the counter reached the allowed memory limit. On some instances of Table 1 **MO*** is shown because it was the translator of pseudo-Boolean to SAT that reached the memory limit instead of the counter. Observe that **MO*** can only take place with the BBR encoding.

The experimental results are shown in Table 1, and are organized in three parts, according to the source of the problem instances. The first part of Table 1 presents results for instances of finding the *Frobenius* number plus 1 in knapsack problems [11]. For these instances, *Latte* is the only solver able to count the number of models. The underline numerical problem proved to be very difficult for pseudo-Boolean or SAT counters. Nevertheless, our solver *pb_counter* was able to find partial solutions for some instances, while both *cachet* and *relsat* (in both encodings) were unable to do so.

The second part of Table 1 presents results for ILP benchmarks, generated from well-known graph coloring benchmarks. Since the optimum number of colors is known for these instances, a constraint was added in order to count the number of optimum solutions. For these instances *Latte* performed poorly. In fact, our experience is that *Latte* performs better for numerical problems with very few variables. On the other hand, both *relsat* and *pb_counter* were able to solve some instances, as well as providing partial solutions for most instances. It can also be observed from the number of partial solutions found, that better results were obtained using *relsat* with the BBR encoding. Like *relsat*, *cachet* was also able to solve completely 3 instances (using the BBR encoding), but on all others no solution was found because *cachet* exhausted the available memory.

Finally, the third part of Table 1 presents results for benchmarks which encode the problem of finding the minimum-size prime implicant [15] for several instances from the DIMACS [10] benchmark suite. As in the graph coloring instances, the optimum value for these instances is also known. For each instance, a constraint was added in order to only allow models with the optimum value. Hence, for each instance, the number of models corresponds to number of different minimum-size prime implicants. Observe that *Latte* was unable to solve any of these instances, while the other model counters were able to find the total number of models for most instances. In fact, it was surprising that *pb_counter* was able to outperform the other counters on these instances, since most of the constraints are originally propositional clauses.

Table 1. Results on several benchmark instances

Benchmark	# Models	Latte	relat Warners	BBR	cachet Warners	BBR	pb_counter
cuww1	1	0.11	TO	MO	MO	MO	TO
cuww2	1	0.40	TO	MO*	MO	MO*	TO
cuww3	2	0.35	TO	MO*	MO	MO*	TO
cuww4	1	0.55	TO	MO*	MO	MO*	TO
cuww5	1	4.51	TO	MO*	MO	MO*	TO
prob1	8.592e8	22.14	TO	MO*	MO	MO*	TO
prob2	2.047e6	13.51	TO	MO*	MO	MO*	TO
prob3	0	27.01	TO	TO	MO	MO	TO
prob4	6.319e7	19.68	TO	MO*	MO	MO*	TO
prob5	2.178e10	18.65	TO	MO*	MO	MO*	≥ 4.337e4
prob6	2.188e5	96.12	TO	MO*	MO	MO*	≥ 7
prob7	4.198e12	81.20	TO	MO*	MO	MO*	≥ 3.812e4
prob8	6.743e6	257.77	TO	MO*	MO	MO*	TO
prob10	1.024e17	145.83	TO	MO*	MO	MO*	≥ 764
1-FullIns_3	5.069e7	–	≥ 8.448e6	2701.56	MO	67.73	≥ 5.194e6
2-FullIns_3	–	–	≥ 3.061e7	≥ 5.381e8	MO	MO	≥ 1.067e7
2-Insertions_3	–	–	≥ 4.156e7	≥ 4.625e8	MO	MO	≥ 9.006e6
3-FullIns_3	–	–	≥ 5.542e6	TO	MO	MO	≥ 5.217e6
3-Insertions_3	–	–	≥ 6.291e7	≥ 2.527e10	MO	MO	≥ 3.094e7
4-Insertions_3	–	–	≥ 6.753e7	≥ 1.307e12	MO	MO	≥ 1.501e7
games120	–	–	≥ 1.296e4	≥ 1.407e6	MO	MO	≥ 1.194e6
mug100_1	–	–	≥ 2.967e13	≥ 2.725e23	MO	MO	≥ 2.476e7
mug88_1	–	–	≥ 9.834e6	≥ 1.138e23	MO	MO	≥ 1.513e7
myciel3	12480	–	4.75	1.27	0.41	0.26	0.96
myciel4	–	–	≥ 2.995e6	≥ 7.215e7	MO	MO	≥ 5.065e6
myciel5	–	–	≥ 1.150e9	≥ 3.637e11	MO	MO	≥ 1.134e7
queen5_5	240	–	1123.32	151.69	71.31	29.78	4.38
queen6_6	–	–	TO	≥ 2	MO	MO	≥ 1.251e4
queen7_7	–	–	TO	TO	MO	MO	≥ 1.447e3
aim-100-1_6-yes1-2	1	–	0.19	9.04	3.58	2.27	0.02
aim-100-2_0-yes1-3	1	–	0.19	9.59	2.73	2.61	0.03
aim-100-3_4-yes1-4	1	–	0.2	9.95	2.37	4.98	0.04
aim-100-6_0-yes1-1	1	–	0.2	9.76	0.70	1.31	0.05
aim-200-1_6-yes1-3	1	–	0.42	185.29	30.67	20.42	0.04
aim-200-2_0-yes1-4	1	–	0.44	181.01	15.89	41.97	0.07
aim-200-3_4-yes1-1	1	–	0.42	172.86	13.16	MO	0.08
aim-200-6_0-yes1-2	1	–	0.51	169.77	6.67	24.55	0.14
ii8a1	1056	–	17.41	45.77	18.67	7.35	7.25
jnh1	12	–	2.26	1079.66	7.52	8.12	0.53
jnh12	1	–	0.21	1.95	0.97	0.81	0.07
jnh17	35	–	0.42	6.13	1.19	2.49	0.20
jnh7	26	–	1.37	6.09	4.18	1.75	0.26
ssa7552-038	–	–	TO	TO	MO	MO	≥ 1.853e4
ssa7552-158	–	–	≥ 1.319e13	TO	MO	MO	≥ 5.183e4

7 Related Work

Besides the work on model counting and enumeration in SAT [6,12,17,18], there has been work on model counting in non-Boolean domains, including Integer Linear Programming (ILP) [5,11] and Linear Integer Arithmetic (LIA) [7,16].

Existing work on model counting in LIA is described in [7,16]. The work of [7] is based binary decision diagrams and does not scale to large number of variables. The work of [16] enumerates a large number of applications for model counting in LIA. The proposed algorithm is also only suitable for a small number of variables, or when most variables have fixed values.

The most well-known work on model counting in ILP is Barvinok's algorithm [5]. An existing implementation, *LattE* [11], which incorporates a number of improvements, has been extensively used. As the results of Section 6 confirm, Barvinok's algorithm is adequate for instances of ILP with a small number of variables which may have larger domains. Observe that the algorithms for model counting in LIA can also be used for ILP (a special case of LIA) but current solutions can only handle small instances.

8 Conclusions

This paper proposes two alternative approaches for counting models in ILP instances. The first approach is based on encoding ILP into Pseudo-Boolean (PB) and using a PB counter. A PB counter was developed for this purpose. A second approach is based on encoding ILP into SAT, using an intermediate encoding into PB. The paper shows that some PB to SAT encoding may overestimate the number of models, whereas others are shown to yield the correct number of models. As a result, counting models in integer domains can be achieved by encoding ILP constraints into SAT and directly using SAT model counters, thus taking advantage of the techniques already incorporated into SAT counters. Experimental results indicate that the PB counter is competitive with the SAT counters. Moreover, an existing alternative to SAT-based model counters, using Barvinok's algorithm [5,11], provides essentially orthogonal results, being more efficient for problem instances having few variables with large domains, and being inadequate for problem instances having many variables with small domains.

Despite the interesting insights, many challenges still remain. The PB counter is a prototype, aiming to prove the concept of counting models for PB constraints. A more sophisticated algorithm is expected to provide significant gains, for example if connected components are identified for PB constraints. There is also a clear gap between *LattE* (the implementation of Barvinok's algorithm) and the SAT-based solutions. Work on closing this gap is also an interesting challenge. Finally, the utilization of the model counter in instances of linear integer arithmetic [7,16], one of the motivations for this work, will require significantly more optimized ILP counters.

Acknowledgments. This work is partially supported by FCT under research projects POSI/SRI/41926/2001 and POSC/EIA/61852/2004.

References

1. F. Aloul, A. Ramani, I. Markov, and K. Sakallah. Generic ILP versus specialized 0-1 ILP: An update. In *International Conference on Computer-Aided Design*, pages 450–457, November 2002.
2. F. Bacchus, S. Dalmao, and T. Pitassi. Algorithms and complexity results for #SAT and bayesian inference. In *Symposium on Foundations of Computer Science*, pages 340–351, 2003.
3. O. Bailleux and Y. Boufkhad. Full CNF encoding: The counting constraints case. In *Seventh International Conference on Theory and Applications of Satisfiability Testing*, 2004.
4. O. Bailleux, Y. Boufkhad, and O. Roussel. A translation of pseudo Boolean constraints to SAT. *Journal on Satisfiability, Boolean Modeling and Computation*, 2, March 2006.
5. A. Barvinok and J. Pommersheim. An algorithmic theory of lattice points in polyhedra. In *New Perspectives in Algebraic Combinatorics*, volume 38, pages 91–147. MSRI Publications, Cambridge University Press, 1999.
6. R. J. Bayardo and J. D. Pehoushek. Counting models using connected components. In *National Conference on Artificial Intelligence*, 2000.
7. B. Boigelot and L. Latour. Counting the solutions of presburger equations without enumerating them. *Theoretical Computer Science*, 313(1):17–29, 2004.
8. D. Chai and A. Kuehlmann. A Fast Pseudo-Boolean Constraint Solver. In *Proceedings of the Design Automation Conference*, pages 830–835, 2003.
9. N. Eén and N. Sörensson. Translating pseudo-Boolean constraints into SAT. *Journal on Satisfiability, Boolean Modeling and Computation*, 2, March 2006.
10. D. S. Johnson and M. A. Trick. Second DIMACS Implementation Challenge. DIMACS Series in Discrete Mathematics and Theoretical Computer Science, 1994.
11. J. A. D. Loera, R. Hemmecke, J. Tauzer, and R. Yoshida. Effective lattice point counting in rational convex polytopes. *J. Symb. Comput.*, 38(4):1273–1302, 2004.
12. K. L. McMillan. Applying SAT methods in unbounded symbolic model checking. In *International Conference on Computer-Aided Verification*, 2002.
13. A. Morgado, P. Matos, V. Manquinho, and J. Marques-Silva. Counting models in integer domains. Technical Report 05/2006, INESC-ID, March 2006.
14. G. L. Nemhauser and L. A. Wolsey. *Integer and Combinatorial Optimization*. John Wiley & Sons, 1988.
15. C. Pizzuti. Computing Prime Implicants by Integer Programming. In *Proceedings of the International Conference on Tools with Artificial Intelligence*, pages 332–336, November 1996.
16. W. Pugh. Counting solutions to presburger formulas: How and why. In *Conference on Programming Language Design and Implementation*, pages 121–134, June 1994.
17. K. Ravi and F. Somenzi. Minimal satisfying assignments for conjunctive normal formulae. In *International Conference on Tools and Algorithms for the Construction and Analysis of Systems*, 2004.
18. T. Sang, F. Bacchus, P. Beame, H. A. Kautz, and T. Pitassi. Combining component caching and clause learning for effective model counting. In *International Conference on Theory and Applications of Satisfiability Testing*, May 2004.
19. J. P. Warners. A linear-time transformation of linear inequalities into conjunctive normal form. *Information Processing Letters*, 68(2):63–69, 1998.

sharpSAT – Counting Models with Advanced Component Caching and Implicit BCP

Marc Thurley

Institut für Informatik, Humboldt-Universität zu Berlin
thurley@informatik.hu-berlin.de

Abstract. We introduce sharpSAT, a new #SAT solver that is based on the well known DPLL algorithm and techniques from SAT and #SAT solvers. Most importantly, we introduce an entirely new approach of coding components, which reduces the cache size by at least one order of magnitude, and a new cache management scheme. Furthermore, we apply a well known look ahead based on BCP in a manner that is well suited for #SAT solving. We show that these techniques are highly beneficial, especially on large structured instances, such that our solver performs significantly better than other #SAT solvers.

Introduction

The appearance of highly optimized SAT solvers [7,5,8] encouraged applying these SAT solvers to the closely related problem of counting the solutions of a propositional formula, known as #SAT. Applying the DPLL algorithm [4] to model counting was proposed in [3]. relsat 2 (cf. [2]) combined clause learning [11,12] with *component decomposition*. Recently, *Cachet* by Sang et al. [9,10] provided *component caching* and new branching heuristics.

We introduce sharpSAT - a new #SAT solver that inherits these techniques, improves upon them and contributes new ideas, such that it is able to outperform the best #SAT solvers (its source code is available at [1]).

After some basic definitions we will give a brief overview of our #SAT solver. Then we will discuss a new way of component caching that differs significantly from the scheme known so far (see [9]). It reduces cache sizes by at least by one order of magnitude. In the course of this, we will propose a cache management scheme which bounds the cache size explicitly and deletes old cache entries by means of a simple utility function.

Section 2 provides a discussion of *implicit BCP* - an adaptation of a well known "look ahead" technique based on boolean constraint propagation (BCP) (cf.[6]). Implicit BCP is built to integrate this technique well with other common #SAT solving techniques. This frequently results a smaller search space and reduces the cache size even further.

Eventually, in section 3, we will compare sharpSAT to the sate-of-the-art #SAT solver Cachet. This will reveal that the new techniques perform exceptionally well especially on very large instances, such as those from bounded model checking, which often contain several thousands of variables.

A. Biere and C.P. Gomes (Eds.): SAT 2006, LNCS 4121, pp. 424–429, 2006.

We consider propositional formulas F in *conjunctive normal form* (*CNF*). Let $F|_\sigma$ denote the *residual* formula under an assignment σ, where satisfied literals (and clauses) evaluate to **1** and unsatisfied ones to **0**. If $\mathbf{0} \in F|_\sigma$, i.e $F|_\sigma$ contains the *empty clause* **0** we say that σ *conflicts* with F. $F|_\sigma$ is *satisfied* if all clauses evaluate to **1**.

A short outline of the basic techniques. From SAT solvers sharpSAT inherits clause learning (cf. [11], [12]) and a fast BCP algorithm, based on the "Two Watched Literal" scheme (see [7]). Recall the notion of BCP: Whenever $F|_\sigma$ contains a *unit clause* $C = \lambda$ then λ must be satisfied, i.e. $\sigma(\lambda) = 1$. BCP performs these assignments until either no unit clause is left or a conflict occurs.

From #SAT solvers we adopted *bounded* component analysis and caching - these techniques and their correctness were discussed in [9]. For selecting *branch* variables sharpSAT applies the VSADS heuristic from Cachet (cf. [10]).

1 Component Caching

As is done in Cachet, components can be identified by strings, omitting satisfied clauses and assigned literals. Let, for example $F = (x_1 \vee x_2 \vee x_3) \wedge (x_1 \vee \bar{x}_4 \vee \bar{x}_5) \wedge (x_6 \vee x_2 \vee x_3) \wedge (x_6 \vee \bar{x}_4 \vee \bar{x}_5)$, then a string coding this would be $(1, 2, 3, 0, 1, -4, -5, 0, 6, 2, 3, 0, 6, -4, -5, 0)$ (zeros denote ends of clauses). Call this the *Standard scheme* (STD).

sharpSAT codes the components differently. First of all, only *sound components* are cached, i.e. those which contain only clauses with at least two unassigned literals. This is reasonable as length zero clauses (i.e. the empty clause) denote conflicts and unit clauses are handled by BCP.

Let $var(F)$ ($cl(F)$) be the set of variables (clauses, resp.) in F and $var_{id}(F)$ ($cl_{id}(F)$) the corresponding sets of indices. Let $id(\lambda)$ ($id(C)$) denote the index of a literal λ (clause C). With F as in the example above we have $cl(F) = \{\{x_1, x_2, x_3\}, \{x_1, \bar{x}_4, \bar{x}_5\}, \{x_6, x_2, x_3\}, \{x_6, \bar{x}_4, \bar{x}_5\}\}$ but $cl_{id}(F) = \{1, 2, 3, 4\}$. We code components G by writing $var_{id}(G)$ and $cl_{id}(G)$ to strings a and b in increasing order of the indices, which yields a *code* (a, b). For F as above we have $a = (1, 2, 3, 4, 5, 6)$ and $b = (1, 2, 3, 4)$. Call this the *Hybrid coding scheme* (HC). The correctness of HC is displayed by the following lemma.

Lemma 1. *Given $F \in CNF$, (partial) assignments σ, τ and components G of $F|_\sigma$ and B of $F|_\tau$ then $(var(B) = var(G)$ and $cl(B) = cl(G))$ iff $(var_{id}(B) = var_{id}(G)$ and $cl_{id}(B) = cl_{id}(G))$.*

Proof. The forward direction is trivial. For the reverse let $var_{id}(B) = var_{id}(G)$ and $cl_{id}(B) = cl_{id}(G)$. Obviously, $var(B) = var(G)$ holds, but suppose for contradiction, that $cl(B) \neq cl(G)$. As $cl_{id}(B) = cl_{id}(G)$, there is a clause $\gamma_B \in cl(B)$ for which $\gamma_G \in cl(G)$ exits with $id(\gamma_B) = id(\gamma_G)$ but $\gamma_B \neq \gamma_G$.

As B and G are components of restrictions of F, there is a clause $\gamma \in F$ with $id(\gamma) = id(\gamma_B) = id(\gamma_G)$. Now, as $\gamma_B \neq \gamma_G$, there is a literal λ such that w.l.o.g. $\lambda \in \gamma_B \setminus \gamma_G$. Since $\lambda \notin \gamma_G$ we have that λ is assigned in G but not in B, contradicting $var(B) = var(G)$.

Note that STD is more general than HC. For example, for F as above, $\sigma = [x_1 \leftarrow 0, x_6 \leftarrow 1]$ and $\tau = [x_1 \leftarrow 1, x_6 \leftarrow 0]$ we have $F|_\sigma = F|_\tau = (x_2 \vee x_3) \wedge (\bar{x}_4 \vee \bar{x}_5)$ which could be recognized by STD but not by HC as $cl_{id}(F|_\sigma) = \{1, 2\}$ and $cl_{id}(F|_\tau) = \{3, 4\}$. However, our experimental results (see sect. 3) show the effectiveness of HC and hence suggest that this case is not very likely.

We can reduce the sizes of the codes even further. As only sound components G are cached, storing the ids of binary clauses is redundant. Why is this so? Consider a formula F, an assignment σ and a sound component G of $F|_\sigma$ with code (a, b). Assume that G contains $C|_\sigma$ for a binary clause $C = (\lambda \vee \kappa) \in F$. Suppose that at least one of κ and λ is assigned in σ. If exactly one is assigned, $C|_\sigma$ is satisfied (otherwise BCP could be applied). If both are assigned, $C|_\sigma$ is satisfied as well, as G is sound. Thus, $C|_\sigma$ occurs in G iff κ and λ are unassigned, and the occurrence of $cl_{id}(C|_\sigma)$ in the code b can be reconstructed by the presence of $id(\lambda)$ and $id(\kappa)$ in a. Hence, we can omit storing the identifiers of binary clauses in the component codes. Call this scheme *Omitting binary clauses*.

We can do even better. Each code (a, b) of a component G is packed before caching and before cache look-up. To obtain the packed form $(\hat{a}, \hat{b})$, we determine $n := \lceil log_2 |var(F)| \rceil$ and $m := \lceil log_2 |cl(F)| \rceil$. Identifiers in a contain information only in the n least significant bits, thus a is packed into $\hat{a}$ by bitshifting. b is treated analogously. Call this the *Packing* scheme.

Table 1 illustrates the coding schemes when applied to formulas from SATLIB. *HCO* is HC omitting binary clauses, and *HCOP* shows this in its packed form.

Table 1. Comaring codes sizes in bytes (*= unit clauses removed via BCP)

Problems	vars	clauses	STD	HC	HCO	HCOP
flat200	600	2237	27644	11348	3200	950
uf200	200	860	13760	4240	4240	1075
logistics.a.cnf	828	6718	98532	30184	16964	5301
logistics.b.cnf	843	7301	105300	32576	16576	6006
bmc-ibm-1.cnf	7085*	35419*	822420	170016	49484	20104
bmc-galileo-8.cnf	43962*	183532*	4079 KB	884 KB	302 KB	151 KB

We compare sizes of the different codes of the input formulas. Experiments show that this gives a good estimate of the relative cache sizes. The starred numbers denote a preprocessing before forming the codes: all unit clauses in the input formula had to be propagated via BCP, as the hybrid scheme does not cache formulas with unit clauses. Observe that the efficiency of HC in comparison to STD increases with clause-to-variable ratio. HCO is futile on formulas without binary clauses (see uf200) but it is highly beneficial for example on the flat200 formulas. Clearly, packing shows the least advantages on large instances (ibm-galileo-8) but still reduces the code size to about 50%.

Cache Management. On hard formulas the cache size quickly exceeds any reasonable bound, which necessitates a good cache management. In our experiments, we observed drawbacks of bounding the cache size by an oldest age bound: a good bound depends highly on the formula size and has to be set manually.

In sharpSAT an absolute bound $maxSize$ in bytes on the cache size is set. Furthermore, we keep scores for each cache entry in a way reminiscent of the VSIDS heuristic (cf. [7]). If an entry is hit its score is increased. All scores are divided periodically. The cache is cleared only if it exceeds a fixed fraction (0.9, say) of $maxSize$, if so all entries with a score lower than $minScore$ are deleted. Directly after cleanup we try to keep the cache size at about $0.5 \cdot maxSize$. To achieve this, we increase or decrease $minScore$ accordingly.

This quickly stabilizes the cache size after cleanup to about the desired value. Furthermore, this scheme is quite fast, as entries are deleted only when necessary and updating scores creates almost no time overhead.

2   Implicit BCP

BCP plays a central role in the performance of SAT and #SAT solvers. Branching heuristics based on BCP, called Unit Propagation (UP) heuristics (cf. [6]) try to maximize the possible effect of BCP by applying a form of "look ahead".

UP heuristics determine branch variables by estimating the effect an assignment has for BCP. To achieve this, for each variable x of a certain set S of free variables the assignments $x \leftarrow 0$ and $x \leftarrow 1$ are made independently and BCP is applied in each case. If any of these cases, say $x \leftarrow 0$, causes a conflict, a *failed literal* $(\bar{x})$ is found and x is chosen directly as branch variable. Otherwise, the variables in S are evaluated according to their effect on BCP and one of these is chosen.

sharpSAT applies an algorithm for finding failed literals. It deviates from the traditional UP heuristics approach, as for example pursued in Cachet, in at least two ways. First, it is applied independently of the branching heuristics and only failed literals are sought. If a failed literal, say $\lambda = \bar{x}$ is found, a conflict clause C^λ is learned directly and the algorithm proceeds as if $x \leftarrow 1$ was found by BCP via C^λ. The process stops either if a conflict occurs, or no failed literals are found anymore. In SAT solvers this might show no big difference to UP heuristics, but in our #SAT solver a large amount of component analysis and cache look up and storing is avoided by this procedure as in the course of implicit BCP these operations are not applied.

Furthermore, the set S of candidates for failed literals is computed differently. We only consider literals from *original* clauses that have become binary in the most recent call of BCP. Thus in instances that allow for few implications only, S is small and thus implicit BCP induces almost no overhead. In cases of many implications S is larger but failed literals are more likely as well.

3   Comparison

We compare sharpSAT with and without implicit BCP and Cachet (version 1.22) on instances from SATLIB, to wit, the flat200, uf200, bmc and logistics suite. Tests were run on a 3GHz Pentium 4 with 1GB of main memory, a time bound of 10 hours and a maximum cache size of 512MB. Table 2 displays the results, entries for bmc-ibm-6,7,10,13 and logistics.d are missing as neither Cachet nor sharpSAT solved them. In contrast to the model counts shown in scientific number form, the solvers used BigNum packages for exact model counting.

Note the effect of implicit BCP. Comparing the savings in terms of the run time to the reduced number of decisions reveals the overhead of implicit BCP, e.g. in the uf200 suite, a reduction by a factor of 7 in the decisions is reflected only in a factor of 1.5 in the actual running time. However, there *is* a benefit in time as a lot of component analysis as well as cache look-up and storing is avoided. Each of these operations is performed once per decision. Hence, less decisions are always beneficial, as irrespective of decreased running times, the cache size is always reduced by about the same factor as the number of decisions.

The footnotes on the running times for Cachet refer to oldest age bounds on the cache entries, which were adjusted manually. Where these are given, Cachet could not solve the instances without them due to out-of-memory errors. For sharpSAT the maximum cache size was set to 512MB for all the instances given.

sharpSAT dominates especially on structured instances. However, a general link between small running time and exactly one of the new techniques is not obvious. On some instances with very low model counts (e.g. galileo-8 and 9) (see table 2) the dominance is clearly due to implicit BCP. On the other hand, for ibm-bmc-5 and logistics.c the dominance is based solely on the new caching

Table 2. Comparing sharpSAT with and without implicit BCP and Cachet (X = time out after 10 hours; [1-5] oldest age bounds: 1 = 500; 2 = 50; 3 = 30; 4 = 15; 5 = 3000)

Problems	vars	clauses	solutions	implicit BCP decisions	secs	w/o implicit BCP decisions	secs	Cachet secs
flat200 avg.	600	2,237	2.22e+13	3,378	**1.77**	14,141	2.18	3.98
uf200 avg.	200	860	1.57e+09	9,597	7.36	70,448	10.9	**6.63**
bmc								
ibm-1	9,685	55,870	7.33e+300	11,991	**16**	37,808	32	47
ibm-2	2,810	11,683	1.33e+19	148	**0.09**	584	0.11	**0.09**
ibm-3	14,930	72,106	2.47e+19	2,657	64.5	14,705	72.2	**58**
ibm-4	28,161	139,716	9.73e+79	59,334	**111**	1.6e+6	1,346	X
ibm-5	9,396	41,207	2.46e+171	191,558	152	198,464	**64.5**	486[1]
ibm-11	32,109	150,027	3.53e+74	429,575	**3,331**	2.3e+6	15,204	26,823[2]
ibm-12	39,598	194,778	2.1e+112	25,456	**833**	–	X	977[3]
galileo-8	58,074	294,821	8.14e+40	12,945	**326**	168,748	1,716	628[3]
galileo-9	63,624	326,999	3.46e+44	15,798	**392**	313,474	3,688	786[4]
logistics								
a	828	6,718	3.78e+14	4,412	**0.8**	19,176	2.14	5.31
b	843	7,301	4.53e+23	15,711	**7.15**	93,885	11.8	17.8
c	1,141	10,719	3.98e+24	2.3e+6	**426**	3.9e+6	480	1,003[5]

scheme. In all other instances we claim that the effect is due to the combination of both techniques.

4 Conclusion

We introduced sharpSAT - a #SAT solver that outperforms the current state-of-the-art solver Cachet on a wide range of structured instances. This is due to new techniques which comprise a highly optimized way of coding the components for caching and the implicit BCP algorithm that performs well in practice.

References

1. www.informatik.hu-berlin.de/~thurley/sharpSAT.
2. Robert J. Bayardo and Joseph D. Pehoushek. Counting models using connected components. *Proceedings, AAAI-00: 17th International Conference on Artificial Intelligence*, pages 157–162, 2000.
3. Elazar Birnbaum and Eliezer L. Lozinskii. The good old davis-putnam procedure helps counting models. *Journal of Artificial Intelligence Research*, 1999.
4. Martin Davis, George Logeman, and Donald Loveland. A machine program for theorem-proving. *Communications of the ACM*, 1962.
5. E. Goldberg and Y. Novikov. Berkmin: A fast and robust sat-solver. *Design, Automation and Test in Europe (DATE'02)*, pages 142–149, 2002.
6. Chu Min Li and Anbulagan. Heuristics based on unit propagation for satisfiability problems. In *IJCAI (1)*, pages 366–371, 1997.
7. Matthew W. Moskewicz, Conor F. Madigan, Ying Zhao, Lintao Zhang, and Sharad Malik. Chaff: Engineering an efficient sat solver. In *Proceedings of the 38th Design Automation Conference (DAC'01)*, June 2001.
8. Lawrence Ryan. Efficient algorithms for clause learning sat solvers. Master's thesis, Simon Fraser University, 2004.
9. Tian Sang, Fahiem Bacchus, Paul Beame, Henry Kautz, and Toniann Pitassi. Combining component caching and clause learning for effective model counting. In *Seventh International Conference on Theory and Applications of Satisfiability Testing*, 2004.
10. Tian Sang, Paul Beame, and Henry Kautz. Heuristics for fast exact model counting. In *Eighth International Conference on Theory and Applications of Satisfiability Testing, Edinburgh, Scotland*, 2005.
11. Joao P. Marques Silva and Karem A. Sakallah. Grasp - a new search algorithm for satisfiability. In *Proceedings of IEEE/ACM International Conference on Computer-Aided Design*, pages 220–227, November 1996.
12. Lintao Zhang, Conor F. Madigan, Matthew W. Moskewicz, and Sharad Malik. Efficient conflict driven learning in a boolean satisfiability solver. In *International Conference on Computer-Aided Design (ICCAD'01)*, pages 279–285, 2001.

A Distribution Method for Solving SAT in Grids

Antti E. J. Hyvärinen, Tommi Junttila, and Ilkka Niemelä

Helsinki University of Technology, Laboratory for Theoretical Computer Science
{Antti.Hyvarinen, Tommi.Junttila, Ilkka.Niemela}@tkk.fi

Abstract. The emerging large-scale computational grid infrastructure is providing an interesting platform for massive distributed computations. In this paper a novel distribution method called scattering is introduced for solving SAT problem instances in grid environments. The key advantages of scattering are that it can be used in conjunction with any sequential SAT solver (including industrial black box solvers), the distribution heuristic is strictly separated from the heuristic used in sequential solving, and it requires no communication between processes solving subproblems but still allows coordination of such processes. An implementation of the method has been developed for NorduGrid, a large widely distributed production-level grid running in Scandinavia.

1 Introduction

We study the *propositional satisfiability problem* (SAT) of determining whether a given propositional formula has a satisfying truth assignment. Decision methods for SAT and their implementation techniques have advanced considerably during the last decade and SAT based techniques have been applied successfully in several areas such as planning [1] and model checking [2].

An interesting approach to boosting the applicability of SAT based problem solving is to exploit parallel computation to solve SAT problem instances. In particular, the emerging large scale computational grids make this approach increasingly attractive. For example, the largest software project currently funded by the European Union is the EGEE project (Enabling Grids for E-sciencE; http://public.eu-egee.org/).

In this paper we study how to exploit the grid infrastructure in solving challenging SAT instances. Compared to more tightly coupled parallel computing architectures, grids have properties that need to be taken into account when designing distributed algorithms. In particular, (i) the available resources can be quite heterogeneous in a grid, (ii) no shared memory is available and communication delays to grid nodes are significant, (iii) inter-node communication is very limited and often not available, (iv) individual jobs executed in grid nodes have non-negligible failure rates. The goal is an approach where we can exploit the best SAT solving techniques and the grid resources.

Several techniques for distributed SAT solving have been proposed [3,4,5,6,7, 8,9,10,11,12]. However, in these approaches it is not possible to exploit a chosen sequential SAT solver directly but they are based on developing a special purpose distributed SAT solver. Moreover, distribution of work and load balancing are

A. Biere and C.P. Gomes (Eds.): SAT 2006, LNCS 4121, pp. 430–435, 2006.

fairly tightly coupled with the decisions made in the SAT solver and a significant amount of inter-process communication is needed.

A straightforward way to use grid computing to solve SAT problems is to employ an approach we call *Simple Distributed SAT* (SDSAT) where a SAT instance is solved by running a number of SAT solvers on the same instance as independent jobs and waiting until one of them solves the problem. If a randomized SAT solver is available, this solver can be used with different seeds. This approach has potential as results, e.g., in [13, 14] indicate.

However, the natural way of running SDSAT by starting jobs without re-source bounds and waiting until one of them succeeds is not available in grids. This is because each process needs to be given resource bounds (CPU time, memory) when the process is sent to a grid. Moreover, grid job management typically takes into account the resource requirements of a job, implying lower priority to jobs with substantial resource demands and, thus longer delays.

In this paper we present a novel distributed SAT solving method called *scattering*. The basic idea is to divide a given SAT instance gradually to increasingly more constrained subproblems which are sent to the grid to be solved using practically any available SAT solver. While this is somewhat similar to SDSAT, there are substantial differences. (i) The subproblems to be solved become easier to solve as the computation proceeds. (ii) Learning techniques used in sequential solvers can be exploited when dividing a problem to subproblems. (iii) The division of a problem to subproblems is based on estimating the computational cost of the subproblems in order to achieve better load balancing.

Scattering has similar modest communication requirements as SDSAT and is thus different from typical guiding path based approaches, like [3], which need inter-process communication and special customized solvers for dynamic load balancing (and failure recovery). In scattering dynamic load balancing (and failure recovery) is handled without inter-process communication by (i) dividing a running job into subproblems and starting jobs on them even before the result of the running job is available and (ii) setting at job construction time resource bounds (time and memory) which when exceeded terminate the job.

Scattering is implemented in SATU, a SAT solver designed to work with Nor-duGrid [15] (http://www.nordugrid.org/), a widely distributed Scandinavian grid. SATU is available at http://www.tcs.hut.fi/Software/satu/.

2   The Algorithm

The goal is to develop an algorithm for solving challenging SAT instances on a wide range of grids. In order to support this, we make minimal assumptions regarding the *Distributed Execution Environment* (DEE) provided by a grid. DEE is assumed to offer a simple interface between the client sending *executions* (executable programs together with their inputs), and the environment receiving and running the executions. The only functionalities available to the client are (i) Send, sending an execution to the environment, (ii) Monitor, reporting the state of the execution, and (iii) Receive, returning the result of an execution.

We assume that the executions are not able to communicate directly with each other. We also assume that DEE has some maximum of simultaneous executions it can hold. If this limit is reached, the environment is *saturated*, and any new executions may fail without a result. The executions must finish when some condition given at construction time is triggered, e.g, a CPU time or memory limit is exceeded.

The simple distributed SAT (SDSAT) scheme is not optimal for solving resource intensive problems in a grid because of the resource bounds and job management policies explained in the introduction. In order to address deficiencies of SDSAT we have developed a distribution method called *scattering*. The basic ideas underlying scattering are quite straightforward.

- A SAT problem instance is divided to a set of subproblems by adding new clauses to the original problem to make the subproblems easier to solve.
- The division of a problem to subproblems (a scattering step) is done so that (i) if all subproblems have been solved, we get a solution to the original problem, and (ii) solution spaces of the subproblems are disjoint.
- Until a computed answer is obtained, subproblems are further divided by scattering steps whenever possible, i.e., when DEE is below the saturation limit. This means that the execution of a subproblem exceeding a given resource bound can be interrupted (or a grid node can suffer a failure) without compromising the completeness of the method.

2.1   The Basic Scattering Rule and the Scattering Tree

The scattering rule takes as input a formula F and a number $sf\,(\geq 2)$ (scattering factor) and constructs sf scattered formulas $F_1, \ldots, F_{sf}$ from F such that

$$F_i = \begin{cases} F \wedge T_1 & \text{if } i = 1 \\ F \wedge \neg T_1 \wedge \cdots \wedge \neg T_{i-1} \wedge T_i & \text{if } 1 < i < sf \\ F \wedge \neg T_1 \wedge \cdots \wedge \neg T_{sf-1} & \text{if } i = sf. \end{cases} \tag{1}$$

Each T_i is a conjunction $l_1^i \wedge \cdots \wedge l_{d_i}^i$ of d_i literals heuristically selected (Sect. 2.3) and each number d_i is selected to yield comparably sized subproblems (Sect. 2.2). The expression $\neg T_i = \bar{l}_1^i \vee \cdots \vee \bar{l}_{d_i}^i$ is the negation of the conjunction T_i. Thus constructed propositional formulas have the properties that (i) the disjunction $F_1 \vee \cdots \vee F_{sf}$ is logically equivalent to the formula F, and (ii) no two formulas F_i, F_j, $i \neq j$, share a satisfying truth assignment.

We solve a SAT instance F_r by performing a distributed search in a *scattering tree* with root F_r. The nodes are formulas obtained by the scattering rule so that the children of a node F are the scattered formulas $F_1, \ldots, F_{sf}$. The search is implemented by sending formulas associated to nodes as jobs to be solved in DEE. A part of a possible scattering tree is given in Fig. 1.

A node of the tree is *computed satisfiable* if the corresponding job returns with this answer, or if at least one of the children is computed satisfiable. The node is *computed unsatisfiable* if the corresponding job returns unsatisfiable, or if all children have been computed unsatisfiable. The scattering tree is constructed incrementally according to the chosen search strategy while sending jobs to DEE whenever DEE is below its saturation limit until F_r is computed satisfiable or unsatisfiable.

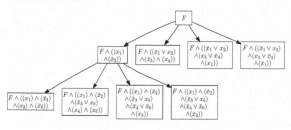

Fig. 1. A part of a scattering tree

The correctness of the computed answer follows from the following observations. Since scattered formulas of F are more constrained than F by (1), a satisfying truth assignment for one of the scattered formulas satisfies F as well. Moreover, since the disjunction of the scattered formulas is logically equivalent to F, if all scattered formulas are unsatisfiable, F is also unsatisfiable. Furthermore, if a formula is computed unsatisfiable, all possible scattered formulas, being more constrained, are also unsatisfiable.

An ancestor of a formula in the scattering tree is less constrained than the child. As a result, all clauses which are logical consequences of the ancestor are also logical consequences of the child. Learned clauses from formula F can thus be included to the formula F' if F is ancestor of F' in the scattering tree.

2.2   Balancing the Subproblems

The goal in a scattering step is to divide a given formula F to scattered formulas $F_1, \ldots, F_{sf}$ that have comparable estimated computational costs by choosing d_i fixed literals for each F_i. This is done by selecting for each $i = 1, \ldots, sf - 1$, the number d_i such that $|2^{-d_i} - r_i|$ is minimized, where $r_i = (sf - i + 1)^{-1}$.

An example for $sf = 7$ is given in Table 1. The first column shows values of r_i for $i = 1, \ldots, 7$, the second shows the values of d_i minimizing $|2^{-d_i} - r_i|$, and the third the resulting estimated fractions of the full problem.

The idea behind this approach is to divide the solution space of F to sf portions of equal size. The solution spaces of different F_is are disjoint by construction. Hence, for each i, if $F_1, \ldots, F_{i-1}$ already cover $i - 1$ of these portions, then an equal size portion is obtained by taking the fraction r_i from the remaining solution space. By fixing d_i literals a fraction 2^{-d_i} of the solution space can be chosen. Hence, for each i, the numbers d_i are chosen so that 2^{-d_i} is as close as possible to r_i.

Table 1.
d_i ($sf = 7$)

r_i	d_i	Est.
$\frac{1}{7}$	3	0.125
$\frac{1}{6}$	3	0.109
$\frac{1}{5}$	2	0.191
$\frac{1}{4}$	2	0.144
$\frac{1}{3}$	2	0.108
$\frac{1}{2}$	1	0.161
1	0	0.161

2.3   The Heuristic

The selection of literals for the scattered formulas in (1) is a heuristic process which we have implemented on top of a SAT solver implementation [16] similar to zChaff [17]. The SAT solver is run first on the formula F until a given number of conflicts is reached to initialize the heuristic scores of the literals as in zChaff and the initialization is continued as long as DEE is above the saturation limit. When selecting the d_i literals for the scattered formula F_i, the procedure takes as input $F \wedge \neg T_1 \wedge \cdots \wedge \neg T_{i-1}$, runs the zChaff type algorithm until it reaches the decision level $d_i + 1$ and then selects as the literals $l_1^i, \ldots, l_{d_i}^i$ the d_i decision literals chosen on levels $1, \ldots, d_i$. If the decision level $d_i + 1$ cannot be reached, this produces a computed solution for $F_i, \ldots, F_{sf}$.

We use a modification of the original zChaff VSIDS heuristic [17] which aims at finding variables that divide the remaining search space into parts of comparable size. In contrast to the greedy VSIDS heuristic, we rank each variable x with the scores of literals x and $\neg x$ and choose the variable which has the best lower score, and finally select the literal corresponding to the higher score of this variable as the decision literal.

3   Experimental Results and Conclusions

Some results, obtained with NorduGrid and SATU 0.2 having zChaff as the SAT solver, are shown in Table 2. The scatter results are with scattering factor $sf = 7$, per job timeout ranging randomly between 60 and 90 minutes, memory limit of 1 GB and saturation limit of 64. The times are compared to the minimum time of 64 runs of randomized zChaff, simulating idealized SDSAT. The fourth column indicates the number of cases where the run time of randomized zChaff was smaller than the wall clock time of SATU. We run randomized zChaff on 2 GHz AMD Athlon(tm) 64 Processor 3200+ with 2 GB of memory. Note that the results of randomized zChaff do not include the very substantial communication and management delays of a widely distributed grid. More experimental results are available in [18].

Scattering differs from other distributed SAT solving methods, such as [3, 4, 5, 6, 7, 8, 9, 10, 11, 12] in a number of ways: (i) any SAT solver, including industrial black box solvers, can be used with no modifications, (ii) it has modest requirements for communication but still allows process coordination, (iii) when the solving of a problem needs to be distributed, it is possible to divide the problem to an arbitrary number of subproblems, and (iv) heuristic for dividing a problem to subproblems is separated from the heuristic for solving individual subproblems.

Table 2. Run times with SATU on formulas from SAT2002 competitions

Name	scatter (s)	zChaff$_{min}$ (s)	#	result
cnt10	1121	540	57	sat
dp10u09	586	149	19	unsat
f2clk_40	8078	4831	34	unsat
lisa20_1_a	240	1	20	sat
lisa21_3_a	1017	206	5	sat
Mat26	1503	4151	0	unsat
vda_gr_rcs_w8	1160	1	5	sat

Interesting topics of future work include the optimization of the scattering algorithm and job management. Also a more thorough comparison between scattering and other methods is needed.

References

1. Kautz, H., Selman, B.: Pushing the envelope: Planning, propositional logic, and stochastic search. In: AAAI/IAAI 1996, AAAI Press (1996) 1194–1201
2. Clarke, E., Biere, A., Raimi, R., Zhu, Y.: Bounded model checking using satisfiability solving. Formal Methods in System Design **19**(1) (2001) 7–34
3. Zhang, H., Bonacina, M., Hsiang, J.: PSATO: A distributed propositional prover and its application to quasigroup problems. J. Symbolic Computation **21**(4) (1996) 543–560
4. Chrabakh, W., Wolski, R.: GridSAT: A chaff-based distributed SAT solver for the grid. In: SC 2003, IEEE (2003)
5. Jurkowiak, B., Li, C., Utard, G.: A parallelization scheme based on work stealing for a class of SAT solvers. Journal of Automated Reasoning **34**(1) (2005) 73–101
6. Sinz, C., Blochinger, W., Küchlin, W.: PaSAT — Parallel SAT-checking with lemma exchange: Implementation and applications. In: SAT 2001. Volume 9 of Electronic Notes in Discrete Mathematics., Elsevier (2001) 12–13
7. Blochinger, W., Sinz, C., Küchlin, W.: Parallel propositional satisfiability checking with distributed dynamic learning. J. Parallel Computing **29**(7) (2003) 969–994
8. Forman, S., Segre, A.: NAGSAT: A randomized, complete, parallel solver for 3-SAT. In: SAT 2002. (2002) Online proceedings at http://gauss.ececs.uc.edu/Conferences/SAT2002/sat2002list.html.
9. Boehm, M., Speckenmeyer, E.: A fast parallel SAT-solver: Efficient workload balancing. Annals of Mathematics and Artificial Intelligence **17**(4-3) (1996) 381–400
10. Okushi, F.: Parallel cooperative propositional theorem proving. Annals of Mathematics and Artificial Intelligence **26**(1-4) (1999) 59–85
11. Speckenmeyer, E., Boehm, M., Heusch, P.: On the imbalance of distributions of solutions of CNF-formulas and its impact on satisfiability solvers. In: Satisfiability Problem: Theory and Applications, DIMACS (1997) 669–676
12. Blochinger, W., Westje, W., Küchlin, W., Wedeniwski, S.: ZetaSAT – Boolean satisfiability solving on desktop grids. In: CCGrid 2005, IEEE (2005) 1079–1086
13. Gomes, C.P., Selman, B., Crato, N., Kautz, H.A.: Heavy-tailed phenomena in satisfiability and constraint satisfaction problems. J. Automated Reasoning **24**(1/2) (2000) 67–100
14. Gomes, C.P., Selman, B.: Algorithm portfolios. Artificial Intelligence **126**(1-2) (2001) 43–62
15. Eerola, P., et al.: Building a production grid in Scandinavia. IEEE Internet Computing **7**(4) (2003) 27–35
16. Zhang, L.: SAT-solving: From Davis-Putnam to Zchaff and beyond, lecture notes (2003) Available online at http://research.microsoft.com/users/lintaoz/SATSolving/satsolving.htm.
17. Moskewicz, M.W., Madigan, C.F., Zhao, Y., Zhang, L., Malik, S.: Chaff: Engineering an efficient SAT solver. In: DAC 2001, ACM (2001) 530–535
18. Hyvärinen, A.E.J.: SATU: A system for distributed propositional satisfiability checking in computational grids. Research Report A100, Helsinki Univ. of Technology, Lab. for Theoretical Comp. Science, Espoo, Finland (2006)

Author Index

Lecture Notes in Computer Science

For information about Vols. 1–4007

please contact your bookseller or Springer

Vol. 4054: A. Horváth, M. Telek (Eds.), Formal Methods and Stochastic Models for Performance Evaluation. VIII, 239 pages. 2006.

Vol. 4053: M. Ikeda, K.D. Ashley, T.-W. Chan (Eds.), Intelligent Tutoring Systems. XXVI, 821 pages. 2006.

Vol. 4052: M. Bugliesi, B. Preneel, V. Sassone, I. Wegener (Eds.), Automata, Languages and Programming, Part II. XXIV, 603 pages. 2006.

Vol. 4051: M. Bugliesi, B. Preneel, V. Sassone, I. Wegener (Eds.), Automata, Languages and Programming, Part I. XXIII, 729 pages. 2006.

Vol. 4049: S. Parsons, N. Maudet, P. Moraitis, I. Rahwan (Eds.), Argumentation in Multi-Agent Systems. XIV, 313 pages. 2006. (Sublibrary LNAI).

Vol. 4048: L. Goble, J.-J.C.. Meyer (Eds.), Deontic Logic and Artificial Normative Systems. X, 273 pages. 2006. (Sublibrary LNAI).

Vol. 4047: M. Robshaw (Ed.), Fast Software Encryption. XI, 434 pages. 2006.

Vol. 4046: S.M. Astley, M. Brady, C. Rose, R. Zwiggelaar (Eds.), Digital Mammography. XVI, 654 pages. 2006.

Vol. 4045: D. Barker-Plummer, R. Cox, N. Swoboda (Eds.), Diagrammatic Representation and Inference. XII, 301 pages. 2006. (Sublibrary LNAI).

Vol. 4044: P. Abrahamsson, M. Marchesi, G. Succi (Eds.), Extreme Programming and Agile Processes in Software Engineering. XII, 230 pages. 2006.

Vol. 4043: A.S. Atzeni, A. Lioy (Eds.), Public Key Infrastructure. XI, 261 pages. 2006.

Vol. 4042: D. Bell, J. Hong (Eds.), Flexible and Efficient Information Handling. XVI, 296 pages. 2006.

Vol. 4041: S.-W. Cheng, C.K. Poon (Eds.), Algorithmic Aspects in Information and Management. XI, 395 pages. 2006.

Vol. 4040: R. Reulke, U. Eckardt, B. Flach, U. Knauer, K. Polthier (Eds.), Combinatorial Image Analysis. XII, 482 pages. 2006.

Vol. 4039: M. Morisio (Ed.), Reuse of Off-the-Shelf Components. XIII, 444 pages. 2006.

Vol. 4038: P. Ciancarini, H. Wiklicky (Eds.), Coordination Models and Languages. VIII, 299 pages. 2006.

Vol. 4037: R. Gorrieri, H. Wehrheim (Eds.), Formal Methods for Open Object-Based Distributed Systems. XVII, 474 pages. 2006.

Vol. 4036: O. H. Ibarra, Z. Dang (Eds.), Developments in Language Theory. XII, 456 pages. 2006.

Vol. 4035: T. Nishita, Q. Peng, H.-P. Seidel (Eds.), Advances in Computer Graphics. XX, 771 pages. 2006.

Vol. 4034: J. Münch, M. Vierimaa (Eds.), Product-Focused Software Process Improvement. XVII, 474 pages. 2006.

Vol. 4033: B. Stiller, P. Reichl, B. Tuffin (Eds.), Performability Has its Price. X, 103 pages. 2006.

Vol. 4032: O. Etzion, T. Kuflik, A. Motro (Eds.), Next Generation Information Technologies and Systems. XIII, 365 pages. 2006.

Vol. 4031: M. Ali, R. Dapoigny (Eds.), Advances in Applied Artificial Intelligence. XXIII, 1353 pages. 2006. (Sublibrary LNAI).

Vol. 4029: L. Rutkowski, R. Tadeusiewicz, L.A. Zadeh, J. Zurada (Eds.), Artificial Intelligence and Soft Computing – ICAISC 2006. XXI, 1235 pages. 2006. (Sublibrary LNAI).

Vol. 4028: J. Kohlas, B. Meyer, A. Schiper (Eds.), Dependable Systems: Software, Computing, Networks. XII, 295 pages. 2006.

Vol. 4027: H.L. Larsen, G. Pasi, D. Ortiz-Arroyo, T. Andreasen, H. Christiansen (Eds.), Flexible Query Answering Systems. XVIII, 714 pages. 2006. (Sublibrary LNAI).

Vol. 4026: P.B. Gibbons, T. Abdelzaher, J. Aspnes, R. Rao (Eds.), Distributed Computing in Sensor Systems. XIV, 566 pages. 2006.

Vol. 4025: F. Eliassen, A. Montresor (Eds.), Distributed Applications and Interoperable Systems. XI, 355 pages. 2006.

Vol. 4024: S. Donatelli, P. S. Thiagarajan (Eds.), Petri Nets and Other Models of Concurrency - ICATPN 2006. XI, 441 pages. 2006.

Vol. 4021: E. André, L. Dybkjær, W. Minker, H. Neumann, M. Weber (Eds.), Perception and Interactive Technologies. XI, 217 pages. 2006. (Sublibrary LNAI).

Vol. 4020: A. Bredenfeld, A. Jacoff, I. Noda, Y. Takahashi (Eds.), RoboCup 2005: Robot Soccer World Cup IX. XVII, 727 pages. 2006. (Sublibrary LNAI).

Vol. 4019: M. Johnson, V. Vene (Eds.), Algebraic Methodology and Software Technology. XI, 389 pages. 2006.

Vol. 4018: V. Wade, H. Ashman, B. Smyth (Eds.), Adaptive Hypermedia and Adaptive Web-Based Systems. XVI, 474 pages. 2006.

Vol. 4017: S. Vassiliadis, S. Wong, T.D. Hämäläinen (Eds.), Embedded Computer Systems: Architectures, Modeling, and Simulation. XV, 492 pages. 2006.

Vol. 4016: J.X. Yu, M. Kitsuregawa, H.V. Leong (Eds.), Advances in Web-Age Information Management. XVII, 606 pages. 2006.

Vol. 4014: T. Uustalu (Ed.), Mathematics of Program Construction. X, 455 pages. 2006.

Vol. 4013: L. Lamontagne, M. Marchand (Eds.), Advances in Artificial Intelligence. XIII, 564 pages. 2006. (Sublibrary LNAI).

Vol. 4012: T. Washio, A. Sakurai, K. Nakajima, H. Takeda, S. Tojo, M. Yokoo (Eds.), New Frontiers in Artificial Intelligence. XIII, 484 pages. 2006. (Sublibrary LNAI).

Vol. 4011: Y. Sure, J. Domingue (Eds.), The Semantic Web: Research and Applications. XIX, 726 pages. 2006.

Vol. 4010: S. Dunne, B. Stoddart (Eds.), Unifying Theories of Programming. VIII, 257 pages. 2006.

Vol. 4009: M. Lewenstein, G. Valiente (Eds.), Combinatorial Pattern Matching. XII, 414 pages. 2006.

Vol. 4008: J.C. Augusto, C.D. Nugent (Eds.), Designing Smart Homes. XI, 183 pages. 2006. (Sublibrary LNAI).